Experimental Models for Renal Diseases: Pathogenesis and Diagnosis

Contributions to Nephrology

Vol. 169

Series Editor

Claudio Ronco Vicenza

Experimental Models for Renal Diseases

Pathogenesis and Diagnosis

Volume Editor

Guillermo A. Herrera Tempe, Ariz.

76 figures, 27 in color, and 13 tables, 2011

KARGER Basel · Freiburg · Paris · London · New York · Bangalore · Bangkok · Shanghai · Singapore · Tokyo · Sydney

Contributions to Nephrology
(Founded 1975 by Geoffrey M. Berlyne)

Guillermo A. Herrera
Associate Director, Nephrocor
1700 N Desert Dr.
Tempe, Arizona, 85 281

Library of Congress Cataloging-in-Publication Data

Experimental models for renal diseases : pathogenesis and diagnosis / volume editor, Guillermo A. Herrera.
p. ; cm. -- (Contributions to nephrology, ISSN 0302-5144 ; v. 169)
Includes bibliographical references and indexes.
ISBN 978-3-8055-9537-7 (hard cover : alk. paper) -- ISBN 978-3-8055-9538-4 (e-ISBN)
1. Kidneys--Diseases--Pathogenesis. I. Herrera, Guillermo A., MD. II. Series: Contributions to nephrology ; v. 169. 0302-5144
[DNLM: 1. Kidney Diseases--diagnosis. 2. Kidney Diseases--etiology. 3. Models, Animal. W1 CO778UN v.169 2011 / WJ 300]
RC903.9.E97 2011
616.6'14--dc22

2010046760

Bibliographic Indices. This publication is listed in bibliographic services, including Current Contents® and Index Medicus.

Drug Dosage. The authors and the publisher have exerted every effort to ensure that drug selection and dosage set forth in this text are in accord with current recommendations and practice at the time of publication. However, in view of ongoing research, changes in government regulations, and the constant flow of information relating to drug therapy and drug reactions, the reader is urged to check the package insert for each drug for any change in indications and dosage and for added warnings and precautions. This is particularly important when the recommended agent is a new and/or infrequently employed drug.

www.karger.com
Printed in Switzerland on acid-free and non-aging paper (ISO 9706) by Reinhardt Druck, Basel
ISSN 0302–5144
ISBN 978–3–8055–9537–7
e-ISBN 978–3–8055–9538–4

Contents

Foreword

Experimental Models for Renal Diseases: Impact on Understanding Pathogenesis and Improving Diagnosis

The need for this book became apparent as discussions with colleagues clearly indicated that our understanding of pathogenesis and ability to accurately classify and diagnose renal diseases has improved considerably in the last two decades thanks to basic and translational research efforts and that this information has not been collated and organized in a proper format making it available to students, diagnosticians, researchers and/or teachers of renal diseases. A single source in which to find this information presented in an up-to-date, succinct, yet comprehensive manner is not currently available. There are excellent textbooks addressing clinical diagnosis and pathology of renal diseases and in these there are, sometimes, incursions into pathogenesis, but generally not an in-depth discussion of how research has impacted our knowledge and understanding of the pathogenesis of renal diseases and/or improved our ability to diagnose and treat them.

I am satisfied to have assembled many of the world leaders in the various renal disorders to contribute to this book. These scientists have themselves contributed with their own research to the advancement of our understanding of renal diseases. Together we have condensed vast amounts of information into relatively short chapters due to the usual space constraints that are imposed by the publisher. This is only possible when the writer's understanding of the subject matter is clear and comprehensive making it possible to synthesize the knowledge in an insightful fashion. In many instances, these collaborators have developed their careers making their own seminal contributions to the literature. To these giants in the field of renal diseases, I am forever indebted for the time they have spent collecting, organizing and presenting the material in a coherent and comprehensive manner while at the same time conveying the concepts in a superb fashion, making the material amenable to be used for didactic purposes.

The famous anthropologist Claude Levi-Strauss has said, 'the scientific mind does not so much provide the right answers as to ask the right questions', emphasizing that our understanding of scientific issues is continuously being challenged. It is only through persistent questioning that the so-called established 'dogmas' are tested and that eventually advances in science occur. If the wrong question is asked in the first place, undoubtedly the wrong answer will be obtained when the research is done. On the other hand, if the right question is asked, an answer of value (though not necessarily unequivocal or final), but one that will likely advance our knowledge will emerge. The authors of the chapters have excelled in their careers by persistently asking the right questions and, as a result, have discovered and clarified information, helping in this way to formulate our vision of the present and the future, as it pertains to the understanding of the pathophysiologic mechanisms involved in renal disorders.

Without their original and seminal contributions to the various subjects, scientific progress would have been thwarted. Undoubtedly, without the efforts of these dedicated scientists, the treatment and management of patients with renal disorders would have been adversely affected. I feel honored to join them in sharing with the readers new knowledge. In doing so, we are able to explore some unique perspectives in the various subjects that I have carefully selected with our readers in mind. In some chapters, some controversial subjects are discussed with views expressed that indicate how these issues stand today. One of the purposes of doing this is to show our readers areas of future potential research interest.

It is the intent of all contributors that this book be used by students of various disciplines, clinicians and investigators alike and by all those trying to correlate basic research information with clinical issues and value the uniqueness of the translational approach to study renal diseases. At the end of many of the chapters, a summary of its contents is provided. This is often accomplished with a chart to facilitate the understanding of the information and to make it most useful for didactic purposes.

The initial chapters in the book discuss basic concepts and principles of renal tissue reactions to injurious agents using a specific cell/compartment approach. Since the patterns of tissue reaction are limited and different injurious agents lead to similar tissue alterations, understanding these is crucial to conceptualize pathogenesis in a logical fashion. I believe this to be a necessary preamble to discussing the specific entities. Although it is impossible to avoid some repetition, I have tried to minimize this during the editing process.

In this the era of molecular biology, disease processes associated with renal diseases have become much better understood. Specific steps in pathogenic cascades that can be controlled or modulated have been discovered. The elucidation and conceptualization of the complexity of cell reactions and tissue responses has advanced considerably and there is now a sophisticated understanding of molecular pharmacokinetics. New therapeutic interventions are now possible as a result of the fact that sound and thorough mechanistic processes have been dissected in many diseases. New therapeutic interventions have been designed with these molecular pathways in mind. New therapies have improved the treatment of these diseases considerably thanks to the basic researchers and clinical scientists that have contributed so much to decipher key pathogenetic events that can be targeted by appropriate drugs. The ultimate beneficiaries of all this hard work are undoubtedly the patients who can now be treated based upon scientifically sound data rather than empirically. As a result, these patients now enjoy improved prognoses with longer, healthier lives.

It should be mentioned that the level of sophistication of the knowledge that patients command today in regards to their diseases is truly commendable. Patients have rather unrestricted access to information in the web to study; they can analyze research data generated by basic and translational research, including those obtained through carefully designed clinical trials conducted in relation to their specific ailments. Today many patients are very well informed and expect health care practitioners to be up to date with the latest advances in the various fields. Therefore, it behooves physicians to keep abreast of developments in the research arena which translate to patient care. This can be a daunting task since pertinent information is proliferating at a very a fast pace, requiring careful analysis of the data and proper evaluation of its meaning and significance in terms of clinical relevance.

I have tried to make this book a concise source where one can obtain most of the valuable information currently available regarding the topics addressed. I hope the readers find this book a useful reference and, above all, a resource for learning, conceptualizing and/or reviewing concepts related to the pathophysiology of renal diseases. My expectation is that this book can also be useful for teaching about the exciting and ever changing field of renal diseases.

Dedication

This book is dedicated to my wife Dr. Elba A. Turbat-Herrera who has been a source of inspiration and encouragement for me in good and bad times, and who has always believed in me. She has insatiably worked at my side as a close collaborator. Her innovative ideas have been crucial for the development and advancement of the research endeavors that we have conducted over the last 30 years. Without her, my career could have never matured. Our relationship has been a unique interactive professional experience that I have valued very much.

I also dedicate this book to my three grandsons: Athanasius Joseph, Linus Cyril and Elias More Brown who may one day read this book and use it as an inspiration to dabble in the world of science and medicine. Perhaps this book will engender or fuel in them the desire to make their own contributions to the scientific world.

Guillermo A. Herrera, MD
Tempe, Arizona, USA
May, 2010

Herrera GA (ed): Experimental Models for Renal Diseases: Pathogenesis and Diagnosis.
Contrib Nephrol. Basel, Karger, 2011, vol 169, pp 1–5

Overview of Models for the Study of Renal Disease

Elba A. Turbat-Herrera

Bostwick Laboratories, Tempe, Ariz., USA

Abstract

This manuscript presents the usefulness of various experimental models in research with the purpose to elucidate mechanisms of disease and help create treatment modalities. The emphasis is on the use of models primarily for the study and treatment of amyloidosis. This is a rare condition that has attracted much attention recently. Amyloidosis is a disease in which the information obtained in the research laboratory has played a key role in advancing our understanding of the mechanisms associated with this group of disorders and in the development of new avenues for therapeutic intervention. Both in-vivo and in-vitro models are addressed. The early animal models and the more recent transgenic models are presented along with their usefulness to study the pathogenesis and dissect gene expression profiles where appropriate. Their pros and cons of the various experimental settings are highlighted. In-vitro models are presented with examples from the literature as to how they have or are being used to study a variety of conditions, renal as well as others. They also serve the purpose of resolving questions primarily relating to pathogenesis and to test the value of novel interventions prior to using in vivo models that are generally more expensive and difficult to work with. Since all models have their limitations, perhaps the best way is to utilize both in-vitro and in-vivo platforms, selecting them depending upon the questions to be answered. Employing a large variety of experimental techniques can serve to obtain and validate the data. The advantages of this approach will become obvious in the book chapters that follow.

Research should not be an end into itself, but ultimately be at the service of the patients, either to elucidate mechanisms of disease, discover how to cure a given condition or stall its progression. The days when research could be performed with human volunteers are over, as human testing has been deemed unethical. Therefore, turning to models of disease for study purposes represents the next logical step. The study models can be either in vivo (animal) or in vitro models, each with its own benefits/advantages.

Recent advances in molecular biology and stem cell biology have allowed animal models to be more useful, permitting researchers to create custom-made animals for their particular needs. Non-human primates have been and are still used for the study of some human diseases. Being phylogenetically closest to humans, it makes sense that these would be the most appropriate animal models. Primates, however, are expensive to purchase and care for, and since their lifespan is almost as long as humans, the studies become long, tedious and expensive. An example of the use of primates as animal models is the vervet monkey (*Cercopithecus aethiops*) which is being used for the study of serum protein transthyretin (TTR) amyloidosis. This is a good model for human TTR related amyloidosis. In the study by Ueda et al. [1], these animals developed systemic amyloidosis, with development of cardiac involvement. Other large mammals have also been employed as animal models, especially dogs which have been preferred for the classic physiologic studies. Rodents, particularly mice, have been increasingly used in studies as models of human disease since the early 1980's. They are excellent models for the different human diseases since they are relatively inexpensive, have a short lifespan and are easy to care for and manipulate. Genetically engineered mouse models are now accepted and widely used as models for the understanding of the genetic basis of human disease [2].

Animal models are essential for medical research. There are many uses for animal models such as the designing and implementation of imaging techniques, invasive procedures, and other studies that can only be performed in animals. In addition, important questions can be answered and new therapies can be developed and implemented for the treatment of human patients. In a similar manner, drug testing is almost always performed in animals first to help evaluate toxicity and efficiency. The toxicity of different compounds and efficiency of drug delivery can also be effectively tested in vitro and even better in treated animals by evaluating tissue injury and the organ/organ system dysfunction that may occur in association with a given therapy.

Animals such as rats, guinea pigs and mice, to name just a few, have been employed as animal models. As a particular species is selected, the inter- and intraspecies variability of these must be taken into account. For example, different strains of mice have variable penetrance for genetic alterations which makes them react differently to environmental conditions. For example, in TTR mice the deposition of amyloid and the extent of such is different from other strains of mice. Some of the characteristics of the different species need to be understood before a certain model is created. Examples are the fact that the induction of serum amyloid A (SAA) is species specific [3] or how barriers such as the central or peripheral nervous system in rodents are different from those in higher animals, especially humans. Some modulating factors are different in different species and even in different strains of mice. For example, reaction to injury, inflammation and the response to heat shock proteins, as well as others vary among the different species and even between strains. Even the aging process can vary in different strains such as in the senescence-accelerated mouse model. These animals have a decreased lifespan allowing for studies where the aging process accelerates a condition or in studies of the effects of the aging process itself.

There are two methods by which animals have been and can be used effectively for research. One is by the injection of chemical and biologic compounds that elicit the desired effect. The other is by utilizing animals that already have a defect that creates the desired effect. This could be a species-specific difference with humans or the product of genetic engineering or manipulation. An example of the first method in which animal models can be used is by injecting mice with either casein, silver nitrate or lipopolysaccharide to ultimately produce amyloid through the stimulation of cytokines such as tumor necrosis factor and other mediators of the inflammatory response: interleukins 1 and 6 (IL-1 and IL-6) stimulating the production of SAA protein [4, 5]. Some of the first experimental models for the study of kidney diseases were conducted at the turn of the 20th century when the hypotheses for the origin of glomerulonephritides were being elucidated. The prevailing thoughts at the time about the pathogenesis of these conditions were that they were probably associated with bacterial infections. In these early studies of the etiology of the glomerulopathies, the investigators attempted to produce disease in the animals by injecting bacterial toxins or chemical compounds [6]. The usefulness of this type of model is limited since it requires continued external stimulation in order to maintain the desired effect with regression of the findings when these stop [7]. The second method is by using transgenic animals as models of disease. The use of transgenic animals has expanded the usefulness of animals in research. Transgenics are animals that have been engineered by the insertion of DNA carrying the genetic modification desired by inserting the appropriate gene/genes into the mouse genome. In this manner, one can produce the desired disease process in order to facilitate the study of a certain condition. The transgenic animal would then express either a nonmutated protein, a dominant-negative form of a protein or a fluorescent-tagged protein. Although the classic definition of transgenesis involves the insertion of DNA from one species into the genome of another species, the definition at the present time encompasses any mouse that has as part of its genome that of either another mouse or from an animal of another species. Knockout and knock-in mice are also encompassed under the term transgenic mice [8]. A good example is the mouse engineered by the human insertion of the IL-6 gene under the control of the metallothionein-1 promoter that increases the levels of SAA protein and develop amyloid deposits in the spleen, liver and kidneys by three months of age [7].

Animal models can also be created by crossbreeding different animals of the same species in order to produce the desired phenotypes. In studies of hypertension, researchers have produced inbred strains, congenic strains, transgenic animals and recombinant inbred strains to produce the genotypes desired to obtain a particular effect. Several models of hypertension have used this approach. Genetically engineered rats are now available, some of them specifically designed for expression of the genes involved in hypertension, for the detection and study of causes of HTN with the advantage that the study of the specific hypertensive phenotypes is now possible [9]. Another powerful reason to explain why mice have been utilized more often than other species is because their gene expression can be manipulated with ease making gene targeting possible in these animals. The creation of knockout animal models or those which overexpress a certain gene (knock-in) to serve as models has been possible, especially in mice, in order to better study different diseases. The study of gene function and regulation can be accomplished by studying models of disease. The creation of gene overexpression in animals can also be useful in studying pathways of injury. Useful mouse models (single knockout or overexpression) of renal disease have reproduced diseases that are very close to the human counterparts, allowing for the understanding and study of these. Doxycycline-inducible SAA is being used in mice to increase the production of amyloid with amyloid production regressing when doxycycline is removed causing a decrease in SAA and resolution [10].

Although it is desirable for animal models to replicate human disease as closely as possible, it must be understood that, at times, the histologic appearance of the animal tissues and the tissue response to injury may lack some of the features exhibited by human tissues. Mesangial expansion and glomerular basement membrane thickening can be found in animal models with impaired glucose tolerance. However, the Kimmelstiel-Wilson nodular glomerulosclerosis lesion of diabetes, and the development of hypertensive lesions do not develop in the animal models [11], but a nodular form of mesangial expansion resembling (but not identical to) human glomerulosclerosis [12].

When working with animals and before selecting a particular species to use in a research, one must be aware that many disease processes do not acquire their full expression or consequences that can be seen in humans. In some cases, since rats can express some of the desired disease processes, scientists should consider using these instead of mice for their studies. In the study of hypertension and all its secondary effects, rats have been preferred over mice, since mice do not normally develop the renal sequelae of hypertension [11, 13]. Although it appears that mice are resistant to some of the more complex disease processes, some researchers have managed to override this. For example: Kirchhoff et al. [14] used a very resistant mouse strain to create a convincing mouse model of hypertensive renal injury. By adding angiotensin II to the required conditions of unilateral nephrectomy, saline in the drinking water and mineralocorticoids necessary to induce renal injury in rats, they managed to create renal injury in mice. In this study, the mice developed proteinuria and glomerular sclerosis in addition to cardiac enlargement and fibrosis.

Molecular mechanisms can be addressed by in vivo studies by utilizing animal models such as in the study of the formation and resolution of amyloid deposits, and in doing so, attempt to elucidate the pathogenic mechanisms involved in the process of amyloid deposition for the ultimate treatment of patients. Activation of metalloproteinase-9 represents another mechanistic approach by which the remodeling of the diseased tissues can be studied. Although animal models are extremely useful in many ways, there are drawbacks to studying disease mechanisms in whole animals. In vitro models offer other alternatives: in working with cell cultures, one has the opportunity of choosing the specific type of cell hypothesized to be affected in the disease process being studied. These cells can then be manipulated genetically, molecularly and biochemically to recreate the desired effect or conditions. In addition, these techniques can be used in combination utilizing the strengths of each one in order to accomplish the desired goal. An example that illustrates this point is a study using a rat model to produce ischemia followed by reperfusion in order to map the temporal and spatial association between caveolin-1 and tubular renal cell epithelial cells and tubular damage. In addition, an in vitro model to study acute renal failure was set up by culture of renal tubular epithelial cells and arterial endothelial cells and then subjecting these to injury initiators for the study of the expression of caveolin proteins in order to learn the role of caveolin-1 in acute renal failure [15]. The collaboration of the different techniques has the advantage of one supplying the deficiencies of another.

In initial experiments, our lab used in vitro models for the study of amyloid formation. Mesangial cells obtained from both rat and human glomeruli can be grown either on coverslips or in culture plates, and when incubated with amyloidogenic proteins, such as light chains obtained from the urine of patients with AL-amyloidosis, they form amyloid [16]. Another advantage of in vitro studies is the ease by which one can dissect a biologic/pathologic process by changing the variables or conditions of the experiment and evaluating the results. For example in another study from our lab, human mesangial cells were grown with AL-LC, light chain deposition disease-LC and tubulopathic LC. Instead of growing them with amyloid enhancing factor which had produced amyloid in previous studies [5], the AEF was omitted in this experiment and amyloid was produced, proving that AEF was not required [16]. In addition, the cells were also grown with chloroquine, known to inhibit lysosomal activity by increasing their pH, since the hypothesis was that lysosomal proteolysis was required for amyloid to be produced. Some of the cells were grown with TGF-β, which is known to mediate formation of extracellular matrix, and amyloid formation decreased. In contrast, thrombospondin increased production of amyloid, likely by a chaperone effect [17].

The limitations of in vitro studies can be compensated by utilizing cell-specific knockout and knock-in systems for genes or mutated genes in mice. By means of these systems, the in vitro data obtained by cell culture studies can be validated by in vivo systems [18]. Using in vivo models to further test data elucidated by in vitro models represents the next logical step.

References

1 Ueda M, Ageyama N, Nakamura S, Chambers JK, Misumi Y, Nakaguma R, Shinriki S, Shono M, Jono H, Obayashi K, Une Y, Ando Y: Analyses of vervet monkeys suffering from ATTR amyloidosis: a model animal for human TTR related amyloidosis. Amyloid 2010;17(suppl 1):47(A).

2 Brayton C, Justice M, Montgomery CA: Evaluating mutant mice: anatomic Pathology. Vet Pathol 2001;38:1–19.

3 Saraiva MJ: State of the art lecture, plenary lecture. XII International Symposium on Amyloidosis: From Molecular Mechanisms Toward the Cure of Systemic Amyloidosis, Rome, 2010.

4 Skinner M, Shirahama T, Benson MD, Cohen AS: Murine amyloid protein AA in casein induced experimental amyloidosis. Lab Invest 1997;36:420–427.

5 Kisilevsky R, Young ID: Reactive amyloidosis and the acute-phase response. Pathogenesis of amyloidosis. Baillieres Clin Rheumatol 1994;8:613–626.

6 Hoedemaeker PJ, Weening JJ: Relevance of experimental models for human nephropathology. Symposium on the pathogenetiic mechanisms in nephritis. Kidney Int 1989;35: 1015–1025.

7 Solomon A, Weiss DT, Schell M, Hrncic R, Murphy CL, Wall J, McGavin MD, Pan HJ, Kabalka GW, Paulus MJ: Transgenic mouse model of AA amyloidosis. Am J Pathol 1992; 154:1267–1272.

8 Cell Migration Consortium: Transgenic and knock-out mouse – approaches. Nature 2007; 359:211–215.

9 Yagil Y, Yagil C: Genetic models of hypertension in experimental animals. Exp Nephrol 2001;9:1–9.

10 Simons P, Al-Shawi R, Ellmerich S, Slamova I, Bodin KG, Minogue S, Pepys MB, Hawkins PN: Controllable expression of SAA in transgenic mice: a refined model of systemic amyloidosis. Plenary session, animal models. XII International Symposium on Amyloidosis: From Molecular Mechanisms Toward the Cure of Systemic Amyloidosis, Rome, 2010.

11 Sustak K, Bitzer M, Meyer TW, Hostetter TH: Animal models of renal disease. Kidney Int 2008;73:526–528.

12 Velazquez MT, Kimmel PL, Michaelis OE 4th: Animal models of spontaneous kidney disease. FASEB J 1990;4:2850–2859.

13 Badyal DK, Lata H, Dadhich AP: Animal models of hypertension and effect of drugs. Indian J Pharm 2003;35:349–362.

14 Kirchhoff F, Krebs C, Abdulhag UN, et al: Rapid development of severe end- organ damage in C57BL/6 mice by combining DOCA salt and angiotensin II. Kidney Int 2008;73:643–650.

15 Mahmouidi M, Willgoss D, Cuttle L, Yang T, Pat B, Winterford C, Endre Z, Johnson DW, Gobe GC: In vivo and in vitro models demonstrate a role for caveolin-1 in the pathogenesis of ischaemic acute renal failure. J Pathol 2003;200:396–405.

16 Tagouri YM, Sanders PW, Picken MM, Siegal GP, Kerby JD, Herrera GA: In vitro AL-amyloid formation by rat and human mesangial cells. Lab Invest 1996;74;290–302.

17 Isaac J, Kerby JD, Russell WJ, Dempsey SC, Sanders PW, Herrera GA: In vitro modulation of AL-amyloid formation by human mesangial cells exposed to amyloidogenic light chains. Amyloid Int J Exp Clin Invest 1998;5:238–246.

18 Schlondorff DO: Overview of factors contributing to the pathophysiology of progressive renal disease. Kidney Int 2008;74: 860–866.

Elba A. Turbat-Herrera, MD
Associate Medical Director
Bostwick Laboratories Inc.
1700 N Desert Dr.
Tempe, AZ 85 281 (USA)
Tel. +1 602 443 6119 ext. 4110, Fax +1 602 443 6102, E-Mail elba.herrera77@gmail.com

Herrera GA (ed): Experimental Models for Renal Diseases: Pathogenesis and Diagnosis.
Contrib Nephrol. Basel, Karger, 2011, vol 169, pp 6–22

Mesangial Homeostasis and Pathobiology: Their Role in Health and Disease

Guillermo A. Herrera[a] · Elba A. Turbat-Herrera[b] · Jiamin Teng[a]

[a]Nephrocor and [b]Bostwick Laboratories, Tempe, Ariz., USA

Abstract

Mesangial homeostasis is an integral component of normal glomerular function. Alterations in mesangial homeostasis occur frequently, not only in primary glomerular disorders, but also in association with primary tubular interstitial and vascular pathology, although generally the disturbances are not as marked in the latter situations. Mesangial changes could be transitory and reversible or permanent and irreversible, depending on the type and degree of damage inflicted and the reparative ability of the mesangium at a given time. Understanding mesangial pathobiology is crucial for comprehending the reactive and pathological processes that occur in glomeruli. The mesangium is usually the first to react to injurious glomerular events and is often the last to return to normal after the pathological insult has ceased and repair mechanisms have been activated. This is obvious in renal biopsy specimens where mesangial hypercellularity and/or matrix expansion are very common findings in primary glomerular disorders and, as a reactive phenomenon, in primary interstitial and vascular diseases. Repairing mesangial damage represents a fundamental process needed for restoring glomerular function. Since a component of the mesangial damage frequently includes the loss of mesangial cells, a way to revamp mesangial cellularity is essential for restoring mesangial homeostasis. This fact should be taken into account when designing therapies aimed at restoring mesangial integrity and homeostasis.

The last glomerular compartment to be discovered and characterized was the mesangium. Zimmermann [1] in 1933 recognized mesangial cells as a distinct cell type. It took several decades to realize that a central glomerular stalk was responsible for maintaining the structural and biological backbone of the glomerulus. The mesangium is essential for regulating many critical glomerular activities (functions). It is clear that when mesangial integrity is lost, proper glomerular function ceases. The most common pathological process responsible for this is referred to as glomerulosclerosis. In this process, the mesangial matrix is expanded and acquires increased quantities of existing matrix components, such as collagen IV and other glycoproteins innate to the mesangium, and it may also accumulate proteins that are not normally found, such as fibrillary collagen (i.e. collagen I and III) [2–4].

Another process is mesangiolysis; however, it is not as common. It may be conceptualized as the opposite of glomerulosclerosis in that the normal mesangial matrix disappears as a result of the excess action of activated metalloproteinases [5]. It appears that mesangiolysis is perhaps easier to repair than glomerulosclerosis, as mesangiolysis only requires rebuilding of the mesangial matrix, while glomerulosclerosis needs breaking down

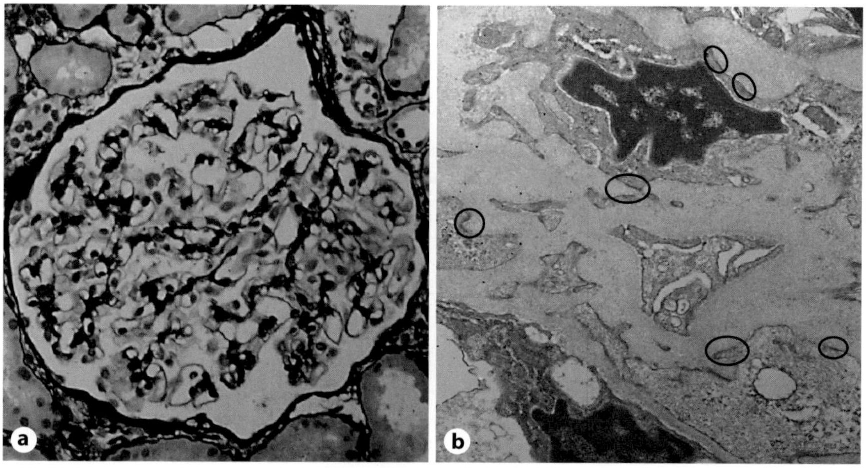

Fig. 1. Normal mesangium. **a** Silver methenamine stain; ×350. **b** Transmission electron microscopy, uranyl acetate and lead citrate; ×8,000. **a** Silver stain highlights normal mesangial areas and peripheral capillary walls in a normal glomerulus. **b** Appearance of a normal mesangium with mesangial cells and surrounding matrix. Cytoplasmic processes detached from main cell bodies and a few mesangial cell bodies are illustrated. The circled areas show attachment plaques on mesangial cells where their filamentous cytoskeletal elements anchor.

of the sclerotic matrix before adequate repair can occur. Data originating from experimental studies and anecdotal human reports support that there is potential for regression after glomerular damage [6, 7].

This chapter will review mechanisms involved in mesangial homeostasis, alterations that may occur due to a variety of injurious agents and conceptual ideas as to how mesangial repair can be accomplished.

Normal and Abnormal Mesangium: Composition and Consequences of Altered Mesangium. A General Overview

The mesangium is composed of mesangial cells and matrix (fig. 1a, b). Mesangial cells constitute 30–40% of the total glomerular cell population. They spread from the hilum in an arboreal pattern. The mesangial cells are embedded in their own matrix, which they manufacture and remodel. Silver stain is routinely used to evaluate glomerular basement membranes and mesangium; the mesangial matrix is silver-positive (fig. 1a). Mesangial cells have continuity and maintain close contact with cells populating the juxtaglomerular apparatus, thus monitoring and regulating its activity [8–11].

In a seminal paper in 1962, Farquhar and Palade provided for the first time evidence that mesangial cells exhibit distinct morphological and functional characteristics that separate them from endothelial and epithelial glomerular cells. The authors speculated that the mesangial cells likely participate in the disposal and degradation of filtration residues and emphasized that these cells closely resemble pericytes morphologically. Both assessments have proven to be accurate [12].

Two types of mesangial cells have been recognized in the normal mesangium. The predominant cell type displays a smooth muscle phenotype with morphological and functional characteristics akin to smooth muscle cells. These cells show branching cytoplasmic processes which are often seen unattached to the main cell bodies (fig. 1b). Their most characteristic cytoplasmic components are filament bundles rich in

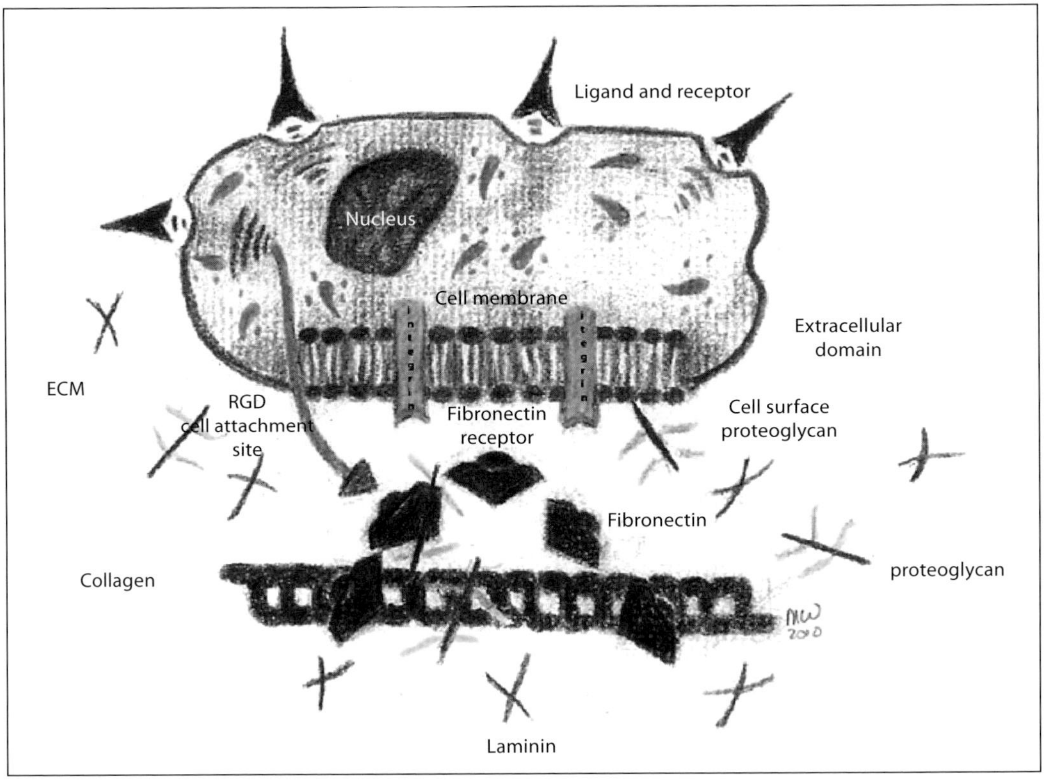

Fig. 2. Schematic representation of normal mesangial cell and matrix. The illustration highlights the presence of receptors at the cell membranes present to interact with various effector molecules either to maintain mesangial homeostasis or to elicit pathologic changes.

actin maintained together by spindle densities or dense bodies. These are present throughout the cytoplasmic matrix. Mesangial cells also have thickening of their cell membranes, known as attachment plaques, where the filaments anchor (fig. 1b). The morphological attributes mentioned above make mesangial cells fit for contractile activities and allow them to maintain glomerular turgor and integrity resulting in wide open glomerular capillaries. The second type of mesangial cells is much less common, accounting for approximately 3–10% of all mesangial cells, and exhibits phagocytic properties. It has been proposed that these cells are derived from bone marrow and they have been shown to express Fc and C3 receptors [7, 8, 13, 14].

The normal mesangial matrix predominantly contains collagen IV, but also has other extracellular matrix glycoproteins, including laminin and fibronectin, as well as proteoglycans such as biglycan and decorin. Ultrastructurally, the mesangial matrix is composed of a dense network of microfibrils 6–8 nm in diameter of variable length, which serve the purpose of anchoring the mesangial cells [8, 9] (fig. 1b). These microfibrils focally connect mesangial cell processes to the glomerular basement membranes, most prominently at mesangial angles. In some areas, these mesangial microfibrils penetrate into the lamina densa of the glomerular basement membrane [10]. A schematic representation of the mesangial cells and matrix is shown in figure 2.

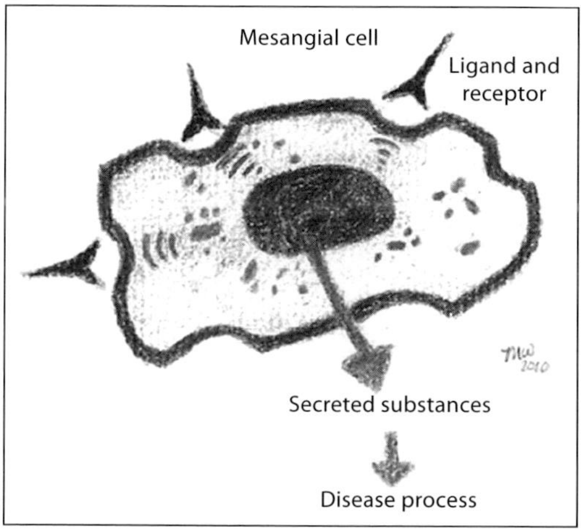

Fig. 3. Schematic representation of the genesis of mesangial sequence of pathologic events. Specific effector molecules interact with cell receptors eliciting the transmission of signals to the nuclei resulting in secretion of a variety of substances that eventually alter the surrounding matrix, creating generic alterations resulting in decreased or increased and biochemically altered matrix.

Plasma percolates through the mesangium under the control of hemodynamic forces. The digestibility of the various molecules reaching the mesangium (i.e. the ability to be cleared by normal cellular mechanisms) determines their clearance and whether they may exert pathologic effects by virtue of remaining in contact with the mesangium for prolonged periods of time. Also as molecules reach the mesangium, they may activate mesangial receptors and initiate effector loops in mesangial cells that may result in pathological mesangial alterations [15] (fig. 3).

Normal mesangial cell functions include regulation of glomerular capillary blood flow, maintenance of glomerular structure, production of vasoactive substances and generation of the mesangial matrix. Mesangial cells help to maintain the structure and function of the glomerular ultrafiltration apparatus and regulate the amount and content of the surrounding extracellular matrix.

Normal mesangial turnover is the result of a tightly controlled and, at the same time, dynamic equilibrium between the synthesis of renovated matrix and removal of old matrix [2, 16, 17]. There are a number of key players that participate in regulating this balance in normal conditions, including growth factors, metalloproteinases (MMPs), tissue inhibitors of metalloproteinases (TIMPs), plasminogen activator inhibitor (PAI), as well as other nonspecific inhibitors such as α2-macroglobulin [5, 18–23]. Changes in the amount and composition of the matrix may have important effects in mesangial homeostasis and produce changes in the mesangium that may or may not be readily identified morphologically [3, 4, 6, 17, 20]. Appreciation of the events involved in mesangial pathobiology is paramount in order to understand normal mesangial function and the alterations that occur in renal disorders, predominantly those centered in the glomerulus [24].

Mesangial cells possess a variety of receptors localized on the surface of their cell membranes. Some of these receptors are co-shared by various effectors, while others are more specific. Among these receptors are those for growth factors including platelet-derived growth factor-β (PDGF-β) and transforming growth factor-β (TGF-β), mannose receptor, CD71 IgA1 Fc receptor, vascular endothelial growth factor (VEGF) receptor and additional Fc receptors, among others. In contrast, CD89-polymeric immunoglobulin (FcαR1) and cubilin/megalin receptors have been shown to be absent in mesangial cells [25–29]. Other receptors may be weakly expressed or not identifiable at all on mesangial cells in the normal kidney, but become readily detectable in pathologic conditions. For example, the ftl1 and KDR receptors, the former which is impossible to detect and the latter which is weakly expressed in normal mesangial cells, are easily identified in mesangial proliferative glomerulonephritis [30].

Recognition of receptors by various effector molecules, followed by engagement and interaction with them, results in activation and the generation of signals which influence mesangial function and may incite genes that are not normally active in these cells. As a result, mesangial cell activities are affected, various growth factors

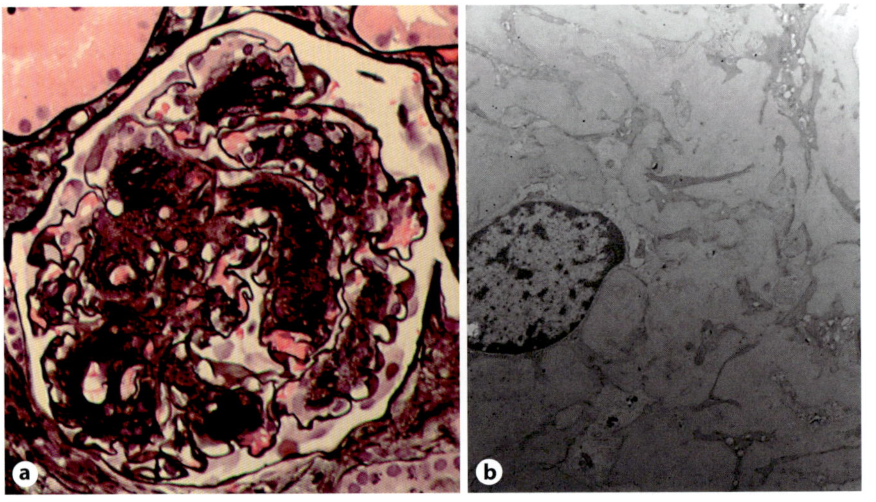

Fig. 4. Nodular glomerulosclerosis with an increased mesangial matrix (accentuated silver staining). **a** Silver methenamine stain; ×350. **b** Transmission electron microscopy, uranyl acetate and lead citrate; ×8,000. Mesangial nodule exhibits intense silver staining indicating increased matrix. **b** Increased mesangial matrix is obvious. Ultrastructurally, the increased mesangial matrix resembles normal matrix morphologically (see fig. 1b) in spite of different biochemical composition. In some instances, other mesangial matrix constituents (such as fibrillary collagen), not normally present in normal mesangium, are identified in the expanded/sclerotic mesangial areas.

and cytokines are secreted by the mesangial cells, and the surrounding extracellular matrix may be altered, further affecting mesangial homeostasis and cellular behavior [31–33] resulting in the morphologic manifestations of the disease process recognizable by light and electron microscopy (fig. 3). This sequence of events could create a self-activating loop that can be perpetuated unless the inciting event is stopped. When mesangial cell homeostasis is affected, it generally results in matrix modifications. These two components (i.e. mesangial cells and matrix) act very much in concert. This is the reason why morphological findings in renal biopsies where the mesangium is altered typically reflect alterations in mesangial cells and surrounding matrix. An expanded matrix results in the deposition of silver-positive material (fig. 4a, b) and is triggered by secretion and activation of TGF-β. In contrast, the disappearance of the normal mesangial matrix leads to loss of mesangial argyrophilia (fig. 5). In some instances, an expanded mesangium with loss of silver positivity is present, indicating that the destroyed mesangial matrix has been replaced by a non-matrix material which can be of variable composition. The pathological mechanisms involved in the disappearance of mesangial matrix may be complex, but conceptually speaking, generally involve secretion and activation of MMPs, which are the main effectors involved in the cleavage and destruction of matrix constituents.

Mesangial cell proliferation mediated by activation and secretion of PDGF-β represents an effective mechanism for bringing more cellular elements to an injured mesangium which could, in turn, accentuate the pathological alterations, result in merely a reactive process without significant consequences, and/or aid in the repair of a damaged mesangium. The factors that regulate the end result are multifactorial, but the nature, intensity and time frame associated with the injurious agents are indeed critical in defining the end result.

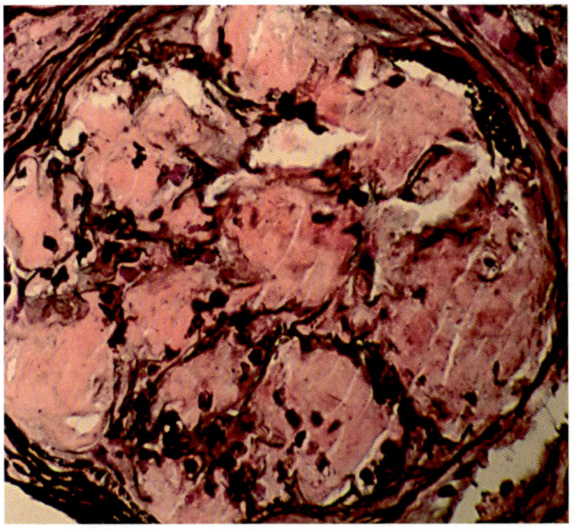

Fig. 5. Amyloidosis with destruction of mesangial matrix, which is replaced by amyloid fibrils. Silver methenamine stain; ×350. Marked loss of mesangial argyrophilia indicating disappearance of matrix.

If the injury is severe enough, mesangial cells may be affected beyond repair and disappear from the mesangium, usually by apoptosis [24, 34–38]. Necrosis occurs less often, but is seen in some forms of glomerulonephritis, most typically represented by those referred to as necrotizing glomerulonephritis. There is often loss of the glomerular integrity in these cases with breaks of capillary walls and leakage of fibrin into the urinary space. These conditions are often associated with crescent formation as a result of cellular proliferation, mostly parietal epithelial cells admixed with fibrin and inflammatory cells. Most of these are therefore necrotizing crescentic glomerulonephritis. In these instances the glomerular injury characteristically extends beyond the mesangium. Necrosis tends to be rapidly replaced by sclerosis and the mesangium likely plays a significant role in the replacement of necrotic tissue by sclerotic matrix.

Mesangial proliferation, mediated by PDGF-β, albeit sometimes mild, is a universal finding in the early stages of most, if not all, primary and secondary disorders that involve the mesangium [39, 40]. Even those diseases that progress to a hypocellular mesangium, such as amyloidosis and nodular glomerulosclerosis, regardless of specific etiology, demonstrate mesangial cell proliferative activity in their early stages.

Mesangial sclerosis is a common mesangial response to injury and is often accompanied by apoptosis [41]. In glomerulosclerosis, the mesangial cells secrete an excess of the normal matrix (i.e. collagen IV) and the activation of some normally quiescent genes results in the production of other matrix proteins that are not normal components of the mesangial matrix. One example of the latter situation is the production of tenascin (not a normal constituent of the mesangial matrix) in some diseases characterized by nodular mesangial glomerulosclerosis, such as in diabetic nephropathy and light chain deposition disease [42–45]. Tenascin has an octopus-like appearance and interacts tightly with surrounding extracellular matrix proteins and mesangial cells creating a stiff matrix. This pathologic matrix protein is not easy to cleave. Its breakdown requires the action of MMP-7 predominantly for its catabolism with minor contributions by MMP-1 and -3. The sclerotic mesangium often contains large quantities of tenascin [40, 43–46] (fig. 6). It is well documented that cell proliferation, differentiation and apoptosis are significantly influenced by the surrounding matrix [41], so an altered mesangial matrix has a great impact on these key cellular processes.

The Role of Integrins in Mesangium

Integrins are transmembrane cell adhesion glycoproteins consisting of αβ heterodimers that serve as anchoring molecules and play a role as receptors for extracellular matrix proteins (fig. 2). They also regulate various cellular behaviors, such as adhesion, spreading and cytoskeletal organization. Their main functions include mediating cell interactions with the environment and transducing signals from within a given cell and from other cells to the surrounding matrix and vice versa. Therefore, integrins are actively involved in signaling pathways. The ability of the integrins to transduce signals from other

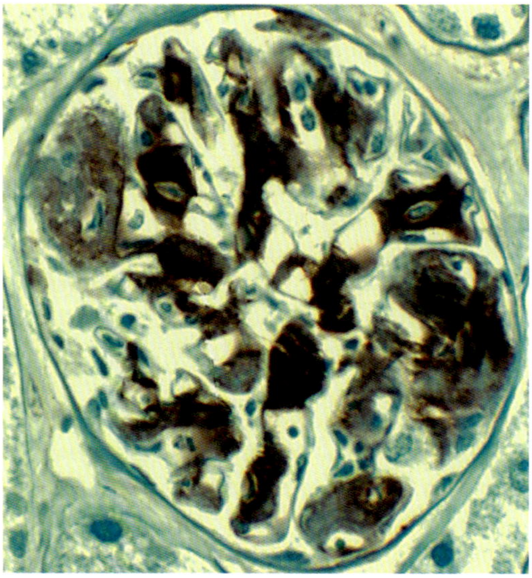

Fig. 6. Nodular glomerulosclerosis with tenascin in mesangium. Immunohistochemical stain for tenascin, with diaminobenzidine as a marker. Methyl green as a counterstain. Intense staining for tenascin in mesangial nodules. Tenascin is the main mesangial matrix constituent of the expanded mesangium in nodular glomerulosclerosis, regardless of the etiology for the nodular glomerulosclerosis.

cells allows them to function as a survival factor for many cell types. These functions are essential for maintaining mesangial homeostasis and play a crucial role in modulating pathological alterations that involve the mesangium [43–46]. Integrins have the capacity to modify transiently or permanently their specificity and affinity for a particular ligand or ligands, and they can do this quite rapidly.

Cross-talk between mesangial cells, extracellular matrix and soluble mediators affecting mesangial function and composition is regulated by integrins. Matrix and cytokines modulate mesangial cell behavior in vitro and in vivo after binding to specific integrin-associated cell surface receptors on these cells. Elucidation of the potential biologic and clinical relevance of cell-matrix interactions in the glomerular mesangium is a crucial issue in renal research, and will impact the quest for new therapeutic interventions aiming to control, modulate, ameliorate and potentially inhibit mesangial alterations that lead to glomerulosclerosis and the irreversibility of renal damage.

Integrins characteristically recognize a common motif, the arginine-glycine-aspartic acid tripeptide (RGD) site on cell membranes for interacting with extracellular matrix proteins (fig. 2). Most integrins recognize one or more matrix proteins as ligands. For example, the $\alpha 9\beta 1$ integrin is a ligand for tenascin and plays a key role in the tight and complex attachments to the mesangial matrix characteristic of this protein [43]. This is partly responsible for the difficulty to remove this protein from the sclerotic mesangium in nodular glomerulosclerosis. Interestingly, $\alpha 9\beta 1$ does not appear to require the RGD site for binding. The integrin $\alpha 9\beta 1$ mediates cell attachment to a non-RGD site in the third fibronectin type III repeat of tenascin [46].

Mesangial Homeostasis: The Role of Growth Factors/Cytokines

Growth factors merely represent a type of cytokines, although in the literature they are often kept separate from the other cytokines. In this chapter they will be separated from other cytokines, as their functions and those of other cytokines, although intertwined, can be reasonably separated in the normal and abnormal mesangium, thus allowing a better understanding of how the mesangial alterations occur in various pathologic settings. The interplay of the various growth factors involved in the pathogenesis of glomerular disorders involving the mesangium are important for understanding what happens when pathological events occur and what may be required to repair the damage inflicted. Normally, cytokines act to maintain homeostasis. Deregulation of their functions is responsible for a number of disease processes, such as some autoimmune, degenerative and fibrotic diseases.

PDGF-β is mitogenic for mesangial cells. It governs mesangial cell proliferation, a crucial

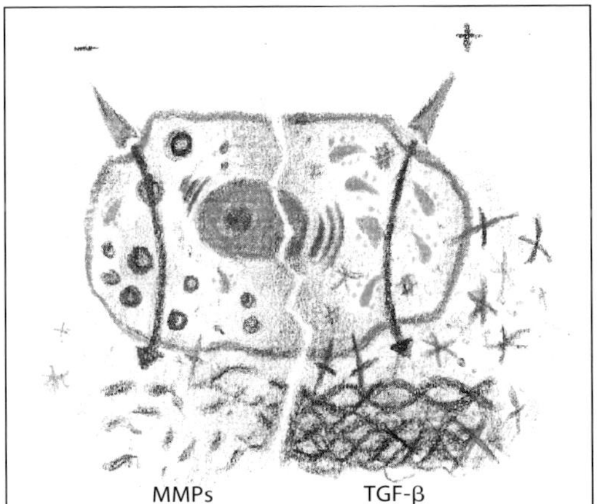

Fig. 7. Schematic representation of main factors participating in degradation and production of mesangial matrix. The diagram pictorially highlights consequences of the action of various factors on mesangial cells, resulting in either mesangial matrix increase (right) or mesangiolysis (left). MMPs are mainly responsible for matrix degradation in maintaining mesangial homeostasis and in pathological processes (mesangiolysis), while TGF-β is the main factor responsible for rebuilding damaged mesangium and producing increased mesangial matrix in pathological conditions (see fig. 8).

initial event in most (if not all) pathological processes involving the mesangium [47–49]. In contrast, TGF-β is a key factor governing matrix production by mesangial cells. Transformation of mesangial cells from their normal smooth muscle to a myofibroblastic phenotype (with increased rough endoplasmic reticulum) is an important step for priming these mesangial cells to engage in the active production of extracellular matrix proteins (fig. 7, right). This growth factor is responsible for the production of extracellular matrix components that are either normally present in the mesangium or other matrix protein that are not typically present in the normal mesangium and only seen in pathologic conditions.

TGF-β is unique among the growth factors in that it performs three fundamental functions which are clearly exemplified in the mesangium: (1) stimulating the synthesis of the matrix, (2) inhibiting matrix degradation by controlling the balance of MMPs/TIMPs and (3) modulating matrix receptors to increase adhesion of cells to the matrix and strengthen the mesangium [49–56]. One of the most important controlling loops is that once TGF-β is activated and mesangial matrix production is increased to a certain point, there is inhibition of PDGF-β. However, this important mechanism to maintain homeostasis is lost in some glomerulonephritis, resulting in concomitant mesangial cell proliferation and overproduction of matrix, entering a 'vicious' cycle that results in irreversible damage if not controlled. Although PDGF-β and fibroblast growth factor act synergistically in the presence of TGF-β, their predominant actions are mitogenesis and angiogenesis, respectively. Neither of these two growth factors stimulates matrix production, as TGF-β controls this activity, albeit in complex ways.

Mesangial cells are extremely active metabolically and synthesize an array of substances, such as cytokines, which result in stimulation of cellular proliferation, cell shape changes, recruitment of inflammatory cells (including macrophages and monocytes), etc. The interactions between cytokines and mesangial cells result in many of the alterations of the mesangial matrix that occur during the process of immune and non-immune complex-mediated renal diseases. Both resident and non-resident glomerular cells secrete factors that may alter glomerular basement membrane permeability and regular blood flow. The native mesangial matrix is expanded as a result of deposition of increasing amounts of normally produced mesangial matrix components with the addition of other extracellular matrix components never present in the normal mesangium. The abnormal mesangium enhances further mesangial matrix expansion by stimulating mesangial cells to continue behaving abnormally, resulting in further departure from its normal composition. Sclerosis is the hallmark of irreversible mesangial injury, which essentially develops as a result of long-term uncontrolled activation of TGF-β signaling and matrix production. In this situation the usual feedback controlling mechanisms fail.

Other cytokines with important mesangial activities include interleukins (primarily IL-1, -4, -6, -10 and -13), tumor necrosis factor (TNF)-α and eicosanoids (lipoxins), among others. Glomerular inflammatory processes are controlled and made quiescent by anti-inflammatory cytokines, including IL-4 and IL-13, antagonists of pro-inflammatory cytokines such as IL-1 receptor antagonist and neuropoietic cytokines (for example, IL-6), and deactivators of inflammatory macrophages such as TGF-β and IL-10 [57]. These control inflammatory reactions by suppressing key mediators and functions (including their production, stability and actual secretion) necessary for the process to continue and generally become damaging to the mesangium and the glomerulus. Cytokines may also upregulate receptors on mesangial cells.

Some of these systems are redundant and complex. For example, VEGF is produced by mesangial cells and receptors on mesangial cells are upregulated by VEGF as well. Also PDGF-β and hypoxia induce the VEGF receptor. VEGF/VEGF receptor systems are frequently upregulated in glomerular diseases and play an important role in mesangial cell pathobiology in general.

Among the important inflammatory mediators released by mesangial cells is nitric oxide (NO), both an instigator and a target of inflammatory processes [58–62]. Inducible NO synthase is expressed in mesangial cells, as well as in resident and infiltrating macrophages. In glomerular injury, mesangial cells are exposed to endogenously produced NO and NO generated from local macrophages. NO and protein targets modified by NO may exert important regulatory control on mesangial cell function. NO has been shown to play important roles in a variety of mesangial cell functions, including DNA replication, transcription, energy metabolism and apoptosis.

The participation of MMPs and TIMPs is obviously intertwined with the activities of growth factors and cytokines. MMPs have been linked to pro-inflammatory cytokines, such as IL-1 and TNF-α, which induce transcription and expression of several MMPs in inflammatory cells (polymorphonuclear cells and monocytes) and in intrinsic glomerular cells, including mesangial cells. In contrast, TGF-β acts as an anti-inflammatory and profibrotic factor that downregulates expression of MMPs and enhances expression of TIMPs [53].

MMPs/TIMPs in Mesangial Homeostasis

MMPs are essential zinc-dependent enzymes that degrade normal mesangial matrix components playing a critical role in the continuous remodeling/turnover of the mesangial milieu [2, 63–65]. MMPs have several domains: a pro-domain, a catalytic domain, a hinge region and a homopexin-like domain. Most MMPs are secreted as pro-MMPs and their activation requires cleavage of the pro-domain by plasmin and other MMPs. There are approximately two dozen MMPs that generally function in concert to maintain general body homeostasis, but some MMPs are more important in some organs than others. MMPs are classified into six groups based on substrate and sequence homology: collagenases, gelatinases, stromelysins, membrane type of MMPs, and other MMPs. MMPs have a high degree of sequence homology, are synthesized as zymogens, and secreted either as free entities or transmembranely attached [66–75].

The most important mesangial MMPs are MMP-2, -5, -7 and -9. MMP-1, -8 and -13 play important roles in other renal compartments and their effects may indirectly or secondarily result in mesangial alterations. Although MMPs are considered to have some substrate specificity, it has been shown that more than one MMP may degrade the same proteins. In fact, there appears to be much cross-activity in protein degradation among them. Since the normal mesangium contains predominantly collagen IV, MMP-2 and -9 (also known as gelatinases or collagenases) play crucial roles in mesangial homeostasis in normal conditions. One of the complexities of the matrix regulatory system is that MMPs may exert different, and sometimes even opposite effects, during the various phases of the evolution of renal diseases. Closely related to the function of the MMPs are the activities of enzymes that stabilize ECM (transglutaminase) and cell receptors for the various ECM components, including integrins and discoidin domain receptor-1, as they also play important roles in the response of mesangial cells to matrix alterations.

In pathological states, other MMPs assume important roles. For example, it has been shown in an experimental model of membranous nephropathy that there is increased expression of MMP-9 by podocytes. This finding has been implicated in the increased basement membrane permeability which typically occurs in this condition [75]. In a mouse model of focal segmental glomerulosclerosis, a lack of *Mpv17* gene expression has been noted, leading to persistently high levels of MMP-2 resulting in heavy proteinuria and, ultimately, glomerulosclerosis [76]. In experimental lupus nephritis in mice, MMP-1, -2 and -3 have been shown to be increased [77]. In contrast, in both the anti-Thy1.1 and passive Heymann nephritis models, MMP-2 and -9 are the ones elevated [78]. Increased mesangial expression of MMP-9, but not MMP-2, has been reported in IgA nephropathy, mesangial proliferative glomerulonephritis and lupus nephritis [79]. Metalloproteinases also play an important pathogenetic role in the acute phase of crescentic glomerulonephritis [80].

Oxidative stress contributes to matrix accumulation, either directly if due to inhibition of MMP-2, or by inducing a cytokine response. Hyperglycemia has a detrimental effect on the mesangium. Glucose levels may indirectly influence matrix turnover by altering MMP function [81–83]. It has been proposed that advanced glycation end products decrease mesangial cell MMP-7 and this may be a mechanism of importance in matrix accumulation in nodular glomerulosclerosis, as seen in diabetic nephropathy [82]. In light chain deposition disease, it has been shown that mesangial cells, although producing MMP-7, have a defect in releasing this MMP into the matrix, resulting in accumulation of tenascin in the expanded mesangial areas [66].

MMPs have also been shown to play an important pathogenetic role in the acute phase of necrotizing crescentic glomerulonephritis. MMP-2, -3 and -9, as well as TIMP1, are active participants in the inflammatory process in ANCA-associated glomerulonephritis, particularly with MMP-9 expression, which correlates with active glomerular lesions, mesangiolysis and tubular interstititial inflammation [80]. MMP activation can also result in the degradation of unwanted material, such as amyloid fibrils, once they have been deposited in the glomerulus [84]. MMPs may also influence matrix turnover via the regulation of some growth factors, as previously alluded to.

Interstitial (fibrillary) collagens (collagens I, II and III) represent important, difficult to degrade, components of the sclerotic mesangium in many situations. MMP-1, -8 and -13 have been demonstrated to play an important role in degrading these collagens. One particular interesting mechanism involves MMP-1, which partially degrades interstitial collagens by making single site-specific cleavages within the triple helix of these complex matrix proteins, allowing for other MMPs to fully degrade them.

Although all this information has surfaced, the role of MMPs in many renal disorders is still not completely defined. Further studies using overexpression and knockout models of different MMPs and TIMPs will further define their roles in various forms of glomerulonephritis.

Decreased overall MMP activity results in accumulation of matrix proteins and renal scarring. Progressive glomerulosclerosis in many diseases is characterized by a profound shift in ECM turnover toward increased matrix accumulation as a result of inhibition of the action of MMPs (fig. 7, right). In contrast, excessive secretion and/or activation of MMPs results in loss of mesangial matrix manifested in tissue sections as decreased or complete loss of mesangial argyrophilia (fig. 5). In essence, glomerular MMP levels determine the slope of progression and degree of glomerular and interstitial scarring and loss of normal matrix constituents [85].

TIMPs are endogenous, specific inhibitors that bind and inhibit MMPs. Four TIMPs have been identified in vertebrates and at least three of them (TIMP-1, -2 and -3) are expressed in the kidney. The overall structure of the TIMPs permits these to fit into the active site of the catalytic domain of metalloproteinases. TIMPs inhibit all MMPs. A number of other inhibitors of MMPs have also been identified, including RECK, α2-macroglobulin and tissue factor pathway inhibitor-2. Some MMPs are secreted in a complex with TIMP-2 [86].

The transcription of TIMPs is regulated by similar cytokines and growth factors that control MMP expression (i.e. TGF-β, TNF-α, IL-1 and IL-6), although the actual regulatory functions may be carried out in distinctive ways. Other endogenous inhibitors include plasma protein α2-macroglobulin and a surface inhibitor of MMPs, the RECK inhibitor.

Activation of MMPs and/or inhibition of TIMPs leads to matrix destruction. In some (not all) instances where this occurs, mesangial cells transform from their normal to a macrophage phenotype, as is the case in amyloidosis (fig. 7, left).

MMPs and TIMPs may be targets of therapeutic intervention in the near future. Temporal control or modulation of these may be important in the management of some renal disorders in order to avoid irreversible glomerular damage.

Mesangial Cell Apoptosis: The Mechanism Involved in Cell Deletion in Mesangial Glomerulosclerosis and in Mesangiolysis

Glomerular diseases with mesangial sclerosis or mesangiolysis are typically associated with significant deletion of mesangial cells, primarily in their advanced stages. The sclerotic areas are characteristically hypocellular with the sclerotic matrix representing the main component of the altered mesangium in nodular glomerulosclerosis. The few remaining cells tend to be localized to the periphery of the sclerosed areas (fig. 8a). It has been shown that mesangial cell deletion occurs by apoptosis and this correlates with the lack of inflammatory components associated with the sclerotic foci. For example, a hyperglycemic mesangium in patients with diabetic nodular glomerulosclerosis leads to mesangial cell apoptosis, and glomerulopathic light chains induce mesangial cell apoptosis [37]. In both cases, mesangial glomerulosclerosis mediated through TGF-β activation is the final morphologic expression of the mesangial damage observed in these two diseases (fig. 8a, b). The same is true of many other non-inflammatory glomerular disorders. The mesangial matrix in these situations is equally altered and these alterations may be seen morphologically as some unexpected matrix constituents appear (i.e. fibrillary collagen) in the expanded/sclerotic mesangial areas. In inflammatory glomerular conditions, cell deletion may not be associated with apoptosis, but may occur as a result of necrosis.

Mesangial Interactions with Other Glomerular Components

Obviously, mesangial cells have to interact with other glomerular components in health and disease. A complicated, yet well-orchestrated, interaction is needed to maintain glomerular homeostasis. Also, as individual glomerular compartments are affected by injurious agents, they must communicate among themselves to coordinate an effective reaction with the purpose of protection from the injurious process, and if necessary, turn on healing mechanisms at the appropriate moment.

Growth factors/cytokines are also produced by glomerular cell types other than mesangial cells, and their activation may indirectly affect mesangial homeostasis leading to detectable changes [85]. Glomerular disorders that are predominantly targeted to other glomerular compartments almost always recruit mesangial cells to be part of the morphological and functional derangements that accompany such processes. Growth factor signaling pathways are altered as a consequence of mutation of genes expressed by podocytes (i.e. transcription factors) and this results in profound effects on endothelial and mesangial cells.

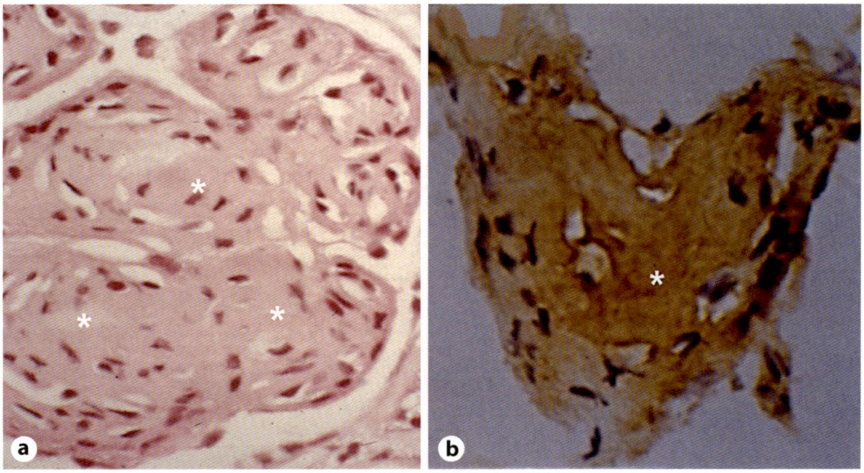

Fig. 8. Nodular glomerulosclerosis. Light chain deposition disease. **a** Hematoxylin and eosin stain; ×500. **b** Immunohistochemical stain for TGF-β, with diaminobenzidine (DAB) as a marker; ×500. **a** Mesangial nodules are clearly shown in this example of nodular glomerulosclerosis, with co-localization of TGF-β in the mesangial nodules (**b**). TGF-β activation and secretion by mesangial cells is responsible for the matrix increase seen in mesangial nodules. *Expanded mesangial areas with increased matrix.

Gene targeting experiments in mice and genetic studies in humans have shown a crucial role for molecular cross-talk among different glomerular cell types during development, maintenance of homeostasis and in disease processes.

Primary tubular interstitial and vascular renal disorders are associated with production and secretion of cytokines that may alter mesangial homeostasis and result in detectable changes, functionally and morphologically. These changes can be conceptualized as reactive, as they are a response to primary non-glomerular pathological events. The most common identifiable changes are those of mesangial cell proliferation and matrix accentuation. In some instances, it is challenging to determine with certainty whether the mesangial alterations are primary or secondary.

Another important area where interaction among different glomerular cell components is essential is regression of glomerulosclerosis. While mesangial cells can readily proliferate and replenish the mesangial areas, interactions with podocytes and endothelial cells are essential for rebuilding functional glomeruli. Podocytes are known to participate in the secretion of specific growth factors, including VEGF-A and angiopoietin-1, which are crucial for maintaining normal glomerular endothelial function. Without interaction of mesangial cells with podocytes and endothelial cells, glomerular healing cannot occur satisfactorily [16, 86].

Mesangial Proteomics

Proteomics has been defined as the systematic analysis for identity, quantity and function of proteins. Proteomics aims at studying proteins quantitatively and qualitatively using various gel-based and gel-free methods. Mass spectroscopy, a type of gel-free proteomics technology, is a powerful methodology used to identify proteins by either direct sequence analysis or by matching peptide fragments using patterns that allow recognition of particular proteins that permit the identification of specific proteins. For example, mass spectroscopy is emerging as the technique of choice for characterizing amyloid precursor proteins in tissues, thus permitting accurate classification of amyloidosis and proper treatment, as there are therapeutic interventions targeted to specific types of amyloidosis [87, 88].

Another important use of proteomics is to identify the target molecule(s) for drug design and produce compounds for therapy of renal diseases. For this purpose, understanding mesangial pathobiology is very important and characterization of mesangium-specific or dominant genes is essential as the great majority of glomerular disorders target the mesangium directly or indirectly. One such gene is megsin (a novel mesangial cell-specific gene) which is upregulated in IgA and diabetic nephropathy when compared with its expression in minimal change disease, membranous nephropathy and normal kidneys, where mesangial participation is not necessarily part of the disease process. It is not surprising that megsin has also been shown to be upregulated in anti-Thy1 nephritis. In essence, megsin appears to be upregulated in disorders characterized by mesangial alterations, such as mesangial cell proliferation and matrix expansion, suggesting a key biological role for this gene in the pathogenesis of glomerulonephritis. However, the exact biological role of megsin remains unknown [89–94].

Proteomics is in its early developmental phase and in the future a number of new applications for this technique to better understand mesangial pathobiology are likely to be developed.

Conclusion

Our improved understanding of mesangial cell pathobiology has cleared the way to start devising new therapeutic interventions aimed at controlling, inhibiting, modulating and activating molecular targets to treat diseases and promote renal recovery. While abolishing the injurious agent(s) will remain paramount, especially when systemic in nature, trying to regulate pathological processes locally may provide a reasonable approach to prevent damage and/or promote or enhance healing. Stem cell therapy in one of its possible forms may soon regain strength and provide new exciting avenues for repairing what is currently felt to be apparently irreversible renal damage [95–102].

Stem cell approaches to heal mesangial injury are only at an early stage of development in the research laboratory, but may soon provide an untapped resource for potentiating rapid recovery. It is exciting that advances in mesangial cell pathobiology are permitting exploration of these new technologies and translation of laboratory findings to the patient's bedside.

Recently, there has been much interest in potential stem cell-based therapies. One of these possibilities is that existing stem cells located in the different renal compartments could replenish injured areas, thus healing the particular compartments that have been damaged. Another possibility is that introduction of progenitor stem cells could be integrated into existing structures and mature in those particular locations.

Although the final word is not in yet, several conclusions appear to be valid at this time. There is no evidence that a renal master stem cell exists. It is more likely that there are progenitors of specific cell types or groups of cells (i.e. epithelial, mesangial, interstitial, tubular, etc.). Delivery of 'support stem cells' to damaged renal compartments may become a form of therapy. A third possible scenario is that progenitor stem cells may be re-stimulated in various locations to enhance repair.

Cell delivery directly into the renal parenchyma could be employed so that restricted regions of the kidney would be targeted. Elucidation of specific 'homing' mechanisms to attract cells to specific compartments and to create the required milieu for proper development and maturity of these cells for various functions is crucial for further development of this important field.

While progenitor stem cells are different than native cells, the end result obtained by enhancing their participation in the repair process, with the result of additional cellular elements engaged in the process, is essentially the same. Depending on the type of injury that occurs in the kidney and the specific compartment(s) affected, one or the other mechanism may be most effective. The future appears exciting and new 'healing' approaches will be integrated into therapeutic algorithms in the near future.

References

1 Zimmerman KW: Über den Bau des Glomerulus der Säugerniere: Weitere Mitteilungen. Z Mikrosk Anat Forsch 1933;32:176–278.
2 Stetler-Stevenson WG: Dynamics of matrix turnover during pathologic remodeling of extracellular matrix. Am J Pathol 1996;148:1345–1350.
3 Morel-Maroger SL, Killen PD, Chi E, Striker GE: The composition of glomerulosclerosis. I. Studies in focal sclerosis, crescentic glomerulonephritis, and membranoproliferative glomerulonephritis. Lab Invest 1984;51:181–192.
4 Thoenes W, Rumpelt HJ: The obsolescent renal glomerulus – collapse, sclerosis, hyalinosis, fibrosis. A light and electron microscopical study of renal biopsies. Virchows Arch A Pathol Anat Histol 1977;377:1–15.
5 Keeling J, Herrera GA: Human metalloprotinases: characteristics and pathologic role in altering mesangial homeostasis. Microsc Res Tech 2008;71:371–379.
6 Fogo A: Progression and potential regression of glomerulosclerosis. Kidney Int 2001;59:804–819.
7 Fogo AB: Can glomerulosclerosis be reversed? Nat Clin Practice Nephrol 2006;2:290–291.
8 Michael F, Keane WF, Vernier RL, Mauer SM: The glomerular mesangium. Kidney Int 1980;17:141–154.
9 Kriz W, Elger M, Lemley R, Sakai T: The structure of the glomerular mesangium: a biochemical interpretation. Kidney Int 1990; 38(Suppl):52–59.
10 Sakai T, Kriz W: The structural relationship between mesangial cells and basement membrane of the renal glomerulus. Anat Embryol (Berl) 1987;176:373–386.
11 Latta H, Maunsbach AB, Madden SC: The centrolobular region of the renal glomerulus studied by electron microscopy. J Ultrastruct Res 1960;4:455–472.
12 Farquhar MG, Palade GE: Functional evidence for the existence of a third cell type in the renal glomerulus. Phagocytosis of filtration residues by a distinctive 'third' type. J Cell Biol 1962;13:55–87.
13 Schlondorff D: The glomerular mesangial cell: an expanding role for a specialized pericyte. FASEB J 1987;1:272–281.
14 Schreiner GF: The mesangial phagocyte and its regulation of contractile cell biology. J Am Soc Nephrol 1992;2:S74–S82.
15 Lopez-Armada MJ, Gomez-Guerrero C, Egido J: Receptors for immune complexes activate gene expression and synthesis of matrix proteins in cultured rat and human mesangial cells. J Immunol 1996;157:2136–2142.
16 El Nahas A-M: Plasticity of kidney cells: role in kidney remodeling and scarring. Kidney Int 2003;64:1553–1563.
17 Herrera GA: Plasticity of mesangial cells: a basis for understanding pathological alterations. Ultrastruct Pathol 2006;30:471–479.
18 Gomez DE, Alonso DF, Yoshiji H, Thorgeirsson UP: Tissue inhibitors of metalloproteinases: structure, regulation, and biochemical functions. Eur J Cell Biol 1997;74:111–122.
19 Couchman JR, Beavan LA, McCarthy KJ: Glomerular matrix: synthesis turn over and role in mesangial expansion. Kidney Int 1994;45:328–335.
20 Ronco P, Chatziantoniou C: Matrix metalloproteinases and matrix receptors in progression and reversal of kidney disease: therapeutic perspectives. Kidney Int 2008;74:873–878.
21 Martin J, Eynstone L, Davies M, et al: Induction of metalloproteinases by glomerular mesangial cells stimulated by proteins of the extracellular matrix. J Am Soc Nephrol 2001;12:88–96.
22 Lenz O, Elliot SJ, Stetler-Stevenson WG: Matrix metalloproteinases in renal development and disease. J Am Soc Nephrol 2000;11:574–581.
23 Schnaper HW, Kleinman HK: Regulation of cell function by extracellular matrix. Pediatr Nephol 1993;7:96–104.
24 Sugiyama H, Kashihara N, Maesshima Y, Okamoto K, Kanao K, Sekikawa T, Makino H: Regulation of survival and death of mesangial cells by extracellular matrix. Kidney Int 1998;54:1188–1196.
25 Terman BI, Dougher-Vermazen M, Carrion ME, Dimitrov D, Armellino DC, Bohlen P: Identification of the KDR tyrosine kinase as a receptor for vascular endothelial cell growth factor. Biochem Biophys Res Commun 1992;187:1579–1586.
26 Haddad E, Moura IC, Arcos-Fajardo M, Macher M-A, Baudoin V, Alberti C, Loirat C, Monteiro RC, Peuchmaur M: Enhanced expression of the CD71 mesangial IgA1 receptor in Berger disease and Henoch-Schönlein nephritis: association between CD71 expression and IgA deposits. J Am Soc Nephrol 2003;14:327–337.
27 Gomez-Guerrero C, Gozalez E, Egido J: Evidence for a specific IgA receptor in rat and human mesangial cells. J Immunol 1993;151:7172–7181.
28 Gomez-Guerrero C, Duque N, Egido J: Mesangial cells possess an asialoprotein receptor for affinity for human immunoglobulin A. J Am Soc Nephrol 1998;9:568–576.
29 Roberts AB, McCune BK, Sporn MB: TGF-β: regulation of extracellular matrix. Kidney Int 1992;41:557–559.
30 Thomas S, Vanuystel J, Gruden G, Rodríguez V, Burt D, Gnudi L, Hartley B, Viberti G: Vascular endothelial growth factor receptors in human mesangium in vitro and in glomerular disease. J Am Soc Nephrol 2000;11:1236–1243.
31 Teng J, Russell WJ, Gu X, Cardelli J, Jones ML, Herrera GA: Different types of glomerulopathic light chains interact with mesangial cells using a common receptor but exhibit different intracellular trafficking patterns. Lab Invest 2004;84:440–451.

32 Moura IC, Arcos-Fajardo M, Sadaka C, et al: Glycosylation and size of IgA1 are essential for interaction with mesangial transferrin receptor in IgA nephropathy. J Am Soc Nephrol 2004;15:622–634.
33 Saito K, Shimizu F, Sato T, et al: Modulation of human mesangial cell behaviour by extracellular matrix components – the possible role of interstitial type III collagen. Clin Exp Immunol 1993;91:510–515.
34 Shimizu F, Kawachi H, Orikasa M: Role of mesangial cell change in progressive renal disease. Kidney Blood Press Res 1999;22:5–12.
35 Ayo SH, Radnik RA, Glass WF, Garoni JA, Rampt ER, Appling DR, Kreisberg JI: Increase in extracellular matrix synthesis and mRNA in mesangial cells grown in high glucose medium. Am J Physiol 1991;260:F185–F191.
36 Liu ZH, Striker GE, Settler-Stevenson M, Fukushima P, Patel A, Striker LJ: TNF-α and IL-1-α induce mannose receptors and apoptosis in glomerular mesangial but not endothelial cells. Am J Physiol 1996;270: C1595–C1601.
37 Tagouri Y, Sanders PW, Zhu I, Picken MM, Winokur TS, Soong S-J, Herrera GA: Glomerulopathic monoclonal light chains potentiate mesangial cell apoptosis in vitro. Lab Invest 1995;92:161A.
38 Khera T, Martin J, Riley S, Steadman R, Phillips AO: Glucose enhances mesangial cell apoptosis. Lab Invest 2006;86:566–577.
39 Silver BJ, Jaffer FE, Abboud HE: Platelet-derived growth factor synthesis in mesangial cells: induction by multiple peptide mitogens. Proc Natl Acad Sci USA 1989;86:1056–1060.
40 Russell WJ, Cardelli J, Harris E, Baier RJ, Herrera GA: Monoclonal light chain-mesangial cells interactions: early signaling events and subsequent pathologic effects. Lab Invest 2001;81:889–703.
41 Razzaque M, Ahsan N, Taguchi T: Role of apoptosis in fibrogenesis. Nephron 2002;90:365–372.
42 Truong LD, Majesky MW, Pindur J: Tenascin is synthesized and secreted by rat mesangial cells in culture and is present in the extracellular matrix in human glomerular disease. J Am Soc Nephrol 1994;4:1771–1777.
43 Turbat-Herrera EA, Isaac J, Sanders PW, Truong LD, Herrera GA: Integrated expression of glomerular extracellular matrix proteins and β1 integrins in monoclonal light chain-related renal diseases. Mod Pathol 1997;10:485–495.
44 Teng J, Zhang PL, Russell WJ, Zheng LP, Herrera GA: Insights into mechanisms responsible for mesangial alterations associated with fibrogenic glomerulopathic light chains. Nephron Physiol 2003;94:28–38.
45 Truong LD, Pundur J, Barrios R, D'Agatti V, Lechago J, Suki W, Majesky M: Tenascin is an important component of the glomerular extracellular matrix in normal and pathologic conditions. Kidney Int 1994;45:201–210.
46 Yokosaki Y, Palmer EL, Prieto AL, Crossini KL, Bourdon MA, Pytela R, Sheppard D: The integrin α9β1 mediates cell attachment to a non-RGD site in the third fibronectin repeat of tenascin. J Biol Chem 2994;43:26691–26696.
47 Ooi BS, Cohen DJ, Veis JH: Biology of the mesangial cell in glomerulonephritis – role of cytokines. Proc Soc Exp Biol Med 1996;213:230–237.
48 Sedor JR, Konieczkowski M, Huang S, et al: Cytokines, mesangial cell activation and glomerular injury. Kidney Int 1993;43:S65–S70.
49 Johnson RJ, Iida H, Alpers CE, et al: Expression of smooth muscle cell phenotype by rat mesangial cells in immune complex nephritis. α-smooth muscle actin is a marker of mesangial cell proliferation. J Clin Invest 1991;87:847–858.
50 Johnson RJ, Floege J, Yoshimura A, et al: The activated mesangial cell: a glomerular 'myofibroblast'? J Am Soc Nephrol 1992;2: S190–S197.
51 Herrera GA, Schultz J, Soong S, Sanders PW: Growth factors in monoclonal light chain-related renal diseases. Hum Pathol 1994;25:883–889.
52 Laiho M, DeCapiro JA, Ludlow JW, Livingston DM, Massague J: Growth inhibition by TGF-β linked to suppression of retinoblastoma protein phosphorilation. Cell 1990;62:175–185.
53 Moses HL, Yang EY, Pietenpol JA: TGF-β stimulation and inhibition of cell proliferation: new mechanistic insights. Cell 1990;63:245–247.
54 Border WA, Ruoslahti E: Transforming growth factor-β in disease: the dark side of tissue repair. J Clin Invest 1992;90:1–7.
55 Roberts AB, Sporn MB: Transforming growth factor-β. Adv Cancer Res 1988;51:107–145.
56 Kim SI, Han DC, Lee HB: Lovastatin inhibits transforming growth factor-β 1 expression in diabetic rat glomeruli and cultured rat mesangial cells. J Am Soc Nephrol 2000;11:80–87.
57 Edwards DR, Murphy G, Reynolds JJ, Whithan SE, Docherty AJ, Angel P, Heath JK: Transforming growth factor β modulates the expression of collagenase and metalloproteinase inhibitor. EMBO J 1987;6:1899–1904.
58 Zhu L, Herrera GA, Murphy-Ullrich JE, Huang Z-Q, Sanders PW: Pathogenesis of glomerulosclerosis in light chain deposition disease: role for transforming growth factor β. Am J Pathol 1995;147:375–385.
59 Baud L, Fouqueray B, Suberville S, Doublier S: Interleukin 10: a logical candidate for suppressing glomerular inflammation? Exp Nephrol 1998;6:22–27.
60 Baud L, Perez J, Ardaillou R: Dexamethasone and hydrogen peroxide production by mesangial cells during phagocytosis. Am J Physiol 1986;250:F596–F604.
61 Wang JS, Tseng HH, Shih DF, Jou HS, Ger LP: Expression of inducible nitric oxide synthase and apoptosis in human lupus nephritis. Nephron 1997;77:404–411.
62 Cattel V: Nitric oxide and glomerulonephritis. Semin Nephrol 1999;19:277–287.
63 Kone BC: Nitric oxide in renal health and disease. Am J Kidney Dis 1997;30:311–333.
64 Kone BC, Baylis C: Biosynthesis and homeostatic role of nitric oxide in the normal kidney. Am J Physiol 1997;272:F561–F578.
65 Davies M, Coles GA, Thomas GJ, Martin J, Lovett DH: Proteinases and the glomerulus: their role in glomerular diseases. Klin Wochenschr 1990;68:1145–1149.
66 Keeling J, Herrera GA: An in-vitro model of light chain deposition disease. Kidney Int 75: 634–645.

67 Keeling J, Herrera GA: Matrix metalloproteinases and mesangial remodeling in light chain-related glomerular damage. Kidney Int 2005;68:1590–1603.
68 Nagase H: Activation mechanisms of matrix metalloproteinases. Biol Chem 1997;378:151–160.
69 Norman JT, Lewis MP: Matrix metalloproteinases (MMPs) in renal fibrosis. Kidney Int Suppl 1997;54:S61-S63.
70 Heidland A, Sebekova K, Paczek L, Teshner M, Dammrich J, Gaciong Z: Renal fibrosis: Role of impaired proteolysis and potential therapeutic strategies. Kidney Int Suppl 1997;62:S32–S35.
71 El Nahas A-M, Muchaneta-Kubara EC, Essawy M, Soylemezoglu O: Renal fibrosis: insights into pathogenesis and treatment. Int J Biochem Cell Biol 1997;29:55–56
72 Imai K, Kusakabe M, Sakakura T, Nakanishi I, Okada Y: Susceptibility to degradation by matrix metalloproteinases and serine proteinases. FEBS Lett 1994;352:216–218.
73 McLennan SV, Yue DK, Turtle JR: Effect of glucose on matrix metalloproteinase activity in mesangial cells. Nephron 1998;79:293–298.
74 Leehey DJ, Song RH, Alavi N, Singh AK: Decreased degradative enzymes in mesangial cells cultured in high glucose media. Diabetes 1995;44:929–935.
75 McMillan JI, Riordan JW, Couser WG, Pollock AS, Lovett DH: Characterization of a glomerular epithelial cell metalloproteinase as matrix metalloproteinase-9 with enhanced expression in a model of membranous nephropathy. J Clin Invest 1996;97:1094–1101.
76 Reuter A, Nestl A, Zwaka RM, Tuckermann J, Waldherr R, Wagner EM, Hoyhtya M, Meyer zum Gottesberger AM, Angel P, Weiher H: Expression of the recessive glomerulosclerosis gene Mpv 17 regulates MMP-2 expression in fibroblasts, the kidney, and the inner ear of mice. Mol Biol Cell 1998;9:1675–1682.
77 Ahmed AK, El Nahas A-M, Johnson TS: Localization of matrix metalloproteinases and their inhibitors in experimental progressive kidney scarring. Kidney Int 2007;71:755–763.
78 Steinmann-Niggli K, Ziswiler R, Kung M, Marti HP: Inhibition of matrix metalloproteinases attenuates anti-Thy 1.1 nephritis. J Am Soc Nephrol 1998;9:397–407.
79 Urushihara M, Kagami S, Kuhara T, Tamaki T, Kuroda Y: Glomerular distribution and gelatinolytic activity of matrix metalloproteinases in human glomerulonephritis. Nephrol Dial Transplant 2002;17:1189–1196.
80 Sanders J-S, van Goor H, Hanemaaijer R, Kallenberg CGM, Stegeman CA: Renal expression of matrix metalloproteinases in human ANCA-associated glomerulonephritis. Nephrol Dial Transplant 2004;19:1412–1419.
81 Salahudeen AK, Kanji V, Reckelhoff JF, Schmidt AM: Pathogenesis of diabetic nephropathy: a radical approach. Nephrol Dial Transplant 1997;12:664–668.
82 McLennan SV, Kelly DJ, Schache M, Waltham M, Dy V, Langham RG, Yue DK, Gilbert RE: Advanced glycation end products decrease mesangial MMP-7: a role in matrix accumulation of diabetic nephropathy? Kidney Int 2007;72:481–488.
83 Nakamura T, Fukui M, Ebihara I, Osada S, Tomino Y, Koide H: Abnormal gene expression of matrix metalloproteinases and their inhibitor in glomeruli from diabetic rats. Diabetes 1993;42:1425–1432.
84 Stix B, Kahne T, Sletten K, Raynes J, Roessner A, Rocken C: Proteolysis of AA amyloid fibril proteins by matrix metalloproteinases- 1,-2, and -3. Am J Pathol 2001;159:561–570.
85 Lenz O, Elliot SJ, Stetler-Stevenson WG: Matrix metalloproteinases in renal development and disease. J Am Soc Nephrol 2000;11:574–581.
86 Chen J, Park H-C, Addabbo F, Ni J, Pelger E, Li H, Plotkin M, Goligorsky MS: Kidney-derived mesenchymal stem cells contribute to vasculogenesis, angiogenesis, and endothelial repair. Kidney Int 2008;74:879–889.
87 Murphy CL, Eulitz M, Hrncic R, Sletten K, Westermark P, Williams T, Macy SD, Wooliver C, Wall J, Weiss DT, Solomon S: Chemical typing of amyloid protein contained in formalin-fixed paraffin-embedded biopsy specimens. Am J Clin Pathol 2001;116:135–142.
88 Picken MM, Herrera GA: The burden of 'sticky' amyloid. Arch Pathol Lab Med 2007;131:850–851.
89 Miyata T, Inagi R, Nangaku M, Imasawa T, Sato M, Izuhara Y, Suzuki D, Yoshino A, Onogi H, Kimura M, Sugiyama S, Kurokawa K: Overexpression of the serpin megsin induces progressive mesangial cell proliferation and expansion. J Clin Invest 2002;109:583–593.
90 Nangaku M, Kurokawa K: Specific tissue distribution of megsin, a novel serpin, in the glomerulus and its up-regulation in IgA nephropathy. Biochem Biophys Res Commun 2001;286:1098–1106.
91 Nangaku M, Miyata T, Suzuki D, Umezono T, Hashimoto T, Wada T, Yagi M, Nagano N, Inagi R, Kurokawa K: Cloning of rodent megsin revealed its up-regulation in mesangioproliferative nephritis. Kidney Int 2001;60:641–652.
92 Suzuki D, Miyata T, Nangaku M, Takano H, Saotome N, Inagi R, Endoh M, Kurokawa K, Sakai H: Expression of megsin mRNA, a novel mesangium-predominant gene, in the renal tissue of various glomerular diseases. J Am Soc Nephrol 1999;10:2606–2613.
93 Inagi R, Miyata T, Imasawa T, Nangaku M, Kurokawa K: Mesangial cell-predominant gene, megsin. Nephrol Dial Transplant 2002;17:32–33.
94 Onogi H, Inagi R, Miyata T, Nangaku M, Kurokawa K: Proteomics and mesangial cell: serpin, megsin, and plasmin; in Thongoboonkerd V, Klein JB (eds): Proteomics in Nephrology. Contrib Nephrol. Basel, Karger, pp 212–220.
95 Kunter U, Rong S, Djuric Z, Boor P, Muller-Newen G, Yu D, Floege J: Transplanted mesenchymal stem cells accelerate glomerular healing in experimental glomerulonephritis. J Am Soc Nephrol 2006;17:2202–2212.
96 Rabelink TJ, van Kooten C: Stem cell therapy for glomerular disease. J Am Soc Nephrol 2006;17:2086–2088.

97 Imai E, Takabatake Y, Mizui M, Isaka Y: Gene therapy in renal diseases. Kidney Int 2004;65:1551–1555.
98 Koh CJ, Atala A: Tissue engineering, stem cells, and cloning: opportunities for regenerative medicine. J Am Soc Nephrol 2004;15:1113–1125.
99 Ito T, Suzuki A, Imai E, et al: Bone marrow is a reservoir of repopulating mesangial cells during glomerular remodeling. J Am Soc Nephrol 2001;12:2625–2635.
100 Imasawa T, Utsunomiya Y, Kawamura T, et al: The potential of bone marrow-derived cells to differentiate to glomerular mesangial cells. J Am Soc Nephrol 2001;12:1401–1409.
101 Poulsom R, Alison MR, Cook T, Jeffery R, Ryan E, Forbes S, Hunt T, Wyles S, Wright NA: Bone marrow stem cells contribute to healing of the kidney. J Am Soc Nephrol Suppl 2003;14:S48–S54.
102 Zerbini G, Piemonti L, Maestroni A, Dell'Antonio G, Bianchi G: Stem cells and the kidney: a new therapeutic tool? J Am Soc Nephrol Suppl 2006;17:S123–S126.

Guillermo A. Herrera, MD
Nephrocor
1700 North Desert Drive
Tempe, AZ 85 281 (USA)
Tel. +1 602 443 6119, Fax +1 602 443 6102 ext. 3106, E-Mail nephronpath@gmail.com

Herrera GA (ed): Experimental Models for Renal Diseases: Pathogenesis and Diagnosis.
Contrib Nephrol. Basel, Karger, 2011, vol 169, pp 23–36

Parietal Epithelial Cells: Their Role in Health and Disease

Paola Romagnani

Excellence Center for Research, Transfer and High Education DENOthe, University of Florence, Florence, Italy

Abstract

Parietal epithelial cells of Bowman's capsules were first described by Sir William Bowman in 1842 in his paper *On the Structure and Use of the Malpighian Bodies of the Kidney* [London, Taylor, 1842], but since then their functions have remained poorly understood. A large body of evidence has recently suggested that parietal epithelial cells represent a reservoir of renal progenitors in adult human kidney which generate novel podocytes during childhood and adolescence, and can regenerate injured podocytes. The discovery that parietal epithelial cells represent a potential source for podocyte regeneration suggests that podocyte injury can be repaired. However, recent results also suggest that an abnormal proliferative response of renal progenitors to podocyte injury can generate hyperplastic glomerular lesions that are observed in crescentic glomerulonephritis and other types of glomerular disorders. Taken together, these results establish an entirely novel view that changes the way of thinking about renal physiology and pathophysiology, and suggest that understanding how self-renewal and fate decision of parietal epithelial cells in response to podocyte injury may be perturbed or modulated will be crucial for obtaining novel tools for prevention and treatment of glomerulosclerosis.

Parietal epithelial cells (PEC) of Bowman's capsules are polygonal flat cells that cover the inner aspect of Bowman's capsules. They were first described by Sir William Bowman in 1842 in his paper *On the Structure and Use of the Malpighian Bodies of the Kidney*, but since then their functions have remained poorly understood and seem limited to the flow of the glomerular ultrafiltrate into the proximal tubule [1]. Nonetheless, morphologic studies have shown that the parietal epithelium is not homogeneous and contains cuboidal cells with prominent microvilli localized at the urinary pole [2–7]. In addition, other studies have shown that there are PEC analogous in size and shape to visceral podocytes, which are localized at the vascular pole in continuity with the podocytes of the glomerular tuft and cover up to 25% of the Bowman's capsule [8–12]. Thus, PEC appear as a heterogeneous population of cells localized within the Bowman's capsule. Their phenotype and function under physiological as well as pathological conditions has remained mostly unknown, thus making it difficult to make a clear definition of them.

During nephrogenesis, PEC, podocytes and proximal tubular cells share a common phenotype in the early stages of nephron development. Indeed, the cortical nephron develops from the metanephric mesenchyme. This development passes through several stages [13]: vesicle, comma and S-shaped, glomerular capillary loop, and mature glomerulus. The distal extremities of the dividing branches of the ureteral bud (the future collecting ducts) penetrate into the loose undifferentiated mesenchyme and induce aggregates of mesenchymal cells that gradually acquire an epithelial phenotype, express cytokeratins (CK) and form renal vesicles [13].

The renal vesicle then undergoes a series of invaginations and elongations to generate the comma-shaped and, afterwards, the S-shaped bodies. At this stage, the proximal end of the S-shaped body becomes invaded by blood vessels, differentiates into podocytes and PEC, and generates the glomerular tuft [13, 14]. Simultaneously, the middle and the distal segments of the S-shaped body that had remained in contact with the ureteric bud epithelium fuse to form a single continuous epithelial tube, and begin to express proteins that are characteristic of tubular epithelia [13, 14]. The phenotype of PEC diverges from podocytes at later stages. At the lower end of the S-shaped bodies, Bowman's space begins to form, limited outside by a narrow band of PEC and inside by the crown of visceral epithelial cells, the future podocytes [13, 14].

When the glomerulus passes from the S-shaped body to the capillary loop stage, podocytes acquire their definitive phenotype, arriving at the mature glomerulus stage, which is associated with the formation of pedicels along the glomerular basement membrane, loss of their mitotic activity and the expression of proliferation markers [13–15]. In contrast, these cells express de novo the cyclin-dependent kinase inhibitors p27 and p57 [16, 17], which prevent cellular division by blocking the cyclin-cyclin-dependent kinase complexes [18], and they also start to express the proteins that are specific to mature podocytes [19, 20]. During nephrogenesis and in mature glomeruli, the common progenitors of PEC and podocytes express PAX2. The disappearance of PAX2 from podocytes corresponds to their definitive loss of ability to divide. PAX2 induces cellular proliferation [21] and is also anti-apoptotic [22, 23]. The downregulation of PAX2 by WT1 seems to be a prerequisite that permits the WT1-controlled differentiation of podocytes [24].

The close relationships between PEC and podocytes has allowed a clear distinction between these two cell types only on the basis of anatomical criteria and has delayed the understanding of the function and role of PEC in renal physiology. In addition, during glomerular disorders these anatomic distinctions are often lost within glomeruli, making it difficult to understand the contribution of PEC and podocytes to the pathogenesis of glomerular injury. However, the recent discovery of specific markers of these cells, as well as the advent of genetic tagging, has allowed the characterization of their phenotype and has shed new light onto the important role of PEC in renal physiology and pathology.

The Role of PEC in Healthy Conditions: A Renal Progenitor Niche

The role of PEC in renal physiology has acquired importance in recent years as the exponentially growing prevalence of renal disorders has focused attention on the mechanisms of kidney repair. The possibility of treatment of renal disorders has been severely impaired by our poor knowledge of the regenerative properties of the kidney [25, 26]. Indeed, the mammalian kidney shares with the majority of organs the ability to repopulate and at least partially repair structures that have sustained some degree of injury [25–27].

In most adult tissues, the process of wound healing by replacing the damaged or dead cells is maintained through the presence of stem/progenitor cells [28, 29]. Thus, a great effort was focused on the attempt to identify putative renal stem cells in adult kidneys. Stem cells are functionally defined by their ability to self-renew and to differentiate into the cell lineages of their tissue of origin [28, 29]. Once activated, epithelial stem cells can generate a proliferating progeny, which is often referred to as transiently amplifying cells. In their normal environment, transiently amplifying cells actively divide for a limited period of time, thus expanding the cellular pool that will then differentiate along a particular cell lineage to make the tissue. In addition, stem cells are critically involved in regeneration in response to wounds. Unless the epithelial stem/progenitor cells are permanently damaged, most epithelia are able to repair their tissues following injury [28, 29]; when epithelial stem cells are depleted, fibrotic responses occur [28, 29].

In our search for putative renal stem cells in adult human kidneys, we used a combination of markers, such as CD24 (a surface molecule that has been used to identify different types of human stem cells) [30, 31], CD133 (a marker of hematopoietic and other types of adult tissue stem cells) [32, 33], the polycomb group protein Bmi-1 (a transcription factor that is critical for maintenance of stem cell self-renewal) [34, 35] and the stem cell-specific transcription factors Oct-4 [36], to identify and recover a subset of PEC of the Bowman's capsule that displayed self-renewal and multidifferentiation potential [14, 25, 26, 37–40]. Differently from all other types of renal cells, CD24+CD133+ PEC also expressed CD106, a surface molecule that together with CD105, CD54 and CD44 is usually co-expressed by adult SC types that grow as adherent cells, such as mesenchymal SC [41]. CD24+CD133+ PEC isolated from adult human kidneys regenerated tubular structures in different portions of the nephron and reduced the morphologic and functional kidney damage in mice affected by acute renal failure, suggesting that these cells can participate in tubular regeneration in adult human kidneys [25, 26, 38–40].

Interestingly, several previous studies had shown that the proximal tubule arises at a variety of angles from Bowman's capsule and that at least one part of the tubuloglomerular junction has an area of intermediate appearance, with prominent microvilli on parietal cells in humans, mammals and fish. These findings suggest that the parietal epithelium may be able to change to tubular and that this might particularly occur as the kidney grows, during severe renal disorders [2–7] or during ageing [8]. In most cases, however, repair of epithelial cells does not depend on cells generated from multipotent progenitors, but directly derives from the migration of epithelial cells from the neighboring epithelia, as previously reported for the skin [42, 43].

Several studies have suggested that the tubular epithelium can be self-renewing after acute kidney injury and that differentiated tubular cells proliferate and migrate to replace the neighboring dead cells [44–46]. Thus, renal stem/progenitor cells might contribute to tubular epithelium repair, but this probably occurs only when a renal tubular injury cannot be spontaneously repaired through the migration of neighboring unwounded tubular cells. However, CD24+CD133+ progenitors are physically located within the Bowman's capsule, the only place in the kidney which appears to be contiguous with both tubular cells and glomerular podocytes [37, 38, 40], suggesting that CD24+CD133+ progenitors might also be responsible for podocyte regeneration after injury.

Unlike tubular cells, podocytes are post-mitotic cells that cannot undergo complete cell division and are therefore unable to regenerate themselves [47, 48]. Previous studies have suggested the existence of transitional cells exhibiting a mixed phenotype between the PEC and the podocyte, predominating around the vascular pole where they are in continuity with the visceral podocytes [8–12]. Parietal podocytes were originally described or illustrated by TEM in the normal rat kidney several decades ago [49], and the close relationship between the JGA and parietal podocytes in the rat and other laboratory animals has been described [50, 51]. CD24+CD133+ progenitors also represent common progenitors of tubular cells and podocytes during renal development [37] as shown by the observation that CD133 and CD24 co-expression identifies a subset of cells in the primordial nephron that display self-renewal and multidifferentiation potential [29] (fig. 1).

CD24+CD133+ progenitors are located in condensed mesenchyme-derived primordial structures, primary vesicles, comma-shaped bodies and S-shaped bodies (fig. 1a–d). In S-shaped bodies, CD24+CD133+ progenitors are located in the proximal loop, which give rise

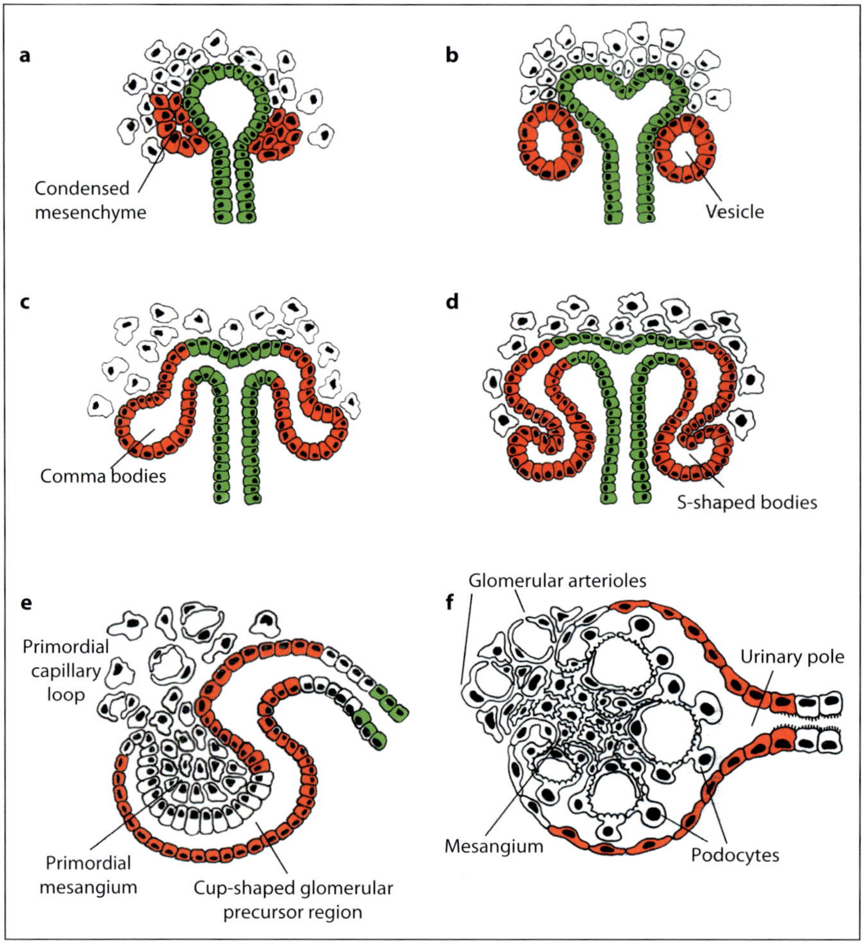

Fig. 1. Series of schematics depicting the morphological events and the localization of CD24+CD133+ renal progenitors during the different phases of nephron development. Reproduced with permission from AlphaMed Press and Wiley-Blackwell [14]. Mesenchymal cells near the tips of the branching ureteric bud are induced and differentiate through a series of forms: aggregate (**a**), renal vesicle (**b**), comma-shaped bodies (**c**) and S-shaped bodies (**d**). Also shown is the developing vasculature within the glomerular cleft of an S-shaped body (**e**) and the glomerulus in a more mature nephron (**f**). Cells of the ureteric bud are stained in green. CD24+CD133+ renal progenitors are stained in red. **a** CD24+CD133+ renal progenitors (red) localize in the condensed mesenchyme, but not in the un-induced mesenchyme (white) or in the ureteric bud (green). **b** CD24+CD133+ renal progenitors localize in a primary vesicle, but not in the un-induced mesenchyme (white) or in the ureteric bud (green). **c** CD24+CD133+ renal progenitors localize in the comma bodies, but not in the un-induced mesenchyme (white) or in the ureteric bud (green). **d** CD24+CD133+ renal progenitors localize in the S-shaped body, in the proximal loop, as well as in the distal loop, but not in the un-induced mesenchyme (white) or in the ureteric bud (green). **e** CD24+CD133+ renal progenitors (red) localize in an S-shaped body after colonization by primordial capillaries and mesangium. **f** In maturing glomeruli, CD24+CD133+ renal progenitors (red) selectively persist as a subset of cells of the Bowman's capsule localized opposite to the vascular pole.

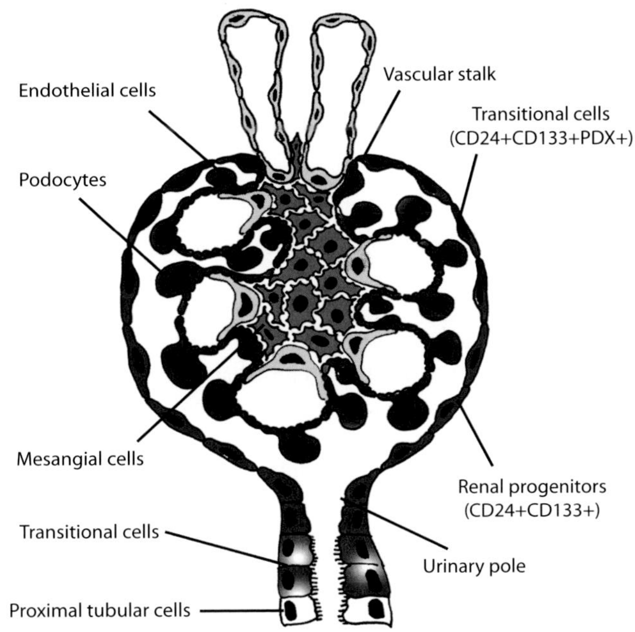

Fig. 2. Hierarchical distribution of renal progenitors in adult human glomeruli. Reproduced with permission from AlphaMed Press and Wiley-Blackwell [14]. CD24+ CD133+ renal progenitors (red) are localized at the urinary pole and are in close contiguity with podocytes (blue) at one extremity (the vascular stalk) and with tubular renal cells (white) at the other extremity. A transitional cell population (CD24+ CD133+PDX+, red/blue) displays features of either renal progenitors (red) or podocytes (blue), and localizes between the urinary pole and the vascular pole. At the vascular stalk of the glomerulus, the transitional cells are localized in close continuity with cells that lack progenitor markers, but exhibit the podocyte markers and the phenotypic features of differentiated podocytes (blue). Endothelial cells: yellow; mesangial cells: green; podocytes: blue; proximal tubular cells: white; renal progenitors: red.

to both podocytes and the Bowman's capsule of the primitive glomerulus, as well as in the distal loop (fig. 1d); however, when a primitive vascular tuft is evident within the cup-shaped glomerular precursor region, CD24+CD133+ progenitors are detected only in PEC of the Bowman's capsule (fig. 1e). In maturing, as well as adult, glomeruli, CD24+CD133+ progenitors are selectively localized at the urinary pole of the Bowman's capsule (fig. 1f).

In adult human kidneys, CD24+CD133+ cells consist of a hierarchical population of progenitors that are arranged in a precise sequence within the Bowman's capsule and exhibit heterogeneous potential for differentiation and regeneration [14, 40]. Cells localized to the urinary pole that express CD133 and CD24, but not podocalyxin (PDX) or other podocyte markers (CD24+CD133+PDX– cells), could regenerate both tubular cells and podocytes (fig. 2). By contrast, cells localized between the urinary pole and vascular pole that expressed both progenitor and podocyte markers (CD24+CD133+PDX+) could regenerate only podocytes [14, 40] (fig. 2). Finally, cells localized to the vascular pole did not exhibit progenitor markers, but did display phenotypic features of fully differentiated podocytes (CD24–CD133–PDX+ cells) [14, 40] (fig. 2).

These findings obtained in humans were consistent with a parallel study performed in rodents, which showed that transitional cells with morphologic and immunohistochemical features of both PEC and podocytes could be detected at the glomerular vascular stalk [52]. More importantly, the same study, by using an elegant model of genetic tagging of PEC in a triple-transgenic doxycycline-inducible mouse line, unequivocally demonstrated that podocytes are recruited from PEC which proliferate and differentiate from the urinary to the vascular stalk generating novel podocytes [52]. Thus, there is convincing evidence that podocytes can be recruited from a progenitor population localized at the urinary pole of the Bowman's capsule, thus explaining how the growing glomerulus is covered by podocytes despite the inability of these cells to undergo cell division.

The Role of PEC in Diseases: A Response to Renal Injury

Glomerular injury can involve excessive proliferation of resident glomerular epithelial cells, resulting in the formation of crescents and obliteration of Bowman's space. These abnormalities are typically seen in the hypercellular lesions of crescentic glomerulonephritis, but can also be observed in other types of glomerular disorders. So far, theories addressing the origin of aberrant epithelial cells in crescentic glomerulonephritis have been controversial [53–97].

While one explanation would be that these cells exclusively originate from PEC, another possibility is that some de-differentiated podocytes can acquire markers of PEC [53–97] (tables 1 and 2). Initially, based on ultrastructural observations, it was postulated that crescents result from the proliferation of glomerular PEC [67–72]. Subsequent studies demonstrated the presence of macrophages in crescents [73, 74]. Nonetheless, the majority of cells in crescents exhibit staining for PEC markers (e.g. CK, PGP9.5, pan-cadherin), indicating that these are the predominant cell type in crescents [70–72]. In addition, PEC play an active role in the progression from cellular to fibrous crescents by producing chemoattractants for monocytes, neutrophil infiltration and transmigration into periglomerular and intraglomerular spaces [75, 76], and by producing extracellular matrix molecules such as fibronectin and collagens I, III and IV [77, 78].

Despite the large series of studies showing the involvement of PEC in inflammatory glomerular diseases, PEC have generally been considered to be 'innocent bystanders' in non-inflammatory glomerular disease. However, epithelial hyperplasia and sometimes pseudo-crescent formation are found in a number of non-inflammatory conditions, most notably in focal segmental glomerulosclerosis (FSGS). In fact, lesions in FSGS with prominent epithelial hyperplasia may resemble cellular crescents, with the result that these cases may even be inappropriately diagnosed as crescentic glomerulonephritis.

Collapsing glomerulopathy is described as a clinically and pathologically distinct variant of FSGS characterized by black racial predominance, massive proteinuria and relatively rapid progressive renal insufficiency [79]. Collapsing glomerulopathy is pathologically characterized by segmental-to-global collapse of the capillary tuft and pronounced epithelial cell hyperplasia. Such a pattern has been described in HIV-associated nephropathy [80], parvovirus B19 infection [81], pamidronate toxicity [82] and in some patients with idiopathic FSGS [83, 84].

In a study on collapsing idiopathic FSGS and HIV-associated nephropathy, proliferating cells that did not express podocyte markers were described as podocytes based on their localization on top of the glomerular tuft, without an apparent connection to Bowman's capsule [53]. However, these cells expressed markers that are not normally seen in podocytes, such as CK [85], PAX2 [86] and CD68 [86, 87], suggesting the concept of the 'dysregulated podocyte'. Cells with a 'dysregulated podocyte phenotype' were also observed in a transgenic mouse (Tg26), in which several genes of the HIV genome are ectopically expressed in tubular and glomerular epithelial cells. Therefore, it was proposed that, the expression of HIV genes leads to de-differentiation and dysregulation of podocytes that is observed in HIV-related collapsing glomerulopathy.

It was also proposed that dysregulated podocytes can regain the ability to proliferate, causing epithelial hyperplasia. The observation that podocytes transfected with specific HIV genes (*Tat* or *Nef-1*) proliferate in vitro appeared consistent with this hypothesis [88, 89]. However, the existing data in the literature are conflicting. In contrast to in vitro studies, although podocyte-

Table 1. Role and distribution of PEC and podocytes in glomerular hyperplastic lesions of human biopsies

Reference	Renal disorder	Types of cells within the lesion
Thorner et al. [98]	crescentic glomerular disease (35 cases)	– ≤50% nestin-positive – negative for other podocyte markers (podocin and WT1) – the majority of Ki-67+ cells in crescents are negative for nestin
Kihara et al. [97]	collapsing glomerulopathy (3 cases)	hyperplastic epithelial cells are: – negative for the podocyte markers (PHM5 and synaptopodin)
Ohtaka et al. [85]	primary FSGS (35 cases)	cells in the lesion showed: – low expression of WT1 – significantly high expression of Pax2 and CK (PEC markers)
Nagata et al. [62]	collapsing glomerulopathy (2 cases)	hyperplastic and hypertrophic epithelial cells are: – mostly CK+Ki-67+ – negative for PHM-5 (marker of podocyte) – small number of hyperplastic epithelial cells expressed both PHM-5 and CK
Nagata et al. [61]	FSGS (21 cases)	cells in the lesions are: – 82.4% CK+Ki-67+ – negative for podocyte markers (PHM-5 and synaptopodin)
Dijkman et al. [63]	collapsing glomerulopathy (3 cases)	The proliferating cells Ki-67+: – negative for the podocyte markers (synaptopodin, vascular endothelial growth factor and CD10) – positive for the PEC markers (CK8 and PAX2)
Dijkman et al. [96]	collapsing glomerulopathy (1 case, 137 glomeruli)	Epithelial cells in Bowman's space were: – negative for podocyte markers (synaptopodin, CD10, vimentin), – negative for macrophage markers (CD68) – positive for PEC markers (CK8, pan-cadherin and PAX2)
Barisoni et al. [52]	collapsing idiopathic FSGS (10 cases), HIV-associated nephropathy (8 cases), minimal change disease (5 cases), membranous glomerulopathy (5 cases)	– 11% of glomeruli in collapsing idiopathic FSGS and 10% of glomeruli in HIV-nephropathy showed from one to five Ki-67+ podocytes – 89% of glomeruli do not have Ki-67+ podocytes.
Smeets et al. [99]	crescentic glomerulonephritis (25 cases), collapsing glomerulopathy (6 cases), FSGS with intraglomerular hyperplasia (7 cases), tip lesion (5 cases)	50–80% of cell in intraglomerular hyperplastic lesions of patients with crescentic glomerulonephritis, collapsing glomerulopathy, FSGS and the tip lesion are derived from CD24+CD133+PEC (immature progenitors), 10–40% of the cells are derived from podocytes and 2–25% of the cells are CD24+CD133+PDX+ podocyte-committed progenitor; only PEC proliferate.

Table 2. Role and distribution of PEC and podocytes in glomerular hyperplastic lesions of animal models of glomerular disorders

Reference	Renal disorder	Types of cells within the lesion
Moeller et al. [55]	doubly transgenic 2.5P-Cre/ROSA26 mice with constitutive expression of β-galactosidase; crescentic glomerulonephritis was induced with a rabbit anti-GBM antiserum	in 52% of cellular crescents: – approximately one-half of the cells or more were β-galactosidase-positive (podocytes) Cells were negative for podocyte-specific antigens, including WT1, synaptopodin, nephrin and podocin, – 48% of the glomeruli did not contain podocytes
Smeets et al. [65]	Thy-1.1 transgenic mice that express the Thy-1.1 antigen on podocytes; injection of anti-Thy-1.1 mAb induces an acute albuminuria and FSGS lesions that resemble human collapsing FSGS	cells in the lesion: – strongly express CD10 (marker for PEC) – are negative for Thy-1.1 (transgenic tagging of podocytes) proliferating cells were: – Ki-67+Cd10+ (PEC) – negative for desmin and all podocyte markers (WT1, synaptopodin, ASD33, Thy-1.1 antigen)
Asano et al. [59]	transgenic mouse model equipped with a podocyte-selective injury induction system	proliferating cells within the lesion were: – lacZ-negative – podocytes are absent
Thorner et al. [98]	mouse models of crescentic glomerulonephritis (anti-GBM model and Von Hippel-Lindau mouse model)	– nestin-positive cells were more often Ki-67-negative than positive
Suzuki et al. [95]	FSGS model in p21-deficient mice with podocyte lineage identified by genetic tagging	hypercellular and monolayer extracapillary lesions: – X-gal-negative and nestin-negative – positive for PEC markers (CK, Pax-2 and claudin-1). – expressed CK and Pax-2 (PEC markers) – negative for synaptopodin or macrophage markers
Smeets et al. [100]	mouse models of crescentic glomerulonephritis (Anti-GBM model) and of collapsing glomerulopathy (podocyte specific Thy-1.1 transgenic mice treated with anti-Thy-1.1 mAb) obtained in genetically tagged PEC	85±8% of the crescentic lesions are derived from genetically-tagged PEC. Almost the totality of the cells generating the lesion in collapsing glomerulopathy are derived from PEC. Only PEC proliferate

specific expression of HIV-1 Nef in mice caused podocytes to enter the cell cycle, mitosis was not completed and there was no proliferation [90, 91]. Furthermore, recent studies in animal models have suggested that the epithelial cells that proliferate in collapsing FSGS are not podocytes, but PEC. This hypothesis was supported by data obtained in a model of FSGS developed in transgenic mice carrying the mouse-human chimeric transgene that causes ectopic expression of the mouse Thy-1.1 antigen on podocytes [92, 93]. Injection of a monoclonal antibody directed against the Thy-1.1 antigen induces FSGS lesions with intraglomerular epithelial cell hyperplasia, which resembles human collapsing glomerulopathy [92, 93]. The proliferating epithelial cells in Bowman's space were negative for all podocyte markers tested, including Thy-1.1, and expressed CD10, which is a PEC marker in the mouse, which was expressed from the pre-capillary stage onwards.

In a recent study where transgenic mice with podocyte-specific expression of CD25 and LacZ, driven by the Nphs-1 promoter, were generated, injection of LMB2 (an immunotoxin against the podocyte CD25 antigen) caused podocyte injury characterized by vacuolar degeneration of podocytes, hyalinosis and glomerular sclerosis with proliferation of epithelial cells in the Bowman's space [94]. No evidence for proliferation or transdifferentiation of podocyte-derived cells was found; however, proliferating PEC migrating onto the glomerular tuft were observed.

Finally, in another study performed in HIV13FBV mice (which express HIV-1 genes in podocytes), prominent cell proliferation adjacent to the parietal epithelium lining Bowman's capsule was observed in the Bowman's space and were often positive for Ki-67 and CK [94]. In contrast, the cells adjacent to the glomerular basement membrane were rarely positive for Ki-67 and no cells were found to co-express synaptopodin and Ki-67, suggesting that most proliferating cells in Bowman's space were derived from the parietal epithelium [94].

Similar results were also obtained in another model of collapsing glomerulopathy [95]. Indeed, podocyte injury in a model of p21-deficient mice, where podocytes were identified by podocyte lineage by genetic tagging, developed a complete histological picture of collapsing glomerulopathy [95]. X-gal staining and immunohistochemistry for podocyte and PEC markers revealed progressive podocytopenia with capillary collapse accompanied by PEC hyperplasia leading to FSGS. Non-tagged cells expressed neither WT1 nor nestin, while Ki-67 (a proliferation marker) was rarely associated with podocytes, but was expressed at high levels in PEC. Proliferating epithelial cells in Bowman's space from a patient with recurrence of idiopathic FSGS after transplantation also stained positively for all PEC markers without expressing podocyte markers, thus arguing in favor of their origin from the parietal epithelium [96].

Obviously, the validity of marker analysis to determine cell origin can be questioned in view of possible de-differentiation accompanied by neo-expression by podocytes of antigens commonly found on PEC, especially since PEC and podocytes share a common embryonic origin. However, in the latter study [96], an additional argument was provided by three-dimensional analysis of the glomeruli. It was indeed shown that, in the absence of cell bridges between Bowman's capsule and the glomerular tuft, PEC covering Bowman's capsule acquired the CK8-positive phenotype before proliferation. These data are in agreement with the findings of other studies suggesting that PEC identified by their CK expression contribute to the hypercellular lesions in patients with idiopathic FSGS [63, 64, 97].

Taken all together, these studies allow the conclusion that PEC proliferate and play a role in the progression of human FSGS. Although the recent discovery that a relevant percentage of cells within crescents are stained by nestin, an intermediate filament protein specifically associated with podocytes in mature glomeruli and expressed during all stages of podocyte development, renewed the concept of the de-differentiated podocyte [66, 98].

However, we have recently explored the possibility that hyperplastic epithelial cells present in human biopsies of different glomerular disorders might derive from distinct populations of CD133+CD24+ renal progenitors normally present in the Bowman's capsule of the adult human kidney, i.e. more undifferentiated cells lacking podocyte markers (CD133+CD24+PDX–nestin–) and transitional cells (CD133+CD24+PDX+nestin+) [40]. By using confocal microscopy, laser capture microdissection and real-time quantitative RT-PCR, we could show that the majority of cells present in the hyperplastic lesions of collapsing glomerulopathy or crescentic glomerulonephritis display the features of renal progenitors of the Bowman's capsule, with heterogeneous expression of podocyte markers. In addition, virtually all the proliferating cells (Ki-67+) within the hyperplastic lesions exhibited the renal progenitor markers CD133 and CD24, with or without the podocyte marker nestin.

Podocytes which were included within these lesions did not proliferate. Therefore, we suggest that the glomerular hyperplastic lesions are the consequence of an aberrant proliferative response of the renal progenitors of the Bowman's capsule that may be considered as an attempt to replace the damaged podocytes. It might be hypothesized that in the chain of events that leads to glomerular sclerosis, the podocyte is initially injured and that renal progenitors from the Bowman's capsule proliferate to replace damaged podocytes [99].

A further confirmation of this hypothesis came from lineage tracing experiments performed in transgenic mice with genetically labeled parietal epithelial cells in the nephrotoxic nephritis model of inflammatory crescentic glomerulonephritis as well as in the Thy-1.1 transgenic mouse model of collapsing glomerulopathy [64, 100]. In both models, genetically labeled PEC constituted the majority of cells that composed early extracapillary proliferative lesions and almost all of the proliferating cells [100].

Lineage-tracing experiments have demonstrated that following massive podocyte injury, PEC generate cell bridges with the glomerular tuft in several areas of the glomerulus, thus probably providing a slide for a quick replacement of lost podocytes. However, numerous areas of podocyte injury and glomerular epithelial stem cell proliferation heavily distort glomerular structural integrity, thus altering the polarity of PEC division. A polarized proliferation is a critical determinant of a correct stem cell differentiation [101–105]. This might explain why a disruption in PEC polarity initiates their abnormal proliferation and the development of hyperplastic glomerular lesions, thus impairing repair. Taken altogether, the results of these recent studies suggest that PEC represent a crucial player in glomerular physiology and in the pathogenesis of glomerular disorders.

Conclusion

Since they were first described almost 200 years ago, the function of PEC has remained poorly understood. A large body of evidence has recently suggested that PEC represent a reservoir of renal progenitors in adult human kidney which generate novel podocytes during childhood and adolescence and can regenerate injured podocytes. The discovery of a hierarchical population of renal progenitors among PEC of the Bowman's capsule provides a new point of view for the understanding of glomerular physiology. In addition, the observation that renal progenitors generate hyperplastic glomerular lesions opens a novel view of the pathogenesis of different types of glomerular disorders, which may result from abnormal regenerative responses to podocyte injury.

References

1 Sir William Bowman (1816–1892): On the Structure and Use of the Malpighian Bodies of the Kidney, with Observations on the Circulation. London, R. and J. E. Taylor, 1842.

2 Finckh ES, Joske RA: The occurrence of columnar epithelium in Bowman's capsule. J Pathol Bacteriol 1954;68:646–648.

3 Nachman RL: Metaplasia of parietal capsular epithelium of renal glomerulus. Arch Pathol 1962;73:48–52.

4 Kanel GC, Peters RL: Glomerular tubular reflux – a morphologic renal lesion associated with the hepatorenal syndrome. Hepatology 1984;4:242–246.

5 Valdes AJ, Zhang JM: Intraglomerular tubular epithelial cells. A marker of glomerular hematuria. Arch Pathol Lab Med 1987;111:189–191.

6 Andrews PM: The presence of proximal tubulelike cells in the kidney parietal epithelium in response to unilateral nephrectomy. Anat Rec 1981;200:61–65.

7 Castelletto L, Goya RG: Sex-related incidence of tubular metaplasia in Bowman's capsule of aging rats. Virchows Arch B Cell Pathol Incl Mol Pathol 1990;59:79–82.

8 Haensly WE, Lee JC: Metaplasia of the parietal layer of Bowman's capsule: a histopathological survey of the human kidney. Histol Histopathol 1986;1:363–367.
9 Ryan GB, Goghlan JP, Scoggins BA: The granulated peripolar epithelial cells: a potential secretory component of the renal juxtaglomerular complex. Nature 1979;277:655–656.
10 Gall JA, Alcorn D, Butkus A, Coghlan JP, Ryan GB: Distribution of glomerular peripolar cells in different mammalian species. Cell Tissue Res 1986;244:203–208.
11 Gibson IW, Downie I, Downie TT, Han SW, More IAR, Lindop GB: The parietal podocyte: a study of the vascular pole of the human glomerulus. Kidney Int 1992;41:211–214.
12 Bariety J, Mandet C, Hill GS, Bruneval P: Parietal podocytes in normal human glomeruli. J Am Soc Nephrol 2006;17:2770–2780.
13 Dressler GR: The cellular basis of kidney development. Annu Rev Cell Dev Biol 2006;22:509–529.
14 Romagnani P: Toward the identification of a 'renopoietic system'? Stem Cells 2009;27: 2247–2253.
15 Kreidberg JA: Podocyte differentiation and glomerulogenesis. J Am Soc Nephrol 2003;14:806–814.
16 Nagata M, Nakayama K, Terada Y, Hoshi S, Watanabe T: Cell cycle regulation and differentiation in the human podocyte lineage. Am J Pathol 1998;153:1511–1520.
17 Hiromura K, Haseley LA, Zhang P, Monkawa T, Durvasula R, Petermann AT, Alpers CE, Mundel P, Shankland SJ: Podocyte expression of the CDK-inhibitor p57 during development and disease. Kidney Int 2001;60:2235–2246.
18 Shankland SJ, Eitner F, Hudkins KL, Goodpaster T, D'Agati V, Alpers CE: Differential expression of cyclin-dependent kinase inhibitors in human glomerular disease: role in podocyte proliferation and maturation. Kidney Int 2000;58:674–683.
19 Barisoni L, Kriz W, Mundel P, D'Agati V: The dysregulated podocyte phenotype: a novel concept in the pathogenesis of collapsing idiopathic focal segmental glomerulonephritis and HIV-associated nephropathy. J Am Soc Nephrol 1999;10:51–61.
20 Bariety J, Bruneval P, Hill G, Irinopoulou T, Mandet C, Meyrier A: Posttransplantation relapse of FSGS is characterized by glomerular epithelial cell transdifferentiation. J Am Soc Nephrol 2001;12:261–274.
21 Mundlos S, Pelletier J, Darveau A, Bachmann M, Winterpacht A, Zabel B: Nuclear localization of the protein encoded by the Wilms' tumor gene WT1 in embryonic and adult tissues. Development 1993;119:1329–1341.
22 Yang Y, Jeanpierre C, Dressler GR, Lacoste M, Niaudet P, Gubler MC: WT1 and PAX-2 podocyte expression in Denys-Drash syndrome and isolated diffuse mesangial sclerosis. Am J Pathol 1999;154:181–192.
23 Torban E, Eccles MR, Favor J, Goodyer PR: PAX2 suppresses apoptosis in renal collecting duct cells. Am J Pathol 2000;157:833–842.
24 Ryan G, Steele-Perkins V, Morris JF, Rauscher F Jr, Dressler GR: Repression of PAX-2 by WT1 during normal kidney development. Development 1995;121:867–875.
25 Romagnani P, Kalluri R: Possible mechanisms of kidney repair. Fibrogenesis Tissue Repair 2009;26:2–3.
26 Sagrinati C, Ronconi E, Lazzeri E, et al: Stem-cell approaches for kidney repair: choosing the right cells. Trends Mol Med 2008;14:277–285.
27 Remuzzi G, Benigni A, Remuzzi A: Mechanisms of progression and regression of renal lesions of chronic nephropathies and diabetes. J Clin Invest 2006;116:288–296.
28 Blanpain C, Horsley V, Fuchs E: Epithelial stem cells: turning over new leaves. Cell 2007;128:445–458.
29 Gurtner GC, Werner S, Barrandon Y, et al: Wound repair and regeneration. Nature 2008;453:314–321.
30 Shackleton M, Vaillant F, Simpson KJ, et al: Generation of a functional mammary gland from a single stem cell. Nature 2006;439:84–88.
31 Kubota H, Avarbock MR, Brinster RL: Spermatogonial stem cells share some, but not all, phenotypic and functional characteristics with other stem cells. Proc Natl Acad Sci USA 2003;100:6487–6492.
32 Coskun V, Wu H, Blanchi B, et al: CD133+ neural stem cells in the ependyma of mammalian postnatal forebrain. Proc Natl Acad Sci USA 2008;105:1026–1031.
33 Kania G, Corbeil D, Fuchs J, et al: Somatic stem cell marker prominin-1/CD133 is expressed in embryonic stem cell–derived progenitors. Stem Cells 2005;23:791–804.
34 Rajasekhar VK, Begemann M: Roles of polycomb group proteins in development and disease: a stem cell perspective. Stem Cells 2007;25:2498–2510.
35 Kozakowski N, Soleiman A, Pammer J: BMI-1 expression is inversely correlated with the grading of renal clear cell carcinoma. Pathol Oncol Res 2008;14:9–13.
36 Pesce M, Schöler HR: Oct-4: gatekeeper in the beginnings of mammalian development. Stem Cells 2001;19:271–278.
37 Lazzeri E, Crescioli C, Ronconi E, et al: Regenerative potential of embryonic renal multipotent progenitors in acute renal failure. J Am Soc Nephrol 2007;18:3128–3138.
38 Sagrinati C, Netti GS, Mazzinghi B, et al: Isolation and characterization of multipotent progenitor cells from the Bowman's capsule of adult human kidneys. J Am Soc Nephrol 2006;17:2443–2456.
39 Mazzinghi B, Ronconi E, Lazzeri E, et al: Essential but differential role for CXCR4 and CXCR7 in the therapeutic homing of human renal progenitor cells. J Exp Med 2008;205: 479–490.
40 Ronconi E, Sagrinati C, Angelotti ML, et al: Regeneration of glomerular podocytes by human renal progenitors. J Am Soc Nephrol 2009;20:322–332.
41 Pittenger MF, Mackay AM, Beck SC, Jaiswal RK, Douglas R, Mosca JD, Moorman MA, Simonetti DW, Craig S, Marshak DR: Multilineage potential of adult human mesenchymal stem cells. Science 1999;284:143–147.
42 Ito M, Liu Y, Yang Z, et al: Stem cells in the hair follicle bulge contribute to wound repair but not to homeostasis of the epidermis. Nat Med 2005;11:1351–1354.
43 Levy V, Lindon C, Harfe BD, et al: Distinct stem cell populations regenerate the follicle and interfollicular epidermis. Dev Cell 2005;9:855–861.
44 Liu KD, Brakeman PR: Renal repair and recovery. Crit Care Med 2008;36(Suppl 4):187–192.

45 Vogetseder A, Picard N, Gaspert A, et al: Proliferation capacity of the renal proximal tubule involves the bulk of differentiated epithelial cells. Am J Physiol Cell Physiol 2008; 294:C22–C28.
46 Vogetseder A, Palan T, Bacic D, et al: Proximal tubular epithelial cells are generated by division of differentiated cells in the healthy kidney. Am J Physiol Cell Physiol 2007;292:C807–C813.
47 Wiggins RC: The spectrum of podocytopathies: a unifying view of glomerular diseases. Kidney Int 2007;71:1205–1214.
48 Marshall CB, Shankland SJ: Cell cycle and glomerular disease: a minireview. Nephron Exp Nephrol 2006;102:e39–e48.
49 Marcus PB: Podocytic 'metaplasia' of parietal Bowman's capsular epithelium. Arch Pathol Lab Med 1977;101:664.
50 Rosivall L: Morphology and function of the distal part of the afferent arteriole. Kidney Int 1990;38(Suppl):S10–S15.
51 Appel D, Kershaw DB, Smeets B, et al: Recruitment of podocytes from glomerular PECs. J Am Soc Nephrol 2009;20:333–343.
52 Barisoni L, Kriz W, Mundel P, D'Agati V: The dysregulated podocyte phenotype: a novel concept in the pathogenesis of collapsing idiopathic focal segmental glomerulosclerosis and HIV-associated nephropathy. J Am Soc Nephrol 1999;10:51–61.
53 Barisoni L, Schnaper HW, Kopp JB: A proposed taxonomy for the podocytopathies: a reassessment of the primary nephrotic diseases. Clin J Am Soc Nephrol 2007;2:529–542.
54 Bariety J, Nochy D, Mandet C, Jacquot C, Glotz D, Meyrier A: Podocytes undergo phenotypic changes and express macrophagic-associated markers in idiopathic collapsing glomerulopathy. Kidney Int 1998;53:918–925.
55 Moeller MJ, Soofi A, Hartmann I, Le Hir M, Wiggins R, Kriz W, Holzman LB: Podocytes populate cellular crescents in a murine model of inflammatory glomerulonephritis. J Am Soc Nephrol 2004;15:61–67.
56 Ding M, Cui S, Li C, Jothy S, Haase V, Steer BM, Marsden PA, Pippin J, Shankland S, Rastaldi MP, Cohen CD, Kretzler M, Quaggin SE: Loss of the tumor suppressor Vhlh leads to upregulation of Cxcr4 and rapidly progressive glomerulonephritis in mice. Nat Med 2006;12: 1081–1087.
57 Kriz W, Gretz N, Lemley KV: Progression of glomerular diseases: is the podocyte the culprit? Kidney Int 1998;54:687–697.
58 Albaqumi M, Barisoni L: Current views on collapsing glomerulopathy. J Am Soc Nephrol 2008;19:1279–1281.
59 Asano T, Niimura F, Pastan I, Fogo AB, Ichikawa I, Matsusaka T: Permanent genetic tagging of podocytes: fate of injured podocytes in a mouse model of glomerular sclerosis. J Am Soc Nephrol 2005;16:2257–2262.
60 Matsusaka T, Xin J, Niwa S, Kobayashi K, Akatsuka A, Hashizume H, Wang QC, Pastan I, Fogo AB, Ichikawa I: Genetic engineering of glomerular sclerosis in the mouse via control of onset and severity of podocyte-specific injury. J Am Soc Nephrol 2005;16:1013–1023.
61 Nagata M, Horita S, Shu Y, Shibata S, Hattori M, Ito K, Watanabe T: Phenotypic characteristics and cyclin-dependent kinase inhibitors repression in hyperplastic epithelial pathology in idiopathic focal segmental glomerulosclerosis. Lab Invest 2000;80:869–880.
62 Nagata M, Hattori M, Hamano Y, Ito K, Saitoh K, Watanabe T: Origin and phenotypic features of hyperplastic epithelial cells in collapsing glomerulopathy. Am J Kidney Dis 1998;32:962–969.
63 Dijkman HB, Weening JJ, Smeets B, Verrijp KC, van Kuppevelt TH, Assmann KK, Steenbergen EJ, Wetzels JF: Proliferating cells in HIV and pamidronate-associated collapsing focal segmental glomerulosclerosis are PECs. Kidney Int 2006;70:338–344.
64 Smeets B, Dijkman HB, Wetzels JF, Steenbergen EJ: Lessons from studies on focal segmental glomerulosclerosis: an important role for PECs? J Pathol 2006;210:263–272.
65 Smeets B, Te Loeke NA, Dijkman HB, Steenbergen ML, Lensen JF, Begieneman MP, van Kuppevelt TH, Wetzels JF, Steenbergen EJ: The parietal epithelial cell: a key player in the pathogenesis of focal segmental glomerulosclerosis in Thy-1.1 transgenic mice. J Am Soc Nephrol 2004;15:928–939.
66 Perry J, Ho M, Viero S, Zheng K, Jacobs R, Thorner PS: The intermediate filament nestin is highly expressed in normal human podocytes and podocytes in glomerular disease. Pediatr Dev Pathol 2007;10: 369–382.
67 Morita T, Suzuki Y, Churg J: Structure and development of the glomerular crescent. Am J Pathol 1973;72:349–368.
68 Cattell V, Jamieson SW: The origin of glomerular crescents in experimental nephrotoxic serum nephritis in the rabbit. Lab Invest 1978;39: 584–590.
69 Magil AB: Histogenesis of glomerular crescents. Immunohistochemical demonstration of cytokeratin in crescent cells. Am J Pathol 1985;120:222–229.
70 Harrison DJ, MacDonald MK: The origin of cells in the glomerular crescent investigated by the use of monoclonal antibodies: Histopathology 1986;10:945–952.
71 Shirato I, Asanuma K, Takeda Y, Hayashi K, Tomino Y: Protein gene product 9.5 is selectively localized in PECs of Bowman's capsule in the rat kidney. J Am Soc Nephrol 2000;11: 2381–2386.
72 Usui J, Kanemoto K, Tomari S, et al: Glomerular crescents predominantly express cadherin-catenin complex in pauci-immune-type crescentic glomerulonephritis. Histopathology 2003;43:173–179.
73 Ophascharoensuk V, Pippin JW, Gordon KL, Shankland SJ, Couser WG, Johnson RJ: Role of intrinsic renal cells versus infiltrating cells in glomerular crescent formation. Kidney Int 1998;54:416–425.
74 Boucher A, Droz D, Adafer E, Noel LH: Relationship between the integrity of Bowman's capsule and the composition of cellular crescents in human crescentic glomerulonephritis. Lab Invest 1987;56:526–533.

75 Cockwell P, Brooks CJ, Adu D, Savage CO: Interleukin-8: a pathogenetic role in antineutrophil cytoplasmic autoantibody-associated glomerulonephritis. Kidney Int 1999;55:852–863.

76 Cockwell P, Howie AJ, Adu D, Savage CO: In situ analysis of C-C chemokine mRNA in human glomerulonephritis. Kidney Int 1998;54:827–836.

77 He JS, Hayashi K, Horikoshi S, Funabiki K, Shirato I, Tomino Y: Identification of cellular origin of type I collagen in glomeruli of rats with crescentic glomerulonephritis induced by anti-glomerular basement membrane antibody. Nephrol Dial Transplant 2001;16:704–711.

78 Barnes JL, Torres ES, Mitchell RJ, Peters JH: Expression of alternatively spliced fibronectin variants during remodeling in proliferative glomerulonephritis. Am J Pathol 1995;147:1361–1371.

79 Detwiler RK, Falk RJ, Hogan SL, Jennette JC: Collapsing glomerulopathy: a clinically and pathologically distinct variant of focal segmental glomerulosclerosis. Kidney Int 1994;45:1416–1424.

80 D'Agati V, Suh JI, Carbone L, Cheng JT, Appel G: Pathology of HIV-associated nephropathy: a detailed morphologic and comparative study. Kidney Int 1989;35:1358–1370.

81 Moudgil A, Nast CC, Bagga A, et al: Association of parvovirus B19 infection with idiopathic collapsing glomerulopathy. Kidney Int 2001;59:2126–2133.

82 Markowitz GS, Appel GB, Fine PL, et al: Collapsing focal segmental glomerulosclerosis following treatment with high-dose pamidronate. J Am Soc Nephrol 2001;12:1164–1172.

83 Valeri A, Barisoni L, Appel GB, Seigle R, D'Agati V: Idiopathic collapsing focal segmental glomerulosclerosis: a clinicopathologic study. Kidney Int 1996;50:1734–1746.

84 Nagata M, Tomari S, Kanemoto K, Usui J, Lemley KV: Podocytes, parietal cells, and glomerular pathology: the role of cell cycle proteins. Pediatr Nephrol 2003;18:3–8.

85 Ohtaka A, Ootaka T, Sato H, Ito S: Phenotypic change of glomerular podocytes in primary focal segmental glomerulosclerosis: developmental paradigm? Nephrol Dial Transplant 2002;17(Suppl 9):11–15.

86 Yang Y, Gubler MC, Beaufils H: Dysregulation of podocyte phenotype in idiopathic collapsing glomerulopathy and HIV-associated nephropathy. Nephron 2002;91:416–423.

87 Bariety J, Bruneval P, Hill G, Irinopoulou T, Mandet C, Meyrier A: Posttransplantation relapse of FSGS is characterized by glomerular epithelial cell transdifferentiation. J Am Soc Nephrol 2001;12:261–274.

88 Conaldi PG, Bottelli A, Baj A, Serra C, Fiore L, Federico G, Bussolati B, Camussi G: Human immunodeficiency virus-1 tat induces hyperproliferation and dysregulation of renal glomerular epithelial cells. Am J Pathol 2002;161:53–61.

89 Husain M, Gusella GL, Klotman ME, Gelman IH, Ross MD, Schwartz EJ, Cara A, Klotman PE: HIV-1 Nef induces proliferation and anchorage-independent growth in podocytes. J Am Soc Nephrol 2002;13:1806–1815.

90 Husain M, D'Agati VD, He JC, Klotman ME, Klotman PE: HIV-1 Nef induces dedifferentiation of podocytes in vivo: a characteristic feature of HIVAN. AIDS 2005;19:1975–1980.

91 Petermann AT, Pippin J, Hiromura K, Monkawa T, Durvasula R, Couser WG, Kopp J, Shankland SJ: Mitotic cell cycle proteins increase in podocytes despite lack of proliferation. Kidney Int 2003;63:113–122.

92 Kollias G, Evans DJ, Ritter M, Beech J, Morris R, Grosveld F: Ectopic expression of Thy-1 in the kidneys of transgenic mice induces functional and proliferative abnormalities. Cell 1987;51:21–31.

93 Smeets B, Dijkman HB, Te Loeke NA, van Son JP, Steenbergen EJ, Assmann KJ, Wetzels JF, Groenen PJ: Podocyte changes upon induction of albuminuria in Thy-1.1 transgenic mice. Nephrol Dial Transplant 2003;18:2524–2533.

94 Zhong J, Zuo Y, Ma J, Fogo AB, Jolicoeur P, Ichikawa I, Matsusaka T: Expression of HIV-1 genes in podocytes alone can lead to the full spectrum of HIV-1-associated nephropathy. Kidney Int 2005;68:1048–1060.

95 Suzuki T, Matsusaka T, Nakayama M, Asano T, Watanabe T, Ichikawa I, Nagata M: Genetic podocyte lineage reveals progressive podocytopenia with parietal cell hyperplasia in a murine model of cellular/collapsing focal segmental glomerulosclerosis. Am J Pathol 2009;174:1675–1682.

96 Dijkman H, Smeets B, van der Laak J, Steenbergen E, Wetzels J: The parietal epithelial cell is crucially involved in human idiopathic focal segmental glomerulosclerosis. Kidney Int 2005;68:1562–1572.

97 Kihara I, Yaoita E, Kawasaki K, Yamamoto T, Hara M, Yanagihara T: Origin of hyperplastic epithelial cells in idiopathic collapsing glomerulopathy. Histopathology 1999;34:537–547.

98 Thorner PS, Ho M, Eremina V, Sado Y, Quaggin S: Podocytes contribute to the formation of glomerular crescents. J Am Soc Nephrol 2008;19:495–502.

99 Smeets B, Angelotti ML, Rizzo P, Dijkman H, Lazzeri E, Mooren F, Ballerini L, Parente E, Sagrinati C, Mazzinghi B, Ronconi E, Becherucci F, Benigni A, Steenbergen E, Lasagni L, Remuzzi G, Wetzels J, Romagnani P: Renal progenitor cells contribute to hyperplastic glomerular lesions of different types of podocytopathies and in crescentic glomerulonephritis. J Am Soc Nephrol 2009;20:2593–2603.

100 Smeets B, Uhlig S, Fuss A, Mooren F, Wetzels JF, Floege J, Moeller MJ: Tracing the origin of glomerular extracapillary lesions from parietal epithelial cells. J Am Soc Nephrol 2009;20:2604–2615.

101 Chambers SM, Shaw CA, Gatza C, Fisk CJ, Donehower LA, Goodell MA: Aging hematopoietic stem cells decline in function and exhibit epigenetic dysregulation. PLoS Biol 2007;5:e201.

102 Brack AS, Conboy MJ, Roy S, Lee M, Kuo CJ, Keller C, Rando TA: Increased Wnt signaling during aging alters muscle stem cell fate and increases fibrosis. Science 2007;317:807–810.

103 Liang Y, Van Zant G, Szilvassy SJ: Effects of aging on the homing and engraftment of murine hematopoietic stem and progenitor cells. Blood 2005;106:1479–1487.

104 Conboy IM, Conboy MJ, Wagers AJ, Girma ER, Weissman IL, Rando TA: Rejuvenation of aged progenitor cells by exposure to a young systemic environment. Nature 2005;433:760–764.

105 Rizvi AZ, Wong MH: Stem cell niche: there's no place like home. Stem Cells 2005;23:150–165.

Paola Romagnani
Department of Clinical Pathophysiology, Nephrology Section, University of Florence
Viale Pieraccini 6
IT–50139 Firenze (Italy)
Tel. +39 554271356, Fax +39 554271357, E-Mail p.romagnani@dfc.unifi.it

Herrera GA (ed): Experimental Models for Renal Diseases: Pathogenesis and Diagnosis.
Contrib Nephrol. Basel, Karger, 2011, vol 169, pp 37–50

Role of Proximal Tubules in the Pathogenesis of Kidney Disease

Nazih Nakhoul · Vecihi Batuman

Nephrology Section, Tulane University Health Sciences Center and VA Medical Center, New Orleans, La., USA

Abstract

The proximal tubules make up a significant portion of the kidneys; proximal tubule epithelial cells are the most populous cell type in the kidney, and carry out diverse regulatory and endocrine functions where numerous transporters are located. Under normal circumstances, more than two thirds of filtered salt and water, and all filtered bicarbonate is reabsorbed in the proximal tubule. A number of inherited and acquired acid-base and tubule disorders are linked to impaired transporters in the proximal tubule cells. Equally important is the intrinsic immune characteristics of proximal tubule cells that give them the ability to also function as immune responders to a wide range of immunologic, ischemic or toxic injury. It is therefore not surprising that proximal tubule-related phenomena are closely related to the pathogenesis of a vast array of kidney diseases. Many kidney diseases, acute and chronic, first manifest with proximal tubule disorders. Recent insight into molecular characteristics of transport functions in the proximal tubules, and the recognition that proximal tubule cells possess intrinsic immune responses have contributed to an improved understanding of important areas in nephrology, such as Fanconi's syndrome, renal tubular acidosis, phosphate wasting syndromes, Dent's disease, cystinuria and other amino acid transport disorders, acute kidney injury, and the role of proximal tubules in progressive kidney disease. Megalin/cubilin-mediated endocytosis by proximal tubule cells of increased quantities of filtered proteins (protein overloading) in glomerular diseases appears to evoke cell stress responses resulting in increased inflammatory cytokines leading to tubulointerstitial inflammation and fibrosis. Finally, the proximal tubule may be the site of both active vitamin D synthesis through the action of 1-α-hydroxylase, and the site where erythropoietin synthesis takes place. Thus, proximal tubule injury also contributes to two distressing consequences of chronic kidney disease: mineral-bone disorder and anemia.

The proximal tubule is the nephron segment that starts at the urinary pole of the glomerulus and ends at the loop of Henle. Based on morphologic-functional properties, the proximal tubule is viewed in three segments. The first two are denoted as S1 and S2, and are populated with cells of high complexity. The straight segment, S3, is made up of cells with relatively lower complexity [1]. Proximal tubules carry out diverse regulatory and endocrine functions where numerous transporters are located. Equally important is their intrinsic immune characteristics that give them the ability to also function as immune responders to a wide range of immunologic, ischemic or toxic injury. It is therefore not surprising that proximal tubule-related phenomena are closely related to the pathogenesis of a vast array of kidney diseases.

The role of proximal tubules in kidney disease can be viewed in four broad categories: (1)

proximal tubule transport disorders that lead to electrolyte or acid-base abnormalities, glucosuria, phosphaturia, Fanconi's syndrome, and amino-aciduria including cystinuria; (2) overt proximal tubule cell injury associated with apoptosis or necrosis as a result of ischemic or toxic insult giving rise to the clinical syndrome of acute renal failure, or acute kidney injury; (3) proximal tubules' contribution to the pathophysiology of chronic kidney disease through pro-inflammatory and inflammatory cytokines produced by the injured proximal tubule cells, or by epithelial-to-mesenchymal transformation of proximal tubule cells, leading to renal interstitial fibrosis [2]; and (4) vitamin D and erythropoietin deficiencies as the basis for mineral-bone disorder and anemia in chronic kidney disease [3, 4].

Proximal Tubular Disorders Linked to Transporter Dysfunctions

The proximal tubule is the most reabsorptive segment of the nephron. It is responsible for the bulk reabsorption of the largest fraction of glomerular filtrate, including NaCl and $NaHCO_3$, as well as equivalent volume of osmotically obligated water. The proximal tubule reabsorbs major fractions of filtered nutrients such as glucose and amino acids, and divalent ions such as sulfates (SO_4^{-2}), phosphates (PO_4^{-2}) and Ca^{+2}. It also has the capacity to reabsorb a variety of small molecular weight proteins and peptides by a process of endocytosis. These functions of the proximal tubule underscore its major role in regulating water and electrolyte transport as well as acid-base balance in the body.

Solute and electrolyte reabsorption in the proximal tubule occurs via both paracellular and transcellular pathways. In the paracellular pathway, ions and solutes move through a rather leaky tight junction entirely through an extracellular route. In the transcellular pathway, ions and solutes sequentially traverse the apical and basolateral membranes before reaching the blood. Transcellular transport is achieved by many transport mechanisms that occupy both membranes; including channels, co-transporters and exchangers, as well as primary active transporters (ATP-dependent pumps). The coordinated functions of these membrane proteins are responsible for vectorial transepithelial reabsorption or secretion of major electrolytes, solutes and acid-base equivalents in the proximal tubule.

More than 50 years ago, a variety of fluid and electrolyte disorders were described in which the main disturbance appeared to be specific functional defects in renal epithelial transport functions [5, 6]. Recent genetic and molecular studies helped to elucidate the molecular mechanisms of most, but not all, of these disorders that were initially proposed by the physiologic studies. Here, we address several diseases known to be caused by defects in the proximal tubule transport.

Fanconi's Syndrome

Fanconi's syndrome, acquired or hereditary (discussed in detail in a separate chapter), is manifested by a global disruption in sodium-coupled transporter systems and not necessarily a disturbance in a specific transporter. The clinical presentation includes polyuria, dehydration, hypokalemia, acidosis (in adults) and impaired growth and rickets (in children). The general dysfunction is in the proximal tubule and leads to impaired net reabsorption of HCO_3^-, phosphate, glucose and amino acids (fig. 1). In the apical membrane, several co-transporters mediate the downhill uptake of Na^+ to the uphill uptake of solutes such as glucose, amino acids, phosphates, sulfate and mono- and dicarboxylic acids.

The lumen-to-cell Na^+ gradient is maintained by the low intracellular Na^+ concentration $[Na^+]$ i and drives the secondary active uptake of the other solutes. Many of these co-transporters are electrogenic and carry a net positive charge; therefore, they are further driven by the negative potential of the apical membrane. Also, apical Na^+ entry is coupled to H^+ efflux through the electroneutral Na-H exchange (NHE-3). Thus the downhill Na^+ gradient across the apical membrane is the primary driving force for uptake of

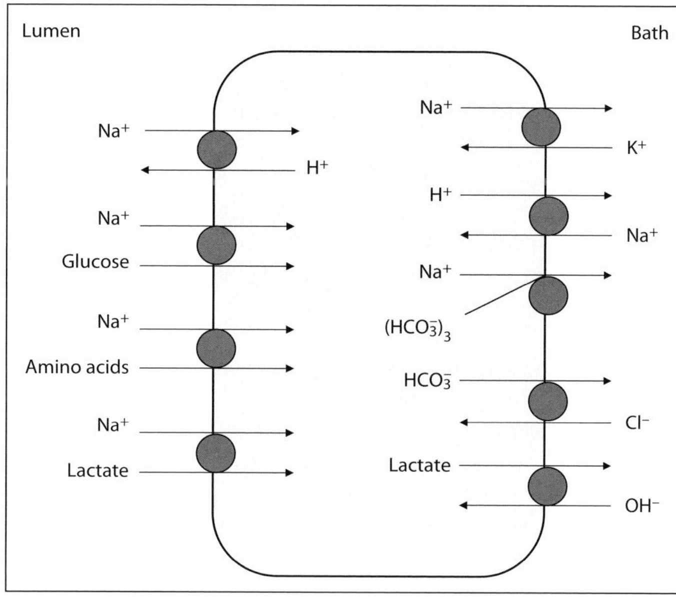

Fig. 1. Proximal epithelial cell. Sodium-dependent transport in the proximal tubule cell.

many solutes. The critical step in maintaining this favorable Na^+ gradient is the low intracellular Na^+ concentration, which is established by the Na-K pump (Na-KATPase) in the basolateral membrane and to a lesser extent the electrogenic Na/HCO_3^- co-transporter (NBC), both of which extrude Na^+ from the cell towards the blood side.

A disruption of this Na^+ gradient is the likely underlying mechanism for the general dysfunction observed in Fanconi's syndrome. Bergeron et al. [7] proposed a transport model to explain Fanconi's syndrome. According to this model, the apical Na^+ gradient that drives the uptake of most solutes is altered by a change in the activity of the Na-K ATPase. Another factor that could act independently or in sync with the above model is a change in the ATP-ADP ratio in the cells, which is maintained by a balance of the activities of various membrane-bound ATPase (e.g. Na-KATPase and H^+-ATPase) that consume ATP and mitochondrial re-phosphorylation of ADP that generates ATP. A small decrease in the ability of mitochondria to maintain the ATP-ADP ratio will also lead to a decrease in the activity of Na-K ATPase causing an increase in $[Na^+]i$, a subsequent decrease in the apical Na^+ gradient and inhibition of solute uptake. Although, the underlying cause of Fanconi's syndrome has yet to be determined, both mechanisms seem plausible.

Renal Glucosuria

Abnormal renal handling of glucose occurs in a number of inherited and acquired diseases [8]. In the absence of increased filtered glucose, there are two main causes of renal glucosuria. Primary (familial) renal glucosuria (FRG) is when glucose is detected in urine in the absence of hyperglycemia and (usually) in the absence of general tubular dysfunction (as in Fanconi's syndrome). The second disorder, glucose-galactose malabsorption, presents early in life, mostly by intestinal symptoms including severe diarrhea and dehydration. Patients with glucose-galactose malabsorption may present with or without glucosuria, which is usually mild. Both disorders are caused by impaired glucose transport. Nearly all the filtered glucose is absorbed in the early segments (S1 and S2) of the proximal tubule, and the remainder in the late proximal segment (S3). At the apical membrane, a Na/Glucose co-transporter couples the uptake of D-glucose (but non L-glucose) to Na^+. In the early proximal tubule (S1 segment), a member of this family (SGLT2) mediates apical glucose uptake. This is a high-capacity/low-affinity transporter and has a stoichiometry of Na-to-glucose of 1:1 and is therefore electrogenic.

In the late proximal tubule (S3 segment), SGLT1 is responsible for apical Na^+ uptake. SGLT1 is a high-affinity/low capacity co-transporter and has a stoichiometry of Na^+:glucose of 2:1 and, therefore, an electrogenic transporter with more ability to generate a larger glucose gradient. Glucose reabsorption in the proximal tubule is mostly transcellular; paracellular transport is minimal and diminishes significantly in later segments of the tubule.

The exit of glucose across the basolateral membrane is mediated by another family of glucose transporters (GLUT), which are Na^+-independent and transport glucose by facilitated diffusion. In the early proximal tubule, GLUT2 is responsible for basolateral glucose transport, whereas GLUT1 is the main transporter in the late proximal tubule. SGLT and GLUT can be inhibited by phloridzin; however, the basolateral GLUTs are significantly less sensitive to this inhibitor. Glucose appears in urine when the rate of glucose reabsorption reaches its maximal transport capacity, which occurs when the apical glucose transporters become fully saturated. Inherited FRG can be autosomal recessive or autosomal dominant and occurs as a result of different mutations of SGLT2 [9, 10]. The most severe form of FRG, classified as type 0, is caused by nonfunctioning mutations of SGLT2 causing complete absence of renal tubular reabsorption of glucose. Patients with less severe forms of FRG (types A and B) also have mutations in SGLT2 (Gene SLC5A2). FRG patients, including those with the most severe form type 0, are generally asymptomatic. Inherited glucose-galactose malabsorption syndrome is an autosomal recessive disease caused by mutation of SGLT1 (gene SLC5A1) [11, 12]. Although this disease is dominated by intestinal symptoms that are generally severe, some patients present with glucosuria. However, given that the role of SGLT1 is minor in renal reabsorption of glucose, glucosuria in these patients is usually mild or absent [13].

Loss of function mutations of the gene encoding GLUT2 (SLC2A2) usually result in a rare autosomal recessive glycogen storage disease known as Fanconi-Bickel syndrome [11]. Dysfunction of GLUT2 also results in glucose-galactose intolerance with varying symptoms. Glucosuria is also a manifestation of GLUT2-induced nephropathy because it is involved in basolateral uptake of glucose. There are no known mutations of GLUT1 that cause renal glucosuria. All three transporters, SGLT1, SGLT2 and GLUT2, affect transport of glucose in many organs. SGLT2, however, is present almost exclusively in renal proximal tubules; therefore, malfunction of this transporter is a primary cause of renal tubular impairment of glucose reabsorption.

Renal Hypophosphatemic Tubular Disorders

Phosphate (PO_4^{-2}) plays a critical role in many cellular processes, particularly in cell energy metabolism (as part of ATP) and in activation and deactivation of enzymes. Phosphate metabolism depends on bone, where most PO_4^{-2} is present, intestines, which absorb typically more than half of the dietary intake of PO_4^{-2}, and the kidneys where PO_4^{-2} is freely filtered. Under normal conditions, the kidneys reabsorb most of the filtered PO_4^{-2} and excrete an amount equivalent to the net amount absorbed by the gastrointestinal tract. Inorganic phosphates are very abundant in the diet; consequently, PO_4^{-2} deficiency is unusual unless significant PO_4^{-2} wasting occurs.

The renal proximal tubule reabsorbs more than 80% of the freely filtered phosphate predominantly through the transcellular pathway. Apical uptake of PO_4^{-2} is coupled to Na^+ and driven by the Na^+ electrochemical gradient. An apical Na-phosphate co-transporter (NaPi type 2; NaPi-2) transports 3 Na^+ with 1 phosphate ion, which could be either HPO_4^{-2} or HPO_4^- making pH an important factor in regulating apical phosphate uptake. Luminal, as well as intracellular, pH affect PO_4^{-2} transport with intracellular H^+ stimulating NaPi-2 (probably by allosteric modification), whereas luminal H^+ inhibits NaPi-2. Luminal H^+ seems to compete with Na^+ and also shifts HPO_4^{-2} to $H_2PO_4^-$ equilibrium, both negatively influencing transport.

At the basolateral membrane, phosphate exits the cell passively; however, the exact mechanism is not well understood. The basolateral membrane has a NaPi Type 3 that electrogenically transports 3 Na^+ with each PO_4^{-2}, which mediates cellular uptake of PO_4^{-2} and not exit. Several important factors affect PO_4^{-2} transport. Parathyroid hormone (PTH) is probably the most important regulator and it inhibits PO_4^{-2} reabsorption rapidly and independently of an effect on protein synthesis. PTH binding to its receptor (1R receptor) leads to endocytosis of NaPi-2 through a cellular cascade involving both PKA and PKC signaling. A decrease in PTH has the opposite effect: insertion of NaPi-2 in the membrane and an increase in PO_4^{-2} absorption. Other factors also have a role: phosphate depletion and 1,25 $(OH)_2$-Vit D_3 stimulate, but glucocorticoids and low pH inhibit. Thus, a main mechanism of PO_4^{-2} regulation appears to involve insertion or retrieval of NaPi-2 into the apical membrane.

There are several well-characterized inherited hypophosphatemic disorders due to renal wasting of PO_4^{-2} caused by impaired renal PO_4^{-2} transport [14]. All are characterized by bone disease and (often) abnormal vitamin D metabolism. X-linked hypophosphatemia (XLH), an X-linked disorder with dominant mode of inheritance is characterized by malabsorption of renal PO_4^{-2}, osteomalacia, rickets and growth retardation. Autosomal dominant hypophosphatemic rickets (ADHR) is very similar, but less common, and has a variable age of onset. Unlike these two inherited forms, oncogenic hypophosphatemic osteomalacia is an acquired and rare form of renal PO_4^{-2} wasting, but clinically resembles XLH and ADHR. Another disorder, hereditary hypophosphatemic rickets with hypercalciuria is distinguished from the others by elevated serum levels of 1,25 $(OH)_2$-Vit D_3 and hypercalciuria, although other clinical symptoms are similar.

The transport defect in XLH is a decrease in renal reabsorption of PO_4^{-2} with a decrease in the maximal velocity of NaPi-2 secondary to a decrease in the abundance of NaPi-2 mRNA and protein in the renal brush border membranes [15–17]. The gene responsible for XLH is known as PHEX, 'phosphate regulating gene with homology to endopeptidases on the X chromosome', where various mutations (currently 166) have been identified [18]. Interestingly, PHEX is expressed in osteoblasts, osteocytes and odontoblasts, but not in kidney. However, loss of function of PHEX leads to a decrease in renal PO_4^{-2} reabsorption.

It has been proposed that the product of PHEX, a neutral endopeptidase, inactivates phosphatonin. This phosphaturic factor presumably inhibits NaPi-2, and inhibits 1-α-hydroxylase activity leading to inhibition of 1,25-$(OH)_2$-Vit D_3 activity, which in turn inhibits NaPi-2 [19, 20]. The features of ADHR are similar to those of XLH. The gene responsible for ADHR is *FGF23* and encodes fibroblast growth factor (FGF-23), a 251 amino acid peptide [21]. Studies in vitro on the effect of FGF-23 on NaPi-2 are not conclusive, with some reporting no effect [22] or an inhibition [23, 24]. In oncogenic hypophosphatemic osteomalacia, secretion of phosphatonin or possibly other phosphaturic factors by the tumor has been implicated in inhibition of renal PO_4^{-2} reabsorption because removal of the tumor completely corrects the phenotype [25].

The last disorder, hypophosphatemic rickets with hypercalciuria, shares similar features as XLH and ADHR. The abnormality can be corrected by PO_4^{-2} supplementation alone, implying that the primary defect was in renal PO_4^{-2} reabsorption [26]. A mutation in the gene encoding NaPi-2 *(NPT2)* was suggested; however, genetic studies thus far are inconclusive [14].

Chloride Channel Renal Disease

Over a span of several decades, several syndromes including Dent's disease, X-linked recessive nephrolithiasis with renal failure and X-linked recessive hypophosphatemic rickets were all recognized to involve renal proximal tubular dysfunction. An important discovery later showed that all were caused by mutations that affected a chloride channel [27]. Acknowledging Charles Dent who first described two cases of hypercalciuric rickets with proximal tubular dysfunction, these X-linked diseases were collectively referred to as Dent's disease [28, 29]. Cl^- channels are ubiquitous in most cells and their function is necessary for electrical excitability in nerve and muscle cells, salt and water transport in epithelia, cell volume regulation, and acidification of intracellular organelles. In the kidney, different types of Cl^- channels are present in every segment of the nephron and they play significant roles in salt and water homeostasis, and the pathophysiology of several renal diseases.

The proximal tubule reabsorbs Cl^- both by paracellular and transcellular routes. In the early proximal tubule, passive paracellular Cl^- transport is predominant and driven through the tight junction by the lumen-negative electrical gradient established by various Na-solute co-transporters (e.g. Na-glucose, Na-amino acids, etc.). Solvent drag is another factor that favors passive Cl^- reabsorption in the S1 segment. In the late proximal tubule (S2, and S3 segments) the electrical gradient is dissipated because the lumen becomes slightly positive. However, the preferential reabsorption of HCO_3^- over Cl^- in the S1 segment raises Cl^- concentration in the lumen creating a favorable chemical gradient for passive Cl^- reabsorption that overcomes the now unfavorable electrical gradient.

With the decreasing paracellular driving force for Cl^- reabsorption in the late proximal tubule, the transcellular Cl^- route becomes more dominant. This route involves uphill uptake of Cl^- across the apical membrane via a 'tertiary' transporter that transports Cl^- in exchange for cellular anions (mainly formate and oxalate). The secreted anions are neutralized in the lumen by H^+ secreted by the Na-H exchanger. The basolateral exit of Cl^- occurs via Cl^- channels and possibly other transporters such as K-Cl co-transport. Different types of Cl^- channels are present in different segments of the nephron and some have been implicated in various renal disorders. However, in most cases these disorders of Cl^- (salt) reabsorption are caused by mutations of Cl^- channels in later segments of the nephron such as Bartter's syndrome type III (CLC-Kb in the thick ascending limb of Henle) or Bartter's type IV (Barttin in TAL). On the other hand, in the proximal tubule, the predominant Cl^- channel is ClC-5 and mutations of this channel (more than 30) are known to cause Dent's disease [30, 31]. However, Dent's disease is not caused by a primary defect in proximal Cl^- reabsorption, but by a disruption of the ability of the proximal tubule to carry on renal endocytosis [32, 33].

The proximal tubule is responsible for the uptake of low molecular weight proteins from the glomerular filtrate via a receptor-mediated endocytosis and recycling of various membrane proteins and transporters. In the proximal tubule, ClC-5 is present in a sub-apical region rich with endocytic vesicles and is co-localized with V-type H-ATPases. The endosomes are acidified by H-ATPases where the influx of H^+ into the vesicles is balanced by the influx of Cl^-, likely via ClC-5 channels. This step is essential; otherwise the buildup of intravesicular positive potential (due to H^+ uptake) would hinder further proton pumping. Intravesicular acidification is necessary for endosomal fusion and interference with this process impairs endocytosis and expression of various membrane transporters [34].

Consequently, the expression of two major receptors for endocytosis, megalin and cubilin, decreases in proximal tubule cells lacking ClC-5 [35, 36]. Impaired endocytosis can readily explain the clinical manifestations of Dent's disease, such as low molecular weight proteinuria that is observed early in patients with Dent's disease. Similarly, megalin-dependent uptake of vitamin D and its binding protein (DBP), which are essential for recycling or reactivating this

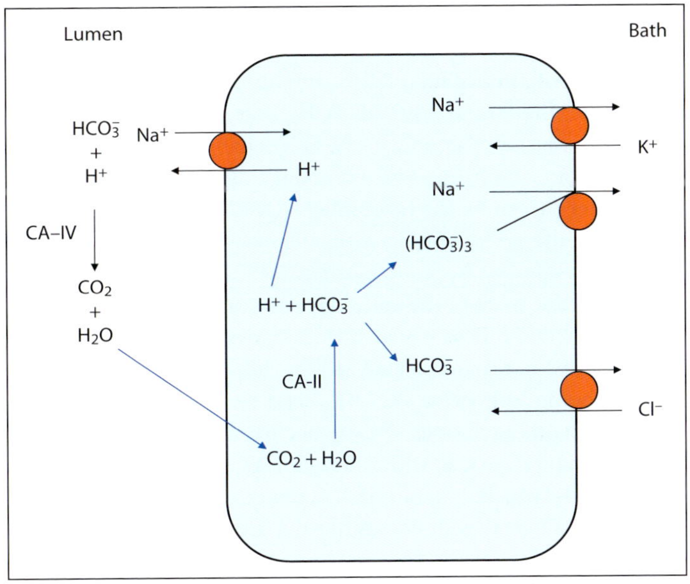

Fig. 2. Proximal epithelial cell. Bicarbonate reabsorption (reclamation) and the role of carbonic anhydrases.

vitamin, are also affected. Disruption of ClC-5 leads to decreased uptake of the inactive precursor, 25(OH)$_2$-Vit D$_3$ and its subsequent conversion by 1-α-hydroxylase to the active vitamin D3, 1,25(OH)$_2$-Vit D$_3$ as well as loss of the active form in the urine. However, this defect is balanced by decreased PTH uptake, also a megalin-mediated endocytic process, which is impaired in Dent's disease leading to accumulation of PTH in the lumen. This activates luminal PTH receptors, which in turn stimulate production of 1-α-hydroxylase and increased production of active Vit D$_3$ (if the substrate is available). Active vitamin D$_3$ stimulates intestinal Ca^{+2} reabsorption and indirectly causes hypercalciuria.

Finally in Dent's disease, impaired PTH endocytosis also causes phosphaturia. In ClC-5 knockout mice, the increased luminal PTH causes a decrease in the amount of NaPi-2 that was found mainly in intracellular vesicles resulting in reduced uptake of filtered PO_4^{-2} and phosphaturia [36]. Hence in the proximal tubule, loss of function mutations of CLC-5 inhibits vesicular acidification and endocytosis, but not necessarily Cl^- reabsorption. Impaired endocytosis is the main cause of Dent's disease.

Proximal Renal Tubular Acidosis

Renal tubular acidosis (RTA) is metabolic acidosis that occurs when the kidneys' ability to secrete acid, reabsorb filtered HCO_3^- and generate new HCO_3^- is compromised (fig. 2). RTA does not always imply a primary defect in the acid secreting mechanisms of the kidney, but could be secondary to generalized tubule transport impairment as observed in Fanconi's syndrome. There are three types: type 1 or classical distal RTA, type 2 or proximal RTA (pRTA), and type 4 which is also called hyperkalemic RTA. Type 3 is not used anymore and is considered a mixed disorder of types 1 and 2. This classification is based on our understanding of renal transport of acid-base equivalents in different segments of the nephron.

pRTA often presents as an acquired disease, but isolated pRTA occurs rarely and is caused by a breakdown in the process of reabsorption of HCO_3^-. This process includes several mechanisms, the dysfunction of any one of them could lead to pRTA. Filtered HCO_3^- is almost completely reabsorbed (reclaimed) by the proximal tubule. This process involves at least three steps.

First, H^+ is secreted from the cell into tubule fluid across the apical membrane resulting in the production of intracellular OH^-. In the proximal tubule, the largest fraction of H^+ secretion is carried out by the apical Na-H exchanger (NHE-3), an electroneutral secondary active transporter driven by the favorable Na gradient. Although other H^+ secreting mechanisms (such as H-ATPase) may be contributing, NHE-3 is responsible for most of the apical H^+ secretion.

The generated intracellular OH^- is converted to HCO_3^- by a reaction catalyzed by intracellular carbonic anhydrase (CA-II). The second step in reclaiming filtered HCO_3^- occurs when the secreted H^+ reacts with filtered HCO_3^- to form H_2CO_3, which is then rapidly converted to CO_2 and H_2O catalyzed by another CA isoform (CA-IV) present in the luminal brush border membranes. CO_2 and H_2O then diffuse into the cell, and the net result is removal of filtered HCO_3^- and its regeneration in the cell.

The third step in reabsorbing HCO_3^- is the transport of newly formed intracellular HCO_3^- from the cell to the blood side. In the proximal tubule, HCO_3^- efflux across the basolateral membrane takes place predominantly by the electrogenic Na/HCO_3^- co-transporter (NBC) [37, 38]. NBC carries 3 HCO_3^- and 1 Na^+, and is therefore an electrogenic transporter that is most abundant in the S1 segment. In the S3 segment, the Cl^- HCO_3^- exchanger belonging to the anion exchangers family of transporters is present in the basolateral membrane and contributes to efflux of remaining HCO_3^-. Both transporters are inhibited by the stilbene derivative DIDS. Thus, the hallmarks of net acid-base transport in the proximal tubule are apical H^+ secretion, conversion of luminal HCO_3^- by CA-IV and regeneration of cellular HCO_3^- by CA-II, and basolateral efflux of HCO_3^- by NBC and to a lesser extent by Cl^- HCO_3^-.

A dysfunction in any one of these processes could cause pRTA [38]. NBC was investigated as a candidate for inherited isolated pRTA [39]. Two mutations in SLC4A4, the gene encoding electrogenic NBC, R289S and R510H, were shown to significantly reduce the activity of the transporter [40]. Patients with these mutations had autosomal recessive pRTA. A patient with a mutation in the same gene, Q29X, had pRTA and glaucoma [41]. Another mutation, S427L, was reported to cause pRTA, glaucoma, cataracts and short stature [42]. NBC or one of its variants is expressed in corneal endothelium, lens epithelium, pancreas and other tissues as well. The expression of NBC in the eye could explain the ocular phenotype. More recently, Lemann et al. [43] reported autosomal dominant pRTA in a large Costa Rican family, but the molecular identity of this disease has not been identified yet.

As described above, both CA-II and CA-IV are necessary for HCO_3^- reclamation in the proximal tubule. Soon after the gene for CA-II was cloned, several mutations were shown to cause CA-II deficiency [44–47]. CA-II is present only in kidney and brain, and patients with mutations of this enzyme had impaired ability to secrete acid. Mutations that affect function of NBC or CA can relatively easily explain the effect on acid secretion by the kidneys. Moreover, because pH affects the function of many enzymes and metabolic processes, pH variations can have a more systemic effect with variable clinical manifestations.

Amino Acid Transport Disorders

The kidney reabsorbs almost all of the filtered amino acids in the proximal tubule. Absorption of amino acids both in kidney and intestine involves a variety of transporters for different groups of amino acids. During the past decade, most amino acid transporters in the proximal tubule have been identified on a molecular level. This has considerably improved our understanding of rare inherited aminoacidurias, such as Hartnup disorder and iminoglycinuria [48, 49]. Although these disorders are rare and clinically not important, they have led to a better understanding of more complex disorders that affect the kidneys, such as Fanconi's syndrome and maturity-onset diabetes of the young.

Among aminoacidurias, cystinuria is the best known and is relatively common, occurring in

1:7,000 patients. Cystinuria is an autosomal recessive defect of the transport of cystine and dibasic amino acids ornithine, arginine and lysine from the renal proximal tubules and small intestines. It is also associated with increased urinary excretion of these four amino acids. Of the four, only cystine has low solubility at physiologic urine pH levels predisposing to stone formation.

Cystinuria is classified as type 1 and non-type 1 (types 2 and 3) diseases. Type 1 is completely recessive and involves mutations in the gene *SLC3A1* (solute carrier family 3) on chromosome 2. This gene encodes a type II membrane glycoprotein, which is one of the components of the renal amino acid transporter that transports cystine, dibasic and neutral amino acids in the renal proximal tubule and intestinal tract. The non-type 1 forms are incompletely recessive and involve mutations in the gene *SLC7A9* (solute carrier family 7) on chromosome 19. This gene encodes for proteins that mediate the high-affinity and sodium-independent transport of cystine, and neutral and dibasic amino acids in the proximal tubule.

Homozygotes with all three types excrete a high concentration of cystine in the urine and often present with symptomatic cystine stones. Heterozygotes with type 1 cystinuria are asymptomatic carriers and have normal urine cystine excretion. In contrast, heterozygotes with type 2 and 3 may excrete high concentrations of cystine and are at risk of stone formation. The diagnosis is made by microscopic examination of the urine sediment and by stone analysis. Urine microscopy may show typical hexagonal cystine crystals, and the spot urine cystine concentration may be significantly elevated.

Quantitative urine amino acid analysis yields high levels of cystine, ornithine, arginine and lysine. Stone analysis confirms that cystine is the major stone component [50, 51]. Treatment is difficult, but could be rewarding in reducing stone formation and preventing renal injury. In addition to promoting diuresis to reduce cystine concentration in the urine, agents such as penicillamine, captopril and dimercaptosuccinic acid, and urinary alkalinization agents, have been used to increase solubility of cystine in the urine.

Proximal Tubules in Disease States

Proximal tubule cells are vulnerable to ischemic and toxic insult, and the clinical syndrome of acute kidney injury is predominantly a proximal tubule event. A whole host of toxins and ischemic events may cause proximal tubule cell injury, which lead to alterations in cytoskeletal organization of the cell including their integrin repertory, loss of cell polarity, disruption in tubule basement membranes, anoikis, apoptosis and outright necrosis of tubule cells.

The severity of tubule injury is generally related to the intensity of the insult, for example the dose of the nephrotoxic drug or agent, such as calcineurin inhibitors, aminoglycoside antibiotics, contrast dye, antineoplastic agents (e.g. cisplatin), and nucleotide and nucleoside reverse transcriptase inhibitors used in HIV disease. Ischemic injury can occur in severe hypovolemia and hypovolemic shock, sepsis and septic shock, and ischemia reperfusion injury seen after cardiopulmonary bypass. Among other toxins important ones are myoglobin, which can be released into circulation after metabolic or traumatic muscle injury, and another pigment, hemoglobin, which can occur during severe hemolysis, such as mismatched blood transfusion. Acute kidney injury, especially in hospitalized patients, has a worse prognosis, and severe acute kidney injury seen in critically ill patients often requires dialysis and is associated with a mortality rate of 50–60%. Remarkably, patients who survive often recover their kidney function attesting to the intrinsic regenerative capability of the proximal tubule cells.

Many kidney diseases, acute and chronic, first manifest with disorders of proximal tubule transporters described above. Events that result in proximal tubule injury can result in manifestation of either the full complement or some components of Fanconi's syndrome, which is described above. In many cases, Fanconi's syndrome is suspected from an incidental finding of glucosuria in the urine of a patient without diabetes, or abnormalities in plasma chemistry, such as low bicarbonate, low phosphorus, low magnesium or low uric acid. Interestingly, many cases of Fanconi's syndrome are also accompanied by the excretion of low molecular weight proteins, such as immunoglobulin light chains, β_2-microglobulin, retinol binding protein, lysozyme, etc., reflecting impaired proximal tubule absorption of such filtered proteins [52].

Proteins small enough to pass through the glomerular barrier are reabsorbed by receptor-mediated endocytosis after binding to megalin/cubilin. The cubilin/megalin tandem endocytic receptor belongs to a unique class of giant glycoprotein receptors related to the LDL receptor super family, is abundantly expressed in the proximal tubule brush borders and can bind multiple ligands [53, 54]. Many of the low molecular weight proteins that are relatively unhindered in the glomerulus are endocytosed after binding to cubilin/megalin. Among these proteins are α_1-, and β_2-microglobulin, cytochrome c, retinol binding protein, peptide hormones (e.g. insulin, PTH, etc.), and κ- and λ-light chains [52–56]. Megalin deficiency as seen in selective knock-out models, or blocked receptor-mediated endocytosis as seen in ClC5 deficiency (Dent's disease), and cubilin deficiency as seen in Imerslund-Grasbeck syndrome [57, 58] have generally been associated with low molecular weight proteinuria in addition to a variety of other disorders, such as vitamin B12 deficiency, thyroid deficiency, etc. [58–63].

Role of Proximal Tubules in Glomerular Diseases

Many kidney diseases originate in the glomerulus, and through a variety of immunologic or metabolic injuries impair the glomerular barrier against plasma proteins, resulting in proteinuria. While some glomerular diseases are reversible, many progress to chronic kidney disease and end-stage kidney disease requiring dialysis or transplantation. Prognosis of chronic kidney disease correlates with the magnitude of proteinuria and antiproteinuric therapy preserves kidney function [64, 65]. It is now clear that proximal tubule cells have an innate capability to mount a cascade of immune responses against various insults.

These responses include activation of nuclear transcription factors that drive nuclear transcription of cytokines, such as interleukin (IL)-6, IL-8, RANTES, monocyte chemoattractant protein-1, TNF-α and TGF-β, which may be released to the interstitium and mediate the recruitment of inflammatory cells and activate myofibroblasts resulting in the destruction of the components of kidney interstitium. A spectrum of signaling mechanisms involving Toll-like receptors, Rho and mitogen-activated protein kinases (MAPKs) are involved [52, 66–71]. Additionally, in response to a whole host of insults ranging from viral infection, urinary tract obstruction, transplant rejection, ischemia-reperfusion, metabolic diseases including diabetes, inherited disorders such as autosomal dominant polycystic kidney disease, etc., the proximal tubule cells may undergo epithelial-mesenchymal transition, transform into myofibroblasts and contribute to tubulointerstitial fibrosis [2, 70–75].

Cytokine production in response to filtered serum proteins appears as an important mechanism in the progression of kidney diseases in proteinuric states. It has been hypothesized that excessive endocytosis of filtered proteins including albumin causes 'protein overloading' and endoplasmic reticulum stress, activating nuclear transcription factors NF-κB that result in increased transcription of inflammatory cytokines [64, 65]. These responses are mediated through MAPKs. For example, albumin endocytosis in proximal tubule cells may trigger activation of ERK and JNK, and these events may explain the inflammatory consequences of excessive albumin endocytosis [65, 76].

Among the major classes of MAPKs, p38 MAPKs appear particularly important in regulating endocytic trafficking in response to extracellular stimuli. P38 MAPKs are also critical in cytokine responses [77, 78]. Interestingly, both p38 MAPK and JNK are also involved in EMT pathways [2, 79]. We observed similar responses in proximal cells exposed to myeloma light chains, implicating a role for these signaling mechanisms in the pathogenesis of the characteristic tubulointerstitial kidney disease in myeloma [52, 80–82]. These observations point to a pivotal role by activation of MAPKs in proximal tubule cells in protein overload nephropathy, and other renal diseases resulting in chronic injury.

Epithelial Mesenchymal Transition

Recent research has shown that mesenchyme-derived proximal tubule epithelial cells can transform to their embryonic phenotype and acquire fibroblast characteristics under a variety of injuries [2, 72, 73]. EMT has emerged as a key event in the pathogenesis of tubulointerstitial fibrosis and progression of kidney disease. Transformed epithelia lose expression of cadherins and begin expressing FSP-1, MMP2, vimentin, type I collagen and α-SMA. These cells then acquire increased migratory capability and move into the interstitium depositing collagen and comprising of up to 36% of the interstitial fibroblasts [73].

TGF-β and other growth factors such as EGF, VEGF and FGF-2 play a critical role in driving EMT, while BMP7 and HGF oppose and may even reverse EMT [83, 84]. Although the best characterized signaling pathway that results in EMT is through Smad proteins after TGF-β receptors are activated, in cell stress responses that result in phosphorylation of MAPKs (particularly p38 MAPKs), nuclear translocation of NF-κB and cytokines including IL-8 and MCP1 are also involved [2, 72, 73, 85]. Certain immunoglobulin light chains can induce EMT in cultured proximal tubule epithelial cells, which may be an important mechanism in the pathogenesis of myeloma kidney as well as other proteinuric diseases associated with increased quantities of polyclonal light chains in the urine because of impaired glomerular permselectivity [52, 80, 86].

Anemia and Mineral-Bone Disorder of Chronic Kidney Disease

Anemia of renal disease and uremic mineral bone disorder are among the most distressing features of chronic kidney disease and both are major determinants of increased cardiovascular morbidity and mortality among kidney patients [3, 4]. Although there is some controversy, most investigators point to the proximal tubule cells as the site where erythropoietin is synthesized [87]. Similarly, while some studies locate vitamin D synthesis in the distal tubule [88], more recent studies identify the proximal tubule as the site of active vitamin D synthesis [89] through the action of the enzyme 25-hydroxyvitamin D 1-alpha-hydroxylase or CYP27B1, the key enzyme in the two-step activation process of vitamin D to 1,25-dihydroxyvitamin D (1,25D). Therefore, tubular injury and damage to proximal tubule cells may also be the origin of these two major abnormalities seen in chronic kidney patients.

Conclusions

The proximal tubules are a major component of the kidney and are responsible for the reabsorption of the majority of the glomerular ultrafiltrate, along with the reabsorptive or secretory transport of a vast number of solutes regulating the fluid-electrolyte and acid-base regulation of the body. Proximal tubule injury or acquired or inherited disorders of these transporters may manifest with a spectrum of kidney diseases such as Fanconi's syndrome, renal tubular acidosis, phosphate wasting disorders, glucosuria and aminoaciduria including cystinuria. Proximal tubule cell damage caused by viruses, ischemia, or drugs and other toxins may lead to anoikis, apoptosis and necrosis, and give rise to the syndrome of acute renal failure or acute kidney injury. The intrinsic immune responsiveness of the proximal tubule cells result in activation of various inflammatory cytokines, EMT and other inflammatory pathways in response to a spectrum of injuries, such as protein overloading, ischemic or toxic injury and contribute to the progressive tubulointerstitial destruction in chronic kidney disease. Finally, proximal tubule cell damage may be responsible for the most distressing features of chronic kidney disease, anemia and mineral-bone disorder of uremia.

References

1 Maunsbach AB, Madden SC, Latta H: Light and electron microscopic changes in proximal tubules of rats after administration of glucose, mannitol, sucrose, or dextran. Lab Invest 1962;11:421–432.

2 Kalluri R, Neilson EG: Epithelial-mesenchymal transition and its implications for fibrosis. J Clin Invest 2003;112:1776–1784.

3 Pendse S, Singh AK: Complications of chronic kidney disease: anemia, mineral metabolism, and cardiovascular disease. Med Clin North Am 2005;89:549–561.

4 Zabetakis PM, Nissenson AR: Complications of chronic renal insufficiency: beyond cardiovascular disease. Am J Kidney Dis 2000;6: S31–S38.

5 Guay-Woodford LM: Molecular insights into the pathogenesis of inherited renal tubular disorders. Curr Opin Nephrol Hypertens 1995;4:121–129.

6 Scheinman SJ, Guay-Woodford LM, Thakker RV, Warnock DG: Genetic disorders of renal electrolyte transport. N Engl J Med 1999;340:1177–1187.

7 Bergeron M, Gougoux A, Vinay P: The Renal Fanconi Syndrome; in: The Metabolic and Molecular Bases of Inherited Diseases. New York, McGraw Hill, 1995.

8 Marsenic O: Glucose control by the kidney: an emerging target in diabetes. Am J Kidney Dis 2009;53:875–883.

9 Magen D, Sprecher E, Zelikovic I, Skorecki K: A novel missense mutation in SLC5A2 encoding SGLT2 underlies autosomal-recessive renal glucosuria and aminoaciduria. Kidney Int 2005;67:34–41.

10 Santer R, et al: Molecular analysis of the SGLT2 gene in patients with renal glucosuria. J Am Soc Nephrol 2003;14:2873–2882.

11 Brown GK: Glucose transporters: structure, function and consequences of deficiency. J Inherit Metab Dis 2000;23:237–246.

12 Turk E, Zabel B, Mundlos S, Dyer J, Wright EM: Glucose/galactose malabsorption caused by a defect in the Na+/glucose cotransporter. Nature 1991;350:354–356.

13 Wright EM, Hirayama BA, Loo DF: Active sugar transport in health and disease. J Intern Med 2007;261:32–43.

14 Tenenhouse HS, Murer H: Disorders of renal tubular phosphate transport. J Am Soc Nephrol 2003;14:240–248.

15 Tenenhouse HS: X-linked hypophosphataemia: a homologous disorder in humans and mice. Nephrol Dial Transplant 1999;14:333–341.

16 Tenenhouse HS, Beck L: Renal Na(+)-phosphate cotransporter gene expression in X-linked Hyp and Gy mice. Kidney Int 1996;49:1027–1032.

17 Tenenhouse HS, et al: Renal Na(+)-phosphate cotransport in murine X-linked hypophosphatemic rickets. Molecular characterization. J Clin Invest 1994;93:671–676.

18 Sabbagh Y, Jones AO, Tenenhouse HS: PHEXdb, a locus-specific database for mutations causing X-linked hypophosphatemia. Hum Mutat 2000;16:1–6.

19 Beck L, et al: Pex/PEX tissue distribution and evidence for a deletion in the 3′ region of the Pex gene in X-linked hypophosphatemic mice. J Clin Invest 1997;99:1200–1209.

20 Econs MJ, Francis F: Positional cloning of the PEX gene: new insights into the pathophysiology of X-linked hypophosphatemic rickets. Am J Physiol 1997;273:F489–F498.

21 Zhang SL, et al: High levels of glucose stimulate angiotensinogen gene expression via the P38 mitogen-activated protein kinase pathway in rat kidney proximal tubular cells. Endocrinology 2000;141:4637–4646.

22 Shimada T, et al: Cloning and characterization of FGF23 as a causative factor of tumor-induced osteomalacia. Proc Natl Acad Sci USA 2001;98:6500–6505.

23 Bowe AE, et al: FGF-23 inhibits renal tubular phosphate transport and is a PHEX substrate. Biochem Biophys Res Commun 2001;284:977–981.

24 Yamashita T, Konishi M, Miyake A, Inui K, Itoh N: Fibroblast growth factor (FGF)-23 inhibits renal phosphate reabsorption by activation of the mitogen-activated protein kinase pathway. J Biol Chem 2002;277: 28265–28270.

25 Econs MJ, Drezner MK: Tumor-induced osteomalacia – unveiling a new hormone. N Engl J Med 1994;330:1679–1681.

26 Tieder M, et al: 'Idiopathic' hypercalciuria and hereditary hypophosphatemic rickets. Two phenotypical expressions of a common genetic defect. N Engl J Med 1987;316:125–129.

27 Scheinman SJ: X-linked hypercalciuric nephrolithiasis: clinical syndromes and chloride channel mutations. Kidney Int 1998;53:3–17.

28 Frymoyer PA, et al: X-linked recessive nephrolithiasis with renal failure. N Engl J Med 1991;325:681–686.

29 Wrong OM, Norden AG, Feest TG: Dent's disease; a familial proximal renal tubular syndrome with low-molecular-weight proteinuria, hypercalciuria, nephrocalcinosis, metabolic bone disease, progressive renal failure and a marked male predominance. QJM 1994;87:473–493.

30 Birkenhager R, et al: Mutation of BSND causes Bartter syndrome with sensorineural deafness and kidney failure. Nat Genet 2001; 29:310–314.

31 Simon DB, et al: Mutations in the chloride channel gene, CLCNKB, cause Bartter's syndrome type III. Nat Genet 1997;17:171–178.

32 Jentsch TJ: Chloride transport in the kidney: lessons from human disease and knockout mice. J Am Soc Nephrol 2005;16:1549–1561.

33 Jentsch TJ, Maritzen T, Zdebik AA: Chloride channel diseases resulting from impaired transepithelial transport or vesicular function. J Clin Invest 2005;115:2039–2046.

34 Gekle M, Mildenberger S, Freudinger R, Silbernagl S: Endosomal alkalinization reduces Jmax and Km of albumin receptor-mediated endocytosis in OK cells. Am J Physiol 1995;268:F899–F906.
35 Christensen EI, et al: Loss of chloride channel ClC-5 impairs endocytosis by defective trafficking of megalin and cubilin in kidney proximal tubules. Proc Natl Acad Sci USA 2003;100:8472–8477.
36 Piwon N, Gunther W, Schwake M, Bosl MR, Jentsch TJ: ClC-5 Cl--channel disruption impairs endocytosis in a mouse model for Dent's disease. Nature 2000;408:369–373.
37 Boron WF, Boulpaep EL: Intracellular pH regulation in the renal proximal tubule of the salamander. Basolateral HCO_3- transport. J Gen Physiol 1983;81:53–94.
38 Laing CM, Toye AM, Capasso G, Unwin RJ: Renal tubular acidosis: developments in our understanding of the molecular basis. Int J Biochem Cell Biol 2005;37:1151–1161.
39 Igarashi T, et al: Mutations in SLC4A4 cause permanent isolated proximal renal tubular acidosis with ocular abnormalities. Nat Genet 1999;23:264–266.
40 Romero MF, Fulton CM, Boron WF: The SLC4 family of HCO 3 - transporters. Pflugers Arch 2004;447:495–509.
41 Igarashi T, et al: Novel nonsense mutation in the Na^+/HCO_3- cotransporter gene (SLC4A4) in a patient with permanent isolated proximal renal tubular acidosis and bilateral glaucoma. J Am Soc Nephrol 2001; 12:713–718.
42 Dinour D, et al: A novel missense mutation in the sodium bicarbonate cotransporter (NBCe1/SLC4A4) causes proximal tubular acidosis and glaucoma through ion transport defects. J Biol Chem 2004;279:52238–52246.
43 Lemann J Jr, Adams ND, Wilz DR, Brenes LG: Acid and mineral balances and bone in familial proximal renal tubular acidosis. Kidney Int 2000;58:1267–1277.
44 Fathallah DM, Bejaoui M, Sly WS, Lakhoua R, Dellagi K: A unique mutation underlying carbonic anhydrase II deficiency syndrome in patients of Arab descent. Hum Genet 1994;94:581–582.
45 Hu PY, et al: Carbonic anhydrase II deficiency: single-base deletion in exon 7 is the predominant mutation in Caribbean Hispanic patients. Am J Hum Genet 1994;54:602–608.
46 Roth DE, Venta PJ, Tashian RE, Sly WS: Molecular basis of human carbonic anhydrase II deficiency. Proc Natl Acad Sci USA 1992;89:1804–1808.
47 Shah GN, Bonapace G, Hu PY, Strisciuglio P, Sly WS: Carbonic anhydrase II deficiency syndrome (osteopetrosis with renal tubular acidosis and brain calcification): novel mutations in CA2 identified by direct sequencing expand the opportunity for genotype phenotype correlation. Hum Mutat 2004;24:272.
48 Broer S: Apical transporters for neutral amino acids: physiology and pathophysiology. Physiology (Bethesda) 2008;23:95–103.
49 Broer S: Amino acid transport across mammalian intestinal and renal epithelia. Physiol Rev 2008;88:249–286.
50 Chiu PK, Chan ES, Hou SS, Ng CF: Cystinuria: a rare diagnosis that should not be missed. Hong Kong Med J 2008;14:399–401.
51 Martens K, Jaeken J, Matthijs G, Creemers JW: Multi-system disorder syndromes associated with cystinuria type I. Curr Mol Med 2008;8:544–550.
52 Batuman V: Proximal tubular injury in myeloma. Contrib Nephrol 2007;153:87–104.
53 Christensen EI, Birn H: Megalin and cubilin: synergistic endocytic receptors in renal proximal tubule. Am J Physiol Renal Physiol 2001;280:F562–F573.
54 Verroust PJ, Birn H, Nielsen R, Kozyraki R, Christensen EI: The tandem endocytic receptors megalin and cubilin are important proteins in renal pathology. Kidney Int 2002; 62:745–756.
55 Klassen RB, Allen PL, Batuman V, Crenshaw K, Hammond TG: Light chains are a ligand for megalin. J Appl Physiol 2005;98:257–263.
56 Batuman V, et al: Myeloma light chains are ligands for cubilin (gp280). Am J Physiol 1998;275:F246–F254.
57 Hauck FH, Tanner SM, Henker J, Laass MW: Imerslund-Gräsbeck syndrome in a 15-year-old German girl caused by compound heterozygous mutations in CUBN. Eur J Pediatr 2008;167:671–675.
58 Gräsbeck R: Imerslund-Gräsbeck syndrome (selective vitamin B(12) malabsorption with proteinuria). Orphanet J Rare Dis 2006;1:17.
59 Souza-Menezes J, Morales MM, Tukaye DN, Guggino SE, Guggino WB: Absence of ClC5 in knockout mice leads to glycosuria, impaired renal glucose handling and low proximal tubule GLUT2 protein expression. Cell Physiol Biochem 2007;20:455–464.
60 Tanuma A, et al: Functional characterization of a novel missense CLCN5 mutation causing alterations in proximal tubular endocytic machinery in Dent's disease. Nephron Physiol 2007;107:p87–p97.
61 Pollock CA, Poronnik P: Albumin transport and processing by the proximal tubule: physiology and pathophysiology. Curr Opin Nephrol Hypertens 2007;16:359–364.
62 Diwakar R, Pearson AL, Colville-Nash P, Brunskill NJ, Dockrell ME: The role played by endocytosis in albumin-induced secretion of TGF-beta1 by proximal tubular epithelial cells. Am J Physiol Renal Physiol 2007;292: F1464–F1470.
63 Nielsen R, et al: Endocytosis provides a major alternative pathway for lysosomal biogenesis in kidney proximal tubular cells. Proc Natl Acad Sci USA 2007;104:5407–5412.
64 Abbate M, et al: In progressive nephropathies, overload of tubular cells with filtered proteins translates glomerular permeability dysfunction into cellular signals of interstitial inflammation. J Am Soc Nephrol 1998;9: 1213–1224.
65 Sidaway JE, et al: Inhibitors of 3-hydroxy-3-methylglutaryl-CoA reductase reduce receptor-mediated endocytosis in opossum kidney cells. J Am Soc Nephrol 2004;15:2258–2265.
66 Peng F, et al: RhoA/Rho-kinase contribute to the pathogenesis of diabetic renal disease. Diabetes 2008;57:1683–1692.
67 Fu P, et al: Signaling mechanism of renal fibrosis in unilateral ureteral obstructive kidney disease in ROCK1 knockout mice. J Am Soc Nephrol 2006;17:3105–3114.
68 Rangan GK, Wang Y, Tay YC, Harris DC: Inhibition of nuclear factor-kappaB activation reduces cortical tubulointerstitial injury in proteinuric rats. Kidney Int 1999;56:118–134.

69 Tian W, Zhang Z, Cohen DM: MAPK signaling and the kidney. Am J Physiol Renal Physiol 2000;279:F593–F604.
70 Anders HJ, et al: Activation of Toll-like receptor-9 induces progression of renal disease in MRL-Fas(lpr) mice. FASEB J 2004;18: 534–536.
71 Anders HJ, Banas B, Schlondorff D: Signaling danger: Toll-like receptors and their potential roles in kidney disease. J Am Soc Nephrol 2004;15:854–867.
72 Harris RC, Neilson EG: Toward a unified theory of renal progression. Annu Rev Med 2006;57:365–380.
73 Liu Y: Epithelial to mesenchymal transition in renal fibrogenesis: pathologic significance, molecular mechanism, and therapeutic intervention. J Am Soc Nephrol 2004;15:1–12.
74 Strutz FM: EMT and proteinuria as progression factors. Kidney Int 2009;75:475–481.
75 Liu Y: New insights into epithelial-mesenchymal transition in kidney fibrosis. J Am Soc Nephrol 2010;21:212–222.
76 Dixon R, Brunskill NJ: Albumin stimulates p44/p42 extracellular-signal-regulated mitogen-activated protein kinase in opossum kidney proximal tubular cells. Clin Sci (Lond) 2000;98:295–301.
77 Cavalli V, et al: The stress-induced MAP kinase p38 regulates endocytic trafficking via the GDI: rab5 complex. Mol Cell 2001;7:421–432.
78 Abbate M, Remuzzi G: Proteinuria as a mediator of tubulointerstitial injury. Kidney Blood Press Res 1999;22:37–46.
79 Bakin AV, Rinehart C, Tomlinson AK, Arteaga CL: p38 mitogen-activated protein kinase is required for TGFbeta-mediated fibroblastic transdifferentiation and cell migration. J Cell Sci 2002;115:3193–3206.
80 Li M, Hering-Smith KS, Simon EE, Batuman V: Myeloma light chains induce epithelial mesenchymal transition in human renal proximal tubule epithelial cells. Nephrol Dial Transplant 2008;23:860–870.
81 Sengul S, Zwizinski C, Batuman V: Role of MAPK pathways in light chain-induced cytokine production in human proximal tubule cells. Am J Physiol Renal Physiol 2003;284:F1245–F1254.
82 Sengul S, et al: Endocytosis of light chains induces cytokines through activation of NF- kappaB in human proximal tubule cells. Kidney Int 2002;62:1977–1988.
83 Zeisberg M, et al: BMP-7 counteracts TGF-beta1-induced epithelial-to-mesenchymal transition and reverses chronic renal injury. Nat Med 2003;9:964–968.
84 Zeisberg M, et al: Bone morphogenic protein-7 inhibits progression of chronic renal fibrosis associated with two genetic mouse models. Am J Physiol Renal Physiol 2003; 285:F1060–F1067.
85 Huber MA, et al: NF-kappaB is essential for epithelial-mesenchymal transition and metastasis in a model of breast cancer progression. J Clin Invest 2004;114:569–581.
86 Sengul S, Li M, Batuman V: Myeloma kidney: toward its prevention–with new insights from in vitro and in vivo models of renal injury. J Nephrol 2009;22:17–28.
87 Mujais SK, Beru N, Pullman TN, Goldwasser E: Erythropoietin is produced by tubular cells of the rat kidney. Cell Biochem Biophys 1999;30:153–166.
88 Kumar R, Schaefer J, Grande JP, Roche PC: Immunolocalization of calcitriol receptor, 24-hydroxylase cytochrome P-450, and calbindin D28k in human kidney. Am J Physiol 1994;266:F477–F485.
89 Hendrix I, Anderson PH, Omdahl JL, May BK, Morris HA: Response of the 5′-flanking region of the human 25-hydroxyvitamin D 1alpha-hydroxylase gene to physiological stimuli using a transgenic mouse model. J Mol Endocrinol 2005;34:237–245.

Nazih Nakhoul, PhD
Nephrology Section, Tulane University Health Sciences Center and VA Medical Center
1430 Tulane Avenue
New Orleans, LA 70112 (USA)
Tel. +1 504 988-7819, E-Mail nakhoul@tulane.edu

Herrera GA (ed): Experimental Models for Renal Diseases: Pathogenesis and Diagnosis.
Contrib Nephrol. Basel, Karger, 2011, vol 169, pp 51–72

Kidney Microvasculature in Health and Disease

Kuang-Yu Jen[a] · Lukas Haragsim[b] · Zoltan G. Laszik[a]

[a]University of California San Francisco, San Francisco, Calif., and [b]University of Oklahoma, Oklahoma City, Okla., USA

Abstract

The renal microvasculature is composed of the glomerular and peritubular capillary beds which supply the cellular constituents of the kidney with oxygen and nutrients as well as maintain renal function by providing an adequate glomerular filtration rate. As a result, endothelial dysfunction within the kidney can lead to devastating consequences. Recently, a plethora of information regarding the molecular players involved in kidney microvasculature development and disease has emerged. Many of these studies focus on intricate signaling pathways within the local microenvironment of endothelial cells. Here, we highlight some of these studies and relate them to the molecular pathogenesis of glomerular and peritubular endothelial cells in both native and transplant renal diseases.

The kidney is an extraordinarily well-vascularized organ, receiving on average approximately 20–25% of the total cardiac output. The high volume of blood flow is required to sustain a sufficient glomerular filtration rate (GFR) in order to adequately clear circulating waste products through urine production. Additionally, the highly specialized function of the kidney demands a considerable metabolic load. While the larger caliber arteries and arterioles are essential for delivering blood into the kidneys from the systemic circulation; ultimately, the renal microvasculature is responsible not only for supplying nutrients and oxygen to the renal parenchyma, but also for determining GFR. Thus, microvascular endothelial dysfunction can be exceedingly detrimental to renal function and is intimately involved in the pathogenesis of a wide spectrum of renal diseases.

Functionally and compartmentally, the renal microvasculature can be divided into two major categories: the glomerular capillaries and the peritubular capillaries (PTCs). The glomerular capillaries receive incoming blood flow from the afferent arterioles and facilitate the filtration functions of the nephron. Post glomerular capillary blood flow coalesces and exits the glomerulus through the efferent arteriole and feeds the PTCs, which enshroud and interact with the tubules of the nephron. In this chapter, we will focus on experimental models that specifically examine the kidney microvasculature in both normal and disease states.

Glomerular Endothelial Cells

Glomerular endothelial cells (GEnCs) are highly specialized endothelial cells with a distinct morphologic appearance characterized by an extremely flattened cell shape that is densely populated with transcytoplasmic perforating fenestrations. They line the inner aspect of the glomerular capillary walls and are an essential component of the glomerular filtration barrier (GFB). These cells are exceptionally dynamic and continuously interact with their local microenvironment. Not only are appropriate external signals necessary for guiding glomerular vasculogenesis, but they are also required to maintain the distinctively specialized morphology and function of the GEnC in the mature state. Therefore, in addition to direct endothelial damage, destruction or perturbation of the GEnC microenvironment plays a crucial role in glomerular disease. In the following sections, we focus our discussion on the molecular pathways involved in the development, maturation and maintenance of the glomerular endothelium.

Glomerular Endothelial Cell Origin

Vascular development can proceed through: (1) vasculogenesis in which endothelial cell precursors (angioblasts) aggregate to form vessels de novo, (2) angiogenesis in which new vessels sprout from the existing vasculature, or (3) a combination of both processes. For the most part, the glomerular capillaries are generated through vasculogenesis rather than angiogenesis. During mammalian glomerular development, angioblasts migrate and invade into the vascular/proximal cleft of the S-shaped body to give rise to the eventual glomerular endothelium. Whether these cells arise from the local metanephric mesenchyme or at least in part from an extrarenal source has been long debated. This controversy largely stems from conflicting data reported from a number of inter- and intraspecific graft experiments.

Early organ culture studies revealed that embryonic kidneys grown in vitro form avascular glomeruli without evidence of capillary in-growth and suggested that the glomerular microvasculature originates from an extrarenal source [1]. This hypothesis was further supported by a series of studies using mouse/quail and quail/chick interspecific grafts [2–4]. In this model, undifferentiated embryonic kidneys from one species were grafted in ovo onto host chorioallantoic membranes. Although the majority of the glomeruli in the resultant grafts were avascular, a few well-developed well-vascularized glomeruli were present and contained podocytes of graft origin, while the glomerular endothelium expressed host nuclear markers. Comparable results were reported more recently in interspecific grafting models utilizing other animal species combinations [5, 6]. Taken together, these data suggest that the glomerular capillaries are derived from an infiltrating extrarenal source of angioblasts rather than from local differentiation of cells originating within the metanephric mesenchyme.

On the other hand, contrary results were observed by Hyink et al. [7] and Robert et al. [8] using an in oculo transplant model. In this approach, interspecific grafts between rats and mice were established in which fetal kidneys were implanted into the anterior eye chambers of adult hosts. In contrast to the in ovo grafting model, the in oculo grafting procedure yielded relatively abundant well-vascularized glomeruli. Species-specific monoclonal antibodies revealed that the glomerular microvascular extracellular matrices within grafts were typically of donor origin [7]. Furthermore, using reverse-orientation splice acceptor 26 (ROSA26) murine hosts, which constitutively express β-galactosidase in every cell, these investigators showed that although β-galactosidase activity was present in peripheral vessels, the majority of GEnCs lacked β-galactosidase activity and, therefore, were of donor origin [7]. At embryonic day (E) 12, scattered cells through out the metanephric mesenchyme exhibited positive staining for the mouse vascular endothelial growth factor receptor Flk1 (designated KDR or VEGFR2 in humans), a marker for angioblasts and endothelial cells [8]. At later points in kidney development, these presumed angioblasts became concentrated in the glomerular vascular clefts and glomerular tufts. Therefore, these findings indicate that cells endogenous to the developing kidney give rise to the glomerular microvasculature.

These disparate results have generated uncertainty regarding the origin of GEnCs. Some of these differences may arise from the limitations of surgical graft animal models. A variety of factors, including the site of graft placement, the species used as graft and host, the age of the host and/or the developmental stage of the engrafted embryonic kidney, may lead to the variability of donor as opposed to host contribution to the eventual makeup of the GEnCs. Another major limitation may arise from the grafting procedure itself. Presumably, graft survival relies on sufficient revascularization of the engrafted fetal kidney, which may interfere with the normal in situ mechanisms of GEnC development. Interestingly, when fetal kidneys are engrafted into adult host kidneys, the glomerular endothelium originates solely from the graft [8]. On the contrary, when fetal kidneys are engrafted into newborn hosts, the GEnCs are derived from an admixture of host and graft-derived cells. Abrahamson and Robert [9] point out that these findings may explain the differences seen in the in ovo versus in oculo interspecific graft experiments. They suggest that host chorioallantoic membranes may contain angioblasts capable of responding to the inductive signals within the graft and therefore contribute to glomerular capillary formation in an artificial manner.

Overview of Glomerular Development

The glomerular tuft is largely composed of three cell types: visceral epithelial cells (podocytes), mesangial cells and GEnCs. Additionally, extracellular matrix produced by these cells [e.g. glomerular basement membrane (GBM) and mesangial matrix] act as a scaffold for the glomerular tuft architecture. Since the genesis and maturation of these individual cellular and extracellular components are highly interdependent, GEnC development must be viewed in the context of glomerulogenesis and nephrogenesis (reviewed in [10–12]).

The developing nephron proceeds through several distinct morphologic stages and begins with the condensation of the metanephric mesenchyme in response to inductive signals from the ureteric bud (fig. 1a). The resultant peritubular aggregate undergoes mesenchymal-to-epithelial transition and generates the renal vesicle (fig. 1b). Subsequently, the renal vesicle fuses with the ureteric bud and undergoes segmentation to give rise to the comma- and S-shaped stages. The proximal end of the S-shaped body displays the first sign of glomerular specification with the appearance of a layer of columnar epithelia representing early developing podocytes. These cells are responsible for establishing a number of crucial spatial and temporal cues that provide a permissive microenvironment for angioblast recruitment into the adjacent vascular cleft and subsequent glomerular capillary formation (fig. 1c). Following invasion into the vascular cleft of the S-shaped body, the angioblasts undergo proliferation and homotypic aggregation to form a single solid pre-capillary cord of undifferentiated endothelial cells [10, 13]. Initially, these structures are without identifiable lumina. Lumen formation occurs only after a subset of endothelial cells within the pre-capillary cord undergoes selective apoptosis in a transforming growth factor (TGF)-β-dependent manner [14]. The remaining surviving endothelial cells differentiate by flattening out and covering the inner surface of the developing capillary wall, which results in lumen formation. Shortly thereafter, mesangial cells, recruited in response to platelet-derived growth factor B signals produced by GEnCs, generate and stabilize the capillary tuft by dividing the capillary into six to eight individual loops [15–18]. Concurrently, the early GBM is laid down by both the podocytes and GEnCs as two adjacent separate structures, which subsequently fuses to give rise to the mature GBM [4, 19, 20]. It should be kept in mind that the components of the glomerular tuft are exceedingly interdependent for proper glomerulogenesis since this process is largely driven by cell-cell and cell-matrix interactions. Some of these interactions will be highlighted later in this chapter.

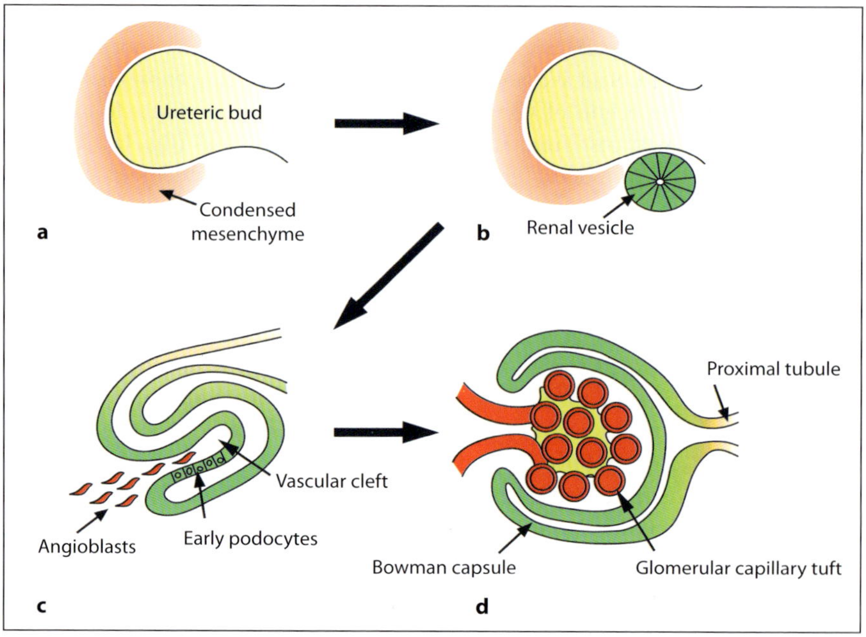

Fig. 1. Glomerular development. **a** Inductive signals from the ureteric bud cause condensation of the metanephric mesenchyme. **b** A portion of the condensed mesenchyme forms the peritubular aggregate which undergoes mesenchymal-to-epithelial transition to form the renal vesicle. **c** The renal vesicle fuses with the ureteric bud and eventually forms the S-shaped body. Signals produced from early podocytes lining the vascular cleft attract the migration of angioblasts. **d** The proximal portion of the S-shaped body develops into the mature glomerulus.

Glomerular Endothelial Fenestrations

GEnCs constitute the inner-most aspect of the glomerular capillary wall and possess distinct morphologic characteristics. The GEnC body tends to be located at the base of glomerular capillary loops, adjacent to the mesangium; whereas the peripheral cytoplasm extends along the GBM and completely covers the glomerular capillary lumen circumferentially. Consequently, these cells assume a remarkably flattened shape, measuring only approximately 50–150 nm in thickness away from the cell body [13]. Most notably, the attenuated peripheral cytoplasm of GEnCs is densely populated with perforating fenestrations. These transcytoplasmic holes represent approximately 30–50% of the luminal surface of the glomerular endothelium and can be easily appreciated on transmission electron microscopy [13, 21] (fig. 2).

Glomerular endothelial fenestrations are unique structures unlike other types of endothelial fenestrations. In general, endothelial fenestrations can be classified into three categories [22, 23]. First, some endothelial fenestrations contain a bridging diaphragm such as those seen in endocrine glands, gastrointestinal mucosa and renal PTCs. These fenestrations are relatively similar in size measuring 60–70 nm in diameter and are organized in a regular pattern. Their diaphragms are 3–5 nm thick and are formed by the fusion of plasmalemmal vesicles (caveolae), which contain the cationic endothelial glycoprotein PV-1. In contrast, discontinuous endothelia, such as those of the liver sinusoids, spleen or bone marrow, contain fenestrations that are much larger (100–

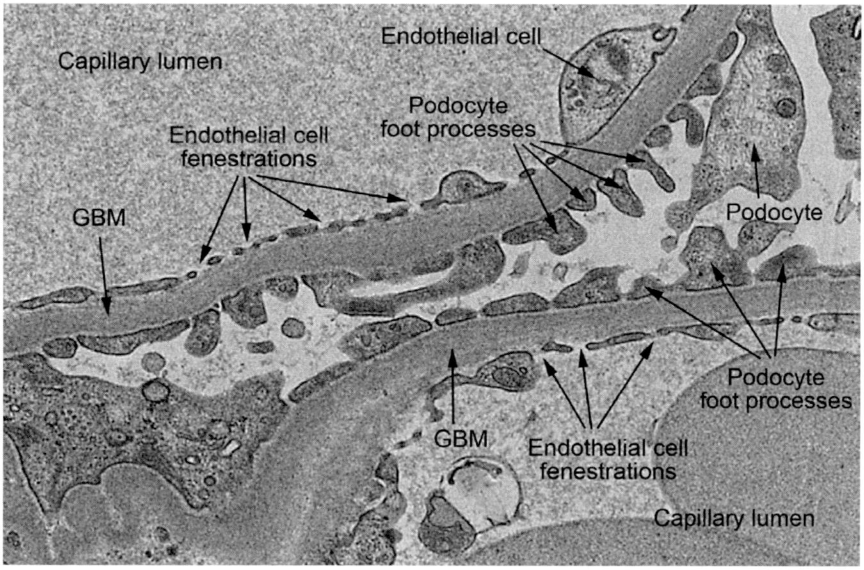

Fig. 2. Transmission electron micrograph showing a cross-sectional view of adjacent glomerular capillary walls. Several components of the GFB, including the foot processes of the podocytes, GBM and glomerular endothelial cell fenestrations, are readily identified.

175 nm) and more variable in diameter. Unlike the first type of endothelial fenestration, these do not contain diaphragms. The third type of endothelial fenestration is unique to GEnCs and morphologically resembles the first type; however, they lack diaphragms. Interestingly, during glomerular endothelial development, PV-1 expression associated with bridging diaphragms is seen in early GEnC fenestrations, but is subsequently lost as GEnC maturation proceeds [23].

The glomerular capillary tuft initiates the process of urine production by filtering circulating blood. Essentially, the glomerular capillary wall constitutes the GFB by allowing the passage of water and low molecular weight waste products into the Bowman space while retaining cells and macromolecules circulating within the vasculature. This filtration process relies mainly on size and charge selection. Much of the literature has concentrated on the importance of the bridging slit diaphragms of podocyte foot processes and the GBM in maintaining GFB integrity. It is clear that disruption of either of these structures results in GFB breakdown and proteinuria. However, the degree to which the glomerular capillary endothelium contributes to establishing a functional GFB has been greatly debated. Much of this disagreement stems from contrary views regarding GEnC fenestration structure and its functional consequences.

It is widely accepted that glomerular endothelial fenestrations affect hydraulic conductivity of the glomerular capillary wall. As a result, GFR directly correlates with the size and density of glomerular endothelial fenestrations. This idea is supported by the fact that fenestrated capillaries are a great deal more permeable to water and small solutes than non-fenestrated capillaries. Also, reduced density or loss of fenestrations in both human diseases and animal models leads to reduced GFR [24–26]. However, conflicting views exist regarding the permeability of GEnC fenestrations to macromolecules. One standpoint is that GEnC fenestrations contribute little to the GFB; instead, they act as a relatively porous structure, allowing for high hydraulic conductivity as well as offering little resistance to the flux of macromolecules across the glomerular endothelium. Consistent with this hypothesis, macromolecules are able to penetrate the glomerular endothelium under conventional fixation conditions [27]. A cross-sectional view under electron microscopy of GEnC fenestrations reveals transcytoplasmic perforating 'holes' devoid of a bridging diaphragm or any other substance (fig. 2). Furthermore, the absence of PV-1 in mature GEnC fenestrations implies that they do not possess the diaphragms seen in other fenestrated endothelia, although this observation does not exclude the possibility of diaphragms lacking PV-1 [28]. Hence, these findings suggest that GEnCs allow a direct communication of the glomerular capillary luminal contents with the GBM.

In contrast, other studies have proposed that although glomerular endothelial fenestrations do not possess PV-1-associated diaphragms, labile molecules lining the GEnC surface and fenestrations detected by special techniques may contribute to establishing a functional GFB. The ability for glomerular endothelium to retain macromolecules under experimental conditions appears to depend on the method of perfusion/fixation [29, 30]. Glomerular capillary endothelia are able to retain macromolecules when renal tissue is perfused with oxygenated solutions during labeling and/or fixation, whereas this property is lost under conventional fixation techniques. Structural evaluation following fixation with oxygenated solutions reveal the presence of a delicate filamentous surface coat in addition to the 10–20 nm thick negatively-charged endothelial glycocalyx. Depending on the fixation and labeling technique, this filamentous layer measures between 60 and 300 nm in thickness and may represent the endothelial surface layer [31, 32].

The endothelial surface layer is formed from the adsorption of a layer of proteins onto the plasma membrane-anchored glycocalyx. Functionally, it is detected by the observation of an exclusion zone, which extends over 200 nm above the glomerular endothelium [33]. Infusion with glycosaminoglycan-degrading enzymes diminished the thickness of the endothelial surface layer and significantly increased the permeability of albumin across the GFB [34]. Notably, filamentous 'sieve plugs' (also named fascinae fenestrae) occluding the fenestrations were also detected [32]. Whether any of these endothelial surface molecules play a substantial role in the permselectivity barrier has yet to be convincingly proven. Critics question whether some of the findings from structural studies are artifacts from fixation [23].

Part of the problem in assigning a substantial role of GEnCs to the GFB lies in the difficulty of interpreting functional results due to the extraordinary interdependence seen between podocytes, GBM, mesangial cells and GEnCs. Although there is no doubt that glomerular endothelial injury can lead to fenestration loss and proteinuria both in the clinical setting and in animal models, whether the proteinuria is a direct consequence of fenestration loss or a secondary effect of glomerular endothelial damage is not well established. Interestingly, recent studies have shown that proteinuria following glomerular endothelial injury may arise from podocyte dysfunction due to altered endothelial-podocyte interaction. For example, sera from women with preeclampsia altered podocyte barrier-forming capacity by promoting nephrin loss from podocytes and was mediated by GEnC-derived endothelin-1 [35, 36]. Nephrin loss was due to protease cleavage and cytoskeleton redistribution rather than reduced synthesis [36]. Although these studies do not address the importance of GEnC fenestrations to the GFB, they provide a possible mechanism by which glomerular endothelial injury may influence GFB integrity through effects on podocyte health.

Molecular Signaling Pathways in Glomerular Endothelial Cell Development

As mentioned previously, during glomerulogenesis, chemoattractive signals (many of which consist of podocyte-secreted growth factor ligands) bind to their respective cognate receptors, which are expressed on presumed angioblasts. Consequently, the angioblasts initiate capillary in-growth by migrating towards the podocytes and settling within the vascular cleft to give rise to the developing GEnCs. Additional cell-cell and cell-matrix interactions are necessary for subsequent endothelial cell survival, proliferation, differentiation, maturation and eventual glomerular capillary assembly. The continued expression and activation of many of these signaling pathways are required to maintain mature glomerular endothelial morphology (including fenestrations) and function.

Dissecting the precise roles of these signaling pathways both in glomerular development and mature glomerular function presents a major challenge due to the high degree of interdependence between the constituents of the glomerular tuft. For instance, by specifically manipulating GEnCs, one often observes disruption of podocyte and mesangial cell function (the vice versa also often occurs). Therefore, a key lesson to take note of is that if a gene appears essential for GEnC development and/or function, it does not necessarily mean that the gene is operating in a GEnC-specific manner. Instead, the gene could be affecting the other constituents within the glomerular tuft, which provide the necessary microenvironment for the GEnCs.

Another point to mention concerns animal models using targeted gene inactivation (also termed gene 'knockout'). This method is one of the most useful and widely applied approaches for studying the effects of gene-specific function in vivo. Unfortunately, many of the knockout animal models of genes expressed during glomerulogenesis result in early embryonic lethality prior to nephrogenesis, owing to their indispensable nature in early vascular development. Therefore, in order to examine glomerular vascular development, conditional gene inactivation models are usually required.

This strategy attempts to overcome early embryonic lethality by specifically inactivating the gene of interest only in a subset of cells within the transgenic animal. In most cases, this approach allows the mouse to reach a more mature developmental stage permitting examination of glomerulogenesis and nephrogenesis. Commonly, a cell type-specific promoter is used to drive expression of the Cre recombinase transgene. At the same time, the transgenic mouse genome contains loxP sites within the gene of interest. The cell type-specific expression of Cre in combination with available loxP sites within the genome will cause the excision and inactivation of the target gene only in the cells that are able to activate the cell type-specific promoter while the remainder of the mouse remains wild-type. Most often, studies examining glomerular development use podocyte-specific knockout mice due to the availability of podocyte-specific promoters. As a result, the literature may over-represent podocyte-specific contribution to GEnC development and health, although there is no doubt that podocytes are essential to GEnCs. In the following sections, we highlight some of the important genes and signaling pathways in the development and maintenance of the glomerular endothelium.

Vascular Endothelial Growth Factors

Perhaps the best studied growth factor signaling pathway in glomerular capillary development is vascular endothelial growth factor A (VEGFA; also designated VEGF). VEGFA is the founding member of a family of homodimeric signaling glycoproteins structurally similar to platelet-derived growth factors. Additional family members include VEGFB, VEGFC, VEGFD and placenta growth factor (PlGF). VEGFA plays an essential role in mammalian vascular development, the importance of which is highlighted by the fact that mouse embryos lacking functional VEGFA do not survive past E10.5, largely due to failure of vascular development [37, 38]. Mice with only a single functional copy of VEGFA are haploinsufficient and display similar defects leading to early embryonic lethality.

VEGFA binds with high affinity to two type III receptor tyrosine kinases: FLT1 (also designated VEGFR1) and KDR (also designated FLK1 or

VEGFR2), whose expression are mainly restricted to endothelial cells and their precursors. Results from several studies have shed some light on the roles of FLT1 and KDR in VEGFA-mediated signaling. First, targeted disruption of *Kdr* leads to embryonic lethality between E8.5 and E9.5 due to defects in hematopoietic and endothelial cell development, confirming its importance in vascular development [39]. In contrast, mice lacking functional FLT1 show increased numbers of endothelial progenitor cells, which result in vascular disorganization and early embryonic lethality [40]. This increase in endothelial progenitor cells appears to be due to an alteration in cell fate determination among mesenchymal cells leading to increased numbers of hemangioblasts and has been attributed to VEGFA hyperactivity resulting from loss of FLT1 [41]. Additionally, KDR is activated more efficiently than FLT1 upon VEGFA binding in endothelial cells [42–44]. Finally, the FLT1 tyrosine kinase domain appears to be dispensable for vascular development [45]. These data suggest that VEGFA primarily exerts its biological actions through KDR, whereas FLT1 may act as a nonfunctional decoy receptor to sequester excess extracellular VEGFA. This property of FLT1 may allow for precise control of overall VEGFA signaling.

Overwhelming evidence implicates the importance of VEGFA in the developing kidney. During glomerulogenesis, VEGFA is produced at high levels by the developing podocytes and continued podocyte-specific VEGFA production is seen in the mature glomerulus [46, 47]. Moreover, angioblasts expressing the VEGFA receptor, KDR, are present within the metanephric mesenchyme [8]. These same cells migrate towards the developing podocytes and form the eventual glomerular endothelium. Exogenous administration of antibodies directed against VEGFA in newborn mice inhibits angioblast migration into the developing glomerulus and yields numerous abnormal glomeruli lacking capillary tufts [48]. Similarly, another group showed that anti-VEGFA antibodies and soluble FLT1 (an inhibitor of VEGFA signaling seen in preeclampsia; see below) are able to cause GEnC detachment and hypertrophy leading to proteinuria [49]. Given that VEGFA null mouse embryos die prior to kidney organogenesis, Eremina et al. [50] generated mice with podocyte-specific disruption of *VEGFA* by employing a conditional Cre-loxP targeting strategy. Interestingly, the severity of glomerular vascular defects follows a VEGFA dose-dependent manner. Podocyte-specific deletion of both copies of *VEGFA* results in perinatal lethality. Examination of mutant kidneys reveals complete glomerular capillary developmental failure due to defective endothelial cell migration, differentiation and survival. In contrast, mice possessing a single functional copy of *VEGFA* in their podocytes initially demonstrate normal glomerular morphology with functional GFBs [50]. However, by 2.5 weeks of age, these mice develop endotheliosis (endothelial cell swelling), reminiscent of the typical glomerular lesion seen in preeclampsia. Proteinuria soon follows with subsequent endothelial cell necrosis and progression to end-stage renal failure by 9–12 weeks. Likewise, mice with a single hypomorphic *VEGFA* allele in podocytes show normal in-growth of glomerular capillaries and proper glomerulogenesis; however, GEnCs do not survive and disappear by 1–3 days postnatally [51]. Notably, increased VEGFA expression is also detrimental to glomerular function. Podocyte-specific overexpression of VEGFA led to capillary loop collapse and hyperpermeability of the GFB with morphologic features resembling the collapsing variant of focal segmental glomerulosclerosis.

To add another level of complexity, the overall contribution of VEGFA to vascular development may depend on the various isoforms present. The *VEGFA* gene consists of eight exons, which give rise to three main isoforms: VEGFA121, VEGFA165 and VEGFA189 in humans and VEGFA120, VEGFA164 and VEGFA188 in mice, each number representing the protein product size [54]. Additionally, for each major isoform, a corresponding 'b' isoform also exists in which a differential splice site in exon 8 produces a protein of the same size, but with an alternate carboxyl terminus [55].

Although the functional capabilities of these isoforms are not fully delineated, some evidence suggests that the various isoforms possess varying degrees of angiogenic and anti-angiogenic capabilities. For example, the VEGFA165b isoform has been reported to be anti-angiogenic and capable of inhibiting VEGFA165-mediated signaling [55–57]. In another study, investigators showed that mice possessing only the VEGFA120 isoform display defects in GEnC development and survival [58]. Furthermore, remodeling of PTCs, vascular coverage by pericytes and branching of renal arteries were impaired. In contrast, although extrarenal defects are present, mice expressing only the VEGFA188 isoform demonstrate no glomerular phenotype [59]. Similarly, mice that express only the VEGFA164 isoform appear to exhibit no abnormalities. Part of these differences may be due to VEGFA164-specific binding of neuropilin-1, a co-receptor of KDR which is also expressed in GEnCs [60–62]. These various isoforms of VEGFA may provide yet another mechanism for precise control of overall VEGFA-mediated signaling.

Modifiers of VEGFA Signaling

In addition to the various VEGFA isoforms and VEGFA receptors, other mechanisms have been proposed by which net VEGFA signaling may be modified. Since the GBM contains abundant heparin sulfate and since certain isoforms of VEGFA can bind to heparin sulfate with high affinity [63], it has been proposed that the GBM may act as a sink for VEGFA, resulting in storage and modified bioavailability and/or turnover of VEGFA. Several genes have also been hypothesized to modify VEGFA signaling. For instance, semaphorin3a (SEMA3A) is a secreted chemorepellent important in axonal and endothelial migration. In the kidney, it is expressed in podocytes and collecting tubules. In addition to abnormal podocytes, mice lacking SEMA3A show defects in renal vascular patterning and excess endothelial cells within glomerular capillaries while podocyte-specific SEMA3A overexpression resulted in GEnC apoptosis and glomerular hypoplasia [64]. These findings indicate that SEMA3A may act as a negative regulator of endothelial cell survival in developing glomeruli. Interestingly, the podocyte and endothelial damage seen with exogenous administration of SEMA3A can be prevented by concurrent VEGFA165 administration, suggesting that SEMA3A affects VEGFA signaling [65].

Another example is cysteine-rich motor neuron protein 1 (CRIM1). A transmembrane cysteine-rich repeat-containing protein related to chordin, CRIM1 has been shown to tether growth factors to the cell surface [66]. In the developing kidney, CRIM1 is co-expressed with VEGFA in podocytes, among other places [67]. Wilkinson et al. [68] report that mice lacking fully functional CRIM1 undergo perinatal death due to defects in multiple organ systems. The kidneys of these mice exhibit abnormal glomerulogenesis, characterized by enlarged capillary loops, podocyte effacement and mesangiolysis. These investigators also show that CRIM1 interacts directly with VEGFA. Furthermore, CRIM1-deficient mice display mislocalized VEGFA within the developing glomerulus and increased activation of KDR in GEnCs, suggesting that CRIM1 regulates the delivery of VEGFA by the podocytes to the endothelial cells.

Transforming Growth Factor-β

TGF-β signaling is important in a wide variety of cellular processes; its role in vascular development is well-established. Abnormalities in blood vessel development and morphology are observed when TGF-β signaling is disrupted in both animal models and humans. With regard to glomerular pathology, infusion of neutralizing antibodies against TGF-β-1 (TGFB1) in neonatal rats clearly attenuated angioblast invasion into the comma- and S-shaped bodies and inhibited glomerular capillary assembly [69]. In addition, capillary lumen formation and endothelial cell fenestration were abrogated in developing glomeruli, whereas mature glomeruli were unaffected.

In GEnCs, Choi et al. [70] showed that in vitro capillary morphogenesis is associated with apoptosis. Moreover, they demonstrated that using dominant negative TGF-β type II receptors to interfere with TGF-β signaling abolishes capillary formation. This defect is likely due in part to the requirement of TGF-β dependent apoptosis for glomerular capillary lumen formation [14]. Recently, some studies suggest that TGF-β may interact with VEGFA-mediated signaling during angiogenesis. Although VEGFA promotes endothelial cell survival while TGFB1 induces endothelial cell apoptosis, it appears that in the presence of TGFB1, VEGFA-mediated signaling can be transiently converted into an apoptotic cue [71, 72].

Angiopoietins

The angiopoietin family of vascular growth factors includes angiopoietin-1 (ANG1) and its naturally occurring antagonist angiopoietin-2 (ANG2). ANG1 signaling through its cognate receptor tyrosine kinase TEK (also designated TIE-2) provides crucial endothelial cell-specific actions effecting vascular proliferation, development, maturation and permeability. Both ANG2 and TIE1 are able to antagonize ANG1-mediated TEK signaling and may provide additional controls for this pathway [73]. In terms of the kidney, *Ang1*, *Ang2* and *Tek* are all expressed in both the embryonic and adult mouse kidney [74–76]. In the embryonic kidney, *Ang1* is expressed in the metanephric mesenchyme, differentiating tubular epithelia and maturing podocytes. In contrast, *Tek* is present in clusters of presumed angioblasts within the metanephric mesenchyme and may represent the same cells that are KDR-positive [73]. Tek is also seen in the mouse peritubular and glomerular capillaries.

Mouse embryos with targeted inactivation of *Ang1* are able to undergo early vasculogenesis; however, subsequent vascular remodeling is hampered, resulting in embryonic lethality prior to nephrogenesis [77]. To date, a podocyte-specific conditional knockout mouse of Ang-1 has not been reported. However, overexpression of Ang-2 in podocytes, which should in effect cause decreased Ang-1-induced TEK signaling, results in proteinuria and GEnC apoptosis [78]. Curiously, significantly decreased expression of VEGFA and nephrin proteins was observed, although podocytes were intact on electron microscopy. In another study, chimeric mice that possess both wild-type and mutant cells lacking functional Tie genes demonstrate the presence of mutant cells in early nonrenal vessels; however, the mutant cells are absent in mature renal vessels in late gestation [79]. Taken together, these findings suggest that ANG1-mediated TEK signaling is important for endothelial survival and vascular remodeling in glomerular morphogenesis.

Notch

The Notch signaling pathway is a highly evolutionarily conserved mechanism for intercellular communication and has been implicated in a variety of developmental functions, including cell survival, proliferation and differentiation as well as cell-fate determination. The mammalian Notch family of proteins includes four members: NOTCH1, NOTCH2, NOTCH3 and NOTCH4. All of the Notch family members trigger intracellular signaling in a similar manner. Upon ligand binding to the extracellular domain, proteolytic cleavage of the heterodimerization domain ensues, resulting in the release of the intracellular domain. Subsequently, the intracellular domain enters the nucleus and acts as a transcription factor to alter gene expression of target genes.

The various Notch genes are required for multiple steps in kidney organogenesis. Specifically for glomerular development, NOTCH2 appears to be essential since mice homozygous for a hypomorphic variant of NOTCH2 (NOTCH2-del1) died perinatally from defects in glomerulogenesis [80]. These mice display hypoplastic kidneys, which contained glomeruli lacking a normal capillary tuft. Although the early stages of metanephric kidney development proceeded relatively normally, angioblast migration into the vascular cleft of the S-shaped body was attenuated and subsequent glomerular tuft formation was lacking in NOTCH2-del1 mice. During capillary in-growth, NOTCH2 expression was confirmed in developing podocytes while Jagged1, a Notch ligand, was expressed in the GEnCs. Gamma-secretase, one of the proteases capable of cleaving Notch, appears to be required for proximal tubule and podocyte fate determination at the S-shaped stage of glomerulogenesis [81]. These results indicate the importance of Notch in glomerulogenesis and glomerular vascularization and point to a possible role of Notch in podocyte-endothelial interaction. Interestingly, recent evidence suggests that overactivation of Notch signaling may contribute to podocyte damage and subsequent glomerular disease [82, 83].

Additional Signaling Pathways

A myriad of genes have been shown to yield defects in glomerulogenesis when perturbed, of which a substantial number give what appears to be glomerular vascular abnormalities. To cover all of these genes is not within the scope of this chapter. Instead we highlight some recent findings.

The presence of certain GBM proteins at the appropriate developmental stage appears to be essential for glomerular vascularization. During glomerulogenesis, several components of the GBM, including laminins and collagen, transition from one isoform to another as development progresses (reviewed in [84]). One of these shifts in protein expression involves laminin α-1 to laminin α-5 transition. This event appears to be crucial for sustaining glomerular vascularization since mice lacking the laminin α-5 chain fail to form glomerular capillaries [85]. Recently, Abrass et al. [86] reported that mice lacking the laminin α-3 *(Lama3)* gene demonstrate abnormalities in glomerulogenesis. LAMA3 is the α-3 subunit of laminin-5 and is present in the GBM adjacent to GEnCs. It is first detected in the early capillary loop stage. In *Lama3*$^{-/-}$ mice, endothelial cells are able to migrate into the vascular cleft of the developing glomerulus; however, they are unable to flatten out, develop fenestrae or form lumens. Furthermore, mesangial cell recruitment is absent.

The ligand for CXCR4, chemokine (C-X-C motif) ligand 12 (CXCL12; also designated SDF1), has also been implicated in glomerular vasculogenesis. CXCL12 is a strong chemoattractant for lymphocytes and is often induced by pro-inflammatory stimuli. Additionally, CXCL12-induced CXCR4 signaling plays a key role in hematopoiesis and organogenesis. Takabatake et al. [87] examined the role of this pathway in kidney development. They found that CXCL12-secreting podocytes are in close proximity to CXCR4-positive endothelial cells. Furthermore, embryonic kidneys from both CXCL12- and CXCR4-deficient mice demonstrate impaired glomerular vessel formation marked by aneurismal dilatation of the developing glomerular tuft. A similar phenotype was seen in mice with endothelial cell-specific inactivation of CXCR4.

Bone morphogenetic protein 4 (BMP4) also appears to be essential in kidney development. With regard to glomerulogenesis, mice with podocyte-specific inactivation of BMP4 exhibit glomerular microaneurysms, collapsed glomerular capillary tufts, enlarged Bowman's capsules, and fewer normal proximal tubules [88]. In contrast, overexpression of BMP4 in podocytes resulted in defective glomerular capillary formation, but podocytes show no appreciable defects. Most likely, podocyte-derived BMP4 exerts its actions on neighboring endothelial and/or mesangial cells since phosphorylated Smads, the mediators of BMP signaling, are found in these cells rather than in podocytes.

Peritubular Endothelium

PTCs receive post-glomerular blood flow from the efferent arteriole and drain into interlobular veins. PTCs develop in conjunction with the glomerular vasculature; however, the PTC endothelium does not share the unique morphology seen in GEnCs. Most notably, PTC endothelial cells possess fenestrations which are subtended by PV-1-associated diaphragms [28]. By far, PTC development is not as well-studied as glomerular capillary formation, but many of the same signals derived from podocytes in glomerular development are produced by the tubular epithelial cells. Functionally, PTCs deliver oxygen and nutrients to the tubulointerstitium as well as reabsorb and return some of the filtered contents within the Bowman capsule and urinary space. Therefore, PTC dysfunction can lead to direct detriment to the tubulointerstitium, often marked by interstitial edema with variable infiltration of inflammatory cells and tubular damage in the acute setting while chronic irreversible changes are typically characterized by tubular atrophy and interstitial fibrosis. PTCs are also a prime target in allograft pathology as the endothelial cells are most susceptible to immune-mediated attack. We will discuss the pathogenesis and mechanisms of PTC-related diseases in more detail later in this chapter.

Microvascular Injury in Renal Disease

The kidney microvasculature is involved in a wide spectrum of renal diseases. Depending on the mechanism of disease, the endothelial cells may be damaged directly or the endothelial microenvironment may be altered giving rise to endothelial dysfunction. Pathways essential to development and maintenance of the renal microvasculature are often disrupted and either lead to renal disease directly or indirectly. Endothelial damage induces a wide variety of consequential events including leukostasis and inflammatory cell recruitment, vasospasm, pro-coagulation, and loss of structure and function (especially evident in GEnCs). These events lead to release of material into the circulation, which can be used as markers of endothelial damage/dysfunction. Several markers have been used reliably including von Willebrand's factor, intercellular adhesion molecule, vascular cell adhesion molecule, thrombomodulin and endothelial selectin [89].

Since renal microvascular damage and dysfunction plays a major role in a wide range of renal diseases, it is not possible to thoroughly discuss the dauntingly expansive amount of data regarding this subject. Therefore, we have highlighted some specific diseases and conditions. Other major diseases involving microvascular dysfunction are discussed in separate chapters. Here we highlight several renal diseases in which the renal microvasculature is crucial to the disease pathogenesis.

Preeclampsia

Preeclampsia is the most frequent renal complication encountered during pregnancy and is the leading cause of maternal and fetal morbidity and mortality. It is characterized by new-onset hypertension and proteinuria arising after 20 weeks of gestation. Preeclampsia develops in approximately 5% of all pregnancies, making it perhaps the most common glomerular disease worldwide [90]. Severe preeclampsia may progress into eclampsia if the mother develops tonic-clonic seizures. Additionally, HELLP syndrome, a severe life-threatening variant of preeclampsia, may develop if additional laboratory features of hemolysis, elevated liver enzymes and low platelets (thrombocytopenia) are present. Currently, delivery of the placenta is the only definitive cure; however, the crucial problem lies in weighing the benefits of delivery of the placenta for the mother against the risks of pre-term delivery for the child.

Under light microscopic examination, renal biopsies of patients with preeclampsia reveal pathologic glomerular morphology similar to that seen in thrombotic microangiopathy (TMA). In fact, preeclampsia can be considered a subclass of TMA. These diseases are characterized by 'bloodless glomeruli' in which severe GEnC swelling (glomerular endotheliosis) causes glomerular capillary occlusion. Overall, glomerular volume is increased, but glomerular cellularity remains relatively unchanged. Preeclampsia generally does not contain platelet fibrin thrombi within glomerular capillaries, whereas this finding is common in other forms of TMA. Ultrastructurally, the glomerular capillaries lose endothelial fenestrations.

The pathogenesis of preeclampsia may be multifactorial; however, microvascular endothelial injury appears to play a central role in the mechanism of this disease. One hypothesis is that abnormal placentation leads to placental hypoxia, which triggers the release of placenta-derived factors into the maternal circulation. Consequently, these factors cause damage to the microvascular endothelium, possibly through altering local angiogenic and vasodilatory signals. In turn, the GFB is disrupted and clinical signs of proteinuria and hypertension ensue.

Several circulating factors have been implicated in the pathogenesis of preeclampsia. One of the best studied is soluble Flt-1 (sFlt1). This secreted form of VEGFR1 is able to bind and sequester the angiogenic factors VEGFA and PlGF, thus acting as an endogenous inhibitor of VEGF-receptor signaling. In preeclampsia, excessive levels of placenta-derived sFlt1 are believed to disrupt VEGF-receptor signaling in the glomeruli, causing GEnC injury. In support of this idea, women with established preeclampsia display greatly elevated levels of serum sFlt1 [91, 92]. In animal models, excess sFlt1 produces clinical signs and glomerular lesions reminiscent of human preeclampsia [92]. Similar observations have been noted when neutralizing antibodies to VEGFA are administered to rodents [49]. Treatment with the 121 isoform of VEGFA alleviates this preeclampsia-like phenotype in rats without apparent harm to the fetus [93]. Likewise, renal damage induced in mice by adenoviral overexpression of sFlt1 is prevented by co-expression of VEGF [94]. Notably, the development of proteinuria, hypertension and TMA in patients receiving anti-VEGFA therapy for cancer have been reported (reviewed in [95, 96]).

The increased sFlt1 levels in preeclampsia may be due to autoantibodies against angiotensin II type 1 receptor [AT(1)-AAs]. It has long been known that AT(1)-AAs are present in preeclamptic women [97]; however, these antibodies are not specific to preeclampsia and may represent a secondary response to a hypoxic vascular insult [98]. Recently, Zhou et al. [99, 100] induced a preeclampsia-like condition in pregnant mice after injection with AT(1)-AAs from women with preeclampsia. They further report that AT(1)-AAs induce sFlt-1 production through angiotensin receptor activation [100].

Soluble endoglin (sEng) also appears to play a role in the pathogenesis of preeclampsia; and in conjunction with sFlt1, may contribute to the development of HELLP syndrome. Similar to sFlt1, sEng (most likely produced by the placenta) is elevated in the sera of women with preeclampsia and correlates with disease severity [101]. Although overexpression of sEng alone causes only mild proteinuria as compared to animals with sFlt1 overexpression; in combination, these factors lead to severe preeclampsia with evidence of HELLP syndrome. The effects of sEng may be due to disruption of TGFB1 signaling. In vitro studies show that sEng inhibits TGFB1 binding to its receptors with subsequent impairment of TGFB1-mediated activation of eNOS [101]. Since hypoxia upregulates the placental expression of endoglin [102], placental hypoxia due to abnormal placentation may cause concurrent sEng elevation, and in combination with increased sFlt1 levels, may result in preeclampsia or HELLP syndrome.

Although the exact pathogenesis of preeclampsia is still incompletely understood, the studies mentioned here provide extraordinary insight into the mechanism of this disease. It also provides possible diagnostic tools and avenues of therapy. For instance, the levels of preeclampsia-associated circulating factors may provide a useful biomarker for the early detection of preeclampsia. Levine et al. [103] report that the combination of elevated sEng level and increased sFlt1:PlGF ratio was more strongly predictive of preeclampsia than either biomarker alone. Furthermore, targeted therapy specifically inhibiting these placenta-derived circulating factors may provide treatment options for preeclampsia in the future.

Ischemia in Chronic Kidney Disease

Chronic kidney disease (CKD) is a progressive deterioration of renal function over time. Although the etiology of CDK may differ, renal microvascular dysfunction appears to play a central role in propagating kidney damage by creating an ischemic environment. Numerous rodent models have reported an association between hypoxia and CKD (reviewed in [104]). One of the leading hypotheses postulates that glomerular injury may cause restricted post-glomerular blood flow leading to PTC damage and chronic ischemia [105]. As a result, tubulointerstitial scars develop, which may be deleterious to adjacent nephrons; and thus, propagate renal damage. The overall effect is seen as progressive decline in renal function, the hallmark of CKD.

Chronic ischemia arising from damage to the kidney microvasculature may be due to both structural loss (i.e. decreased capillary number) and functional loss (i.e. vasoconstriction) of blood flow. In both human studies and animal models, a decrease in glomerular and PTC number correlates directly with the development of glomerulosclerosis and interstitial fibrosis, respectively [106–109]. Whether renal microvascular injury/loss is a primary cause of renal scarring or secondary to scar formation is unclear. It is likely that both scenarios occur.

Loss of renal microvasculature may be due to altered vascular homeostasis arising from local angiogenic/anti-angiogenic factor imbalance. Animal models of progressive glomerulosclerosis show GEnC apoptosis with insufficient capillary regeneration, resulting in a net decline of glomerular capillary loops and eventual glomerulosclerosis [110, 111]. A decreased Ang-1:Ang-2 ratio is observed in multiple animal models of progressive glomerulosclerosis [73, 112–114]. Using a remnant kidney model, Kang et al. [107] showed that although an initial angiogenic response is observed following subtotal renal ablation, progressive glomerular and PTC loss with associated glomerulosclerosis and interstitial fibrosis ensues. Renal microvascular rarefaction correlated with diminished VEGF and elevated thrombospondin-1 (TSP-1). Furthermore, they showed that loss of VEGF correlated with sites of macrophage infiltration and that macrophage-associated cytokines (IL-1β, TNF-α and IL-6) inhibited VEGF production in medullary thick ascending limb (mTAL) cells in vitro. TSP-1 induction may occur as an upshot of macrophage-derived cytokines as well, since IL-1β and TNF-α stimulated TSP-1 mRNA expression in mTAL cells. Since VEGF is a known angiogenic and endothelial survival factor while TSP-1 displays anti-angiogenic properties, including the ability to directly induce endothelial cell apoptosis, these investigators concluded that impaired angiogenesis leads to excessive capillary loss in the remnant kidney and that macrophage-derived cytokines contribute to this impairment by directly inhibiting local VEGF production while at the same time promoting TSP-1 expression. Furthermore, exogenous administration of VEGF reduced interstitial fibrosis and stabilized renal function in the rat remnant kidney model [115]. In humans, increased microvascular damage combined with an imbalance of angiogenic and anti-angiogenic factors are seen in CKD [116–118]. Interestingly, excess sFlt-1, a VEGFA antagonist, is associated with endothelial dysfunction and poor renal function in CKD [119].

Another major contributor to chronic ischemia is functional loss rather than structural loss of blood flow. Vasoactive mediators within the local microenvironment can cause inadequate delivery of necessary oxygen and nutrients to the renal parenchyma. Several studies have shown that nitric oxide (NO) is reduced in CKD (reviewed in [120]). NO is a gaseous signaling molecule produced by the enzyme nitric oxide synthase (NOS) through the oxidation of L-arginine. Three different subtypes of NOS (neuronal, inducible and endothelial) exist and are encoded by separate genes. In the endothelium, NO is synthesized by endothelial NOS (eNOS), which acts locally to produce vasodilatory and antithrombotic effects. NO is also of prime importance in maintaining endothelial health. Both pharmacological reduction of NO and targeted disruption of eNOS in mice produce hypertension, consistent with its importance in regulating vasoactivity [121–124]. Renal hemodynamics is highly dependent on local NO production. Vasodilatory effects of NO on both afferent and efferent arterioles enhance glomerular perfusion and GFR [89, 125].

Patients with ESRD who are on hemodialysis show impaired NO-mediated vasodilation, which may stem from reduced bioavailability of NO resulting in an attenuated vasodilatory state [126]. Some evidence point to asymmetric dimethylarginine (ADMA) as the culprit. ADMA is a metabolite of arginine and an endogenous inhibitor of eNOS. Elevated ADMA has been associated with increased cardiovascular risk and with progression of renal disease [127]. Infusion of ADMA into healthy volunteers decreased effective renal plasma flow, increased renal and systemic vascular resistance, and increased blood pressure, whereas GFR remained unaffected [128]. Some CKD patients show high levels of ADMA independent of reduced renal clearance, which may cause lower eNOS activity [129]. ADMA levels predict progression of CKD [130, 131].

A study using a mouse remnant kidney model suggests that the elevation of ADMA may be due to a combination of increased synthesis and decreased breakdown [132]. Other studies have pointed to reduced L-arginine availability due to decreased production and/or transport into endothelial cells as a possible mechanism by which NO synthesis is impaired in CKD [120]. Notably, ADMA inhibition of NOS promotes endothelial-mesenchymal transformation and the expression of anti-angiogenic factors such as endostatin [133]. It should be pointed out that reduced NO will not only have detrimental effects on kidney perfusion but also on general cardiovascular health. Consistent with this idea, ESRD is associated with an increased risk of cardiovascular disease. Therefore, dysfunction in vasoactivity extends beyond the kidney microvasculature and may cause CKD patient morbidity and mortality through its general vascular effects.

Angiotensin II (ATII) appears to also be important in the progression of CKD. Although ATII levels are not elevated in CKD, it has been postulated that the overall balance between ATII and NO activity could be a primary factor that contributes to CKD [134, 135]. The benefits of angiotensin-converting enzyme (ACE) inhibitors and ATII receptor antagonists in slowing CKD progression may act not only through the mitigation of glomerular hyperfiltration and glomerular hypertension, but also through the improvement of renal vascular flow by abating ATII-related renal ischemia [136]. In support of this hypothesis, it has been reported that ATII inhibition raises renal interstitial microvascular pO_2 in rats [137].

Another important point to mention is that vasoactive mediators can influence the local expression of angiogenic factors. For example, both NO and ATII modulate VEGF expression in mTAL cells [138]. Therefore, it is likely that structural and functional loss of the renal microvasculature are interconnected.

These pathways by which the kidney may become vulnerable to chronic ischemia and subsequent renal scarring brings up potential avenues of therapy to slow progressive renal disease. Several studies point to the benefits of angiogenic factors such as VEGF in stimulating capillary repair/formation and slowing renal scarring and renal decline (reviewed in [136]). Similarly, increasing vasodilation through pharmacological means may constitute another strategy to combat renal scarring by increasing the oxygen tension in the local renal environment.

Aging Kidney

Even in the absence of renal disease, both human and animal studies indicate that GFR declines with advancing age [139, 140]. This decline appears to be part of the normal aging process and is associated with structural changes. As the kidney ages, the renal cortical mass decreases primarily due to progressive nephron loss. Histologically, a gradual increase in the severity of glomerular and tubulointerstitial scarring in conjunction with loss of glomerular and PTC density is observed [141–144]. Since these observations are surprisingly similar to those seen in progressive renal disease, the leading hypothesis implicates age-related microvascular dysfunction to progressively worsening chronic renal ischemia, eventually giving rise to these age-related histologic and functional changes. Consistent with this idea, elevated numbers of circulating endothelial cells, a marker reflecting microvascular injury/dysfunction, is seen in older individuals [141]. Additionally, studies have pointed to the imbalance of angiogenic and anti-angiogenic factors in contributing to the loss of the renal microvasculature, similar to that seen in CKD [145, 146]. As in the remnant kidney model, TSP-1 is also elevated in the aging kidney [145].

Microvascular vasoactivity perturbations are seen in the aging kidney as well. With advanced age, the endothelial response to vasoconstrictors appears to dominate over vasodilatory actions [141]. Just as in CKD, the NO system declines and is associated with an increased concentration of circulating endogenous NOS inhibitors with advancing age [147–149]. The overall effect of these vasoactive substances may contribute to reduced PTC blood flow in the aging kidney and may be a key contributor to age-associated tubulointerstitial fibrosis as a result of chronic ischemia [141]. Interestingly, the NO system displays sexual dimorphism with elevated NO and eNOS levels in females [142, 150–152]. This observation is likely due to estrogen-dependent effects on NO production and may explain gender preference in renal protective effects.

Glomerular Endothelial Cells in Transplants

Endothelial injury is of prime importance in the mechanism of several immunologic processes directed against renal allografts. Both acute and ongoing/chronic immune-mediated endothelial injury can result in a decline in allograft function. Effective treatments for acute episodes are available, and immediate treatment is necessary to ensure allograft survival. Currently, however, no effective treatment is available to combat chronic endothelial damage, the end result of which is progressive renal dysfunction leading to allograft loss. Although there is no doubt that endothelial damage is a key feature of type II acute T cell-mediated rejection, vascular damage in this setting is generally centered on arteries and arterioles. In contrast, antibody-mediated rejection appears to target the renal microvasculature more prominently. In the following sections, we focus our discussion on antibody-mediated allograft injury.

Acute Antibody-Mediated Rejection

Acute antibody-mediated rejection (AMR) is a distinct form of allograft rejection in which donor-specific antibodies (DSA) directed towards target antigens present within the graft elicit complement activation and subsequent graft damage. Definitive diagnosis of acute AMR includes three criteria: (1) morphologic evidence of acute tissue injury, (2) diffuse C4d staining in PTCs and (3) serologic evidence of DSA [153]. The pathogenesis of AMR is mediated through the actions of DSA, which most commonly are directed towards class I or class II human leukocyte antigens (HLAs) [154, 155].

HLAs are widely expressed in the kidney. However, since they are abundantly expressed in endothelial cells and immunologic mediators have direct access to the endothelial cell surface, the most severe changes in AMR are seen in the renal vasculature. Following DSA interaction with target antigens, components of the complement system are activated, resulting in the production of complement split products, which in turn cause graft injury. Several complement split products can directly affect the renal microvasculature [155]. For example, C5b-9 can cause cytolysis and apoptosis of endothelial cells as well as upregulate endothelial expression of adhesion molecules to allow for inflammatory cell recruitment and further endothelial and tissue damage. Pro-coagulative properties develop in damaged endothelium creating increased susceptibility to intravascular thrombosis. Vasoactive mediators are also perturbed in vascular injury and can cause vasospasm. Although many complement split products are bound to tissue via antigen-antibody complexes, only C4d is reliable for diagnosis of AMR. C4d is both sensitive and specific for AMR if present in the walls of PTCs and is indicative of endothelial injury related to AMR. Recently, some studies suggest that a significant portion of kidney transplant patients experience antibody-mediated endothelial damage despite being C4d-negative [156]. Using expression microarrays, Sis et al. [157] show that high renal endothelial transcript expression in patients with DSA is an indicator of active antibody-mediated allograft injury and poor graft survival. Histologically, features of chronic AMR are much more common in the biopsies from these patients than acute AMR.

Chronic Active Antibody-Mediated Rejection and Transplant Glomerulopathy

Chronic active AMR is likely caused by the presence of circulation DSA directed against renal endothelial cells, resulting in chronic ongoing allograft injury. The diagnosis of chronic active AMR relies on (1) diffuse C4d staining in PTCs, (2) the presence of circulating DSA and (3) morphologic evidence of chronic tissue injury [158]. Chronic tissue injury is usually in the form of transplant glomerulopathy (TG), multilayering of the PTC basement membrane, interstitial fibrosis/tubular atrophy and/or arterial fibrous intimal thickening. All of these features are attributed to endothelial damage, much of which appears to be targeted against the renal microvasculature.

TG is a morphologically distinct form of chronic allograft injury usually arising as a late graft complication, which progressively leads to renal failure. TG is rarely diagnosed prior to 1 year following kidney transplantation. However, the incidence dramatically rises from approximately 4% at 1 year to between 20 and 40% at 5 years post-transplant on protocol biopsies [159, 160]. Clinically, patients usually display nonspecific signs of proteinuria, hypertension and progressively worsening graft function. Histologically, TG is characterized by thickening of the glomerular capillary walls with re-duplication of the GBM. Various degrees of mesangial matrix expansion and mesangial hypercellularity are present. Glomerulitis is often present, usually consisting of mononuclear leukocytes. Ultrastructural analysis reveals circumferential multilayering of the GBM and is often associated with PTC basement membrane multilayering. Subendothelial widening with mesangial cell interposition is often present. Podocyte foot processes usually show widespread effacement consistent with the clinical observation of proteinuria.

TG is strongly associated with humoral immunity given that approximately half of biopsies with TG demonstrate C4d deposition in PTCs. Patients with TG frequently have ongoing or prior episodes of either acute T cell-mediated rejection or acute antibody-mediated rejection [161]. TG likely develops as a consequence of chronic ongoing immune-mediated endothelial damage. The observation that glomerulitis is commonly present is consistent with this hypothesis. Similar to the mechanism of acute antibody-mediated rejection, TG likely results from donor antigens (most often HLAs), which elicit humoral immunity to generate DSA. The target antigens are abundantly expressed on endothelial cell surfaces. Consequently, endothelial cells damage follows.

In this chapter, we have briefly summarized some of the major events in kidney microvascular development, highlighted features of endothelial cell morphology and function, and presented some of the experimental and clinical data which strongly support the essential role of the kidney microvasculature in maintaining renal health and function. In general, significant renal diseases often occur as a result of perturbations of molecular pathways within the endothelial cells themselves and/or their microenvironment. Microvascular endothelial damage and dysfunction play a major role in both diseases of the native kidneys and in renal allograft pathology. A broad understanding of the renal microvasculature is essential for improving our current treatment of renal diseases and opens the possibility of exploiting molecular targets with the hope of providing more efficacious and less toxic therapies in the future.

References

1 Bernstein J, Cheng F, Roszka J: Glomerular differentiation in metanephric culture. Lab Invest 1981;45:183–190.

2 Ekblom P, et al: The origin of the glomerular endothelium. Cell Differ 1982;11:35–39.

3 Sariola H, et al: Differentiation and vascularization of the metanephric kidney grafted on the chorioallantoic membrane. Dev Biol 1983;96:427–435.

4 Sariola H, et al: Dual origin of glomerular basement membrane. Dev Biol 1984;101:86–96.

5 Rogers SA, Hammerman MR: Transplantation of rat metanephroi into mice. Am J Physiol Regul Integr Comp Physiol 2001;280:R1865–R1869.

6 Takeda S, Rogers SA, Hammerman MR: Differential origin for endothelial and mesangial cells after transplantation of pig fetal renal primordia into rats. Transpl Immunol 2006;15:211–215.

7 Hyink DP, et al: Endogenous origin of glomerular endothelial and mesangial cells in grafts of embryonic kidneys. Am J Physiol 1996;270:F886–F899.

8 Robert B, et al: Evidence that embryonic kidney cells expressing flk-1 are intrinsic, vasculogenic angioblasts. Am J Physiol 1996;271: F744–F753.

9 Abrahamson DR, Robert B: Derivation and differentiation of glomerular endothelial cells. Nephrol Dial Transplant 2003;18(Suppl 6):vi2–vi7.

10 Vaughan MR, Quaggin SE: How do mesangial and endothelial cells form the glomerular tuft? J Am Soc Nephrol 2008;19:24–33.

11 Quaggin SE, Kreidberg JA: Development of the renal glomerulus: good neighbors and good fences. Development 2008;135:609–620.

12 Dressler GR, The cellular basis of kidney development. Annu Rev Cell Dev Biol 2006; 22:509–529.

13 Ballermann BJ: Glomerular endothelial cell differentiation. Kidney Int 2005;67:1668–1671.

14 Fierlbeck W, et al: Endothelial cell apoptosis during glomerular capillary lumen formation in vivo. J Am Soc Nephrol 2003;14:1349–1354.

15 Lindahl P, et al: Paracrine PDGF-B/PDGF-Rbeta signaling controls mesangial cell development in kidney glomeruli. Development 1998;125:3313–3322.

16 Leveen P, et al: Mice deficient for PDGF B show renal, cardiovascular, and hematological abnormalities. Genes Dev 1994;8:1875–1887.

17 Soriano P: Abnormal kidney development and hematological disorders in PDGF beta-receptor mutant mice. Genes Dev 1994;8:1888–1896.

18 Kikkawa Y, Virtanen I, Miner JH: Mesangial cells organize the glomerular capillaries by adhering to the G domain of laminin alpha5 in the glomerular basement membrane. J Cell Biol 2003;161:187–196.

19 Sariola H: Incomplete fusion of the epithelial and endothelial basement membranes in interspecies hybrid glomeruli. Cell Differ 1984;14:189–195.

20 St John PL, Abrahamson DR: Glomerular endothelial cells and podocytes jointly synthesize laminin-1 and -11 chains. Kidney Int 2001;60:1037–1046.

21 Ballermann BJ: Contribution of the endothelium to the glomerular permselectivity barrier in health and disease. Nephron Physiol 2007;106:19–25.

22 Satchell SC, Braet F: Glomerular endothelial cell fenestrations: an integral component of the glomerular filtration barrier. Am J Physiol Renal Physiol 2009;296:F947–F956.

23 Ballermann BJ, Stan RV: Resolved: capillary endothelium is a major contributor to the glomerular filtration barrier. J Am Soc Nephrol 2007;18:2432–2438.

24 Avasthi PS, et al: Effect of gentamicin on glomerular ultrastructure. J Lab Clin Med 1981;98:444–454.

25 Deen WM, Lazzara MJ, Myers BD: Structural determinants of glomerular permeability. Am J Physiol Renal Physiol 2001;281:F579–F596.

26 Lafayette RA, et al: Nature of glomerular dysfunction in pre-eclampsia. Kidney Int 1998; 54:1240–1249.

27 Farquhar MG, Wissig SL, Palade GE: Glomerular permeability. I. Ferritin transfer across the normal glomerular capillary wall. J Exp Med 1961;113:47–66.

28 Stan RV, Kubitza M, Palade GE: PV-1 is a component of the fenestral and stomatal diaphragms in fenestrated endothelia. Proc Natl Acad Sci USA 1999;96:13203–13207.

29 Ryan GB, Hein SJ, Karnovsky MJ: Glomerular permeability to proteins. Effects of hemodynamic factors on the distribution of endogenous immunoglobulin G and exogenous catalase in the rat glomerulus. Lab Invest 1976;34:415–427.

30 Ryan GB, Karnovsky MJ: Distribution of endogenous albumin in the rat glomerulus: role of hemodynamic factors in glomerular barrier function. Kidney Int 1976;9:36–45.

31 Rostgaard J, Qvortrup K: Electron microscopic demonstrations of filamentous molecular sieve plugs in capillary fenestrae. Microvasc Res 1997;53:1–13.

32 Rostgaard J, Qvortrup K: Sieve plugs in fenestrae of glomerular capillaries–site of the filtration barrier? Cells Tissues Organs 2002; 170:132–138.

33 Hjalmarsson C, Johansson BR, Haraldsson B: Electron microscopic evaluation of the endothelial surface layer of glomerular capillaries. Microvasc Res 2004;67:9–17.

34 Jeansson M, Haraldsson B: Morphological and functional evidence for an important role of the endothelial cell glycocalyx in the glomerular barrier. Am J Physiol Renal Physiol 2006;290:F111–F116.

35 Henao DE, et al: Preeclamptic sera directly induce slit-diaphragm protein redistribution and alter podocyte barrier-forming capacity. Nephron Exp Nephrol 2008;110:e73–e81.
36 Collino F, et al: Preeclamptic sera induce nephrin shedding from podocytes through endothelin-1 release by endothelial glomerular cells. Am J Physiol Renal Physiol 2008; 294:F1185–F1194.
37 Ferrara N, et al: Heterozygous embryonic lethality induced by targeted inactivation of the VEGF gene. Nature 1996;380:439–442.
38 Carmeliet P, et al: Abnormal blood vessel development and lethality in embryos lacking a single VEGF allele. Nature 1996;380:435–439.
39 Shalaby F, et al: Failure of blood-island formation and vasculogenesis in Flk-1-deficient mice. Nature 1995;376:62–66.
40 Fong GH, et al: Role of the Flt-1 receptor tyrosine kinase in regulating the assembly of vascular endothelium. Nature 1995;376:66–70.
41 Fong GH, et al: Increased hemangioblast commitment, not vascular disorganization, is the primary defect in flt-1 knock-out mice. Development 1999;126:3015–3025.
42 Waltenberger J, et al: Different signal transduction properties of KDR and Flt1, two receptors for vascular endothelial growth factor. J Biol Chem 1994;269:26988–26995.
43 Gille H, et al: Analysis of biological effects and signaling properties of Flt-1 (VEGFR-1) and KDR (VEGFR-2). A reassessment using novel receptor-specific vascular endothelial growth factor mutants. J Biol Chem 2001; 276:3222–3230.
44 Gille H, et al: A repressor sequence in the juxtamembrane domain of Flt-1 (VEGFR-1) constitutively inhibits vascular endothelial growth factor-dependent phosphatidylinositol 3′-kinase activation and endothelial cell migration. EMBO J, 2000;19:4064–4073.
45 Hiratsuka S, et al: Flt-1 lacking the tyrosine kinase domain is sufficient for normal development and angiogenesis in mice. Proc Natl Acad Sci USA 1998;95:9349–9354.
46 Brown LF, et al: Vascular permeability factor mRNA and protein expression in human kidney. Kidney Int 1992;42:1457–1461.
47 Simon M, et al: Expression of vascular endothelial growth factor and its receptors in human renal ontogenesis and in adult kidney. Am J Physiol 1995;268:F240–F250.
48 Kitamoto Y, Tokunaga H, Tomita K: Vascular endothelial growth factor is an essential molecule for mouse kidney development: glomerulogenesis and nephrogenesis. J Clin Invest 1997;99:2351–2357.
49 Sugimoto H, et al: Neutralization of circulating vascular endothelial growth factor (VEGF) by anti-VEGF antibodies and soluble VEGF receptor 1 (sFlt-1) induces proteinuria. J Biol Chem 2003;278:12605–12608.
50 Eremina V, et al: Glomerular-specific alterations of VEGF-A expression lead to distinct congenital and acquired renal diseases. J Clin Invest 2003;111:707–716.
51 Eremina V, et al: Vascular endothelial growth factor a signaling in the podocyte-endothelial compartment is required for mesangial cell migration and survival. J Am Soc Nephrol 2006;17:724–735.
52 Korgaonkar SN, et al: HIV-1 upregulates VEGF in podocytes. J Am Soc Nephrol 2008; 19:877–883.
53 Tufro A: VEGF spatially directs angiogenesis during metanephric development in vitro. Dev Biol 2000;227:558–566.
54 Eremina V, Baelde HJ, Quaggin SE: Role of the VEGF–a signaling pathway in the glomerulus: evidence for crosstalk between components of the glomerular filtration barrier. Nephron Physiol 2007;106:p32–p37.
55 Bates DO, et al: VEGF165b, an inhibitory splice variant of vascular endothelial growth factor, is down-regulated in renal cell carcinoma. Cancer Res 2002;62:4123–4131.
56 Woolard J, et al: VEGF165b, an inhibitory vascular endothelial growth factor splice variant: mechanism of action, in vivo effect on angiogenesis and endogenous protein expression. Cancer Res 2004;64:7822–7835.
57 Bevan HS, et al: The alternatively spliced anti-angiogenic family of VEGF isoforms VEGFxxxb in human kidney development. Nephron Physiol 2008;110:p57–p67.
58 Mattot V, et al: Loss of the VEGF(164) and VEGF(188) isoforms impairs postnatal glomerular angiogenesis and renal arteriogenesis in mice. J Am Soc Nephrol 2002;13:1548–1560.
59 Stalmans I, et al: Arteriolar and venular patterning in retinas of mice selectively expressing VEGF isoforms. J Clin Invest 2002;109:327–336.
60 Soker S, et al: Neuropilin-1 is expressed by endothelial and tumor cells as an isoform-specific receptor for vascular endothelial growth factor. Cell 1998;92:735–745.
61 Whitaker GB, Limberg BJ, Rosenbaum JS: Vascular endothelial growth factor receptor-2 and neuropilin-1 form a receptor complex that is responsible for the differential signaling potency of VEGF(165) and VEGF(121). J Biol Chem 2001;276:25520–25531.
62 Robert B, Zhao X: Abrahamson DR: Coexpression of neuropilin-1, Flk1, and VEGF(164) in developing and mature mouse kidney glomeruli. Am J Physiol Renal Physiol 2000;279:F275–F282.
63 Robinson CJ, et al: VEGF165-binding sites within heparan sulfate encompass two highly sulfated domains and can be liberated by K5 lyase. J Biol Chem 2006;281:1731–1740.
64 Reidy KJ, et al: Semaphorin3a regulates endothelial cell number and podocyte differentiation during glomerular development. Development 2009;136:3979–3989.
65 Tapia R, et al: Semaphorin3a disrupts podocyte foot processes causing acute proteinuria. Kidney Int 2008;73:733–740.
66 Wilkinson L, et al: CRIM1 regulates the rate of processing and delivery of bone morphogenetic proteins to the cell surface. J Biol Chem 2003;278:34181–34188.
67 Pennisi DJ, et al: Crim1KST264/KST264 mice display a disruption of the Crim1 gene resulting in perinatal lethality with defects in multiple organ systems. Dev Dyn 2007;236:502–511.
68 Wilkinson L, et al: Crim1KST264/KST264 mice implicate Crim1 in the regulation of vascular endothelial growth factor-A activity during glomerular vascular development. J Am Soc Nephrol 2007;18:1697–1708.
69 Liu A, Dardik A, Ballermann BJ: Neutralizing TGF-beta1 antibody infusion in neonatal rat delays in vivo glomerular capillary formation 1. Kidney Int 1999;56:1334–1348.

70 Choi ME, Ballermann BJ: Inhibition of capillary morphogenesis and associated apoptosis by dominant negative mutant transforming growth factor-beta receptors. J Biol Chem 1995;270:21144–21150.
71 Ferrari G, et al: Transforming growth factor-beta 1 (TGF-beta1) induces angiogenesis through vascular endothelial growth factor (VEGF)-mediated apoptosis. J Cell Physiol 2009;219:449–458.
72 Ferrari G, et al: VEGF, a prosurvival factor, acts in concert with TGF-beta1 to induce endothelial cell apoptosis. Proc Natl Acad Sci USA 2006;103:17260–17265.
73 Woolf AS, Gnudi L, Long DA: Roles of angiopoietins in kidney development and disease. J Am Soc Nephrol 2009;20:239–244.
74 Yuan HT, et al: Expression of angiopoietin-1, angiopoietin-2, and the Tie-2 receptor tyrosine kinase during mouse kidney maturation. J Am Soc Nephrol 1999;10:1722–1736.
75 Kolatsi-Joannou M, et al: Expression and potential role of angiopoietins and Tie-2 in early development of the mouse metanephros. Dev Dyn 2001;222:120–126.
76 Satchell SC, et al: Human podocytes express angiopoietin 1, a potential regulator of glomerular vascular endothelial growth factor. J Am Soc Nephrol 2002;13:544–550.
77 Suri C, et al: Requisite role of angiopoietin-1, a ligand for the TIE2 receptor, during embryonic angiogenesis. Cell 1996;87:1171–1180.
78 Davis B, et al: Podocyte-specific expression of angiopoietin-2 causes proteinuria and apoptosis of glomerular endothelia. J Am Soc Nephrol 2007;18:2320–2329.
79 Partanen J, et al: Cell autonomous functions of the receptor tyrosine kinase TIE in a late phase of angiogenic capillary growth and endothelial cell survival during murine development. Development 1996;122:3013–3021.
80 McCright B, et al: Defects in development of the kidney, heart and eye vasculature in mice homozygous for a hypomorphic Notch2 mutation. Development 2001;128:491–502.
81 Cheng HT, Kopan R: The role of Notch signaling in specification of podocyte and proximal tubules within the developing mouse kidney. Kidney Int 2005;68:1951–1952.
82 Niranjan T, et al: The Notch pathway in podocytes plays a role in the development of glomerular disease. Nat Med 2008;14:290–298.
83 Waters AM, et al: Ectopic notch activation in developing podocytes causes glomerulosclerosis. J Am Soc Nephrol 2008;19:1139–1157.
84 Miner JH, Building the glomerulus: a matricentric view. J Am Soc Nephrol 2005;16:857–861.
85 Miner JH, Li C: Defective glomerulogenesis in the absence of laminin alpha5 demonstrates a developmental role for the kidney glomerular basement membrane. Dev Biol 2000;217:278–289.
86 Abrass CK, et al: Abnormal development of glomerular endothelial and mesangial cells in mice with targeted disruption of the lama3 gene. Kidney Int 2006;70:1062–1071.
87 Takabatake Y, et al: The CXCL12 (SDF-1)/CXCR4 axis is essential for the development of renal vasculature. J Am Soc Nephrol 2009; 20:1714–1723.
88 Ueda H, et al: Bmp in podocytes is essential for normal glomerular capillary formation. J Am Soc Nephrol 2008;19:685–694.
89 Zoccali C: The endothelium as a target in renal diseases. J Nephrol 2007;20(Suppl 12): S39–S44.
90 Stillman IE, Karumanchi SA: The glomerular injury of preeclampsia. J Am Soc Nephrol 2007;18:2281–2284.
91 Levine RJ, et al: Circulating angiogenic factors and the risk of preeclampsia. N Engl J Med 2004;350:672–683.
92 Maynard SE, et al: Excess placental soluble fms-like tyrosine kinase 1 (sFlt1) may contribute to endothelial dysfunction, hypertension, and proteinuria in preeclampsia. J Clin Invest 2003;111:649–658.
93 Li Z, et al: Recombinant vascular endothelial growth factor 121 attenuates hypertension and improves kidney damage in a rat model of preeclampsia. Hypertension 2007;50:686–692.
94 Bergmann A, et al: Reduction of circulating soluble Flt-1 alleviates preeclampsia-like symptoms in a mouse model. J Cell Mol Med 2009, E-pub ahead of print.
95 Izzedine H, et al: Angiogenesis inhibitor therapies: focus on kidney toxicity and hypertension. Am J Kidney Dis 2007;50:203–218.
96 Gurevich F, Perazella MA: Renal effects of anti-angiogenesis therapy: update for the internist. Am J Med 2009;122:322–328.
97 Wallukat G, et al: Patients with preeclampsia develop agonistic autoantibodies against the angiotensin AT1 receptor. J Clin Invest 1999; 103:945–952.
98 Grill S, et al: Potential markers of preeclampsia – a review. Reprod Biol Endocrinol 2009; 7:70.
99 Zhou CC, et al: Angiotensin receptor agonistic autoantibodies induce preeclampsia in pregnant mice. Nat Med 2008;14:855–862.
100 Zhou CC, et al: Autoantibody from women with preeclampsia induces soluble Fms-like tyrosine kinase-1 production via angiotensin type 1 receptor and calcineurin/nuclear factor of activated T-cells signaling. Hypertension 2008;51:1010–1019.
101 Venkatesha S, et al: Soluble endoglin contributes to the pathogenesis of preeclampsia. Nat Med 2006;12:642–649.
102 Yinon Y, et al: Severe intrauterine growth restriction pregnancies have increased placental endoglin levels: hypoxic regulation via transforming growth factor-beta 3. Am J Pathol 2008;172:77–85.
103 Levine RJ, et al: Soluble endoglin and other circulating antiangiogenic factors in preeclampsia. N Engl J Med 2006;355:992–1005.
104 Fine LG, Norman JT: Chronic hypoxia as a mechanism of progression of chronic kidney diseases: from hypothesis to novel therapeutics. Kidney Int 2008;74:867–872.
105 Fine LG, Orphanides C, Norman JT: Progressive renal disease: the chronic hypoxia hypothesis. Kidney Int Suppl 1998; 65:S74–S78.
106 Bohle A, Mackensen-Haen S, Wehrmann M: Significance of postglomerular capillaries in the pathogenesis of chronic renal failure. Kidney Blood Press Res 1996;19:191–195.
107 Kang DH, et al: Impaired angiogenesis in the remnant kidney model: I. Potential role of vascular endothelial growth factor and thrombospondin-1. J Am Soc Nephrol 2001; 12:1434–1447.

108 Ohashi R, Kitamura H, Yamanaka N: Peritubular capillary injury during the progression of experimental glomerulonephritis in rats. J Am Soc Nephrol 2000;11:47–56.
109 Ohashi R, et al: Peritubular capillary regression during the progression of experimental obstructive nephropathy. J Am Soc Nephrol 2002;13:1795–1805.
110 Shimizu A, et al: Rare glomerular capillary regeneration and subsequent capillary regression with endothelial cell apoptosis in progressive glomerulonephritis. Am J Pathol 1997;151:1231–1239.
111 Kitamura H, et al: Apoptosis in glomerular endothelial cells during the development of glomerulosclerosis in the remnant-kidney model. Exp Nephrol 1998;6:328–336.
112 Rizkalla B, et al: Temporal renal expression of angiogenic growth factors and their receptors in experimental diabetes: role of the renin-angiotensin system. J Hypertens 2005; 23:153–164.
113 Yuan HT, et al: Angiopoietin correlates with glomerular capillary loss in anti-glomerular basement membrane glomerulonephritis. Kidney Int 2002;61:2078–2089.
114 Lu YH, et al: Changes in angiopoietin expression in glomeruli involved in glomerulosclerosis in rats with daunorubicin-induced nephrosis. Acta Pharmacol Sin, 2006;27:579–587.
115 Kang DH, et al: Impaired angiogenesis in the remnant kidney model: II. Vascular endothelial growth factor administration reduces renal fibrosis and stabilizes renal function. J Am Soc Nephrol 2001;12:1448–1457.
116 Futrakul N, Butthep P, Futrakul P: Altered vascular homeostasis in chronic kidney disease. Clin Hemorheol Microcirc 2008;38:201–207.
117 Futrakul N, et al: A defective angiogenesis in chronic kidney disease. Ren Fail 2008;30:215–217.
118 Choi YJ, et al: Peritubular capillary loss is associated with chronic tubulointerstitial injury in human kidney: altered expression of vascular endothelial growth factor. Hum Pathol 2000;31:1491–1497.
119 Di Marco GS, et al: The soluble VEGF receptor sFlt1 contributes to endothelial dysfunction in CKD. J Am Soc Nephrol 2009;20:2235–2245.
120 Baylis C: Nitric oxide deficiency in chronic kidney disease. Am J Physiol Renal Physiol 2008;294:F1–F9.
121 Mattson DL, et al: Effect of chronic renal medullary nitric oxide inhibition on blood pressure. Am J Physiol 1994;266: H1918–H1926.
122 Gardiner SM, et al: Nitric oxide synthase inhibitors cause sustained, but reversible, hypertension and hindquarters vasoconstriction in Brattleboro rats. Eur J Pharmacol 1992;213:449–451.
123 Shesely EG, et al: Elevated blood pressures in mice lacking endothelial nitric oxide synthase. Proc Natl Acad Sci USA 1996;93:13176–13181.
124 Huang PL, et al: Hypertension in mice lacking the gene for endothelial nitric oxide synthase. Nature 1995;377:239–242.
125 Herrera M, Garvin JL: Recent advances in the regulation of nitric oxide in the kidney. Hypertension 2005;45:1062–1067.
126 Passauer J, et al: Reduced agonist-induced endothelium-dependent vasodilation in uremia is attributable to an impairment of vascular nitric oxide. J Am Soc Nephrol 2005;16:959–965.
127 Zoccali C: Asymmetric dimethylarginine (ADMA): a cardiovascular and renal risk factor on the move. J Hypertens 2006;24:611–619.
128 Kielstein JT, et al: Cardiovascular effects of systemic nitric oxide synthase inhibition with asymmetrical dimethylarginine in humans. Circulation 2004;109:172–177.
129 Xiao S, et al: Circulating endothelial nitric oxide synthase inhibitory factor in some patients with chronic renal disease. Kidney Int 2001;59:1466–1472.
130 Fliser D, et al: Asymmetric dimethylarginine and progression of chronic kidney disease: the mild to moderate kidney disease study. J Am Soc Nephrol 2005;16:2456–2461.
131 Ravani P, et al: Asymmetrical dimethylarginine predicts progression to dialysis and death in patients with chronic kidney disease: a competing risks modeling approach. J Am Soc Nephrol 2005;16:2449–2455.
132 Matsuguma K, et al: Molecular mechanism for elevation of asymmetric dimethylarginine and its role for hypertension in chronic kidney disease. J Am Soc Nephrol 2006;17:2176–2183.
133 O'Riordan E, et al: Chronic NOS inhibition actuates endothelial-mesenchymal transformation. Am J Physiol Heart Circ Physiol 2007;292:H285–H294.
134 Singh P, et al: The balance of angiotensin II and nitric oxide in kidney diseases. Curr Opin Nephrol Hypertens 2008;17:51–56.
135 Joles JA, Willekes-Koolschijn N, Koomans HA: Hypoalbuminemia causes high blood viscosity by increasing red cell lysophosphatidylcholine. Kidney Int 1997;52:761–770.
136 Kang DH, et al: Role of the microvascular endothelium in progressive renal disease. J Am Soc Nephrol 2002;13:806–816.
137 Norman JT, et al: Angiotensin II blockade augments renal cortical microvascular pO2 indicating a novel, potentially renoprotective action. Nephron Physiol 2003;94:p39–p46.
138 Kang DH, et al: Nitric oxide modulates vascular disease in the remnant kidney model. Am J Pathol 2002;161:239–248.
139 Zhou X, Frohlich ED: Ageing, hypertension and the kidney: new data on an old problem. Nephrol Dial Transplant 2003;18:1442–1445.
140 Glassock RJ, Winearls C: Ageing and the glomerular filtration rate: truths and consequences. Trans Am Clin Climatol Assoc 2009;120:419–428.
141 Futrakul N, Futrakul P: Renal microvascular disease in an aging population: a reversible process? Ren Fail 2008;30:353–356.
142 Baylis C: Changes in renal hemodynamics and structure in the aging kidney; sexual dimorphism and the nitric oxide system. Exp Gerontol 2005;40:271–278.
143 Thomas SE, et al: Tubulointerstitial disease in aging: evidence for underlying peritubular capillary damage, a potential role for renal ischemia. J Am Soc Nephrol 1998;9:231–242.
144 McLachlan MS: The ageing kidney. Lancet 1978;2:143–145.
145 Kang DH, et al: Impaired angiogenesis in the aging kidney: vascular endothelial growth factor and thrombospondin-1 in renal disease. Am J Kidney Dis 2001;37:601–611.
146 Shimada T, et al: Angiogenesis and vasculogenesis are impaired in the precocious-aging klotho mouse. Circulation 2004;110:1148–1155.

147 Xiong Y, et al: Elevated serum endogenous inhibitor of nitric oxide synthase and endothelial dysfunction in aged rats. Clin Exp Pharmacol Physiol 2001;28:842–847.
148 Lu R, et al: Effect of age on bone mineral density and the serum concentration of endogenous nitric oxide synthase inhibitors in rats. Comp Med 2002;52:224–228.
149 Kielstein JT, et al: Asymmetric dimethylarginine, blood pressure, and renal perfusion in elderly subjects. Circulation 2003;107:1891–1895.
150 Forte P, et al: Evidence for a difference in nitric oxide biosynthesis between healthy women and men. Hypertension 1998;32:730–734.
151 Baylis C: Sexual dimorphism in the aging kidney: differences in the nitric oxide system. Nat Rev Nephrol 2009;5:384–396.
152 Reckelhoff JF, et al: Gender differences in the renal nitric oxide (NO) system: dissociation between expression of endothelial NO synthase and renal hemodynamic response to NO synthase inhibition. Am J Hypertens 1998;11:97–104.
153 Racusen LC, et al: Antibody-mediated rejection criteria – an addition to the Banff 97 classification of renal allograft rejection. Am J Transplant 2003;3:708–714.
154 Colvin RB, Smith RN: Antibody-mediated organ-allograft rejection. Nat Rev Immunol 2005;5:807–817.
155 Truong LD, et al: Acute antibody-mediated rejection of renal transplant: pathogenetic and diagnostic considerations. Arch Pathol Lab Med 2007;131:1200–1208.
156 Sis B, Halloran PF: Endothelial transcripts uncover a previously unknown phenotype: C4d-negative antibody-mediated rejection. Curr Opin Organ Transplant 2010;15:42–48.
157 Sis B, et al: Endothelial gene expression in kidney transplants with alloantibody indicates antibody-mediated damage despite lack of C4d staining. Am J Transplant 2009;9:2312–2323.
158 Solez K, et al: Banff 07 classification of renal allograft pathology: updates and future directions. Am J Transplant 2008;8:753–760.
159 Gloor JM, et al: Transplant glomerulopathy: subclinical incidence and association with alloantibody. Am J Transplant 2007;7:2124–2132.
160 Nankivell BJ, et al: Evolution and pathophysiology of renal-transplant glomerulosclerosis. Transplantation 2004;78:461–468.
161 Fotheringham J, Angel CA, McKane W: Transplant glomerulopathy: morphology, associations and mechanism. Nephron Clin Pract 2009;113:p. c1–c7, discussion c7.

Zoltan G. Laszik, MD, PhD
Clinical Professor of Pathology
University of California San Francisco
513 Parnassus Avenue, S-566
San Francisco, CA 94143-0102 (USA)
Tel. +1 415-502-8230, Fax +1 415-514-3403, E-Mail zoltan.laszik@ucsfmedctr.org

Herrera GA (ed): Experimental Models for Renal Diseases: Pathogenesis and Diagnosis.
Contrib Nephrol. Basel, Karger, 2011, vol 169, pp 73–93

Renal Interstitial Fibrosis: A Critical Evaluation of the Origin of Myofibroblasts

Jeffrey L. Barnes[a,b] · William F. Glass II[c]

[a]The Medical Research Service, Audie Murphy Memorial Veterans Administration Hospital, South Texas Veterans Health Care System, [b]The Department of Medicine, Division of Nephrology, The University of Texas Health Science Center, San Antonio, Tex., and [c]Nephrocor® and Bostwick Laboratories Inc., Glen Allen, Va., USA

Abstract

Interstitial fibrosis is a common feature of chronic kidney disease, regardless of the etiology of the primary renal syndrome. Moreover, interstitial fibrosis is the strongest morphologic predictor of clinical outcome and is most tightly linked to progression of disease, even though the primary disease may be of glomerular origin. Also, the presence of an interstitial component in chronic kidney disease often indicates an ominous outcome. The incidence of chronic kidney disease has reached epidemic proportions; thus, understanding the pathophysiology of interstitial renal disease is paramount for the development of new therapeutic approaches to slowing progression. Experimental models of renal fibrosis have been essential in determining the mechanisms and sequence of progression of fibrogenesis, including the roles of endothelium, infiltrating leukocytes and myofibroblasts; of profibrogenic cytokine and growth factor release; of tubular injury and atrophy; and of interstitial extracellular matrix accumulation. Over the last decade, considerable attention has been paid to the origin of the myofibroblast, the cell type most responsible for interstitial matrix accumulation. However, the origin(s) of myofibroblasts and how they gain access to the peritubular interstitium has become a matter of controversy and debate. Interstitial myofibroblasts have been proposed to originate from one or more of five sources: resident fibroblasts (pericytes), adventitial fibroblasts, circulating fibrocytes, tubular epithelial-mesenchymal transition (EMT) or endothelial/mesenchymal transition. To date, EMT has become the pre-eminent theory of the origin of myofibroblasts; however, a role for EMT in renal fibrosis is rigorously challenged by anecdotal and recent scientific evidence. This chapter will briefly discuss the experimental models used to explore interstitial renal disease in general and will then focus on the controversy related to the origin of myofibroblasts.

Experimental Models of Renal Interstitial Fibrosis

Over the last four decades, numerous experimental models of interstitial fibrosis have been developed and utilized as tools to gain knowledge about the pathophysiological mechanisms of this devastating complication of diverse renal diseases. Interstitial fibrosis in experimental animals, as in humans, is a common feature of chronic renal disease, regardless of causality. Experimental models of renal fibrosis have been essential in determining the mechanisms and the sequence of progression of fibrogenesis including roles of endothelium, infiltrating leukocytes and myofibroblasts; of profibrogenic cytokine and growth factor release; of tubular injury and atrophy; and of interstitial extracellular matrix accumulation (fig. 1a, b). Investigations using myriad models have allowed the general

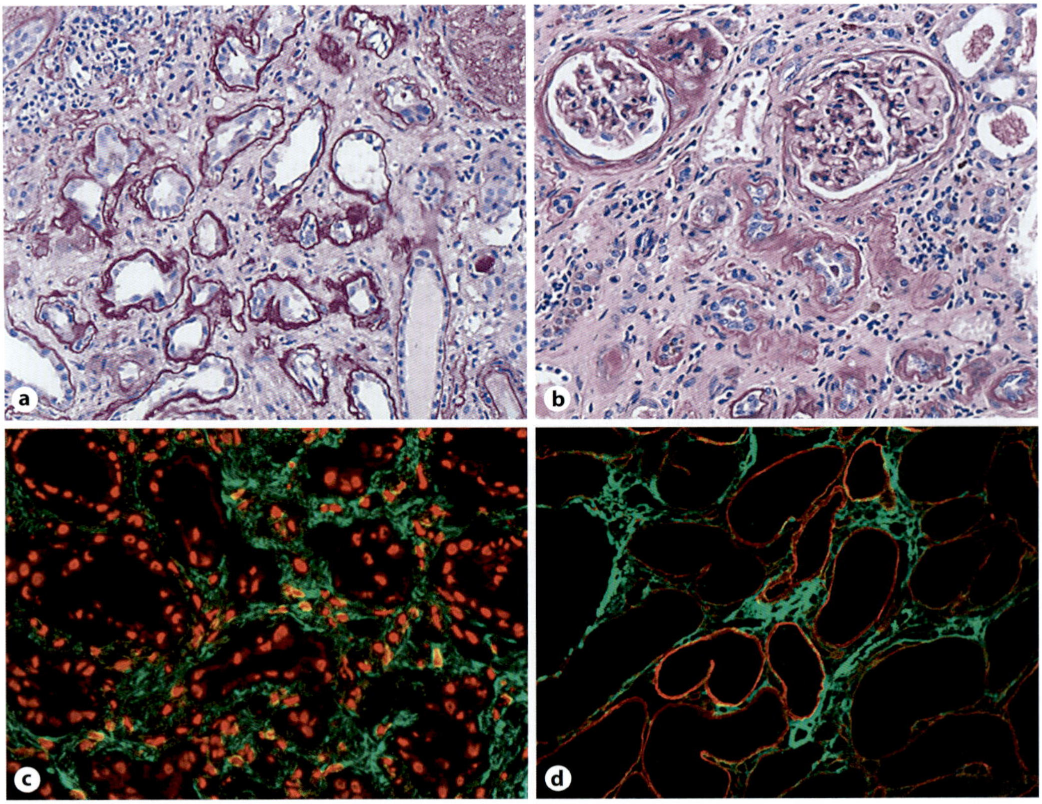

Fig. 1. Features of renal interstitial fibrosis in human (**a**, **c**) and experimental kidney disease in the rat (**b**, **d**). PAS-stained sections of interstitial fibrosis in human focal segmental glomerulosclerosis (**a**) and adriamycin-induced renal disease in the rat (**b**) show very similar pathologic changes including increased interstitial cellularity and matrix expansion, tubular atrophy, and basement membrane thickening. Myofibroblasts (green) are illustrated by immunofluorescence microscopy to be exclusively in the interstitial compartment in human (lupus nephritis, **c**) and 4 days after UUO in the rat (**d**). Nuclei and the tubular basement membrane are depicted as reference structures in red fluorescence in **c** and **d**, respectively. Tubular epithelium does not stain for the myofibroblast marker in either case.

description of a common sequence of progression of fibrogenesis in three phases [1]. An induction phase characterized by infiltration of mononuclear cells and release of profibrogenic cytokines is followed in the second phase by activation of fibroblasts to myofibroblasts and synthesis of extracellular matrix. The third phase is an ongoing matrix synthesis despite resolution of the primary stimulus. The models listed in tables 1 and 2 were adopted to provide a means to study renal fibrosis as a consequence of a particular disease type or examine specific scientific problems. Although, no one model can address all the scientific questions concerning this disease process, the knowledge gained from the composite use of these models has provided incalculable information for our understanding of the mechanisms of the progression of renal disease in the laboratory and clinical settings.

A large number of experimental models of renal disease have interstitial fibrosis as a primary consequence of an initiating event or genetic sequelae, or secondarily as a result of glomerular injury and proteinuria (table 1). 'Natural or spontaneous' chronic renal disease with consequent interstitial fibrosis is a well-known age-related occurrence in humans and animals [2–4]. Age-related renal disease is protracted, but accelerated models have become available by the development of strains naturally predisposed to hypertension and glomerular disease, two common factors responsible for secondary interstitial renal disease as observed in humans. Also, glomerular disease and interstitial fibrosis in most models can be accelerated by superimposing an additional disease or by the administration of angiotensin II, dietary salt, or reduction of renal mass via uninephrectomy or surgical ablation (table 1 and 2).

Proteinuria associated with glomerular diseases has also been utilized regularly as a means to induce renal fibrosis. Such models include direct nephrotoxic antibodies targeted to glomerular basement membrane or cellular targets including the mesangium or epithelium (tables 1). Also, agents that are cytotoxic to glomerular epithelial cells, such as aminoglycosides, lead to severe proteinuria and marked interstitial fibrosis (fig. 1b). Such models with marked proteinuria helped lead to the creation of the tubulocentric theory, which postulates that protein is toxic to tubular epithelial cells. This has resulted in fibrogenic responses and has spawned models of fibrosis using albumin overload with minimal glomerular involvement [5].

Models of primary interstitial fibrosis with little glomerular involvement generally involve administration of nephrotoxic chemicals including heavy metals, or analgesics; a few of which are listed in table 1. However, the most extensively used model to date is the unilateral ureteral obstruction (UUO) model involving ligation of a single ureter as first described in rabbits by Nagle et al. [6] and comprehensively reviewed by Chevalier et al. [7].

More specific interstitial fibrosis models rely on approaches to disease mechanisms that utilize genetic manipulation via breeding which mimic phenotypic and pathologic features of diabetes, lupus nephritis and polycystic renal disease, all of which have interstitial fibrosis as a natural history of the disease. More modern transgenic mouse gene deletion or overexpression provides tools to assess the roles of specific gene products in pathogenic pathways leading to renal fibrosis (table 2). Valuable data will continue to be garnered from genetic models that target selected fibrogenic pathways, such as RAS, growth factors, matrix turnover, etc. Fibrosis in these models is implemented through the natural course of the disease or may require acceleration by superimposing selected additional models, generally the UUO model (table 2).

Mechanisms

Interstitial fibrosis may occur diffusely [104], with or without atrophic tubules, or focally, associated with atrophic tubules (complicated fibrosis). Simple fibrosis may potentially be reversible since the underlying tubular and glomerular architecture are preserved. However atrophic tubules are generally associated with local irreversible injury to the associated nephrons. Tubular atrophy can be due either to glomerular injury or to discrete injury along the tubule. Renal injury can progress to the point at which it is best described as cortical scarring characterized by fibrosis with destruction, loss and replacement of tubular elements by fibrosis. Any of these states can occur accompanied by or preceded by interstitial inflammation with a variety of cells including myofibroblasts and inflammatory cells, such as macrophages and lymphocytes. Viewed in this manner, tubulointerstitial fibrosis is analogous to classic wound healing by primary and secondary intention whereby the former can heal with little or no scarring or functional loss, but the latter is associated with both scarring and loss of local functional adnexal units.

The roles inflammatory, tubular and vascular cells play in renal fibrosis are complex and experimental models have been instrumental in the development of a fund of knowledge on mechanisms of renal fibrosis as described above. An enormous amount of data on roles for macrophages, myofibroblasts, profibrogenic mediators such as TGF-β1, matrix accumulation, plasminogen activators/inhibitors and metalloproteinase

Table 1. Models of Renal Interstitial Fibrosis

Model	Reference	Accelerated model	Reference
Spontaneous disease			
Age-related fibrosis	[2-4]	uninephrectomy	[8]
		dietary protein	[9]
Milan normotensive, aged	[10]		
Munich Wistar Frompter	[11]		
CD1 strain mice (T1 diabetes)		streptozotocin	[12]
db/db mice (T2 diabetes)		uninephrectomy	[13]
MRL/*lpr* mice (lupus)	[14]	L-arginine supplementation	[15]
Han:SPRD-cy rats polycystic kidney	[16]		
diphenylthiazole-induced PKD	[17]		
Surgically induced			
Ablation (5/6 nephrectomy)	[18, 19]	Heymann nephritis	[20]
		anti-HGF neutralizing antibody	[21]
Two-kidney, one-clip	[22]		
UUO	[6]	transgenic models (see below)	
Warm ischemia		uninephrectomy	[23]
Micro-embolism		uninephrectomy	[24]
Experimental renal infection	[25]		
Primary glomerular (proteinuria-inducing)			
Anti-GBM (NTS)	[26]		
Heymann nephritis	[27]	ablation	[20]
Puromycin aminoglycoside	[28]	protamine	[29]
		uninephrectomy	[30]
Adriamycin	[31]	uninephrectomy	[32]
		high-dose, one-kidney	[33]
		dietary salt	[34]
Protein overload	[5]	uninephrectomy	[5]
Anti-Thy1		multiple doses	[35]
		uninephrectomy	[36]
		L-arginine supplementation	[37]

Table 1. Continued

Model	Reference	Accelerated model	Reference
Hypertensive			
Goldblatt			
DOCA/salt		angiotensin II	[38]
Angiotensin II infusion	[39]	habu venom + uninephrectomy	[40]
Spontaneous hypertensive rat (SHR)	[41]	uninephrectomy adriamycin dietary salt	[42] [43] [44]
Stroke-prone spontaneously hypertensive rats, subset of SHR	[45]	all infusion, hypertension, dietary salt aldosterone	[45]
Dahl S rats		dietary salt	[46]
Aldosterone		uninephrectomy, dietary salt	[47]
NaCl overload	[48]		
Growth factor-induced			
TGF-2 (mouse and rat)	[49, 50]		
PDGF BB	[51]		
Nephrotoxic chemicals			
Cadmium	[52]		
Lithium (rabbit)	[53]		
Gold (guinea pig)	[54]		
Uranyl nitrate	[55]		
Analgesics		Gunn strain of rats	[56]
Cyclosporine	[57]	uninephrectomy + salt depletion	[58]
Aristolochic acid	[59]	aristolochic acid + salt depletion	[60]
Diet-induced hypercholesterolemia		uninephrectomy	[61]
Bromoethylamine hydrobromide	[62]		

Models first developed or applied to the investigation of renal fibrosis. Models are of rats unless noted.

Table 2. Genetically engineered models of renal interstitial fibrosis

Transgenic model	Reference	Unmasked or accelerated disease	Reference
Stathmin-deficient (OP18–/–)		+ ischemia	[63]
OVE26 T1 diabetic calmodulin	[64]		
Renin TGR (mRen2) overexpression (rats)	[65]	+ diabetes	[66]
COL4A3-deficiency	[67, 68]		
PAI-1 overexpression		+ UUO	[69]
TGF-β overexpression	[70, 71]		
Relaxin deficiency	[72]		
SAMP1/Sku overexpression		+ UUO	[73]
α-SMA deficiency		+ UUO	[74]
TIMP-1 overexpression (age-related)	[75]	+ UUO	[76]
TIMP-3 deficiency	[77]	+ UUO	[77, 78]
Snail-1 overexpression	[79]		
PTHrp overexpression		+ UUO + folic acid	[80] [81]
RAGE/iNOS overexpression		+ megsin overexpression	[82]
human VEGF overexpression (rabbits)	[83]		
MMP-2 over expression	[84]		
Hemoxygenase-1 (HO-1) deficiency		+ UUO	[85]
eNOS deficiency		+ diabetes + ablation	[86] [87]
Adiponectin deficiency		+ ablation	[88]
Crry$^{-/-}$*C3*$^{-/-}$ mouse kidneys were transplanted into syngeneic C57BL/6 wild-type mice	[89]		
Calcineurin-α–/– and aged α+/	[90]		
PKC-ε deficiency	[91]	+ streptozotocin	[91]
TNF-α-deficiency		+ UUO	[92]
VHL (HIF-1α) deficiency (age-related)	[93]	+ ablation	[93]
All receptor deficiency		+ UUO	[94]
Cux-1 overexpression	[95]		
Cox-2 deficiency	[96]		

Table 2. Continued

Transgenic model	Reference	Unmasked or accelerated disease	Reference
uPAR deficiency		+ UUO	[97]
Decorin deficiency		+ UUO	[98]
Bradykinin B2 receptor deficiency		+ UUO	[99]
Endothelin-1 over expression	[100]		
mIL-8Rh deficiency		Inoculation with *E. coli*	[101]
C3a receptor deficiency		MRL lupus model	[102]
Nephron-deficient Fvb (FvbROP Os/+)	[103]		

Models first developed or applied to the investigation of renal fibrosis. Models are of mice unless noted.

pathways, etc., has been gleaned from the experimental models listed in tables 1 and 2. These topics are reviewed extensively in several excellent publications [7, 104–106] and are beyond the scope of this chapter.

There is common agreement that the myofibroblast is the cell most responsible for matrix synthesis and accumulation leading to interstitial expansion. To understand mechanisms of renal fibrosis, it is important to first understand the origins of myofibroblasts, or 'from where do kidney fibroblasts come?' The simple concept that myofibroblasts are derived from intrinsic 'resting' fibroblasts residing within the peritubular interstitium has evolved into a much more complex and controversial area of investigation. Because of the developing controversy in this area, particularly concerning the process of epithelial-mesenchymal transition (EMT), the majority of this chapter will be devoted to this topic.

The Origin(s) of Myofibroblasts

The origin of myofibroblasts is currently under extensive investigation and scientific reports indicate the cells may be derived from one or more of at least five sources. These sources include expansion of activated resident fibroblasts, expansion of perivascular adventitial fibroblasts, infiltration of circulating bone marrow-derived fibrocytes, tubular EMT or endothelial-mesenchymal transition, each involving and migrating into the peritubular interstitial space. Evidence for each source will be discussed below followed by a critical evaluation of EMT.

Resident Fibroblasts

In seminal studies, Gabbanni et al. [107, 108] described that after injury, or during wound healing, or other fibrotic events, activated fibroblasts transform into myofibroblasts characterized by the acquisition of smooth muscle features. Myofibroblasts express a variety of smooth muscle proteins, such as desmin, caldesmon and SM-myosin heavy chains, and an actin isoform-specific in normal tissues for the smooth muscle phenotype (α-smooth muscle actin, α-SMA) [107–109]. The introduction of specific antibodies against α-SMA enabled identification in tissue of the myofibroblast as the primary cell type responsible for interstitial matrix accumulation in fibrotic diseases in a variety of organs [110, 111], including the kidney [111–115]. Moreover, α-SMA as a marker for myofibroblasts has been described as a reliable tool for experimental renal fibrosis (fig. 1c, d) and has clinical prognostic value in progressive renal diseases [116–118]. Indeed, a large number of studies using a wide variety of models have reported that myofibroblasts are located in the peritubular interstitial space without providing any descriptive or illustrative evidence of tubular sources [19, 39, 40, 71, 119–133]. In addition, mesenchymal cell-derived extracellular matrix proteins, such as cellular fibronectin and collagens type I or III, have been observed exclusively in interstitial spaces in many of these same of models of interstitial renal disease [40, 71, 121, 122, 124–127, 130, 134, 135] as well as in human nephropathy [114, 121, 136–138].

Several contemporary studies refer location of myofibroblasts exclusively to the interstitial compartment [71, 128, 130, 132, 139–144]. Moreover, a number of studies deliberately focusing on the origin of myofibroblasts report the exclusive expression of myofibroblasts in the peritubular interstitium [40, 71, 123, 130–133, 143]. Transgenic models utilizing eloquent gene reporter strategies to define specific myofibroblast lineages determined that these cells are resident fibroblasts derived from pericytes located exclusively in the interstitium and not derived from an epithelial source [123, 124] (see EMT below). In total, the above studies indicate that matrix secreting interstitial myofibroblasts are derived through activation of resident fibroblasts or pericytes.

Adventitial Fibroblasts

Observations of excess deposition of extracellular matrix proteins in the perivascular spaces early during fibrosis in several organs, including the kidney (below), suggested an adventitial origin for fibroblasts [145, 146]. For example, morphometric analysis indicates that myofibroblasts and the deposition of collagen type I or III, and/or fibronectin, first begin in the perivascular spaces of arteries or arterioles in several experimental models of renal fibrosis, including anti-GBM disease [147–149]; two-kidney, one-clip model of hypertensive nephrosclerosis [150]; after renal infection [25]; reduction in renal mass [151]; UUO [123]; and accelerated fibrosis induced by habu venom plus angiotensin II infusion [40]. Similarly, myofibroblasts and expression of mesenchymal matrix proteins was first observed adjacent to vasculature in human nephropathy [152]. Studies intentionally designed to determine the origin of interstitial cells during fibrosis reported myofibroblasts and cellular fibronectin or collagen type I in pericytes and adventitial cells located within the perivascular space without evidence of tubular epithelial expression [40, 123, 124]. Collectively, the above studies suggest that either vascular injury or vascular factors are the most likely triggers for pericyte migration from perivascular regions and differentiation into myofibroblasts in the interstitium.

Extrarenal Fibrocytes

Accumulating evidence suggests that perivascular and interstitial myofibroblasts may additionally be derived from circulating bone marrow fibrocytes and participate in organ fibrosis, including the kidney (reviewed by Wada et al [153]). Iwano et al. [169] reported that 15% of interstitial collagen-producing cells during UUO were derived from CD34-negative fibroblasts originating from the bone marrow. Similarly, CD45- and type I collagen-dual-positive fibrocytes were shown to localize in the interstitium in this same model, particularly in the corticomedullary region [154, 155]. Larger numbers of interstitial myofibroblasts (up to 30%) were attributed to bone marrow-derived cells after ischemia/reperfusion injury [156, 157] and in adriamycin nephropathy [158]. Similarly, bone marrow-derived fibrocytes have been implicated in interstitial disease in humans [159, 160]. A role for bone marrow-derived myofibroblasts, however, is diminished in other studies where fibrocytes were not reported to have made any significant contribution to interstitial fibrosis in the UUO model [123, 161]; rather, fate-mapping experiments indicate that the major source of interstitial collagen-producing cells is from perivascular fibroblasts that may derive from myofibroblast progenitors (see above [124]).

Epithelial-Mesenchymal Transition

The hypothetical process of EMT is based on reversal of a developmental process, known as mesenchymal-epithelial transition, whereby embryonic mesenchymal cells differentiate into an epithelial phenotype. This process is essential for developmental renal epithelialization during renal tubulogenesis [162]. The EMT hypothesis in renal disease assumes that mechanisms exist to reverse the mesenchymal-epithelial transition process and that during tubular injury, epithelial cells lose their differentiated epithelial phenotype, repress epithelial markers such as E-cadherin and cytokeratins, and lose polarity and adhesion. The cells acquire mesenchymal features and express myofibroblast markers including α-SMA, vimentin, calmodulin and fibroblast-specific protein-1 (FSP-1, see below). The newly transformed cells are assumed to acquire an invasive phenotype where they synthesize matrix proteinases that digest the tubular basement membrane and increase mobility, thereby allowing their migration into the peritubular interstitium. Once in the interstitial compartment, the cells proliferate and synthesize extracellular matrix proteins generally attributed to mesenchymal cells, including collagens I and III and cellular fibronectin (for comprehensive reviews on this topic see [1, 162–167]).

The hypothesis that fibroblasts arise from the local conversion of epithelium was originally suggested by Hay and Zuk [162] and corroborated experimentally by Strutz et al. [168] after the discovery of FSP-1 and its demonstration in epithelial cells adjacent to an inflammatory process during tubulointerstitial nephritis in SJL/J mice. Interest in EMT accelerated after the classic report by Iwano et al. [169], suggesting that FSP-1+ fibroblasts also arise in large numbers by local EMT during UUO-induced renal fibrogenesis. With the aid of transgenic reporter mice programmed to produce a tubular epithelial marker and the use of dual-label fluorescence to detect FSP-1, the authors tracked cells expressing both markers to the tubular and interstitial compartments during fibrosis. LacZ/FSP1-positive cells expressed in 36% of tubules co-expressed tubular FSP-1. Additionally, genetically tagged tubular cells were observed to express FSP-1 as well as HSP47, a surrogate marker of collagen synthesis, in the interstitium. Since these seminal studies, EMT has been implicated as the major contributor to fibrosis in a number of renal diseases, including genetically engineered animals [79, 84, 85, 170–172], UUO [169, 173–181], renal ablation [182–184], diabetic nephropathy [185–187], nephrotoxic serum nephritis [188] and polycystic kidney disease [189]. These studies have explored various aspects of EMT, including myofibroblast origin [169, 182, 189], roles for fibrogenic growth factors [174, 178, 187, 188], integrins [175, 186], metalloproteinases [84, 181, 184], apoptosis [177], advanced glycation products [187], signaling pathways [173] and therapeutic manipulation [174, 177–180, 188]. Also, since the initial description of EMT in renal disease, a plethora of reports have described EMT in vitro, a discussion which is beyond the scope of this chapter. Similar to EMT, recent investigations have reported a process whereby endothelial cells transition to mesenchymal cells and occupy the interstitium in several models of renal fibrosis [190, 191], implicating considerable plasticity in resident renal cells during the course of fibrosis.

Growing Evidence Against the EMT Hypothesis

At the time this chapter was written, a Medline search revealed over 100 journal articles, text book chapters, editorials and reports devoted to EMT in renal fibrosis. Many, if not most, of these publications firmly, if not dogmatically, assert EMT as the pre-eminent mode of renal interstitial encroachment and the process a foregone conclusion. However, of all proposed origins of interstitial fibroblasts, EMT has become the most controversial and challenged mode of interstitial encroachment [192]. Our review of past and current scientific evidence has led us to the strongly held opinion that EMT as a source for interstitial myofibroblasts during kidney fibrosis has been broadly overstated and that the evidence seriously challenges the existence of this mode of interstitial encroachment (see below).

The process of EMT would require a highly orchestrated sequence of several teleologically contradictory events involving loss of epithelial phenotype, acquisition of a mesenchymal phenotype, release of matrix-degrading metalloproteinases, upregulation of basement membrane matrix proteins, and cell migration through a thickening tubular basement membrane and emergence as a myofibroblast with consequent synthesis of mesenchymal matrix proteins in the interstitium. Whereas, aside from what we find to be more compelling evidence, it is mechanistically much simpler for resident fibroblasts and perhaps circulating fibrocytes to assume this role. On a broader scale, the reality of EMT as a mode of cell motility in cellular metastasis during neoplasia has also come under a cloud of uncertainty [193, 194]. A counterpoint assessment of EMT as a mode of interstitial encroachment in renal fibrosis is provided below.

Absence of Evidence for EMT during Fibrosis: Herein the Controversy

Lack of Epithelial Markers in the Interstitial Compartment

Indicators of EMT are the loss of specialized epithelial proteins such as E-cadherin, cytokeratin and the acquisition of mesenchymal markers, including vimentin, α-SMA and calmodulin. Numerous studies show a reduction in cytoplasmic epithelial proteins in tubules during renal fibrosis. However, few reports show cells with residual expression of epithelial markers in the interstitium [169, 170], indicating (at least) that if cells are derived from the tubules, they have completely lost their epithelial phenotype during transition prior to migrating into the interstitium. Even in an accelerated model of fibrosis where peritubular myofibroblasts are abundant in the interstitium within one day of initiation, Na+K+-ATPase-positive epithelial cells were not observed in this same location [40], indicating that EMT may not occur, or if it does, the cells rapidly degrade residual epithelial specific proteins prior to transit. Contrary to a rapid loss of epithelial phenotypic proteins during fibrosis are the observations of a newly discovered proximal tubular protein KIM-1 that is upregulated and enhanced during epithelial injury [195]. Kim-1 is abundant in tubules, but not in interstitial cells during the progression of fibrosis induced by adriamycin [196], polycystic kidney disease [197], protein overload nephropathy [198] and in human kidney disease [199, 200]. Thus, even when synthesis of an epithelial protein is revved-up during injury, KIM-1-positive cells are not observed in the peritubular interstitial compartment. How can EMT explain the complete loss of this injury-induced marker precisely as these cells transit the basement membrane into the interstitium?

Similarly, Texas Red-tagged Dextran, a functional marker of proximal tubular epithelium that is endocytosed and retained by the cells, could not be detected within the interstitium in an accelerated model of fibrosis (referred to above) [40]. A natural corollary to these functional tracking experiments is chronic renal hemosiderosis associated with hemolytic anemia that often progresses into renal interstitial fibrosis. In this disease, large amounts of hemosiderin are filtered by glomeruli and endocytosed by proximal tubular cells. Retrospective examination of the literature indicates that hemosiderin, detected by Prussian blue staining, is exclusively observed in the tubular compartment and absent in interstitial fibroblasts [201–203]. That the proximal tubular cells 'jettison' all residual epithelial phenotypic and functional markers prior to escape into the interstitium is very unlikely. An alternative explanation may be that to the extent that EMT occurs, it is a local and stationary intratubular phenomenon where the cells transition to a mesenchymal phenotype, but do not contribute to the interstitial population. A lack of epithelial migration into the peritubular interstitium during the course of renal fibrosis was recently confirmed by Humphreys et al. [124], who developed transgenic mice that express protein labels exclusively in renal tubular epithelial cells. Tracking experiments in the UUO and ischemia models indicated that epithelial cells derived from proximal tubules or collecting ducts did not migrate into the interstitium, even in the presence of gaps in the tubular basement membrane [124].

Absence of Evidence of an Epithelial Conversion to a Mesenchymal Phenotype

Numerous and careful studies that pre-date the EMT hypothesis indicate that myofibroblasts and their matrix proteins are exclusively located within the interstitial spaces (see section above). Although a number of studies have described α-SMA and/or mesenchymal matrix proteins in tubular epithelium (extensively reviewed in references 163–166), other studies published since EMT was 'established' are contradictory to this process and describe α-SMA-positive myofibroblasts exclusively within the interstitial space [71, 128, 130, 132, 139–144, 196]. Moreover, studies designed to examine the origin of myofibroblasts reported the appearance of myofibroblasts exclusively in the peritubular interstitium without evidence for EMT in several fibrosis models, including nephropathy induced by thiazide [131, 133], uranyl nitrate [142], aristolochic acid [143], UUO [123, 124, 132, 204], angiotensin II [40] and TGF-β overexpression [71]. The above observations were corroborated by Dr. Wilhelm Kriz after evaluation of human biopsies and multiple animal models demonstrating that tubular injury is invariably associated with injury to the attached glomerulus and does not propagate from tubule to tubule or from interstitium to tubule [205–207]. Also, transgenic mice that express protein labels exclusively in renal tubular epithelial cells (see above) do not express mesenchymal markers including α-SMA or FSP-1 in UUO and ischemia models, indicating that epithelial cells do not transition into myofibroblast and migrate into the interstitium [124].

Note added in proof: The authors direct the reader to three important articles published after submission of this chapter: a point-by point debate on the existence of EMT in renal fibrosis [228] and two fate mapping studies showing lack of evidence for EMT in separate models of experimental renal disease [229, 230]. Li et al. [229] followed the fate of proximal tubular epithelial cells using genetic tubular markers and fluorescent dextran loading corroborating our earlier studies [40]. Similarly, Koesters et al. [230] showed no evidence of genetically tagged tubular epithelial cells transgressing the tubular basement membrane utilizing a transgenic model of tubular overexpression of TGF-b1. Interestingly, both of these recent studies indicate that tubular epithelial cells undergo autophagy and do not leave the tubular compartment [229, 230].

FSP-1 as a Marker for EMT

Despite the staining characteristics of α-SMA in renal fibrosis, FSP-1 has been purported to be 'the best, and the most consistent marker of fibroblasts' in tissue because 'not all fibroblasts express α-SMA' [168, 169]. Additionally, α-SMA is not expressed in all FSP-1+ fibroblasts [189, 208], thus suggesting that this contractile protein does not define the 'universe of fibroblasts' and a significant number of interstitial cells are most likely derived from tubular epithelium undergoing EMT [163, 168, 169]. Indeed, since its original description, FSP-1 expression in injured proximal tubules, with or without co-localization of mesenchymal markers, has been taken as evidence of EMT [85, 93, 163, 165, 168, 169, 189, 208–211]. Conversely, other investigators have shown FSP-1 to be a poor marker for collagen I-producing cells [71, 123] and to not be associated with an epithelial cell phenotype during renal fibrosis [71, 123, 124, 131, 132, 204, 212]. This questions the existence of EMT as a mode of origin of myofibroblast encroachment in the pathogenesis of kidney disease. Alternatively, FSP-1 was identified in monocytes and macrophages in the renal interstitial space [123, 124, 131, 132, 204, 212], questioning the validity of this protein as a specific marker for fibroblasts as the name implies.

FSP-1 belongs to the S-100 superfamily of small Ca^{2+}-binding proteins implicated in several cellular and extracellular functions, including cytoskeletal-membrane interactions; Ca^{2+} signal transduction; cell growth, differentiation and motility; cancer metastasis; and angiogenesis. FSP-1 is commonly known as S100A4 which was independently cloned by several groups under a number of additional names, including 18A2, pEL98, p9Ka, 42a, CAPL and calvasculin [213, 214]. S100A4 is additionally known to be expressed by normal mesenchymal, epithelial, endothelial cells, chondrocytes and osteoblasts, lymphoid cells, and leukocytes in a variety of organs including intestinal tract, hair follicle liver, prostate, skin and the kidney [213–216].

S100A4 is associated with a wide number of cell behaviors including metastasis, cell detachment and motility, cell proliferation and apoptosis, etc. [213, 214]. Collecting duct cells do not normally express S100A4 either, but it can be induced by adaption to osmotic hypertonicity [217], indicating that expression of this protein may represent an adaptive response to various physiological events. Thus, S100A4 is not specific to any one cell type and appears to be a cell activation marker that is not restricted to cell behaviors only associated with cell invasiveness. For example, S100A4 is associated with cell proliferation and apoptosis, two cell behaviors frequently associated with tubular injury and atrophy during tubulointerstitial renal fibrosis [24, 140].

Evidence for EMT in Human Tissues

There are few studies exploring a role for EMT in human renal disease and as in animal models, they show contradictory results. For example, tubular epithelial acquisition of α-SMA or S100A4 were described in human biopsies from allograft rejection [218–222], IgA nephropathy [208, 221, 223], obstructive nephropathy [224] and a variety of other renal diseases [221, 223]. Also, α-SMA expression in tubular epithelial cells in membranous nephropathy was correlated with areas of fibrosis where disruption of TBM was evident [225]. Most of the above studies described mesenchymal markers in small numbers of tubular cells. Nevertheless, evidence of EMT was reported to provide a valuable histologic index of progression [208, 222]. On the contrary, Vitalone et al. [226] observed that development of early allograft fibrosis correlated with subclinical immune activity rather than markers for EMT. Also, despite immunoreactivity for vimentin in atrophic tubules in diabetic nephropathy, Essawy et al. [118] was only able to observe tubular staining for α-SMA in one of 25 biopsies. Moreover, Kuusniemi et al. [227] found no evidence of EMT in a comprehensive quantitative examination of biopsies from fibrotic kidneys associated with Finnish-type nephrotic syndrome. In this study, only 5 of 84,442 tubular cross sections showed α-SMA-positive cells with similar values for collagen I, MMP2 and MMP9.

It is difficult to reconcile the above studies where scientific evidence for and against EMT may be compelling, yet contradictory. Development of new experimental models of interstitial fibrosis utilizing tracking techniques will be of help. Ultimately, reappraisal of the historical literature, careful experimental design and execution with stringent evaluation and interpretation of the results, and peer communication and scientific review will be necessary to fully resolve this controversy.

References

1 Strutz F, Muller GA: Mechanisms of renal fibrogenesis; in Neilson EG, Couser WG (eds): Immunologic Renal Diseases. Philadelphia, Lippincott Williams & Wilkins, 2001, pp 73–101.

2 Abrass CK, Adcox MJ, Raugi GJ: Aging-associated changes in renal extracellular matrix. Am J Pathol 1995;146:742–752.

3 Inserra F, Romano LA, de Cavanagh EM, Ercole L, Ferder LF, Gomez RA: Renal interstitial sclerosis in aging: effects of enalapril and nifedipine. J Am Soc Nephrol 1996;7:676–680.

4 Thomas SE, Anderson S, Gordon KL, Oyama TT, Shankland SJ, Johnson RJ: Tubulointerstitial disease in aging: evidence for underlying peritubular capillary damage, a potential role for renal ischemia. J Am Soc Nephrol 1998;9:231–242.

5 Eddy AA: Interstitial nephritis induced by protein-overload proteinuria. Am J Pathol 1989;135:719–733.

6 Nagle RB, Kneiser MR, Bulger RE, Benditt EP: Induction of smooth muscle characteristics in renal interstitial fibroblasts during obstructive nephropathy. Lab Invest 1973;29: 422–427.

7 Chevalier RL, Forbes MS, Thornhill BA: Ureteral obstruction as a model of renal interstitial fibrosis and obstructive nephropathy. Kidney Int 2009;75:1145–1152.

8 Greenfeld Z, Stillman IE, Brezis M, Rosen S: Medullary injury in the ageing rat kidney: functional-morphometric correlations. Eur J Clin Invest 1997;27:346–351.

9 Bertani T, Zoja C, Abbate M, Rossini M, Remuzzi G: Age-related nephropathy and proteinuria in rats with intact kidneys exposed to diets with different protein content. Lab Invest 1989;60:196–204.

10 Brandis A, Bianchi G, Reale E, Helmchen U, Kuhn K: Age-dependent glomerulosclerosis and proteinuria occurring in rats of the Milan normotensive strain and not in rats of the Milan hypertensive strain. Lab Invest 1986;55:234–243.

11 Schulz A, Standke D, Kovacevic L, Mostler M, Kossmehl P, Stoll M, Kreutz R: A major gene locus links early onset albuminuria with renal interstitial fibrosis in the MWF rat with polygenetic albuminuria. J Am Soc Nephrol 2003;14:3081–3089.

12 Sugimoto H, Grahovac G, Zeisberg M, Kalluri R: Renal fibrosis and glomerulosclerosis in a new mouse model of diabetic nephropathy and its regression by bone morphogenic protein-7 and advanced glycation end product inhibitors. Diabetes 2007;56: 1825–1833.

13 Ninichuk V, Kulkarni O, Clauss S, Anders HJ: Tubular atrophy, interstitial fibrosis, and inflammation in type 2 diabetic db/db mice. An accelerated model of advanced diabetic nephropathy. Eur J Med Res 2007;12:351–355.

14 Andrews BS, Eisenberg RA, Theofilopoulos AN, Izui S, Wilson CB, McConahey PJ, Murphy ED, Roths JB, Dixon FJ: Spontaneous murine lupus-like syndromes. Clinical and immunopathological manifestations in several strains. J Exp Med 1978;148:1198–1215.

15 Peters H, Border WA, Ruckert M, Kramer S, Neumayer HH, and Noble NA: L-arginine supplementation accelerates renal fibrosis and shortens life span in experimental lupus nephritis. Kidney Int 2003;63:1382–1392.

16 Cowley BD Jr, Gudapaty S, Kraybill AL, Barash BD, Harding MA, Calvet JP, Gattone VH: Autosomal-dominant polycystic kidney disease in the rat. Kidney Int 1993;43:522–534.

17 Carone FA, Butkowski RJ, Nakamura S, Polenakovic M, and Kanwar YS: Tubular basement membrane changes during induction and regression of drug-induced polycystic kidney disease. Kidney Int 1994;46: 1368–1374.

18 Tapp DC, Wortham WG, Addison JF, Hammonds DN, Barnes JL, Venkatachalam MA: Food restriction retards body growth and prevents end-stage renal pathology in remnant kidneys of rats regardless of protein intake. Lab Invest 1989;60:184–195.

19 Kliem V, Johnson RJ, Alpers CE, Yoshimura A, Couser WG, Koch KM, Floege J: Mechanisms involved in the pathogenesis of tubulointerstitial fibrosis in 5/6-nephrectomized rats. Kidney Int 1996;49: 666–678.

20 Okuda S, Oh Y, Tsuruda H, Onoyama K, Fujimi S, Omae T, Fujishima M: Rapidly progressive renal deterioration in partially nephrectomized rats with experimental membranous nephropathy. Nephron 1985;41:359–364.

21 Gong R, Rifai A, Tolbert EM, Biswas P, Centracchio JN, Dworkin LD: Hepatocyte growth factor ameliorates renal interstitial inflammation in rat remnant kidney by modulating tubular expression of macrophage chemoattractant protein-1 and RANTES. J Am Soc Nephrol 2004;15:2868–2881.

22 Mai M, Geiger H, Hilgers KF, Veelken R, Mann JF, Dammrich J, Luft FC: Early interstitial changes in hypertension-induced renal injury. Hypertension 1993;22:754–765.

23 Azuma H, Nadeau K, Takada M, Mackenzie HS, Tilney NL: Cellular and molecular predictors of chronic renal dysfunction after initial ischemia/reperfusion injury of a single kidney. Transplantation 1997;64:190–197.

24 Kimura M, Suzuki T, Hishida A: A rat model of progressive chronic renal failure produced by microembolism. Am J Pathol 1999;155:1371–1380.

25 Hewitson TD, Darby IA, Bisucci T, Jones CL, Becker GJ: Evolution of tubulointerstitial fibrosis in experimental renal infection and scarring. J Am Soc Nephrol 1998;9:6326–6342.

26 Eddy AA: Tubulointerstitial nephritis during the heterologous phase of nephrotoxic serum nephritis. Nephron 1991:59:304–313.

27 Eddy AA, Ho GC, Thorner PS: The contribution of antibody-mediated cytotoxicity and immune-complex formation to tubulointerstitial disease in passive Heymann nephritis. Clin Immunol Immunopathol 1992;62:42–55.

28 Eddy AA, Michael AF: Acute tubulointerstitial nephritis associated with aminonucleoside nephrosis. Kidney Int 1988;33:14–23.

29 Saito T, Sumithran E, Glasgow EF, Atkins RC: The enhancement of aminonucleoside nephrosis by the co-administration of protamine. Kidney Int 1987;32:691–699.

30 Jones CL, Buch S, Post M, McCulloch L, Liu E, Eddy AA: Pathogenesis of interstitial fibrosis in chronic purine aminonucleoside nephrosis. Kidney Int 1991;40:1020–1031.

31 Bertani T, Rocchi G, Sacchi G, Mecca G, Remuzzi G: Adriamycin-induced glomerulosclerosis in the rat. Am J Kidney Dis 1986;7:12–19.

32 Chagnac A, Korzets A, Ben-Bassat M, Zevin D, Hirsh J, Meckler J, Levi J: Uninephrectomy aggravates tubulointerstitial injury in rats with adriamycin nephrosis. Nephron 1994;66:176–180.

33 Bertani T, Cutillo F, Zoja C, Broggini M, Remuzzi G: Tubulo-interstitial lesions mediate renal damage in adriamycin glomerulopathy. Kidney Int 1986;30:488–496.

34 Okuda S, Tamaki K, Ando T, Nagashima A, Nakayama M, Fukuda K, Higashi H, Fujishima M: Increased expression of Na+/H+ exchanger in the injured renal tissues of focal glomerulosclerosis in rats. Kidney Int 1994;46:1635–1643.

35 Yamamoto T, Noble NA, Miller DE, Border WA: Sustained expression of TGF-beta 1 underlies development of progressive kidney fibrosis. Kidney Int 1994;45:916–927.

36 Narita I, Nakayama H, Goto S, Takeda T, Sakatsume M, Saito A, Nakagawa Y, Arakawa M: Identification of genes specifically expressed in chronic and progressive glomerulosclerosis. Kidney Int 1997;63:S215–S217.

37 Peters H, Border WA, Noble NA: L-Arginine supplementation increases mesangial cell injury and subsequent tissue fibrosis in experimental glomerulonephritis. Kidney Int 1999;55:2264–2273.

38 Kirchhoff F, Krebs C, Abdulhag UN, Meyer-Schwesinger C, Maas R, Helmchen U, Hilgers KF, Wolf G, Stahl RA, Wenzel U: Rapid development of severe end-organ damage in C57BL/6 mice by combining DOCA salt and angiotensin II. Kidney Int 2008;73:643–650.

39 Johnson RJ, Alpers CE, Yoshimura A, Lombardi D, Pritzl P, Floege J, Schwartz SM: Renal injury from angiotensin II-mediated hypertension. Hypertension 1992;19:464–474.

40 Faulkner JL, Szcykalski LM, Springer F, Barnes JL: Origin of interstitial fibroblasts in an accelerated model of angiotensin II-induced renal fibrosis. Am J Pathol 2005;167:1193–1205.

41 Feld LG, Brentjens JR, Van Liew JB: Renal injury and proteinuria in female spontaneously hypertensive rats. Renal Physiol 1981;4:46–56.

42 Geiger H, Fierlbeck W, Mai M, Ruchti H, Schonfeld V, Dammrich J, Hugo C, Neumayer HH: Effects of early and late antihypertensive treatment on extracellular matrix proteins and mononuclear cells in uninephrectomized SHR. Kidney Int 1997;51:750–761.

43 Jovanovic D, Jovovic D, Mihailovic-Stanojevic N, Miloradovic Z, Dimitrijevic J, Maksic N, Djukanovic L: Influence of carvedilol on chronic renal failure progression in spontaneously hypertensive rats with adriamycin nephropathy. Clin Nephrol 2005;63:446–453.

44 Yu HC, Burrell LM, Black MJ, Wu LL, Dilley RJ, Cooper ME, Johnston CI: Salt induces myocardial and renal fibrosis in normotensive and hypertensive rats. Circulation 1998; 98:2621–2628.

45 Shibota M, Nagaoka A, Shino A, Fujita T: Renin-angiotensin system in stroke-prone spontaneously hypertensive rats. Am J Physiol 1979;236:H409–H416.

46 Rapp JP, Dene H: Development and characteristics of inbred strains of Dahl salt sensitive and salt-resistant rats. Hypertension 1985;7:340–349.

47 Blasi ER, Rocha R, Rudolph AE, Blomme EA, Polly ML, McMahon EG: Aldosterone/salt induces renal inflammation and fibrosis in hypertensive rats. Kidney Int 2003;63:1791–1800.

48 Roson MI, Cavallero S, Della PS, Cao G, Gorzalczany S, Pandolfo M, Kuprewicz A, Canessa O, Toblli JE, Fernandez BE: Acute sodium overload produces renal tubulointerstitial inflammation in normal rats. Kidney Int 2006;70:1439–1446.

49 Kelly FJ, Anderson S, Thompson MM, Oyama TT, Kennefick TM, Corless CL, Roman RJ, Kurtzberg L, Pratt BM, Ledbetter SR: Acute and chronic renal effects of recombinant human TGF-beta2 in the rat. J Am Soc Nephrol 1999;10:1264–1273.

50 Ledbetter S, Kurtzberg L, Doyle S, Pratt BM: Renal fibrosis in mice treated with human recombinant transforming growth factor-beta2. Kidney Int 2000;58:2367–2376.

51 Tang WW, Ulich TR, Lacey DL, Hill DC, Qi M, Kaufman SA, Van GY, Tarpley JE, Yee JS: Platelet-derived growth factor-BB induces renal tubulointerstitial myofibroblast formation and tubulointerstitial fibrosis. Am J Pathol 1996;148:1169–1180.

52 Goyer RA, Miller CR, Zhu SY, and Victery W: Non-metallothionein-bound cadmium in the pathogenesis of cadmium nephrotoxicity in the rat. Toxicol Appl Pharmacol 1989;101: 232–244.

53 Walker RG, Escott M, Birchall I, Dowling JP, Kincaid-Smith P: Chronic progressive renal lesions induced by lithium. Kidney Int 1986;29:875–881.

54 Ueda S, Wakashin M, Wakashin Y, Yoshida H, Lesato K, Mori T, Mori Y, Akikusa B, Okuda K: Experimental gold nephropathy in guinea pigs: detection of autoantibodies to renal tubular antigens. Kidney Int 1986;29: 539–548.

55 Haley DP, Bulger RE, Dobyan DC: The long-term effects of uranyl nitrate on the structure and function of the rat kidney. Virchows Arch B Cell Pathol 1982;41:181–192.

56 Axelsen RA: Nephrotoxicity of mild analgesics in the Gunn strain of rat. Br J Clin Pharmacol 1980;10(Suppl 2):309S–312S.

57 Bertani T, Perico N, Abbate M, Battaglia C, Remuzzi G: Renal injury induced by long-term administration of cyclosporin A to rats. Am J Pathol 1987;127:569–579.

58 Elzinga LW, Rosen S, Burdmann EA, Hatton DC, Lindsley J, Bennett WM: The role of renal sympathetic nerves in experimental chronic cyclosporine nephropathy. Transplantation 2000;69:2149–2153.

59 Cosyns JP, Dehoux JP, Guiot Y, Goebbels RM, Robert A, Bernard AM, van Ypersele de SC: Chronic aristolochic acid toxicity in rabbits: a model of Chinese herbs nephropathy? Kidney Int 2001;59:2164–2173.

60 Debelle FD, Nortier JL, De Prez EG, Garbar CH, Vienne AR, Salmon IJ, Schodt-Lanckman MM, Vanherweghem JL: Aristolochic acids induce chronic renal failure with interstitial fibrosis in salt-depleted rats. J Am Soc Nephrol 2002;13:431–436.

61 Eddy AA: Interstitial fibrosis in hypercholesterolemic rats: role of oxidation, matrix synthesis, proteolytic cascades. Kidney Int 1998;53:1182–1189.

62 Axelsen RA: Healed experimental renal papillary necrosis and cortical scarring in the rat from 2-bromoethylamine hydrobromide. Virchows Arch B Pathol Anat Histol 1978;381:63–77.

63 Zahedi K, Revelo MP, Barone S, Wang Z, Tehrani K, Citron DP, Bissler JJ, Rabb H, Soleimani M: Stathmin-deficient mice develop fibrosis and show delayed recovery from ischemic-reperfusion injury. Am J Physiol Renal Physiol 2006;290:F1559–F1567.

64 Powell DW, Bertram CC, Cummins TD, Barati MT, Zheng S, Epstein PN, Klein JB: Renal tubulointerstitial fibrosis in OVE26 type 1 diabetic mice. Nephron Exp Nephrol 2009;111:11–19.

65 Bachmann S, Peters J, Engler E, Ganten D, Mullins J: Transgenic rats carrying the mouse renin gene – morphological characterization of a low-renin hypertension model. Kidney Int 1992;41:24–36.

66 Kelly DJ, Wilkinson-Berka JL, Allen TJ, Cooper ME, Skinner SL: A new model of diabetic nephropathy with progressive renal impairment in the transgenic (mRen-2)27 rat (TGR). Kidney Int 1998;54:343–352.

67 Cosgrove D, Meehan DT, Grunkemeyer JA, Kornak JM, Sayers R, Hunter WJ, Samuelson GC: Collagen COL4A3 knockout: a mouse model for autosomal Alport syndrome. Genes Dev 1996;10:2981–2992.

68 Miner JH, Sanes JR: Molecular and functional defects in kidneys of mice lacking collagen alpha 3(IV): implications for Alport syndrome. J Cell Biol 1996;135:1403–1413.

69 Matsuo S, Lopez-Guisa JM, Cai X, Okamura DM, Alpers CE, Bumgarner RE, Peters MA, Zhang G, Eddy AA: Multifunctionality of PAI-1 in fibrogenesis: evidence from obstructive nephropathy in PAI-1-overexpressing mice. Kidney Int 2005;67:2221–2238.

70 Kopp JB, Factor VM, Mozes M, Nagy P, Sanderson N, Bottinger EP, Klotman PE, Thorgeirsson SS: Transgenic mice with increased plasma levels of TGF-beta 1 develop progressive renal disease. Lab Invest 1996;74:991–1003.

71 Chai Q, Krag S, Chai S, Ledet T, Wogensen L: Localisation and phenotypical characterisation of collagen-producing cells in TGF-beta 1-induced renal interstitial fibrosis. Histochem Cell Biol 2003;119:267–280.

72 Samuel CS, Zhao C, Bond CP, Hewitson TD, Amento EP, Summers RJ: Relaxin-1-deficient mice develop an age-related progression of renal fibrosis. Kidney Int 2004;65:2054–2064.

73 Yabuki A, Maeda M, Matsumoto M, Kamimura R, Masuyama T, Suzuki S: SAMP1/Sku as a murine model for tubulointerstitial nephritis: a study using unilateral ureteral obstruction. Exp Animals 2005;54:53–60.

74 Takeji M, Moriyama T, Oseto S, Kawada N, Hori M, Imai E, Miwa T: Smooth muscle alpha-actin deficiency in myofibroblasts leads to enhanced renal tissue fibrosis. J Biol Chem 2006;281:40193–40200.

75 Zhang X, Chen X, Hong Q, Lin H, Zhu H, Liu Q, Wang J, Xie Y, Shang X, Shi S, Lu Y, Yin Z: TIMP-1 promotes age-related renal fibrosis through upregulating ICAM-1 in human TIMP-1 transgenic mice. J Gerontol Series A Biol Sci Med Sci 2006;61:1130–1143.

76 Cai G, Zhang X, Hong Q, Shao F, Shang X, Fu B, Feng Z, Lin H, Wang J, Shi S, Yin Z, and Chen X: Tissue inhibitor of metalloproteinase-1 exacerbated renal interstitial fibrosis through enhancing inflammation. Nephrol Dial Transplant 2008;23:1861–1875.

77 Kassiri Z, Oudit GY, Kandalam V, Awad A, Wang X, Ziou X, Maeda N, Herzenberg AM, Scholey JW: Loss of TIMP3 enhances interstitial nephritis and fibrosis. J Am Soc Nephrol 2009;20:1223–1235.

78 Kawamoto H, Yasuda O, Suzuki T, Ozaki T, Yotsui T, Higuchi M, Rakugi H, Fukuo K, Ogihara T, Maeda N: Tissue inhibitor of metalloproteinase-3 plays important roles in the kidney following unilateral ureteral obstruction. Hypertension Res 2006;29:285–294.

79 Boutet A, De Frutos CA, Maxwell PH, Mayol MJ, Romero J, Nieto MA: Snail activation disrupts tissue homeostasis and induces fibrosis in the adult kidney. EMBO J 2006;25:5603–5613.

80 Ramila D, Ardura JA, Esteban V, Ortega A, Ruiz-Ortega M, Bosch RJ, Esbrit P: Parathyroid hormone-related protein promotes inflammation in the kidney with an obstructed ureter. Kidney Int 2008;73:835–847.

81 Ortega A, Ramila D, Ardura JA, Esteban V, Ruiz-Ortega M, Barat A, Gazapo R, Bosch RJ, Esbrit P: Role of parathyroid hormone-related protein in tubulointerstitial apoptosis and fibrosis after folic acid-induced nephrotoxicity. J Am Soc Nephrol 2006;17:1594–1603.

82 Inagi R, Yamamoto Y, Nangaku M, Usuda N, Okamato H, Kurokawa K, van Ypersele de SC, Yamamoto H, Miyata T: A severe diabetic nephropathy model with early development of nodule-like lesions induced by megsin overexpression in RAGE/iNOS transgenic mice. Diabetes 2006;55:356–366.

83 Liu E, Morimoto M, Kitajima S, Koike T, Yu Y, Shiiki H, Nagata M, Watanabe T, Fan J: Increased expression of vascular endothelial growth factor in kidney leads to progressive impairment of glomerular functions. J Am Soc Nephrol 2007;18:2094–2104.

84 Cheng S, Pollock AS, Mahimkar R, Olson JL, Lovett DH: Matrix metalloproteinase 2 and basement membrane integrity: a unifying mechanism for progressive renal injury. FASEB J 2006;20:1898–1900.

85 Kie JH, Kapturczak MH, Traylor A, Agarwal A, Hill-Kapturczak N: Heme oxygenase-1 deficiency promotes epithelial-mesenchymal transition and renal fibrosis. J Am Soc Nephrol 2008;19:1681–1691.

86 Kosugi T, Heinig M, Nakayama T, Connor T, Yuzawa Y, Li Q, Hauswirth WW, Grant MB, Croker BP, Campbell-Thompson M, Zhang L, Atkinson MA, Segal MS, Nakagawa T: Lowering blood pressure blocks mesangiolysis and mesangial nodules, but not tubulointerstitial injury, in diabetic eNOS knockout mice. Am J Pathol 2009;174: 1221–1229.

87 Nakayama T, Sato W, Kosugi T, Zhang L, Campbell-Thompson M, Yoshimura A, Croker BP, Johnson RJ, Nakagawa T: Endothelial injury due to eNOS deficiency accelerates the progression of chronic renal disease in the mouse. Am J Physiol Renal Physiol 2009;296:F317–F327.
88 Ohashi K, Iwatani H, Kihara S, Nakagawa Y, Komura N, Fujita K, Maeda N, Nishida M, Katsube F, Shimomura I, Ito T, Funahashi T: Exacerbation of albuminuria and renal fibrosis in subtotal renal ablation model of adiponectin-knockout mice. Arterioscler Thromb Vasc Biol 2007;27:1910–1917.
89 Bao J, Tu Z, Sun H, Luo G, Yang L, Song J, Qin M, Shi Y, Bu H, Li Y: R2: identification of renal potential progenitor/stem cells that participate in the renal regeneration processes of kidney allograft fibrosis. Nephrology 2008;13:500–507.
90 Gooch JL, Roberts BR, Cobbs SL, Tumlin JA: Loss of the alpha-isoform of calcineurin is sufficient to induce nephrotoxicity and altered expression of transforming growth factor-beta. Transplantation 2007;83:439–447.
91 Meier M, Menne J, Park JK, Holtz M, Gueler F, Kirsch T, Schiffer M, Mengel M, Lindschau C, Leitges M, Haller H: Deletion of protein kinase C-epsilon signaling pathway induces glomerulosclerosis and tubulointerstitial fibrosis in vivo. J Am Soc Nephrol 2007;18:1190–1198.
92 Morimoto Y, Gai Z, Tanishima H, Kawakatsu M, Itoh S, Hatamura I, Muragaki Y: TNF-alpha deficiency accelerates renal tubular interstitial fibrosis in the late stage of ureteral obstruction. Exp Mol Pathol 2008;85:207–213.
93 Kimura K, Iwano M, Higgins DF, Yamaguchi Y, Nakatani K, Harada K, Kubo A, Akai Y, Rankin EB, Neilson EG, Haase VH, Saito Y: Stable expression of HIF-1alpha in tubular epithelial cells promotes interstitial fibrosis. Am J Physiol Renal Physiol 2008;295:F1023-F1029.
94 Ma J, Nishimura H, Fogo A, Kon V, Inagami T, Ichikawa I: Accelerated fibrosis and collagen deposition develop in the renal interstitium of angiotensin type 2 receptor null mutant mice during ureteral obstruction. Kidney Int 1998;53:937–944.
95 Brantley JG, Sharma M, Alcalay NI, Heuvel GB: Cux-1 transgenic mice develop glomerulosclerosis and interstitial fibrosis. Kidney Int 2003;63:1240–1248.
96 Norwood VF, Morham SG, Smithies O: Postnatal development and progression of renal dysplasia in cyclooxygenase-2 null mice. Kidney Int 2000;58:2291–2300.
97 Zhang G, Kim H, Cai X, Lopez-Guisa JM, Alpers CE, Liu Y, Carmeliet P, Eddy AA: Urokinase receptor deficiency accelerates renal fibrosis in obstructive nephropathy. J Am Soc Nephrol 2003;14:1254–1271.
98 Schaefer L, Macakova K, Raslik I, Micegova M, Grone HJ, Schonherr E, Robenek H, Echtermeyer FG, Grassel S, Bruckner P, Schaefer RM, Iozzo RV, Kresse H: Absence of decorin adversely influences tubulointerstitial fibrosis of the obstructed kidney by enhanced apoptosis and increased inflammatory reaction. Am J Pathol 2002;160:1181–1191.
99 Schanstra JP, Neau E, Drogoz P, revalo Gomez MA, Lopez Novoa JM, Calise D, Pecher C, Bader M, Girolami JP, Bascands JL: In vivo bradykinin B2 receptor activation reduces renal fibrosis. J Clin Invest 2002;110:371–379.
100 Hocher B, Thone-Reineke C, Rohmeiss P, Schmager F, Slowinski T, Burst V, Siegmund F, Quertermous T, Bauer C, Neumayer HH, Schleuning WD, Theuring F: Endothelin-1 transgenic mice develop glomerulosclerosis, interstitial fibrosis, renal cysts but not hypertension. J Clin Invest 1997;99:1380–1389.
101 Svensson M, Irjala H, Alm P, Holmqvist B, Lundstedt AC, Svanborg C: Natural history of renal scarring in susceptible mIL-8Rh-/- mice. Kidney Int 2005;67:103–110.
102 Wenderfer SE, Wang H, Ke B, Wetsel RA, Braun MC: C3a receptor deficiency accelerates the onset of renal injury in the MRL/lpr mouse. Mol Immunol 2009;46:1397–1404.
103 Wang Y, Heilig K O, Minto AW, Chen S, Xiang M, Dean DA, Geiger RC, Chang A, Pravtcheva DD, Schlimme M, Deb DK, Wang Y, Heilig CW: Nephron-deficient Fvb mice develop rapidly progressive renal failure and heavy albuminuria involving excess glomerular GLUT1 and VEGF. Lab Invest 2010;90:83–97.
104 Hewitson TD: Renal tubulointerstitial fibrosis: common but never simple. Am J Physiol Renal Physiol 2009;296:F1239–F1244.
105 Schnaper HW: Renal fibrosis. Meth Mol Med 2005;117:45–68.
106 Wynn TA: Cellular and molecular mechanisms of fibrosis. J Pathol 2008;214:199–210.
107 Gabbiani G: The biology of the myofibroblast. Kidney Int 1992;41:530–532.
108 Desmouliere A, Chaponnier C, Gabbiani G: Tissue repair, contraction, and the myofibroblast. Wound Repair Regen 2005;13:7–12.
109 Desmouliere A, Gabbiani G: Myofibroblast differentiation during fibrosis. Exp Nephrol 1995;3:134–139.
110 Benzonana G, Skalli O, Gabbiani G: Correlation between the distribution of smooth muscle or non muscle myosins and alpha-smooth muscle actin in normal and pathological soft tissues. Cell Motil Cytoskeleton 1988;11:260–274.
111 Gabbiani G: The myofibroblast in wound healing and fibrocontractive diseases. J Pathol 2003;200:500–503.
112 Alpers CE, Pichler R, Johnson RJ: Phenotypic features of cortical interstitial cells potentially important in fibrosis. Kidney Int 1996;54:S28-S31.
113 Alpers CE, Hudkins KL, Floege J, Johnson RJ: Human renal cortical interstitial cells with some features of smooth muscle cells participate in tubulointerstitial and crescentic glomerular injury. J Am Soc Nephrol 1994;5:201–209.

114 Alpers CE, Seifert RA, Hudkins KL, Johnson RJ, Bowen-Pope DF: PDGF-receptor localizes to mesangial, parietal epithelial, and interstitial cells in human and primate kidneys. Kidney Int 1993;43:286–294.
115 Alpers CE, Hudkins KL, Gown AM, Johnson RJ: Enhanced expression of 'muscle-specific' actin in glomerulonephritis. Kidney Int 1992; 41:1134–1142.
116 Badid C, Mounier N, Costa AM, Desmouliere A: Role of myofibroblasts during normal tissue repair and excessive scarring: interest of their assessment in nephropathies. Histol Histopathol 2000;15: 269–280.
117 Badid C, Vincent M, Fouque D, Laville M, Desmouliere A: Myofibroblast: a prognostic marker and target cell in progressive renal disease. Ren Fail 2001;23:543–549.
118 Essawy M, Soylemezoglu O, Muchaneta-Kubara EC, Shortland J, Brown CB, el Nahas AM: Myofibroblasts and the progression of diabetic nephropathy. Nephrol Dial Transplant 1997;12:43–50.
119 Diamond JR, van GH, Ding G, Engelmyer E: Myofibroblasts in experimental hydronephrosis. Am J Pathol 1995;146:121–129.
120 Floege J, Hackmann B, Kliem V, Kriz W, Alpers CE, Johnson RJ, Kuhn KW, Koch KM, Brunkhorst R: Age-related glomerulosclerosis and interstitial fibrosis in Milan normotensive rats: a podocyte disease. Kidney Int 1997;51:230–243.
121 Taneda S, Hudkins KL, Topouzis S, Gilbertson DG, Ophascharoensuk V, Truong L, Johnson RJ, Alpers CE: Obstructive uropathy in mice and humans: potential role for PDGF-D in the progression of tubulointerstitial injury. J Am Soc Nephrol 2003;14:2544–2555.
122 Kimura M, Asano M, Abe K, Miyazaki M, Suzuki T, Hishida A: Role of atrophic changes in proximal tubular cells in the peritubular deposition of type IV collagen in a rat renal ablation model. Nephrol Dial Transplant 2005;20:1559–1565.
123 Lin SL, Kisseleva T, Brenner DA, and Duffield JS: Pericytes and perivascular fibroblasts are the primary source of collagen-producing cells in obstructive fibrosis of the kidney. Am J Pathol 2008;173:1617–1627.
124 Humphreys BD, Lin SL, Kobayashi A, Hudson TE, Nowlin BT, Bonventre JV, Valerius MT, McMahon AP, Duffield JS: Fate tracing reveals the pericyte and not epithelial origin of myofibroblasts in kidney fibrosis. Am J Pathol 2010;176:185–197.
125 Barnes VL, Musa J, Mitchell RJ, Barnes JL: Expression of embryonic fibronectin isoform EIIIA parallels alpha-smooth muscle actin in maturing and diseased kidney. J Histochem Cytochem 1999;47:787–798.
126 Ahmed A, Huang L, Raftery AT, Ahmed AK, Fahmy H, el Nahas AM, Haylor JL: Cyclosporine A sensitizes the kidney to tubulointerstitial fibrosis induced by renal warm ischemia. Transplantation 2004;77:686–692.
127 Huang L, Haylor JL, Hau Z, Jones RA, Vickers ME, Wagner B, Griffin M, Saint RE, Coutts IGC, el Nahas AM, Johnson TS: Transglugaminase inhibition ameliorates experimental diabetic nephropathy. Kidney Int 2009;76:383–394.
128 Ma FY, Liu J, Kitching AR, Manthey CL, Nikolic-Paterson DJ: Targeting renal macrophage accumulation via c-fms kinase reduces tubular apoptosis but fails to modify progressive fibrosis in the obstructed rat kidney. Am J Physiol Renal Physiol 2009;296: F177–F185.
129 Duymelinck C, Dauwe SE, De Greef KE, Ysebaert DK, Verpooten GA, De Broe ME: TIMP-1 gene expression and PAI-1 antigen after unilateral ureteral obstruction in the adult male rat. Kidney Int 2000;58:1186–1201.
130 Suzuki T, Kimura M, Asano M, Fujigaki Y, Hishida A: Role of atrophic tubules in development of interstitial fibrosis in microembolism-induced renal failure in rat. Am J Pathol 2001;158:75–85.
131 Le Hir M, Hegyi I, Cueni-Loffing D, Loffing J, Kaissling B: Characterization of renal interstitial fibroblast-specific protein 1/S100A4-positive cells in healthy and inflamed rodent kidneys. Histochem Cell Biol 2005;123:335–346.
132 Picard N, Baum O, Vogetseder A, Kaissling B, Le Hir M: Origin of renal myofibroblasts in the model of unilateral ureter obstruction in the rat. Histochem Cell Biol 2008;130:141–155.
133 Kaissling B, Le Hir M: The renal cortical interstitium: morphological and functional aspects. Histochem Cell Biol 2008;130:247–262.
134 Hewitson TD, Wu HL, Becker GJ: Interstitial myofibroblasts in experimental renal infection and scarring. Am J Nephrol 1995;15:411–417.
135 Sharma AK, Mauer SM, Kim Y, Michael AF: Interstitial fibrosis in obstructive nephropathy. Kidney Int 1993;44:774–788.
136 Stokes MB, Holler S, Cui Y, Hudkins KL, Eitner F, Fogo A, Alpers CE: Expression of decorin, biglycan, and collagen type I in human renal fibrosing disease. Kidney Int 2000;57:487–498.
137 Hewitson TD, Becker GJ: Interstitial myofibroblasts in IgA glomerulonephritis. Am J Nephrol 1995;15:111–117.
138 Pedagogos E, Hewitson TD, Walker RG, Nicholis KM, Becker GJ: Myofibroblast involvement in chronic transplant rejection. Transplantation 1997;64:1192–1197.
139 Pat B, Yang T, Kong C, Watters D, Johnson DW, Gobe G: Activation of ERK in renal fibrosis after unilateral ureteral obstruction: modulation by antioxidants. Kidney Int 2005;67:931–943.
140 Zhang G, Oldroyd SD, Huang LH, Yang B, Li Y, Ye R, el Nahas AM: Role of apoptosis and Bcl-2/Bax in the development of tubulointerstitial fibrosis during experimental obstructive nephropathy. Exp Nephrol 2001;9:71–80.
141 Hewitson TD, Mookerjee I, Masterson R, Zhao C, Tregear GW, Becker GJ, Samuel CS: Endogenous relaxin is a naturally occurring modulator of experimental renal tubulointerstitial fibrosis. Endocrinol 2007;148:660–669.
142 Fujigaki Y, Muranaka Y, Sun D, Goto T, Zhou H, Sakakima M, Fukasawa H, Yonemura K, Yamamoto T, Hishida A: Transient myofibroblast differentiation of interstitial fibroblastic cells relevant to tubular dilatation in uranyl acetate-induced acute renal failure in rats. Virchows Arch 2005;446:164–176.

143 Pozdzik AA, Salmon IJ, Debelle FD, Decaestecker C, Van den Branden C, Verbeelen D, Deschodt-Lanckman MM, Vanherweghem JL, Nortier JL: Aristolochic acid induces proximal tubule apoptosis and epithelial to mesenchymal transformation. Kidney Int 2008;73:595–607.

144 Taneda S, Pippin JW, Sage EH, Hudkins KL, Takeuchi Y, Couser WG, Alpers CE: Amelioration of diabetic nephropathy in SPARC-null mice. J Am Soc Nephrol 2003;14:968–980.

145 Nakatsukasa H, Nagy P, Evarts RP, Hsia CC, Marsden E, Thorgeirsson SS: Cellular distribution of transforming growth factor-beta 1 and procollagen types I, III, and IV transcripts in carbon tetrachloride-induced rat liver fibrosis. J Clin Invest 1990;85:1833–1843.

146 Hardie WD, Le Cras TD, Jiang K, Tichelaar JW, Azhar M, Korfhagen TR: Conditional expression of transforming growth factor-alpha in adult mouse lung causes pulmonary fibrosis. Am J Physiol Lung Cell Mol Physiol 2004;286:L741–L749.

147 Wiggins R, Goyal M, Merritt S, Killen PD: Vascular adventitial cell expression of collagen I messenger ribonucleic acid in anti-glomerular basement membrane antibody-induced crescentic nephritis in the rabbit. A cellular source for interstitial collagen synthesis in inflammatory renal disease. Lab Invest 1993;68:557–565.

148 Lloyd CM, Dorf ME, Proudfoot A, Salant DJ, Gutierrez-Ramos JC: Role of MCP-1 and RANTES in inflammation and progression to fibrosis during murine crescentic nephritis. J Leukocyte Biol 1997;62:676–680.

149 Lloyd CM, Minto AW, Dorf ME, Proudfoot A, Wells TN, Salant DJ, Gutierrez-Ramos JC: RANTES and monocyte chemoattractant protein-1 (MCP-1) play an important role in the inflammatory phase of crescentic nephritis, but only MCP-1 is involved in crescent formation and interstitial fibrosis. J Exp Med 1997;185:1371–1380.

150 Xu Y, Appay MD, Heudes D, Lemoine R, Hinglais N, Michel JB, Bariety J: Colocalization of collagen overexpression and inflammatory cell infiltration in the two-kidney one-clip rat model from the early days of hypertension onward. Virchows Arch 1998;432:267–277.

151 Muchaneta-Kubara EC and el Nahas AM: Myofibroblast phenotypes expression in experimental renal scarring. Nephrol Dial Transplant 1997;12:904–915.

152 Liu G, Changsirikulchai S, Hudkins KL, Banas MC, Kowalewska J, Yang X, Wietecha TA, Volpone J, Gilbertson DG, Alpers CE: Identification of platelet-derived growth factor D in human chronic allograft nephropathy. Human Pathol 2008;39:393–402.

153 Wada T, Sakai N, Matsushima K, Kaneko S: Fibrocytes: a new insight into kidney fibrosis. Kidney Int 2007;72:269–273.

154 Sakai N, Wada T, Yokoyama H, Lipp M, Ueha S, Matsushima K, Kaneko S: Secondary lymphoid tissue chemokine (SLC/CCL21)/CCR7 signaling regulates fibrocytes in renal fibrosis. Proc Natl Acad Sci USA 2006;103:14098–14103.

155 Sakai N, Wada T, Matsushima K, Bucala R, Iwai M, Horiuchi M, Kaneko S: The renin-angiotensin system contributes to renal fibrosis through regulation of fibrocytes. J Hypertension 2008;26: 780–790.

156 Broekema M, Harmsen MC, van Luyn MJ, Koerts JA, Petersen AH, van Kooten TG, van GH, Navis G, Popa ER: Bone marrow-derived myofibroblasts contribute to the renal interstitial myofibroblast population and produce procollagen I after ischemia/reperfusion in rats. J Am Soc Nephrol 2007;18:165–175.

157 Broekema M, Harmsen MC, Koerts JA, van Kooten TG, Uges DR, Petersen AH, van Luyn MJ, Navis G, Popa ER: Ciclosporin does not influence bone marrow-derived cell differentiation to myofibroblasts early after renal ischemia/reperfusion. Am J Nephrol 2009;30:73–83.

158 Li J, Deane JA, Campanale NV, Bertram JF, Ricardo SD: The contribution of bone marrow-derived cells to the development of renal interstitial fibrosis. Stem Cells 2007;25:697–706.

159 Okon K, Szumera A, Kuzniewski M: Are CD34+ cells found in renal interstitial fibrosis? Am J Nephrol 2003;23:409–414.

160 Grimm PC, Nickerson P, Jeffery J, Savani RC, Gough J, McKenna RM, Stern E, Rush DN: Neointimal and tubulointerstitial infiltration by recipient mesenchymal cells in chronic renal-allograft rejection. N Engl J Med 2001;345:93–97.

161 Roufosse C, Bou-Gharios G, Prodromidi E, Alexakis C, Jeffery R, Khan S, Otto WR, Alter J, Poulsom R, Cook HT: Bone marrow-derived cells do not contribute significantly to collagen I synthesis in a murine model of renal fibrosis. J Am Soc Nephrol 2006;17:775–782.

162 Hay ED and Zuk A: Transformations between epithelium and mesenchyme: normal, pathological, and experimentally induced. Am J Kidney Dis 1995;26:678–690.

163 Strutz F, Neilson EG: New insights into mechanisms of fibrosis in immune renal injury. Springer Semin Immunopathol 2003;24:459–476.

164 Iwano M, Neilson EG: Mechanisms of tubulointerstitial fibrosis. Curr Opin Nephrol Hypertens 2004;13: 279–284.

165 Kalluri R, Neilson EG: Epithelial-mesenchymal transition and its implications for fibrosis. J Clin Invest 2003;112:1776–1784.

166 Liu Y: Epithelial to mesenchymal transition in renal fibrogenesis: pathologic significance, molecular mechanism, and therapeutic intervention. J Am Soc Nephrol 2004;15:1–12.

167 Liu Y: Renal fibrosis: new insights into the pathogenesis and therapeutics. Kidney Int 2006;69:213–217.

168 Strutz F, Okada H, Lo CW, Danoff T, Carone RL, Tomaszewski JE, Neilson EG: Identification and characterization of a fibroblast marker: FSP1. J Cell Biol 1995;130: 393–405.

169 Iwano M, Plieth D, Danoff TM, Xue C, Okada H, Neilson EG: Evidence that fibroblasts derive from epithelium during tissue fibrosis. J Clin Invest 2002;110:341–350.

170 Yang J, Shultz RW, Mars WM, Wegner RE, Li Y, Dai C, Nejak K, Liu Y: Disruption of tissue-type plasminogen activator gene in mice reduces renal interstitial fibrosis in obstructive nephropathy. J Clin Invest 2002; 110:1525–1538.

171 Zhang G, Kernan KA, Collins SJ, Cai X, Lopez-Guisa JM, Degen JL, Shvil Y, Eddy AA: Plasmin(ogen) promotes renal interstitial fibrosis by promoting epithelial-tomesenchymal transition: role of plasmin-activated signals. J Am Soc Nephrol 2007;18:846–859.
172 Bascands JL, Schanstra JP: Obstructive nephropathy: insights from genetically engineered animals. Kidney Int 2005;68:925–937.
173 Kida Y, Asahina K, Teraoka H, Gitelman I, Sato T: Twist relates to tubular epithelial-mesenchymal transition and interstitial fibrogenesis in the obstructed kidney. J Histochem Cytochem 2007;55:661–673.
174 Yang J, Liu Y: Blockage of tubular epithelial to myofibroblast transition by hepatocyte growth factor prevents renal interstitial fibrosis. J Am Soc Nephrol 2002;13:96–107.
175 Li Y, Dai C, Wu C, Liu Y: PINCH-1 promotes tubular epithelial-to-mesenchymal transition by interacting with integrin-linked kinase. J Am Soc Nephrol 2007;18:2534–2543.
176 Lange-Sperandio B, Trautmann A, Eickelberg O, Jayachandran A, Oberle S, Schmidutz F, Rodenbeck B, Homme M, Horuk R, Schaefer F: Leukocytes induce epithelial to mesenchymal transition after unilateral ureteral obstruction in neonatal mice. Am J Pathol 2007;171:861–871.
177 Mao H, Li Z, Zhou Y, Li Z, Zhuang S, An X, Zhang B, Chen W, Nie J, Wang Z, Borkan SC, Wang Y, Yu X: HSP72 attenuates renal tubular cell apoptosis and interstitial fibrosis in obstructive nephropathy. Am J Physiol Renal Physiol 2008;295:F202–F214.
178 Park SH, Choi MJ, Song IK, Choi SY, Nam JO, Kim CD, Lee BH, Park RW, Park KM, Kim YJ, Kim IS, Kwon TH, Kim YL: Erythropoietin decreases renal fibrosis in mice with ureteral obstruction: role of inhibiting TGF-beta-induced epithelial-to-mesenchymal transition. J Am Soc Nephrol 2007;18:1497–1507.
179 Sato M, Muragaki Y, Saika S, Roberts AB, Ooshima A: Targeted disruption of TGF-beta1/Smad3 signaling protects against renal tubulointerstitial fibrosis induced by unilateral ureteral obstruction. J Clin Invest 2003; 112:1486–1494.
180 Tan X, Li Y, Liu Y: Paricalcitol attenuates renal interstitial fibrosis in obstructive nephropathy. J Am Soc Nephrol 2006;17:3382–3393.
181 Yang J, Liu Y: Dissection of key events in tubular epithelial to myofibroblast transition and its implications in renal interstitial fibrosis. Am J Pathol 2001;159:1465–1475.
182 Ng YY, Huang TP, Yang WC, Chen ZP, Yang AH, Mu W, Nikolic-Paterson DJ, Atkins RC, Lan HY: Tubular epithelial-myofibroblast transdifferentiation in progressive tubulointerstitial fibrosis in 5/6 nephrectomized rats. Kidney Int 1998;54:864–876.
183 Kobayashi M, Sugiyama H, Wang DH, Toda N, Maeshima Y, Yamasaki Y, Masuoka N, Yamada M, Kira S, Makino H: Catalase deficiency renders remnant kidneys more susceptible to oxidant tissue injury and renal fibrosis in mice. Kidney Int 2005;68:1018–1031.
184 Cheng S, Lovett DH: Gelatinase A (MMP-2) is necessary and sufficient for renal tubular cell epithelial-mesenchymal transformation. Am J Pathol 2003;162:1937–1949.
185 Oldfield MD, Bach LA, Forbes JM, Nikolic-Paterson D, McRobert A, Thallas V, Atkins RC, Osicka T, Jerums G, Cooper ME: Advanced glycation end products cause epithelial-myofibroblast transdifferentiation via the receptor for advanced glycation end products (RAGE). J Clin Invest 2001;108:1853–1863.
186 Li Y, Yang J, Dai C, Wu C, Liu Y: Role for integrin-linked kinase in mediating tubular epithelial to mesenchymal transition and renal interstitial fibrogenesis. J Clin Invest 2003;112:503–516.
187 Burns WC, Twigg SM, Forbes JM, Pete J, Tikellis C, Thallas-Bonke V, Thomas MC, Cooper ME, Kantharidis P: Connective tissue growth factor plays an important role in advanced glycation end product-induced tubular epithelial-to-mesenchymal transition: implications for diabetic renal disease. J Am Soc Nephrol 2006;17:2484–2494.
188 Zeisberg M, Hanai J, Sugimoto H, Mammoto T, Charytan D, Strutz F, Kalluri R: BMP-7 counteracts TGF-beta1-induced epithelial-to-mesenchymal transition and reverses chronic renal injury. Nat Med 2003;9:964–968.
189 Okada H, Ban S, Nagao S, Takahashi H, Suzuki H, Neilson EG: Progressive renal fibrosis in murine polycystic kidney disease: an immunohistochemical observation. Kidney Int 2000;58:587–597.
190 Zeisberg EM, Potenta SE, Sugimoto H, Zeisberg M, Kalluri R: Fibroblasts in kidney fibrosis emerge via endothelial-to-mesenchymal transition. J Am Soc Nephrol 2008;19:2282–2287.
191 Li J, Qu X, Bertram JF: Endothelial-myofibroblast transition contributes to the early development of diabetic renal interstitial fibrosis in streptozotocin-induced diabetic mice. Am J Pathol 2009;175:1380–1388.
192 Cook HT: The origin of renal fibroblasts and progression of kidney disease. Am J Pathol 2010;176: 22–24.
193 Tarin D: The fallacy of epithelial mesenchymal transition in neoplasia. Cancer Res 2005;65: 5996–6001.
194 Garber K: Epithelial-to-mesenchymal transition is important to metastasis, but questions remain. J Natl Cancer Inst 2008;100:232–239.
195 Bonventre JV: Kidney injury molecule-1 (KIM-1): a specific and sensitive biomarker of kidney injury. Scand J Clin Lab Invest Suppl 2008;241:78–83.
196 Kramer AB, van Timmeren MM, Schuurs TA, Vaidya VS, Bonventre JV, van GH, Navis G: Reduction of proteinuria in adriamycin-induced nephropathy is associated with reduction of renal kidney injury molecule (Kim-1) over time. Am J Physiol Renal Physiol 2009;296:F1136–F1145.
197 Kuehn EW, Park KM, Somlo S, Bonventre JV: Kidney injury molecule-1 expression in murine polycystic kidney disease. Am J Physiol Renal Physiol 2002;283:F1326-F1336.
198 van Timmeren MM, Bakker SJ, Vaidya VS, Bailly V, Schuurs TA, Damman J, Stegeman CA, Bonventre JV, van Goor H: Tubular kidney injury molecule-1 in protein-overload nephropathy. Am J Physiol Renal Physiol 2006;291:F456-F464.

199 Deniz H, Ogutmen B, Cakalagaoglu F, Tuglular S, Ozener C, Akoglu E: Inhibition of the renin angiotensin system decreases fibrogenic cytokine expression in tacrolimus nephrotoxicity in rats. Transplant Proc 2006;38:483–486.
200 Zhang PL, Rothblum LI, Han WK, Blasick TM, Potdar S, Bonventre JV: Kidney injury molecule-1 expression in transplant biopsies is a sensitive measure of cell injury. Kidney Int 2008;73:608–614.
201 Nair RK, Khaira A, Sharma A, Mahajan S, Dinda AK: Spectrum of renal involvement in paroxysmal nocturnal hemoglobinuria: report of three cases and a brief review of the literature. Intl Urol Nephrol 2008;40:471–475.
202 Nath KA, Vercellotti GM, Grande JP, Miyoshi H, Paya CV, Manivel JC, Haggard JJ, Croatt AJ, Payne WD, Alam J: Heme protein-induced chronic renal inflammation: suppressive effect of induced heme oxygenase-1. Kidney Int 2001;59:106–117.
203 Leonardi P, Ruol A: Renal hemosiderosis in the hemolytic anemias: diagnosis by means of needle biopsy. Blood 1960;16:1029–1038.
204 Le Hir M, Kaissling B: Antibodies against macrophages that overlap in specificity with fibroblasts. Kidney Int 2005;68:2400–2001.
205 Kriz W, Le Hir M: Pathways to nephron loss starting from glomerular diseases-insights from animal models. Kidney Int 2005;67:404–419.
206 Schnaper HW, Kopp JB: Why kidneys fail: report from an American Society of Nephrology Advances in Research Conference. J Am Soc Nephrol 2006;17:1777–1781.
207 Kriz W: Histopathological analysis of interstitial fibrosis. Epithelial mesenchymal transition in fibrosis: whither the controversy? Symposium Panel: Morrissey JJ, Liu Y, Lovett DH, Barnes JL, Kriz W. American Society of Nephrology Renal Week San Diego, 2006.
208 Nishitani Y, Iwano M, Yamaguchi Y, Harada K, Nakatani K, Akai Y, Nishino T, Shiiki H, Kanauchi M, Saito Y, Neilson EG: Fibroblast-specific protein 1 is a specific prognostic marker for renal survival in patients with IgAN. Kidney Int 2005;68:1078–1085.
209 Higgins DF, Kimura K, Bernhardt WM, Shrimanker N, Akai Y, Hohenstein B, Saito Y, Johnson RS, Kretzler M, Cohen CD, Eckardt KU, Iwano M, Haase VH: Hypoxia promotes fibrogenesis in vivo via HIF-1 stimulation of epithelial-to-mesenchymal transition. J Clin Invest 2007;117:3810–3820.
210 Higgins DF, Kimura K, Iwano M, Haase VH: Hypoxia-inducible factor signaling in the development of tissue fibrosis. Cell Cycle 2008;7:1128–1132.
211 Nishida M, Hamaoka K: Macrophage phenotype and renal fibrosis in obstructive nephropathy. Nephron Exp Nephrol 2008;110:e31-e36.
212 Ito K, Chen J, El CM, Stern JM, Seshan SV, Khodadadian JJ, Richardson I, Hyman MJ, Vaughan ED Jr, Poppas DP, Felsen D: Renal damage progresses despite improvement of renal function after relief of unilateral ureteral obstruction in adult rats. Am J Physiol Renal Physiol 2004;287:F1283–F1293.
213 Mazzucchelli L: Protein S100A4: too long overlooked by pathologists? Am J Pathol 2002;160:7–13.
214 Garrett SC, Varney KM, Weber DJ, Bresnick AR: S100A4, a mediator of metastasis. J Biol Chem 2006;281:677–680.
215 Duarte WR, Shibata T, Takenaga K, Takahashi E, Kubota K, Ohya K, Ishikawa I, Yamauchi M, Kasugai S: S100A4: a novel negative regulator of mineralization and osteoblast differentiation. J Bone Miner Res 2003;18:493–501.
216 Gibbs FE, Barraclough R, Platt-Higgins A, Rudland PS, Wilkinson MC, Parry EW: Immunocytochemical distribution of the calcium-binding protein p9Ka in normal rat tissues: variation in the cellular location in different tissues. J Histochem Cytochem1995;43:169–180.
217 Rivard CJ, Brown LM, Almeida NE, Maunsbach AB, Pihakaski-Maunsbach K, Andres-Hernando A, Capasso JM, Berl T: Expression of the calcium-binding protein S100A4 is markedly up-regulated by osmotic stress and is involved in the renal osmoadaptive response. J Biol Chem 2007;282:6644–6652.
218 Carvajal G, Droguett A, Burgos ME, Aros C, Ardiles L, Flores C, Carpio D, Ruiz-Ortega M, Egido J, Mezzano S: Gremlin: a novel mediator of epithelial mesenchymal transition and fibrosis in chronic allograft nephropathy. Transplant Proc 2008;40:734–739.
219 Djamali A, Vidyasagar A, Adulla M, Hullett D, Reese S: Nox-2 is a modulator of fibrogenesis in kidney allografts. Am J Transplant 2009;9:74–82.
220 Tyler JR, Robertson H, Booth TA, Burt AD, Kirby JA: Chronic allograft nephropathy: intraepithelial signals generated by transforming growth factor-beta and bone morphogenetic protein-7. Am J Transplant 2006; 6:1367–1376.
221 Jinde K, Nikolic-Paterson DJ, Huang XR, Sakai H, Kurokawa K, Atkins RC, Lan HY: Tubular phenotypic change in progressive tubulointerstitial fibrosis in human glomerulonephritis. Am J Kidney Dis 2001;38:761–769.
222 Vongwiwatana A, Tasanarong A, Rayner DC, Melk A, Halloran PF: Epithelial to mesenchymal transition during late deterioration of human kidney transplants: the role of tubular cells in fibrogenesis. Am J Transplant 2005;5:1367–1374.
223 Rastaldi MP, Ferrario F, Giardino L, Dell'Antonio G, Grillo C, Grillo P, Strutz F, Muller GA, Colasanti G, D'Amico G: Epithelial-mesenchymal transition of tubular epithelial cells in human renal biopsies. Kidney Int 2002;62:137–146.
224 Inoue T, Okada H, Takenaka T, Watanabe Y, Suzuki H: A case report suggesting the occurrence of epithelial-mesenchymal transition in obstructive nephropathy. Clin Exp Nephrol 2009;13:385–388.
225 Zeisberg M, Maeshima Y, Mosterman B, Kalluri R: Renal fibrosis. Extracellular matrix microenvironment regulates migratory behavior of activated tubular epithelial cells. Am J Pathol 2002;160:2001–2008.
226 Vitalone MJ, O'Connell PJ, Jimenez-Vera E, Yuksel A, Wavamunno M, Fung CL, Chapman JR, Nankivell BJ: Epithelial-to-mesenchymal transition in early transplant tubulointerstitial damage. J Am Soc Nephrol 2008;19:1571–1583.

227 Kuusniemi AM, Lapatto R, Holmberg C, Karikoski R, Rapola J, Jalanko H: Kidneys with heavy proteinuria show fibrosis, inflammation, and oxidative stress, but no tubular phenotypic change. Kidney Int 2005;68:121–132.
228 Zeisberg M, Duffield JS: Resolved: EMT produces fibroblasts in the kidney. Debate. J Am Soc Nephrol 2010;21:1247–1253.
229 Li L, Zepeda-Orozoco D, Black R, Lin F: Autophagy is a component of epithelial cell fate in obstructive uropathy. Am J Pathol 2010;176:1767–1778.
230 Koesters R, Kaissling B, Le Hir M, Picard N, Theilig F, Gebhardt R, Glick AB, Hahnel B, Hosser H, Grone H-J, Kriz W: Tubular overexpression of transforming growth factor-b1 induces autophagy and fibrosis but not mesenchymal transition of renal epithelial cells. Am J Pathol 2010;177:632–643.

Jeffrey L. Barnes, PhD
Department of Medicine, Division of Nephrology
The University of Texas Health Science Center
7703 Floyd Curl Drive
San Antonio, TX 78229–3900 (USA)
Tel. +1 210 949 3691, Fax +1 210 567 0517, E-Mail barnesj@uthscsa.edu

Herrera GA (ed): Experimental Models for Renal Diseases: Pathogenesis and Diagnosis.
Contrib Nephrol. Basel, Karger, 2011, vol 169, pp 94–106

Immunopathogenesis of Idiopathic Nephrotic Syndrome

Shao yu Zhang · Vincent Audard · Qingfeng Fan · Andre Pawlak · Philippe Lang · Djillali Sahali

AP-HP, Groupe Henri-Mondor Albert-Chenevier, Service de Nephrologie, Creteil; Université Paris-Est, Faculte de Medecine, UMRS 955, Creteil; INSERM, Unite U955, France

Abstract

Idiopathic nephrotic syndrome is the most frequent glomerular disease in children. The mechanisms underlying its pathophysiology have been investigated by genetic, cellular and molecular approaches. While genetic analyses have provided new insights into disease pathogenesis through the discovery of several podocyte genes mutated in distinct forms of inherited nephrotic syndrome, the molecular bases of minimal change nephrotic syndrome and focal and segmental glomerulosclerosis with relapse remain unclear. The immune system seems to play a critical role in the active phase of this disease through disturbances involving several cell subsets, mainly T cells. The innate immune system may also contribute to the immune disorders. In this review, we discuss recent insights from the molecular and immunological findings and their significance in the context of the clinical course of the disease.

Idiopathic nephrotic syndrome (INS) is a kidney disease defined by massive proteinuria and hypoalbuminemia. INS is recognized as a common chronic illness in childhood. Primary INS includes three histological variants: minimal change nephrotic syndrome (MCNS), focal segmental glomerulosclerosis (FSGS) and membranous nephropathy. MCNS and FSGS respectively account for 70% and 20% of INS in children, and each accounts for 25% of INS in adults [1, 2]. The hallmark of MCNS is the absence of inflammatory injury or immune complex deposition in the glomerulus, whereas FSGS is characterized by adhesion of the glomerular tuft to Bowman's capsule. Ultrastructural analysis shows glomerular morphological changes in the form of foot process effacement: the podocytes seem fused together, assuming a flattened rather than a foot-like morphology.

Membranous nephropathy, which is rare in children, is a distinct glomerular disease characterized by a nephrotic syndrome associated with prominent immune complex deposits between the podocytes and the glomerular basement membrane [3]. While the mechanisms underlying the formation and the shedding of immune complexes in membranous nephropathy are best understood through studies on Heymann nephritis, a mechanistic explanation of the glomerular abnormalities in MCNS and FSGS with relapse remains elusive. In this review, we will discuss the results of recent investigations that may contribute to our understanding of the pathophysiology of these glomerular diseases. Membranous nephropathy, which is a distinct entity, deserves separate consideration.

Whereas MCNS seems to be a single entity, FSGS appears as a heterogeneous disease, with both immune and non-immune causes. In the latter, recent genetic approaches have elucidated several molecular aspects of FSGS by identifying several genes that play a critical role in the glomerular filtration barrier [4–7]. These studies have opened a large field of investigation that has been summarized in a recent excellent review [8].

The molecular relationship between MCNS and FSGS with relapse is unknown. Undoubtedly, some patients with steroid-sensitive MCNS became steroid-resistant and progress to FSGS and renal failure. Conversely, nearly 50% of patients with FSGS are steroid-sensitive, but relapse. Given the infrequency of kidney biopsies in children with nephrotic syndromes, the relative prevalence of MCNS and FSGS at the time of presentation is unknown. Similar immunological disturbances have been reported in the two diseases [9].

The most critical parameter of prognosis is the response to steroid therapy: steroid-sensitive MCNS and FSGS share the same benign course. Of children with nephrotic syndromes, 80–90% (presumably MCNS) are steroid-responsive. Overall, 60–80% of steroid-responsive nephrotic children will relapse and about 60% of them will have five or more relapses [10]. Around 60% of MCNS children show steroid dependence. Resistance to therapy will occur in 50% of FSGS and 10% of MCNS patients despite various immunosuppressive treatments (cyclosporine, cyclophosphamide, high-dose steroids) [11]. Such patients often progress to renal failure, requiring dialysis and transplantation. Recurrent disease in the transplant occurs in 30–50% of patients with FSGS, often in the first hours or days following engraftment [12]. This percentage is as high as 80% for the second graft if recurrence caused the loss of the first graft [13]. The number of years to achieve end-stage renal failure predicts the risk of recurrence after transplantation: 50% if end-stage renal failure was achieved in less than 3 years and 10–20% if this period was longer. Recurrence of the disease in the graft places medicine in a blind alley with regard to possible treatment.

Given the short time between engraftment and disease recurrence, one may exclude the generation of an immune response against any component of the allogeneic glomerular filtration barrier, but movement of a circulating factor endowed with pathogenic properties across the glomerular structure has been postulated [14]. The absence of inflammatory lesions and immune complex deposits within the glomeruli, as well as the rapid relapses following renal transplantation, support the extrarenal origin of a circulating factor. Indirect evidence comes from observations showing that the nephrotic syndrome disappears when an MCNS or an FSGS kidney is transplanted into an INS-free patient [15, 16].

This review focuses on major findings about the immunopathogenesis of MCNS (and likely FSGS with relapse). We discuss recent findings regarding signal transduction and T cell responses in the context of MCNS relapse versus remission. Finally, an alternative mechanism for the pathophysiology of the disease is proposed, one that may reconcile the perspectives of immunologists and nephrologists.

Cytokine Profile in Nephrotic Syndrome

Various alterations in cytokine production during MCNS have been described by a large number of studies, which, unfortunately, are not all in agreement. This variation might result from the different immunogenetic backgrounds of the patients or from the heterogeneity of studies on stimulated cells in environments that are far from physiological. It might also be due to the diversity of the dosage techniques, which do not always take soluble cytokine receptors into account. Table 1 summarizes the data in the literature concerning the expression levels of different cytokines in MCNS. It can be deduced that TNF-α and IL-13 levels often increase during relapse.

Table 1. Summary of data reported in the literature concerning cytokine production in MCNS

Cytokine	Upregulated	Downregulated	No changes	Reference
IL-1	*		******	[9, 26, 55, 84, 85, 86]
IL-2	**		****	[9, 26, 55, 86, 87, 88]
IL-4			****	[9, 26, 39, 88, 89]
IL-6		*		[90]
IL-8	*	**	*	[9, 87, 88, 91]
IL-10		*		[92]
IL-12	*		*	[38, 93]
IL-13	**			[39, 88]
TNF-α	***			[26, 55, 86]
Interferon gamma (IFN-γ)	*	*	*,*,*	[9, 86, 87, 88, 93, 94]

One asterik corresponds to one published study.

Interestingly, the receptors for the cytokines IL-4 and IL-13 are expressed by podocytes in various glomerular diseases, raising the possibility that IL-13 might contribute to the pathophysiology of proteinuria [17]. In fact, Van Den Berg et al. [17] have demonstrated that IL-4 and IL-13 increase transcellular ion transport, but they do not affect permeability to macromolecules. Moreover, in many pathological conditions in which IL-13 is increased, such as asthma, psoriasis and allergic dermatitis, proteinuria does not occur.

Research on the Permeability Factor

In addition to the clinical observations mentioned above, plasma replacement performed in patients with FSGS who escape drug therapy was successful [18, 19]. Moreover, immediate recurrences after transplantation have been successfully treated by plasma exchanges or plasma immunoadsorption techniques [20–22]. Another line of evidence suggesting secretion of a permeability factor comes from experiments showing that systemic infusion of supernatants of cultured PBMC or T cells of patients with MCNS relapse induces proteinuria in rats [23–27]. Immunochemical analyses of affinity column eluate fractions that induced proteinuria suggested that the molecular weight of the permeability factor is below 150 kDa. Numerous attempts to isolate this factor have been unsuccessful.

T Cell Signal Transduction in MCNS

More than three decades ago, it was suggested that MCNS is a systemic disorder of T cell function and cell-mediated immunity [28]. This hypothesis was supported by several clinical observations such as the rapid occurrence of relapses upon antigen challenge (infections or immunization), hyporesponsiveness of lymphocytes to mitogens and decreased delayed hypersensitivity [29]. Allergic manifestations, such as contact dermatitis, rhinitis and asthma, might be observed, particularly in children, but they are uncommon in adult patients with MCNS. Intriguingly, MCNS is the most common kidney disease associated with primary immunological disorders such as Hodgkin's lymphoma [30], leukemia [31] and thymoma [32]. The sensitivity to steroid and immunosuppressive therapy is the most important clinical argument in favor of the immune origin of the MCNS.

Several T cell populations are expanded during the active phase of the disease, such as CD4+ T cells expressing the CD25 antigen (IL-2 receptor-α chain) [33] and CD8+ T cells expressing the memory T cell marker, CD45RO [34]. The CD25 antigen is mainly expressed by two sets of CD4+ T cells: a large population (approx. 50%) that expresses CD25 following activation by an immunogenic stimulus, and a minor population (approx. 10%) that constitutively expresses CD25 and CTLA 4, a negative regulator of CD28 signaling [35]. The expression of the CD25 antigen during relapses suggests an activation of CD4+ T cells and/or a recruitment of CD4+ CD25+ T cells endowed with T-suppressive functions. The possible recruitment of the minor CD4+ CD25+ T cell subset during relapses seems to agree with some findings showing decreased T cell-mediated immunity and reduced lymphocyte response to mitogens [29], but these studies deserve to be revisited in light of new knowledge. Of note, the immune dysfunction is not restricted to CD4+ T cells. Studies of genetic polymorphism in the variable region of the TCR β-chain have revealed a selective recruitment of some Vβ gene families in peripheral CD8+ T cells from nephrotic patients with frequent relapses [36]. These results suggest a clonal expansion of CD8+ T cells in long-lasting, active disease.

We have recently reported a downregulation of the IL-12 receptor β2 subunit (IL-12R β2) during relapse in MCNS T cells [37]. On the other hand, the IL-12R β1 chain was normally expressed. The downregulation of the IL-12R β2 is compatible with the low IL-12 cytokine level, as previously reported by Stefanovic et al. [38] and the lack of induction of the Th1 transcription factor T-bet that we observed during relapse (unpubl. data). These observations, as well as several published reports, support the hypothesis of Th2 polarization during MCNS. An increase in the production of IL-13, a Th2 cytokine, has been reported by independent groups [39].

Valanciuté et al. [40] have shown that the short form of the proto-oncogene c-maf is highly induced during MCNS relapse. c-maf was identified as the first Th2-specific transcription factor that binds to the IL-4 proximal promoter and, therefore, regulates IL-4 expression [41], and it may be activated by a signaling pathway involving B7h/ICOS [42]. c-maf is transcribed into several mRNA forms: the short form, c-maf-S, encodes a protein of 373 amino acids, whereas the long form, c-maf-L, generated by alternative splicing, encodes a protein of 403 amino acids [43]. c-maf belongs to the large maf family, the members of which are characterized by an N-terminal proline/serine/threonine-rich acidic transactivation domain and a C-terminal basic region containing the DNA-binding domain. c-maf binds as a homo- or heterodimer to the DNA palindromic maf recognition element (Mare) that consists of an extended AP-1 motif [44].

Identified early on as the first Th2-specific transcription factor binding to the IL-4 proximal promoter, c-maf is barely expressed in resting T cells and is mainly induced by signals emanating from proximal events involving TCR activation [45]. Surprisingly, the expression level of IL-4 during relapse is low and even downregulated as compared to its remission phase level in the same patients, as well as to the levels in normal subjects. Only some patients who suffered from active allergic diseases, such as bronchial asthma, concomitantly with MCNS, expressed a significant amount of IL-4, but these cases are infrequent. Moreover, despite the increased frequency of allergic diseases in the world, the incidence of MCNS is remarkably stable, thus excluding close relationships between these diseases. Although unexpected in the context of c-maf activation, the low expression level of IL-4 during the relapse phase has been reported by several groups (table 1). We have observed that c-maf expression shuts down in patients who enter remission, even though the IL-4 transcript level increases. This discrepancy raises two questions: what is the mechanism that counteracts the IL-4 induction by c-maf in MCNS, and what is the target gene of c-maf in this disease? The sudden drop in IL-4 level during remission might predict relapse (unpubl. observations), and suggests that IL-4 monitoring could be relevant to controlling the course of the disease.

The rapid relapse of MCNS following immunogenic challenge led us to postulate the possible involvement of the innate immune system, in which the nuclear factor-κB (NF-κB) signaling pathway plays a central role [46]. Moreover, most cytokines whose levels are thought to be increased during relapses and downregulated during remissions are partly or predominantly regulated by the NF-κB proteins. The NF-κB transcription factors play a central role in the triggering and coordination of both innate and adaptive immune responses by regulating the expression of a wide variety of genes, such as those encoding IL-1, IL-6, IL-2, IL-8, IL-12, Rantes, TNF-α, LT-α and chemokines [47]. The NF-κB family includes five members: RelA (p65), NF-κB1 (p50; p105), NF-κB2 (p52; p110), c-Rel and RelB [48]. They have a structurally conserved 300-amino-acid N-terminal region, which is responsible for DNA-binding, dimerization and association with IκB inhibitory proteins [47]. The p50 and p52 proteins are generated by proteolytic processing of precursor p105 and p110 proteins, respectively [49].

Each member of the NF-κB family can form homodimers as well as heterodimers. Homodimers of p50 lacking a transactivation domain can be detected in resting T cell nuclei. The main activated form is the heterodimer of p65 and either p50 or p52 [49]. In resting T cells, the NF-κB complex is sequestered in the cytoplasm through the masking of the nuclear localization signal on NF-κB by the IκB proteins [50]. Upon activation by a large number of inducers, including byproducts of microbial, viral infections and proinflammatory cytokines, cytosolic IκB-α is phosphorylated by the IκB kinase complex. Phosphorylated IκB-α is recognized by the β-Trcp receptor, which targets it for polyubiquitination by an ubiquitin ligase belonging to the SCF (Skp-1/Cul/F box) family [46]. Subsequently, the IκB-α protein is degraded by the proteasome, thus unmasking the nuclear localization signal and allowing the NF-κB proteins to translocate the nucleus to regulate the transcription of their target genes, which encode cytokines and chemokines.

After activation of NF-κB, IκB-α is rapidly re-synthesized and sequesters the NF-κB complex in the cytoplasm by masking the NF-κB protein's nuclear localization signal, switching off its activity [51]. This negative feedback, resulting from transcriptional activation of the IκB-α gene by NF-κB, occurs despite continuous activation by cytokines or protein kinase activators [52]. NF-κB proteins exert a dual role on Th differentiation: cRel regulates Th1 differentiation by T cell-dependent and APC-dependent mechanisms, whereas the NF-κB p50 plays a crucial role in the expression of Gata-3, the master Th2 transcription factor [53].

We have shown that peripheral mononuclear cells, including T cells, exhibit high NF-κB binding activity involving the p50/p65 complexes during relapses, which returned to basal levels during remissions. The mechanism underlying persistent NF-κB activation during the active phase of the disease is not well understood. It implies, at least, a downregulation of IκB-α gene activity through reduced induction of transcription and increased destabilization of IκB-α mRNA, as well as increased proteasome-mediated IκB-α degradation, as demonstrated by the accumulation of the phosphorylated form of IκB-α in the presence of MG132, a proteasome inhibitor [54]. The downregulation of NF-κB activity during remissions is correlated with decreased production of most cytokines, such as TNF-α, IL-8 and IL-13 [55].

This NF-κB activation, involving the NF-κB RelA/p50 or RelA dimer complexes, seems to be responsible for the inhibition of IL-4 induction during the active phase of the disease. Indeed, overexpression of c-maf in T cells induces IL-4-promoter-driven luciferase activity. On the other hand, co-expression of c-maf with NF-κB

RelA/p50, or with RelA dimers, inhibits c-maf-dependent IL-4 promoter activity. Overexpression of p50 amplifies c-maf-mediated IL-4 gene activation. These results suggest that increased RelA, whether as homodimers or as heterodimers with p50, significantly blocks the induction of IL-4 by c-maf, in contrast to NF-κB p50 homodimers. Moreover, the ex vivo inhibition of the proteasome activity, which blocks NF-κB activation, strongly increases the IL-4 mRNA level in MCNS T cells from relapsing patients [40].

One mechanism underlying transcriptional repression of the IL-4 gene relies on mutual expelling of NF-κB/c-maf from their respective DNA binding sites. The contribution of other mechanisms, such as the recruitment of co-repressor complexes, is not excluded. Casolaro et al. [56] have reported such transcriptional interference by the NFATp and NF-κB transcription factors on IL-4 gene activation. Taken together, these results indicate that NF-κB activity plays a critical role in the modulation of IL-4 gene activation. In particular, the NF-κB proteins, except p50, are involved in Th1 polarization as T cells from mice overexpressing a protease-resistant IκB (that inhibits NF-κB activation), or from c-Rel-deficient mice, are unable to mount a Th1 response [57]. In contrast, p50 is the only known NF-κB protein that plays a determinant role in Th2 polarization through the induction of Gata3 [58].

The relapse-related downregulation of the IL-12R β22 chain, absence of induction of T-bet, overexpression of c-maf and increased IL-13 production are consistent with T cell commitment towards a Th2 phenotype in MCNS, through an IL-4-independent mechanism. These findings might explain why patients with MCNS often display a defect in delayed-type hypersensitivity response, suggesting an abnormal Th1-dependent cellular immunity [29]. The low production of IL-4 suggests that a non-classical ('truncated') Th2 phenotype occurs in MCNS. In order to understand the molecular basis of this process, we screened cDNA arrays with multiplex RNA probes from T cells overexpressing both c-maf and RelA. We identified five new c-maf target genes of unknown function, and these are currently under investigation.

The Signaling Link between c-maf and Extracellular Signals: The ICOS/c-mip Pathways?

Activation of the T cell receptor provides insufficient signals for effective T cell function. A second set of signals mediated by co-receptors is needed to promote T cell proliferation, lymphokine secretion and effector functions. Among the co-receptors, ICOS, a member of the CD28 family, delivers signals upon interaction with its ligand, B7H, which is expressed by variety of cells including B cells, antigen-presenting cells (APCs) and epithelial cells. In contrast to CD28, which is constitutively expressed by T cells, ICOS is expressed only on activated T cells [59]. Following ligation to B7H, ICOS binds to PI3K and activates PDK1 and AKT/PKB. Studies of knockout mice have revealed that ICOS-, or B7H-deficient T cells are functionally associated with a selective deficiency of c-maf and a strong reduction of IL-4 and IL-13 production [60]. Consequently, the T-dependent immune response is impaired and, in particular, the production of IgE is reduced, while IgM and IgG levels appear normal [61]. However, the number and maturation of B cells are normal. In contrast to these findings in KO mice, ICOS-deficient human T cells do not display any defect in cytokine expression, T helper differentiation, or Ig isotype switching. On the other hand, serum immunoglobulin levels, including IgA, IgM and IgG, were severely reduced in ICOS-deficient patients, leading to recurrent infections. The number of B lymphocytes is reduced, and the expression of the antigen CD27 is notably lacking, suggesting that B cells fail to mature toward a memory phenotype. Taken together, these data are consistent with a severe defect in B cell maturation in human ICOS deficiency and point out the role of ICOS in T cell help for B cells. The expression of c-maf in ICOS-deficient patients has not yet been reported.

Several observations argue against an alteration in the B7H-ICOS signaling axis in MCNS and FSGS. First, although IgG levels are reduced, the concentration of IgM is increased in MCNS, whereas IgE levels are normal or increased, especially in patients with associated allergic disease [62]. Second, the number and the maturation of B cells are normal in MCNS. Third, the expression of ICOS protein in MCNS is similar to that in normal controls (our unpubl. data). Moreover, recent findings suggest that ICOS contributes to IL-4 and IL-13 regulation, but is not required for Th2 differentiation [63]. These observations led us to postulate that the upregulation of c-maf in MCNS occurs through an ICOS-independent pathway.

Regardless of the initial molecular events allowing c-maf induction in MCNS, T cell proximal signaling involves the activation of c-mip (c-maf-inducing protein), an 85- to 90-kDa protein encoded by a gene previously identified in the human brain [64]. The predicted structure of c-mip includes an N-terminal region containing a pleckstrin homology domain, a middle region characterized by the presence of several interacting domains, including a 14-3-3 module, a PKC domain, an Erk domain and an SH3 domain similar to the p85 regulatory subunit of phosphatidylinositol 3-kinase (PI3K), and a C-terminal region containing a leucine-rich repeat domain. Grimbert et al. [65] have isolated a shortened form of c-mip, Tc-mip (for truncated c-maf-inducing protein); the shortening was caused by deletion of 30 amino acids from the N-terminal region of the pleckstrin homology domain. It has been shown that Tc-mip is highly expressed in CD4+ T cells during the active phase in MCNS patients, and promotes the c-maf signaling pathway [65]. The expression of c-mip/Tcmip is predominant in a few tissues, including the thymus, fetal liver, T cells and the kidney. Overexpression of Tc-mip in Jurkat T cells was shown to strongly induce c-maf protein, transactivate the IL-4 gene and downregulate IFN-γ, thereby committing the T cell to a Th2 phenotype. Filamin-A, a member of the actin/spectrin/dystrophin family of actin-binding proteins, is upregulated along with Tc-mip in MCNS patients, and was also shown to interact with both c-mip and Tc-mip [66]. These results suggest that Tc-mip plays a critical role in the c-maf-mediated Th2 signaling pathway.

Does the Innate Immune System Play a Role in the Pathophysiology of MCNS?

One clinical characteristic, likely specific for MCNS, is the rapid occurrence of relapses following viral or bacterial infection, particularly in children. This observation might be consistent with a possible dysfunction of the innate immune system. The innate immune response occurs within hours after an infectious challenge and relies on recognition of microbial products by specific germ line-encoded receptors, the Toll-like receptors (TLRs), which are expressed by APCs, macrophages, neutrophils, natural killer cells, epithelial cells and endothelial cells. Engagement of TLR triggers a signaling cascade involving myeloid differentiation factor 88, IL-1 receptor-associated kinase and TNF receptor-associated factor 6, resulting in the activation of the NF-κB and AP-1 pathways, which then induce the activation of the inflammatory genes such as TNF-α, IL-1, IL-6 and IL-8 [67].

The strong activation of NF-κB signaling in non-T cell populations during MCNS relapse suggests the involvement of innate immune cells in this process [54]. This hypothesis is further supported by data showing the upregulation of many genes that are recruited in the innate immune response, such as Rantes, TNF receptor-associated factor 6, IL-1, IFN-αR and 1-8D protein [37]. Activation of APCs leads to membrane expression of co-stimulator molecules, including B7.1, B7.2 and B7H, along with microbial peptides bound to MHC antigens. Upon interaction with APCs, naïve T cells are activated and elicit an adaptive immune response. TLR signaling controls the adaptive response by at least two mechanisms: it triggers co-stimulatory signals and influences T cell polarization. Whereas the

Th1 immune response requires TLR signaling, the component of the innate system that drives activated T cells towards Th2 commitment remains unknown.

Can Podocytes Mimic Immune Cell Signaling in Some Pathological Situations?

In some children with MCNS, relapses occur very rapidly following viral or bacterial infections, suggesting that activation of the innate immune system is associated with podocyte dysfunction leading to proteinuria. The underlying mechanism, assumed but not formally demonstrated, is that the immune system, perhaps through T cell dysfunction, releases a factor that induces pathological changes in podocytes. Recently, Reiser et al. [68] provided evidence that another mechanism might account for the podocyte dysfunction. They showed that infusion of LPS, a microbial byproduct, in mice induces the expression of B7.1 antigen on podocytes and triggers disorganization of the podocyte cytoskeleton through the TLR-4 signaling pathway. The ability of LPS to induce proteinuria in SCID mice excludes the involvement of immune cells in proteinuria. Indeed, mouse podocytes constitutively express the LPS receptor, TLR-4 and its co-receptor CD14. Induction of B7.1 in podocytes precedes the development of proteinuria, since B7-deficient mice are resistant to LPS-induced proteinuria. This model raises several questions. First, whereas viral or bacterial infections are common in childhood, why do only some children develop MCNS? Second, if the induction of B7.1 on podocytes is a major phenomenon, are the nephrin and podocin pathways highly susceptible to B7.1 signaling in podocytes? Third, increased epithelial expression of TLRs, such as in lupus nephritis, is usually associated with inflammatory reactions; however, these are lacking in MCNS. Although this mechanism has been demonstrated in a murine model, the relevance of this finding to the pathophysiology of human nephrotic syndrome remains to be investigated.

Downregulation of Activation Markers in MCNS Is Correlated with Remission of Nephrotic Syndrome

Glucocorticoids, the first-line therapy for INS, are among the most potent anti-inflammatory and immunosuppressive agents. They inhibit synthesis of almost all known cytokines and of several cell surface molecules required for immune function, but the mechanism underlying their activity is complex and many aspects of their activities remain unclear. After binding to glucocorticoids, the glucocorticoid receptor dissociates from the 'chaperone protein', hsp90, and moves into the nucleus, where it downregulates NF-κB activity, essentially by preventing NF-κB from binding to DNA and/or by increasing IκB-α transcription [69].

Most cytokine genes that are upregulated in MCNS do not have a glucocorticoid-responsive element, but their promoter exhibits binding sites for NF-κB, which may explain the fact that many cytokines are downregulated during remission [70]. The downregulation of NF-κB and c-maf activation correlates with the remission of MCNS [54]. Cyclosporine, largely used in patients who exhibit frequent relapses, represses the activation of the NF-κB and CD28 pathways [71]. The NF-κB pathway seems to play an important role in the pathogenesis of MCNS, but the mechanisms involved are far from clear. The failure of anti-TNF therapy to induce remission may suggest that NF-κB activation in MCNS does not result from TNF production (our unpubl. data).

Alterations in Podocyte Structure and Function During Nephrotic Syndrome

Visceral glomerular epithelial cells, also called podocytes, are terminally differentiated cells that line the outer aspect of the glomerular basement membrane. They constitute the final barrier to urinary protein loss by the formation and maintenance of the podocyte foot processes (FPs) and the interposed slit diaphragm. The basolateral portion of the FPs represents the center of podocyte function and is defined by three membrane domains: the apical membrane domain, the slit diaphragm protein complex and the sole plate (i.e. the basal membrane domain) [72]. Podocyte FPs contain a contractile and dynamic apparatus composed of actin, myosin II, α-actinin-4, talin, vinculin and synaptopodin [73, 74]. The FPs are anchored to the glomerular basement membrane via α3β1 integrin [75] and dystroglycans [76]. Neighboring FPs are connected by the slit diaphragm, which represents the main selectively permeable barrier in the kidney [77]. Our knowledge of slit diaphragm structure comes from genetic studies of familial nephrotic syndromes, which led to the identification of several proteins (podocin, nephrin, α-actinine-4 and Trp6) that are mutated in these inherited forms [4–7]. Of note, no mutation has been reported in MCNS in which the expression of these proteins is unchanged or downregulated [78, 79].

Besides their role in the structural organization of the slit diaphragm, podocin, nephrin and CD2AP participate in cell signaling pathways that prevent apoptosis [80]. Nephrin and CD2AP, by binding to a subunit of PI3K, stimulate activation of the serine/threonine kinase, Akt. Hence, afferent signaling coming from the slit diaphragm under physiological conditions seems to prevent podocyte apoptosis. On the other hand, alterations in the structure of the slit diaphragm may lead, via attenuated PI3K/Akt signaling, to the apoptotic cell death of podocytes [81]. It is well established that the podocyte number is a critical determinant of glomerulosclerosis and that a decrease in podocyte number leads to progressive renal failure [82].

A Possible Alternative Hypothesis Unifying T Cell Disorder and Podocyte Dysfunction

The clinical and experimental data compiled so far are difficult to reconcile with a single hypothesis involving either an extra-renal factor alone, or a simple molecular defect in podocyte structure, to account for the pathophysiology of MCNS or FSGS with relapse. Rather, one might consider that functional alterations allowing to nephrotic proteinuria could result from downregulation of transduction pathway(s) playing a key role in slit diaphragm function, such as the nephrin-mediated pathway. Immuno-ultrastructural studies have demonstrated that the expression of nephrin is diminished in active MCNS [78]. The failure to restore normal nephrin signaling could be due to an antagonistic pathway that is abnormally upregulated in podocytes in active MCNS or FSGS with relapse. This negative cross-talk may involve the induction of new genes, which are basically repressed in normal podocytes. The upregulation of NF-κB expression has been reported in kidney biopsies of recurrent FSGS [83] although the cellular localization of this perturbation has not been determined. One may postulate that some innate, extracellular or intracellular receptor(s), could be abnormally induced in immune and podocyte cells. This assumption is supported by clinical observations of frequent relapses occurring in the vicinity of antigen challenge, which is not consistent with a usual secondary immune response. How does this functional antagonism take place in the kidney? Future researches may help to elucidate the pathophysiology of this intriguing disease.

How does this functional antagonism take place in the kidney? We postulate that the circulating factor is somehow linked to the NF-κB pathway. Of note, the factor is not especially secreted by T cells. We have previously reported a sustained NF-κB activation during MCNS relapse, which might suggest that negative regulation of NF-κB is altered in immune cells. As a result, podocytes and immune cells might share the same molecular defect in NF-κB negative regulatory pathways. Finally, two mechanisms might account for the pathogenesis of MCNS and FSGS with relapse: (1) an activation of NF-κB in immune cells, followed by secretion of a circulating factor that might alter the podocyte signaling; and (2) a lack of efficient negative regulation of NF-κB inducing a downregulation of nephrin signaling in podocyte and consequently proteinuria, as well as production of a plethora of cytokines by immune cells.

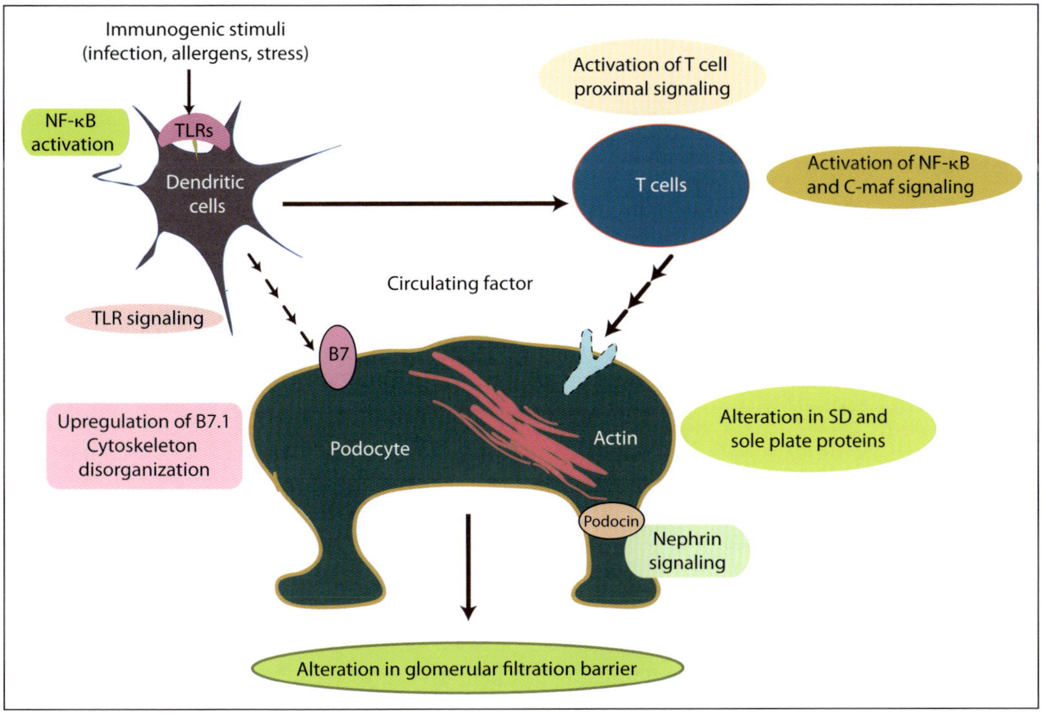

Fig. 1. Potential mechanisms involved in the pathophysiology of INS.

Conclusion

Since the early 1970s, INS has been postulated to be a systemic disorder of T cell function. Clinical observations and recent pathophysiological and experimental data introduce an additional level of complexity, i.e. the potential involvement of the innate immune system. It is conceivable that the innate immune response is triggered by pathogen byproducts that activate Toll receptor-mediated signaling pathways, leading to activation of NF-κB. Alterations in the NF-κB/IκB-α regulatory feedback loop during relapses may be correlated to the degree of recruitment of innate immunity. Subsequently, the innate immune response generates a co-stimulatory signal, which induces an adaptive immune response in which the selective induction of the c-maf pathway contributes to the commitment of T cells towards a Th2 phenotype (fig. 1).

The central role of the podocyte in the induction of the nephrotic syndrome was recently highlighted by the identification of mutations in several podocyte proteins in families with inherited nephrotic syndrome. Moreover, insights from murine models suggest that podocytes may contribute to the activation of innate immunity by expressing the co-stimulatory protein B7-1. How B7 signaling alters podocyte function and disrupts the glomerular barrier remains to be determined. Further studies are required to understand the interplay of immune system disorders and podocyte dysfunction.

Acknowledgments

We thank G. Richet, M. Broyer, P. Niaudet, A. Bensman, P. Ronco, C. Antignac and our numerous colleagues who provided us with blood samples and clinical information as well as for their support and advice.

This work was supported in part by the INSERM institute (the AVENIR program and the APEX program), a grant from the National Academy of Medicine, the association pour l'utilisation du rein artificiel (AURA, Paris) and the University Paris XII.

References

1 Niaudet P: Nephrotic syndrome in children. Curr Opin Pediatr 1993;5:174–179.
2 Nakayama M, Katafuchi R, Yanase T, Ikeda K, Tanaka H, Fujimi S: Steroid responsiveness and frequency of relapse in adult-onset minimal change nephrotic syndrome. Am J Kidney Dis 2002;39:503–512.
3 Ronco P, Allegri L, Brianti E, Chatelet F, Van Leer EH, Verroust P: Antigenic targets in epimembranous glomerulonephritis. Experimental data and potential application in human pathology. Appl Pathol 1989;7:85–98.
4 Kestila M, Lenkkeri U, Mannikko M, et al: Positionally cloned gene for a novel glomerular protein – nephrin – is mutated in congenital nephrotic syndrome. Mol Cell 1998;1:575–582.
5 Boute N, Gribouval O, Roselli S, et al: NPHS2, encoding the glomerular protein podocin, is mutated in autosomal recessive steroid-resistant nephrotic syndrome. Nat Genet 2000;24:349–354.
6 Kaplan JM, Kim SH, North KN, et al: Mutations in ACTN4, encoding alpha-actinin-4, cause familial focal segmental glomerulosclerosis. Nat Genet 2000;24:251–256.
7 Reiser J, Polu KR, Moller CC, et al: TRPC6 is a glomerular slit diaphragm-associated channel required for normal renal function. Nat Genet 2005;37:739–744.
8 Benzing T: Signaling at the slit diaphragm. J Am Soc Nephrol 2004;15:1382–1391.
9 Daniel V, Trautmann Y, Konrad M, Nayir A, Scharer K: T-lymphocyte populations, cytokines and other growth factors in serum and urine of children with idiopathic nephrotic syndrome. Clin Nephrol 1997;47:289–297.
10 Eddy AA, Symons JM: Nephrotic syndrome in childhood. Lancet 2003;362:629–639.
11 Niaudet P: Lipoid nephrosis in childhood. Rev Prat 2003;53:2027–2032.
12 Niaudet P, Gagnadoux MF, Broyer M: Treatment of childhood steroid-resistant idiopathic nephrotic syndrome. Adv Nephrol Necker Hosp 1998;28:43–61.
13 Ramos EL: Recurrent diseases in the renal allograft. J Am Soc Nephrol 1991;2:109–121.
14 Hoyer JR, Vernier RL, Najarian JS, Raij L, Simmons RL, Michael AF: Recurrence of idiopathic nephrotic syndrome after renal transplantation. Lancet 1972;2:343–348.
15 Ali AA, Wilson E, Moorhead JF, et al. Minimal-change glomerular nephritis. Normal kidneys in an abnormal environment? Transplantation 1994;58:849–852.
16 Rea R, Smith C, Sandhu K, Kwan J, Tomson C: Successful transplant of a kidney with focal segmental glomerulosclerosis. Nephrol Dial Transplant 2001;16:416–417.
17 Van Den Berg JG, Aten J, Chand MA, et al: Interleukin-4 and interleukin-13 act on glomerular visceral epithelial cells. J Am Soc Nephrol 2000;11:413–422.
18 Feld SM, Figueroa P, Savin V, Nast CC, Sharma R, Sharma M, Hirschberg R, Adler SG: Plasmapheresis in the treatment of steroid-resistant focal segmental glomerulosclerosis in native kidneys. Am J Kidney Dis 1998;32:230–237.
19 Ginsburg DS, Dau P: Plasmapheresis in the treatment of steroid-resistant focal segmental glomerulosclerosis. Clin Nephrol 1997;48:282–287.
20 Savin VJ, Sharma R, Sharma M, et al: Circulating factor associated with increased glomerular permeability to albumin in recurrent focal segmental glomerulosclerosis. N Engl J Med 1996;334:878–883.
21 Dantal J, Bigot E, Bogers W, et al: Effect of plasma protein adsorption on protein excretion in kidney-transplant recipients with recurrent nephrotic syndrome. N Engl J Med 1994;330:7–14.
22 Dantal J, Godfrin Y, Koll R, et al: Antihuman immunoglobulin affinity immunoadsorption strongly decreases proteinuria in patients with relapsing nephrotic syndrome. J Am Soc Nephrol 1998;9:1709–1715.
23 Boulton Jones JM, Tulloch I, Dore B, McLay A: Changes in the glomerular capillary wall induced by lymphocyte products and serum of nephrotic patients. Clin Nephrol 1983;20:72–77.
24 Yoshizawa N, Kusumi Y, Matsumoto K, et al: Studies of a glomerular permeability factor in patients with minimal-change nephrotic syndrome. Nephron 1989;51:370–376.
25 Tanaka R, Yoshikawa N, Nakamura H, Ito H: Infusion of peripheral blood mononuclear cell products from nephrotic children increases albuminuria in rats. Nephron 1992; 60:35–41.
26 Koyama A, Fujisaki M, Kobayashi M, Igarashi M, Narita M: A glomerular permeability factor produced by human T cell hybridomas. Kidney Int 1991;40:453–460.
27 Lagrue G, Branellec A, Blanc C, et al: A vascular permeability factor in lymphocyte culture supernants from patients with nephrotic syndrome. II. Pharmacological and physicochemical properties. Biomedicine 1975;23: 73–75.
28 Shalhoub RJ: Pathogenesis of lipoid nephrosis: a disorder of T-cell function. Lancet 1974;2:556–560.
29 Fodor P, Saitua MT, Rodriguez E, Gonzalez B, Schlesinger L: T-cell dysfunction in minimal-change nephrotic syndrome of childhood. Am J Dis Child 1982;136:713–717.
30 Peces R, Sanchez L, Gorostidi M, Alvarez J: Minimal change nephrotic syndrome associated with Hodgkin's lymphoma. Nephrol Dial Transplant 1991;6:155–158.
31 Orman SV, Schechter GP, Whang-Peng J, et al: Nephrotic syndrome associated with a clonal T-cell leukemia of large granular lymphocytes with cytotoxic function. Arch Intern Med 1986;146:1827–1829.

32 Karras A, de Montpreville V, Fakhouri F, Grunfeld JP, Lesavre P: Renal and thymic pathology in thymoma-associated nephropathy: report of 21 cases and review of the literature. Nephrol Dial Transplant 2005;20:1075–1082.
33 Neuhaus TJ, Shah V, Callard RE, Barratt TM: T-lymphocyte activation in steroid-sensitive nephrotic syndrome in childhood. Nephrol Dial Transplant 1995;10:1348–1352.
34 Yan K, Nakahara K, Awa S, et al: The increase of memory T cell subsets in children with idiopathic nephrotic syndrome. Nephron 1998;79:274–278.
35 Shevach EM, McHugh RS, Piccirillo CA, Thornton AM: Control of T-cell activation by CD4+ CD25+ suppressor T cells. Immunol Rev 2001;182:58–67.
36 Frank C, Herrmann M, Fernandez S, et al: Dominant T cells in idiopathic nephrotic syndrome of childhood. Kidney Int 2000;57:510–517.
37 Sahali D, Pawlak A, Valanciute A, et al: A novel approach to investigation of the pathogenesis of active minimal-change nephrotic syndrome using subtracted cDNA library screening. J Am Soc Nephrol 2002;13:1238–1247.
38 Stefanovic V, Golubovic E, Mitic-Zlatkovic M, Vlahovic P, Jovanovic O, Bogdanovic R: Interleukin-12 and interferon-gamma production in childhood idiopathic nephrotic syndrome. Pediatr Nephrol 1998;12:463–466.
39 Kimata H, Fujimoto M, Furusho K: Involvement of interleukin (IL)-13, but not IL-4, in spontaneous IgE and IgG4 production in nephrotic syndrome. Eur J Immunol 1995;25:1497–1501.
40 Valanciuté A, le Gouvello S, Solhonne B, et al: NF-kappa B p65 antagonizes IL-4 induction by c-maf in minimal change nephrotic syndrome. J Immunol 2004;172:688–698.
41 Ho IC, Hodge MR, Rooney JW, Glimcher LH: The proto-oncogene c-maf is responsible for tissue-specific expression of interleukin-4. Cell 1996;85:973–983.
42 Nurieva RI, Duong J, Kishikawa H, et al: Transcriptional regulation of th2 differentiation by inducible costimulator. Immunity 2003;18:801–811.
43 Chesi M, Bergsagel PL, Shonukan OO, et al: Frequent dysregulation of the c-maf proto-oncogene at 16q23 by translocation to an Ig locus in multiple myeloma. Blood 1998;91:4457–4463.
44 Kataoka K, Noda M, Nishizawa M: Maf nuclear oncoprotein recognizes sequences related to an AP-1 site and forms heterodimers with both Fos and Jun. Mol Cell Biol 1994;14:700–712.
45 Kim JI, Ho IC, Grusby MJ, Glimcher LH: The transcription factor c-Maf controls the production of interleukin-4 but not other Th2 cytokines. Immunity 1999;10:745–751.
46 Karin M, Ben-Neriah Y: Phosphorylation meets ubiquitination: the control of NF-[kappa]B activity. Annu Rev Immunol 2000;18:621–663.
47 Li Q, Verma IM: NF-kappaB regulation in the immune system. Nat Rev Immunol 2002;2:725–734.
48 Verma IM, Stevenson JK, Schwarz EM, Van Antwerp D, Miyamoto S: Rel/NF-kappa B/I kappa B family: intimate tales of association and dissociation. Genes Dev 1995;9:2723–2735.
49 Silverman N, Maniatis T: NF-kappaB signaling pathways in mammalian and insect innate immunity. Genes Dev 2001;15:2321–2342.
50 Siebenlist U, Franzoso G, Brown K: Structure, regulation and function of NF-kappa B. Annu Rev Cell Biol 1994;10:405–455.
51 Lain de Lera T, Folgueira L, Martin AG, et al: Expression of IkappaBalpha in the nucleus of human peripheral blood T lymphocytes. Oncogene 1999;18:1581–1588.
52 Beg AA, Finco TS, Nantermet PV, Baldwin AS Jr: Tumor necrosis factor and interleukin-1 lead to phosphorylation and loss of I kappa B alpha: a mechanism for NF-kappa B activation. Mol Cell Biol 1993;13:3301–3310.
53 Hilliard BA, Mason N, Xu L, Sun J, Lamhamedi-Cherradi SE, Liou HC, et al: Critical roles of c-Rel in autoimmune inflammation and helper T cell differentiation. J Clin Invest 2002;110:843–850.
54 Sahali D, Pawlak A, Le Gouvello S, et al: Transcriptional and post-transcriptional alterations of IkappaBalpha in active minimal-change nephrotic syndrome. J Am Soc Nephrol 2001;12:1648–1658.
55 Bustos C, Gonzalez E, Muley R, Alonso JL, Egido J: Increase of tumour necrosis factor alpha synthesis and gene expression in peripheral blood mononuclear cells of children with idiopathic nephrotic syndrome. Eur J Clin Invest 1994;24:799–805.
56 Casolaro V, Georas SN, Song Z, et al: Inhibition of NF-AT-dependent transcription by NF-kappa B: implications for differential gene expression in T helper cell subsets. Proc Natl Acad Sci USA 1995;92:11623–11627.
57 Aronica MA, Mora AL, Mitchell DB, Finn PW, Johnson JE, Sheller JR, et al: Preferential role for NF-kappa B/Rel signaling in the type 1 but not type 2 T cell-dependent immune response in vivo. J Immunol 1999;163:5116–5124.
58 Das J, Chen CH, Yang L, Cohn L, Ray P, Ray A: A critical role for NF-kappa B in GATA3 expression and TH2 differentiation in allergic airway inflammation. Nat Immunol 2001;2:45–50.
59 Rudd CE, Schneider H: Unifying concepts in CD28, ICOS and CTLA4 co-receptor signalling. Nat Rev Immunol 2003;3:544–556.
60 Nurieva RI, Mai XM, Forbush K, Bevan MJ, Dong C: B7h is required for T cell activation, differentiation, and effector function. Proc Natl Acad Sci USA 2003;100:14163–14168.
61 Tafuri A, Shahinian A, Bladt F, et al: ICOS is essential for effective T-helper-cell responses. Nature 2001;409:105–109.
62 Lin CY, Chen CH, Lee PP: In vitro B-lymphocyte switch disturbance from IgM into IgG in IgM mesangial nephropathy. Pediatr Nephrol 1989;3:254–258.
63 Dong C, Nurieva RI: Regulation of immune and autoimmune responses by ICOS. J Autoimmun 2003;21:255–260.
64 Nagase T, Kikuno R, Hattori A, Kondo Y, Okumura K, Ohara O: Prediction of the coding sequences of unidentified human genes. XIX. The complete sequences of 100 new cDNA clones from brain which code for large proteins in vitro. DNA Res 2000;7:347–355.

65 Grimbert P, Valanciute A, Audard V, et al: Truncation of C-mip (Tc-mip), a new proximal signaling protein, induces c-maf Th2 transcription factor and cytoskeleton reorganization. J Exp Med 2003;198:797–807.
66 Grimbert P, Valanciute A, Audard V, Lang P, Guellaen G, Sahali D: The Filamin-A is a partner of Tc-mip, a new adapter protein involved in c-maf-dependent Th2 signaling pathway. Mol Immunol 2004;40:1257–1261.
67 Medzhitov R: Toll-like receptors and innate immunity. Nat Rev Immunol 2001;1:135–145.
68 Reiser J, von Gersdorff G, Loos M, et al: Induction of B7-1 in podocytes is associated with nephrotic syndrome. J Clin Invest 2004;113:1390–1397.
69 Ray A, Prefontaine KE: Physical association and functional antagonism between the p65 subunit of transcription factor NF-kappa B and the glucocorticoid receptor. Proc Natl Acad Sci USA 1994;91:752–756.
70 Brattsand R, Linden M: Cytokine modulation by glucocorticoids: mechanisms and actions in cellular studies. Aliment Pharmacol Ther 1996;10(Suppl 2):81–90, discussion 91–92.
71 Meyer S, Kohler NG, Joly A: Cyclosporine A is an uncompetitive inhibitor of proteasome activity and prevents NF-kappaB activation. FEBS Lett 1997;413:354–358.
72 Kerjaschki D: Caught flat-footed: podocyte damage and the molecular bases of focal glomerulosclerosis. J Clin Invest 2001;108:1583–1587.
73 Drenckhahn D, Franke RP: Ultrastructural organization of contractile and cytoskeletal proteins in glomerular podocytes of chicken, rat, and man. Lab Invest 1988;59:673–682.
74 Mundel P, Gilbert P, Kriz W: Podocytes in glomerulus of rat kidney express a characteristic 44 KD protein. J Histochem Cytochem 1991;39:1047–1056.
75 Adler S: Characterization of glomerular epithelial cell matrix receptors. Am J Pathol 1992;141:571–578.
76 Regele HM, Fillipovic E, Langer B, et al: Glomerular expression of dystroglycans is reduced in minimal change nephrosis but not in focal segmental glomerulosclerosis. J Am Soc Nephrol 2000;11:403–412.
77 Tryggvason K, Wartiovaara J: Molecular basis of glomerular permselectivity. Curr Opin Nephrol Hypertens 2001;10:543–549.
78 Wernerson A, Duner F, Pettersson E, et al: Altered ultrastructural distribution of nephrin in minimal change nephrotic syndrome. Nephrol Dial Transplant 2003;18:70–76.
79 Hingorani SR, Finn LS, Kowalewska J, McDonald RA, Eddy AA: Expression of nephrin in acquired forms of nephrotic syndrome in childhood. Pediatr Nephrol 2004;19:300–305.
80 Huber TB, Hartleben B, Kim J, et al: Nephrin and CD2AP associate with phosphoinositide 3-OH kinase and stimulate AKT-dependent signaling. Mol Cell Biol 2003;23:4917–4928.
81 Wolf G, Stahl RA: CD2-associated protein and glomerular disease. Lancet 2003;362:1746–1748.
82 Mundel P, Shankland SJ: Podocyte biology and response to injury. J Am Soc Nephrol 2002;13:3005–3015.
83 Schachter AD, Strehlau J, Zurakowski D, et al: Increased nuclear factor-kappaB and angiotensinogen gene expression in posttransplant recurrent focal segmental glomerulosclerosis. Transplantation 2000;70:1107–1110.
84 Saxena S, Mittal A, Andal A: Pattern of interleukins in minimal-change nephrotic syndrome of childhood. Nephron 1993;65:56–61.
85 Matsumoto K: Decreased production of interleukin-1 by monocytes from patients with lipoid nephrosis. Clin Nephrol 1989;31:292–296.
86 Suranyi MG, Guasch A, Hall BM, Myers BD: Elevated levels of tumor necrosis factor-alpha in the nephrotic syndrome in humans. Am J Kidney Dis 1993;21:251–259.
87 Neuhaus TJ, Wadhwa M, Callard R, Barratt TM: Increased IL-2, IL-4 and interferon-gamma (IFN-gamma) in steroid-sensitive nephrotic syndrome. Clin Exp Immunol 1995;100:475–479.
88 Yap HK, Cheung W, Murugasu B, Sim SK, Seah CC, Jordan SC: Th1 and Th2 cytokine mRNA profiles in childhood nephrotic syndrome: evidence for increased IL-13 mRNA expression in relapse. J Am Soc Nephrol 1999;10:529–537.
89 Hulton SA, Neuhaus TJ, Callard RE, Dillon MJ, Barratt TM: Circulating interleukin 2 receptor (IL2R) in nephrotic syndrome. Kidney Int Suppl 1997;58:S83-S84.
90 Chen WP, Lin CY: Augmented expression of interleukin-6 and interleukin-1 genes in the mesangium of IgM mesangial nephropathy. Nephron 1994;68:10–19.
91 Garin EH, Blanchard DK, Matsushima K, Djeu JY: IL-8 production by peripheral blood mononuclear cells in nephrotic patients. Kidney Int 1994;45:1311–1317.
92 Matsumoto K: Decreased release of IL-10 by monocytes from patients with lipoid nephrosis. Clin Exp Immunol 1995;102:603–607.
93 Matsumoto K, Kanmatsuse K: Increased IL-12 release by monocytes in nephrotic patients. Clin Exp Immunol 1999;117:361–367.
94 Zachwieja J, Bobkowski W, Dobrowolska-Zachwieja A, Lewandowska-Stachowiak M, Zaniew M, Maciejewski J: Intracellular cytokines of peripheral blood lymphocytes in nephrotic syndrome. Pediatr Nephrol 2002;17:733–740.

D. Sahali
INSERM, U955, Hôpital Henri Mondor
51 avenue du Mal de Lattre-de-Tassigny
FR–94010 Créteil (France)
Tel. +33 1 49 81 25 37, Fax +33 1 48 98 25 39, E-Mail dil.sahali@inserm.fr

Herrera GA (ed): Experimental Models for Renal Diseases: Pathogenesis and Diagnosis.
Contrib Nephrol. Basel, Karger, 2011, vol 169, pp 107–125

Membranous Nephropathy

Andrey V. Cybulsky

Department of Medicine, McGill University Health Centre, McGill University, Montreal, Que., Canada

Abstract

The understanding of cellular and molecular mechanisms involved in the pathogenesis of membranous nephropathy (MN) has come from studies carried out in the Heymann nephritis model of MN in the rat, which closely resembles the clinical and pathologic features of the human disease. MN involves the in situ formation of subepithelial immune deposits as a result of circulating antibodies binding to podocyte antigens. Complement activation leads to assembly of C5b-9 on podocyte plasma membranes, and results in sublethal podocyte injury and proteinuria. The podocyte responds to sublethal C5b-9 attack by activating protein kinases, phospholipases, cyclooxygenases, transcription factors, growth factors, NADPH oxidase, stress pathways, proteinases, etc. These signals impact on cell metabolic pathways, the structure/function of lipids and key proteins in the cytoskeleton and slit diaphragm, and on the turnover of extracellular matrix components. Some effects of C5b-9, including dissolution of the actin cytoskeleton, loss of nephrin expression, reduction in F-actin-bound nephrin and loss of slit diaphragm integrity, affect podocytes adversely. Other effects of complement, such as endoplasmic reticulum stress, may limit injury or promote recovery. A number of studies have confirmed the relevance of several experimental insights to the pathogenesis of human MN, including the presence of autoantibodies directed to podocyte antigens in human MN. Increased understanding of nephritogenic antigens, complement activation, and the cellular signaling pathways and targets of C5b-9 will facilitate the design of new approaches to therapy of this important glomerular disease.

Membranous nephropathy (MN) is a common cause of nephrotic syndrome in adults, and while spontaneous remissions occur, MN is an important cause of chronic kidney failure. Morphologically, the disease is characterized by subepithelial immune deposits containing IgG and complement, effacement of visceral glomerular epithelial cell (GEC; podocyte) foot processes, and thickening of the glomerular basement membrane (GBM). Even though a considerable understanding of pathogenic mechanisms in MN has been achieved, treatment of idiopathic (primary) human MN has tended to rely on nonspecific immunosuppression and is often dissociated from the understanding of disease mechanisms. This chapter reviews the mechanisms of glomerular pathophysiology in MN, focusing on complement-mediated podocyte injury, primarily in experimental models, but also in humans (fig. 1). Tubular damage from proteinuria and complement activation, potentially responsible for progressive interstitial inflammation and fibrosis, has been reviewed elsewhere [1, 2].

Experimental Models of MN

The best and most widely employed experimental models to study MN are active and passive

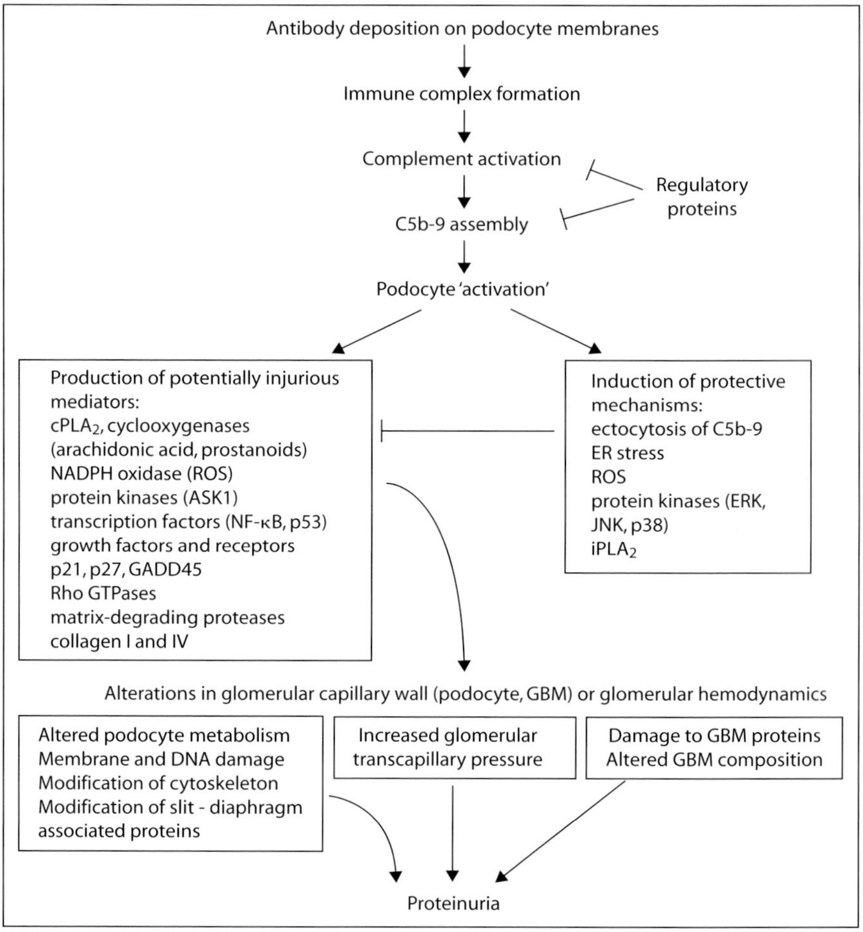

Fig. 1. Pathogenesis of experimental MN. Binding of antibody leads to assembly of C5b-9, which 'activates' podocytes (i.e. alters metabolism and triggers production of mediators) as well as protective mechanisms. The altered metabolism/mediators may contribute to the induction of proteinuria via several targets, including changes in structural components of podocytes, and glomerular hemodynamics. In the longer-term, there are also changes in the GBM.

Heymann nephritis (HN) in rats [3], as the functional and immunohistological features of HN closely resemble the human disease. Studies carried out more than 20 years ago established that immune deposits in HN form in situ as a result of circulating antibodies binding to podocyte antigens. These antibodies may be administered to rats as exogenous heterologous anti-Fx1A in passive Heymann nephritis (PHN), or endogenous autoantibodies may develop after immunization with Fx1A (an antigenic preparation derived from proximal tubular brush border) [4–7]. Furthermore, activation of complement is generally required for the onset of proteinuria, and the effect of complement is due to assembly of C5b-9 and injury to podocytes [4–6].

In contrast to rats, it has been difficult to establish MN models in mice [3]. This is unfortunate since mice provide an opportunity for deletion or overexpression of specific genes, including podocyte-specific expression. A mouse model of MN was induced by administration of cationic bovine serum albumin in three different strains of mice (ICR, BALB/c and C57BL6) [8]. Disease induction was dose-related and strain-specific. ICR and BALB/c mice developed disease, although BALB/c mice required a higher dose. The mice developed features of MN, including subepithelial deposits, proteinuria, diffuse thickening of the GBM and GBM spikes. Glomerular deposition of C3 was reported, but compared with rats, mice generally show more limited glomerular complement activation. Another mouse model of MN was developed by pre-immunizing mice with Freund's adjuvant before administration of a heterologous antibody directed against a murine podocyte cell line [9]. Animals developed albuminuria. Histochemistry revealed diffuse mesangial matrix expansion. Mouse IgG and complement were detected in a linear pattern along the glomerular capillary wall and in the mesangium, but complement depletion did not prevent proteinuria. Glomerular foot processes were flattened. Immunoelectron microscopy localized antibody to podocyte foot processes and the GBM, suggesting that the podocyte was the main target of the antiserum. It should also be noted that a model of MN was developed in rabbits and used to demonstrate that assembly of C5b-9 is required for development of proteinuria [10].

Studies on the structural biology of the podocyte and its role in regulating glomerular permeability have provided additional insights into the pathogenesis of MN [11–14]. The identification of the genes responsible for congenital forms of human proteinuric disorders and proteinuric animal models (e.g. nephrin, neph-1, podocin, cadherins, CD2-AP, α-actinin-4) has led to an appreciation of the structure of the filtration slit diaphragm and its interaction with the podocyte cytoskeleton. It has also established that the slit diaphragm is an important component of the glomerular filtration barrier. Furthermore, proteins of the slit diaphragm have signaling properties, and their signaling may modulate podocyte function.

Nephritogenic Antigens and Immune Deposit Formation: From Experimental to Human MN

Studies in HN have provided the basis for understanding mechanisms of immune deposit formation. Based on work in experimental animal models and cell culture, it can be concluded that subepithelial immune deposits form in situ and are a result of antibody-mediated capping of antigens on the podocyte surface and shedding of the antibody-antigen complexes into the GBM (fig. 1 and 2) [15, 16]. This process occurs independently of preformed circulating immune complexes. The target antigen in HN has been identified as megalin (originally known as gp330), which was shown to reside in coated pits on the soles of podocyte foot processes, where immune complexes are formed [7, 17, 18].

Megalin and receptor-associated protein appear to constitute the target antigen complex in HN. Megalin is a large protein (approx. 600 kDa) and a member of the low-density lipoprotein receptor family [7, 17, 18]. A short fragment of megalin (approx. 80 amino acids) is critical for expression of active disease. This fragment initiates a primary immune response, and subsequently triggers epitope spreading to other domains [19]. Epitope spreading may enhance polyvalent cross-linkage by antibodies, leading to formation of more stable deposits and allowing complement activation and proteinuria. Post-translational modifications and/or conformational determinants of megalin appear to be necessary for its pathogenic potential. Antibodies to receptor-associated protein, which act as a chaperone for megalin, were also detected in HN, but receptor-associated protein by itself does not induce active HN [18].

Megalin is not expressed in human or mouse podocytes, and is not the nephritogenic antigen in idiopathic human MN [18]. Nonetheless, the view that circulating autoantibodies gain access to an antigen on the podocyte foot processes in human MN is a distinct possibility, and is supported by reports from neonatal MN [20]. The mother of an affected child was deficient in neutral endopeptidase (NEP) and developed circulating

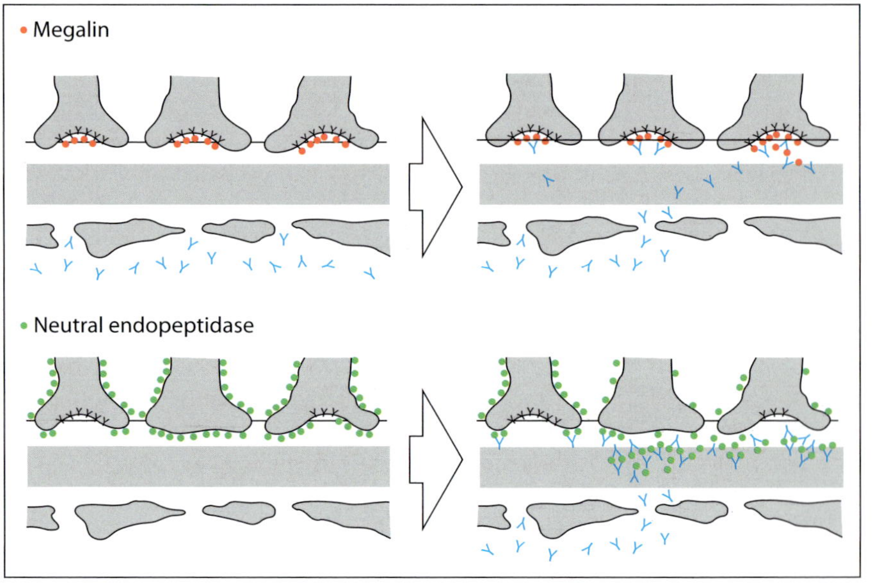

Fig. 2. In situ formation of immune deposits in MN. In PHN and neonatal human MN, nephritogenic antigens are membrane proteins of podocytes. Upper panel: in PHN, the antigen is megalin (red dots), which is concentrated in clathrin-coated pits. Epitope-specific IgG arrives from the circulation to form immune complexes with megalin on the podocyte membrane, followed by shedding and rapid immobilization in the GBM, thus preventing the clearing of complexes via endocytosis by the podocyte. Large immune deposits form by repeated cycles of this mechanism. Lower panel: in neonatal human MN, neutral endopeptidase (NEP; green dots) serves as the pathogenic antigen in the podocyte cell membrane, presumably outside clathrin-coated pits. Antibodies to NEP originate in women who lack NEP because of truncating mutations. Anti-NEP antibody is transported across placenta and causes formation of immune complexes at podocyte membranes, similar to those observed in PHN. Reprinted from Lancet with permission from Elsevier [132].

alloantibodies at the time of a previous miscarriage. Following transplacental transfer, these antibodies caused formation of subepithelial immune deposits in the fetus (fig. 2). Clinical and pathological manifestations of MN were found after birth. The mother's serum contained antibodies to NEP and induced glomerular deposits of IgG when injected into rabbits. Subsequently, additional children from two other families were identified with a similar pathogenesis of MN [21]. It is important to note that four children of one NEP-deficient mother did not develop MN. This mother had low titer anti-NEP antibodies of the noncomplement-fixing IgG4 subclass, but not the complement-fixing IgG1 subclass, whereas mothers of children with MN had high titer IgG1 and IgG4 antibodies [21]. These studies support the concept of in situ immune deposit formation by circulating antibodies to a podocyte antigen in human MN, but NEP may not be the target antigen in idiopathic MN.

Beck et al. [22] demonstrated that serum samples from 26 of 37 patients (70%) with idiopathic, but not secondary, MN specifically identified a 185-kDa glycoprotein in nonreduced glomerular extracts. This protein was identified as M-type phospholipase A_2 receptor (PLA_2R). Anti-PLA_2R autoantibodies in serum samples from patients with MN were mainly IgG4. PLA_2R was expressed in podocytes in normal human glomeruli and co-localized with IgG4 in immune deposits in glomeruli of patients with MN. IgG eluted from kidney biopsies of patients with idiopathic MN, but not in those with lupus MN or IgA nephropathy, also recognized PLA_2R. Therefore, in this cohort of patients with idiopathic MN, a majority had antibodies against a conformation-dependent epitope in PLA_2R, implying that PLA_2R is a major antigen in MN.

Beck et al. [22] also presented preliminary evidence for an association between proteinuria and the presence of the circulating anti-PLA_2R antibodies. Unfortunately, human autoantibodies to PLA_2R apparently do not react with rodent tissue [22], so it may be difficult to prove that these antibodies can induce experimental disease. Perhaps, it may be possible to demonstrate that human MN sera will be able to transfer disease to nonhuman primates that express PLA_2R on podocytes, or that subepithelial deposits develop in a kidney transplanted from a normal donor to a recipient with MN and circulating anti-PLA_2R antibody [23]. In addition, it would be interesting to determine if anti-PLA_2R antibodies can induce capping of the PLA_2R antigen on the podocyte surface, at least in a cell culture model. Another feature of the anti-PLA_2R autoantibodies in human MN is their apparently very narrow specificity for a nonreduced epitope of PLA_2R. This finding appears to be distinct from the process of epitope spreading, which correlates with induction of proteinuria in the animal model of MN [19].

Not all patients with idiopathic MN show anti-PLA_2R autoantibodies. Preliminary observations suggest that low, but significant, levels of anti-NEP antibodies are present in a substantial proportion of patients with MN, and that NEP could be identified in the subepithelial immune deposits by confocal microscopy [24]. There are preliminary reports of other candidate human MN antigens, including cytoplasmic and cytoskeletal proteins [18, 24]. An earlier study showed autoantibodies directed against α-enolase in patients with idiopathic MN [25]. It remains to be determined, however, whether these are primary antigens, or if podocyte injury (see below) exposes intracellular and membrane antigens, thereby inducing a second wave of immunization. In addition, it will be necessary to examine if exogenous antigens trapped in the subepithelial space ('planted' antigens) can induce disease through in situ immune complex formation.

Complement in Experimental and Human MN

The complement system is composed of three pathways, the classical, alternative and mannose-binding lectin pathways, which contain over 30 proteins involved in activation and regulation [26, 27]. The role of complement in experimental MN was overlooked initially because of the absence of glomerular inflammation. However, with the appreciation that the C5b-9 membrane attack complex of complement can injure cells in the absence of inflammatory leukocytes, a role for C5b-9 in inducing podocyte injury was demonstrated in experimental MN [28–30]. Antibodies, which bind to podocyte antigens, lead to complement activation via the classical pathway and subsequent assembly of C5b-9 (fig. 1). The pathogenic role of C5b-9 was demonstrated in both in vivo and ex vivo models of PHN by using sera deficient in terminal complement components (C8 and C6) or by in vivo administration of antibodies to C6. Furthermore, a pathogenic role for C5b-9 was demonstrated in a model of MN in C6-deficient rabbits [10]. In addition, inhibition of complement activation with recombinant human complement receptor 1 (CR1, CD35) led to reduction in proteinuria in PHN [31]. Consistent with these functional data, in PHN, the C5b-9 complex has been localized in immune deposits and GEC plasma membranes on the soles of the foot processes [32]. The complex was endocytosed in GECs, and was released (exocytosed) in membrane vesicles into the urinary space. The latter may represent a mechanism to protect GECs from complement attack (see below).

The alternative pathway of complement is spontaneously active and must be continuously downregulated to prevent bystander injury. Complement regulators are relevant to glomerular diseases [27, 33–35]. The human podocyte expresses CR1 and decay-accelerating factor (CD55), which limit activation of C3 and C5, as well as CD59, an inhibitor of C5b-9 formation. In rodents, Crry supplants the role of CR1 on the podocyte [35]. Neutralizing antibodies to complement regulators are present in the heterologous antisera, which induce PHN [36], and similar antibodies are among the antibody repertoire generated by rats with active HN [37]. By inhibiting podocyte complement regulators, such antibodies can enhance complement activation through the alternative pathway, thereby amplifying the activation of complement induced by other cell-bound antibodies via the classical pathway [35]. In active HN, antibodies that inhibit podocyte Crry are necessary for anti-megalin antibodies to result in sufficient complement activation to induce podocyte injury. Intravenous administration of immunoglobulin to rats with PHN reduced proteinuria and deposition of C3c and C5b-9 [38], suggesting that complement regulation is a possible therapeutic mechanism.

While results of numerous studies support a role for complement in experimental MN, a recent study has demonstrated that proteinuria in PHN may occur without assembly of C5b-9. The authors showed that proteinuria developed in C6-deficient PVG rats to the same extent as in normal PVG rats [39]. Assembly of C5b-9 was evident in the normal rats, but not in the C6-deficient rats. Depletion of CD8+ T cells did not affect proteinuria in the early heterologous phase of the disease, but prevented the increase in proteinuria associated with the later autologous phase. The reason why proteinuria was C5b-9-independent in the PVG rats, but is complement-dependent in the other models is unclear. It is possible that factors including amount of antibody deposition, immune complex size, early complement components or T cells may disrupt glomerular integrity and lead to proteinuria in PVG rats.

Compared with animal models, it is more difficult to implicate complement as being critical to the pathogenesis of human MN. The noncomplement-fixing IgG4 antibody isotype predominates in the immune deposits in idiopathic human MN [5, 18]. Nevertheless, there is significant evidence to support a role for C5b-9. For example, the presence of C3 and C5b-9 [20, 21, 40, 41], as well as C3d [42], has been demonstrated in many cases of human MN. It is important to note that MN associated with anti-NEP antibodies (discussed above) developed in neonates only in the presence of complement-fixing IgG1 antibodies, but not with the noncomplement-fixing IgG4 subclass [18]. Another consideration is that in human MN, complement activation may be initiated through the classical pathway with possible amplification by the alternative pathway. CR1 expression is reduced on podocytes in human MN, and acquired CR1 deficiency could increase the susceptibility of podocytes to complement activation by antibodies in subepithelial immune complexes [43]. However, expression of complement C2 receptor inhibitor trispanning (CRIT), a novel human complement regulatory cell surface receptor, was upregulated on podocytes in MN, and the authors suggested that upregulation of CRIT in activated podocytes might represent a cytoprotective mechanism [44].

C5b-9-Induced Responses in Podocytes

Assembly of C5b-9 in the plasma membrane leads to formation of transmembrane channels or rearrangement of membrane lipids with loss of membrane integrity. Nucleated cells require multiple C5b-9 lesions for lysis, lower doses of C5b-9 induce sublethal (sublytic) injury [45–47], and in vivo, complement-mediated tissue injury is generally associated with sublytic amounts of complement. In sublytic doses, C5b-9 does not simply disrupt the plasma membrane, but activates specific signaling pathways. In GECs, these include phospholipases, protein kinases, transcription factors, growth factors, proteinases, etc. Ultimately, some of these signals may alter metabolic pathways, and the structure or function of lipids and cytoskeleton, slit diaphragm, or other compartments. While these signals injure GECs, it should be noted that other signals activated in parallel limit the complement-induced injury, or promote recovery (fig. 1). As outlined below, multiple pathways are probably activated simultaneously by C5b-9 in podocytes, and there is likely equilibrium among pathways that lead to injury and those that are cytoprotective.

Phospholipase A_2 and Prostanoids

An important pathway activated by C5b-9 in GECs involves release of arachidonic acid (AA) by cytosolic phospholipase A_2-α ($cPLA_2$) [48, 49], and subsequent AA metabolism to prostanoids [50–52]. Assembly of C5b-9 in cultured GECs results in an increase in cytosolic free Ca^{2+} concentration, which is primarily due to calcium influx [53]. $cPLA_2$ activation is regulated by changes in cytosolic Ca^{2+} concentration and by phosphorylation. Activation of $cPLA_2$ and release of AA is compartmentalized to specific GEC organelles, in particular the endoplasmic reticulum (ER). $cPLA_2$ is an important mediator of C5b-9-dependent GEC injury; it provides AA for metabolism via cyclooxygenases (COX)-1 and -2, and may mediate injury more directly, by inducing membrane phospholipid hydrolysis and enhancing the ER stress response (see below). More recently, it has been demonstrated that complement-dependent AA release in cultured GECs may also be mediated calcium-independent phospholipase A_2-γ ($iPLA_2$), and that activation of $iPLA_2$ attenuated complement-induced cytotoxicity [54]. The mechanism requires further elucidation, although it has been suggested that $iPLA_2$ may facilitate metabolism of damaged/oxidized phospholipid acyl chains.

The AA released by $cPLA_2$ may be metabolized to prostanoids by COX enzymes [55]. Resting GECs in culture and glomeruli of normal rats primarily express COX-1. In culture, COX-2 was upregulated by C5b-9, while both isoforms were upregulated in glomeruli of rats with PHN [56]. Upregulation of COX-2 was dependent on activation of protein kinase C and the c-Jun NH_2-terminal kinase (JNK; see below) [57]. Both isoforms of COX were involved in enhanced prostanoid generation. Nonselective COX inhibition, and selective inhibition of COX-2 have been shown to reduce glomerular prostanoid generation and proteinuria in PHN, although inhibition of both COX isoforms was required to achieve a maximum antiproteinuric effect [57–61]. The antiproteinuric effects of COX inhibitors are independent of changes in the glomerular filtration rate. Reduction of proteinuria in PHN was also achieved by a thromboxane synthase inhibitor [62]. A fish oil diet, which shifts production of dienoic prostanoids to inactive trienoic metabolites, reduced production of glomerular thromboxane A_2 and had both protective and therapeutic effects even after the onset of proteinuria in PHN [63]. Thus, a number of distinct experimental approaches support a role for prostanoids in exacerbating proteinuria in PHN, and there may be several mechanisms. First, prostanoids may increase glomerular transcapillary pressure, such that proteinuria is increased on a hemodynamic basis. Second, production of prostanoids may exacerbate complement-induced GEC injury directly [58, 64].

ER Injury and ER Stress

ER stress refers to a pathological state, which results in an accumulation of misfolded proteins in the ER (fig. 3a). To rescue misfolded proteins, the ER has in place quality control machinery, including the unfolded protein response [65, 66]. In GECs, C5b-9 induced damage to the membrane of the ER, as well as ER stress (unfolded protein response), and these effects were dependent (in part) on the activation of $cPLA_2$ [67, 68]. The latter is in keeping with the observation that activation of $cPLA_2$ leads to phospholipid hydrolysis at the membrane of the ER. Thus, complement increased expression of ER chaperones (BiP and GRP94) and stimulated activation of PERK (PKR-like ER kinase), phosphorylation of eukaryotic translation initiation factor-2a subunit (eIF2a), and a reduction in protein synthesis, in a $cPLA_2$-dependent manner. Expression of BiP antisense mRNA in GECs exacerbated complement-mediated GEC injury, confirming a functionally important protective role for BiP.

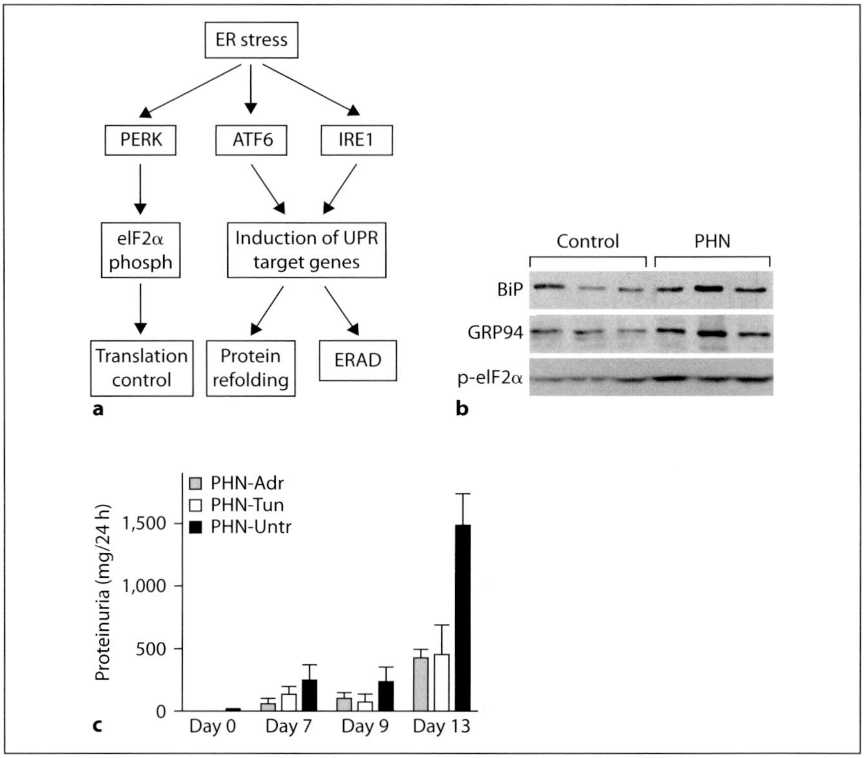

Fig. 3. ER stress in experimental MN. **a** The major UPR signaling pathways are initiated by three protein sensors: PERK, activating transcription factor-6 (ATF6) and inositol requiring-1α (IRE1). Upon accumulation of unfolded proteins in the ER, PERK is activated to phosphorylate eIF2α, which reduces initiation AUG codon recognition. The general rate of translation is reduced. ATF6 moves to the Golgi, where it is cleaved by site-1 and site-2 proteases to yield a cytosolic fragment. This fragment migrates to the nucleus to activate transcription of UPR target genes (e.g. BiP, GRP94), which then facilitate protein folding or ER-associated degradation (ERAD). IRE1 dimerizes and activates its endoribonuclease activity. It cleaves X-box binding protein-1 mRNA and changes the reading frame to yield a potent transcriptional activator. **b** ER stress parameters are increased in PHN. Glomeruli were isolated from normal rats (control) and from rats with PHN on day 14, and lysates were immunoblotted with antibodies to BiP, GRP94 or phospho-eIF2α. **c** Preconditioning of rats with a subnephritogenic dose of adriamycin (Adr) or tunicamycin (Tun) reduces proteinuria in PHN. Rats were untreated (Untr), or were injected with adriamycin or tunicamycin to upregulate ER stress proteins. Four days later, rats were injected with nephritogenic antibody to induce PHN (day 0). Urine protein excretion was measured on days 0, 7, 9 and 13. Adapted from research originally published in Cybulsky et al. [67, 68].

Moreover, fibroblasts from PERK-knockout mice were more susceptible to complement-mediated injury, as compared with wild-type fibroblasts. Thus, induction of ER stress represents a mechanism for protection of cells from complement attack. Furthermore, 'preconditioning' GECs with other stimuli that increased expression of ER stress proteins (e.g. puromycin aminonucleoside) limited the amount of injury during subsequent complement attack.

Compared with normal rats, glomerular BiP and GRP94 proteins were upregulated, and phosphorylation of eIF2α was enhanced in proteinuric rats with PHN (fig. 3b). Administration of a subnephritogenic dose of adriamycin or tunicamycin to rats enhanced glomerular expression of BiP and GRP94. On the basis of these results, rats were treated with a subnephritogenic dose of adriamycin, or tunicamycin, and PHN was then induced in the pretreated rats and in untreated animals. Substantial proteinuria developed in untreated rats with PHN, whereas the amount of proteinuria was significantly lower in the rats with PHN that had been pretreated with the two compounds (fig. 3c) [67]. Thus, in vivo 'preconditioning' to increase ER stress can reduce C5b-9-mediated GEC injury. It should be noted ER stress preconditioning also reduced C5b-9-mediated mesangial cell injury in vivo [69]. These studies provide a rationale for developing non-toxic methods to induce ER stress in vivo, which may eventually have applications to therapy of complement-mediated glomerular disease.

Production of Reactive Oxygen Species

Generation of reactive oxygen species (ROS) in phagocytic cells is catalyzed by a multicomponent enzyme complex, the NADPH oxidase, which in the presence of NADPH, reduces molecular oxygen to the superoxide anion [70]. Superoxide can be further metabolized to other ROS. GECs in culture and in vivo express components of the NADPH oxidase [71, 72], and incubation of cultured GECs with complement stimulated superoxide production via the NADPH oxidase. The mechanism involves activation of $cPLA_2$ and release of AA [73].

Significant amounts of ROS are produced in PHN, and ROS are associated with lipid peroxidation and modification of GEC membrane proteins and GBM components by malondialdehyde adducts [72, 74]. Inhibition of ROS and lipid peroxidation with an antioxidant drug reduced urine protein excretion in PHN, implying that ROS exacerbate proteinuria by damaging the glomerular capillary wall [74]. Although classically, production of ROS is associated with cell or tissue damage, more recently it has been shown that ROS (particularly at low concentrations) function in signaling [70]. In GECs in culture and in vivo, complement-induced superoxide production resulted in the activation of JNK [73]. This process was dependent on $cPLA_2$ and AA release, the latter stimulating JNK activity via the membrane-associated NADPH oxidase and production of ROS. In GEC culture, JNK activation was associated with cytoprotection [73]. Thus, in PHN, ROS may both facilitate signal transduction and contribute to cell injury.

Activation of Protein Kinases

Studies in GEC culture and in PHN showed that assembly of C5b-9 resulted in the transactivation of receptor tyrosine kinases [75]. This was followed by activation of the Ras-extracellular signal-regulated kinase (ERK) pathway (see below), as well as phospholipase C-γ1 [76]. Transactivated tyrosine kinases may serve as scaffolds for assembly and/or activation of proteins, which then lead to activation of downstream effector pathways, e.g. protein kinase C, either independently or in conjunction with increased cytosolic Ca^{2+} concentration.

In addition to the aforementioned protein kinases, C5b-9 can activate the p38 kinase pathway in GECs [77]. The functional role of p38 appears rather complex. In culture, the p38 pathway reduced complement-mediated cytotoxicity, and the cytoprotective effect may involve HSP27, a cytoskeleton-interacting protein. By analogy to cultured GECs, p38 activity was increased in glomeruli of rats with PHN, and treatment of these rats with a p38 inhibitor exacerbated proteinuria. In keeping with experimental data, enhanced phosphorylation of p38 was demonstrated in podocytes in biopsy specimens from human MN [78].

Apoptosis signal-regulating kinase-1 (ASK1) is a newer member of the MAPK kinase kinase family, which is regulated by ROS, and is reported to stimulate both the p38 and JNK pathways [79]. ASK1 was activated in the glomeruli of rats with PHN, and exposure of GECs in culture to C5b-9 stimulated ASK1 activity, in part via the NADPH oxidase and ROS [80]. Complement stimulated p38 and JNK activation, which was amplified in GECs that overexpress ASK1. Complement-induced injury was enhanced in GECs that overexpress ASK1, and was attenuated in GECs that express a dominant negative ASK1 mutant. Inhibition of p38 attenuated complement injury in GECs that overexpress ASK1, but not in control cells. In the latter, generation of ROS restricted complement-mediated GEC injury, but the protective effect of ROS was lost when ASK1 was overexpressed. Thus, the level of ASK1 expression determined the functional effect of p38 activation, i.e. when ASK1 was overexpressed, p38 activation was amplified, and C5b-9 assembly led to GEC injury via ASK1 and p38.

Transcription Factors and Growth Factor Signaling

Cell injury responses may include proliferation and/or apoptosis, and may involve induction of growth factors, growth factor receptors and proteins that regulate the cell cycle [81]. The platelet-derived growth factor-B chain was upregulated in PHN prior to the onset of proteinuria [82]. Platelet-derived growth factor-C expression was not demonstrated in PHN [83], but was prominent in podocytes in human MN [84]. Heparin-binding epidermal growth factor-like factor (HB-EGF) was upregulated in PHN, prior to the onset of proteinuria [85]. At this early time point, HB-EGF was localized in the cytoplasm of GECs; later, HB-EGF showed a nodular pattern within GECs and along the GBM, consistent with secretion and binding to the GBM. Ret, the glial cell-derived neurotrophic factor receptor tyrosine kinase, was upregulated in podocytes in PHN and in cultured mouse podocytes after exposure to sublytic C5b-9 [86]. The functional relevance of these growth factors will require further elucidation.

In vivo, podocytes are believed to be terminally differentiated cells with a low capacity for proliferation under normal circumstances and after injury [14]. In PHN, podocytes do not proliferate, although an increase in proliferating cell nuclear antigen expression has been reported [82]. Various changes in cyclins, cyclin-dependent kinases (CDKs) and CDK inhibitors have also been reported in PHN. There is upregulation of p21 and p27, in association with decreased CDK2 activity [81]. p57 was decreased [87], but Cdc2; cyclins B1, B2 and D1; and phosphorylated histone 3 were increased [88, 89]. Administration of basic fibroblast growth factor to rats with PHN attenuated the increase in p21 levels and increased GEC mitosis and ploidy [81]. The low proliferative capacity of GECs in response to injury may be, at least in part, due to an increase in expression of specific CDK inhibitors.

Complement activated nuclear factor-κB (NF-κB) and NF-κB-dependent gene transcription in cultured GECs and in vivo [57, 90]. In mild PHN (i.e. where rats developed modest proteinuria), administration of the NF-κB inhibitor pyrrolidone dithiocarbamate reduced autologous phase proteinuria, but the drug had no effect in PHN with high levels of urine protein excretion [90]. These results suggest that NF-κB may (in part) contribute to proteinuria in PHN. A recent study has demonstrated that downregulation of nephrin expression may upregulate NF-κB in podocytes [91]. While nephrin downregulation has been reported in PHN (see below), it remains to be determined if the nephrin-NF-κB pathway is active in MN.

In cultured GECs and in PHN, C5b-9 increased DNA damage in association with increases in the tumor suppressor gene, p53, growth-arrest DNA damage-45 (GADD45) gene, p21, and checkpoint kinases-1 and -2 [88, 92]. Inhibition of ERK reduced the increase in p21 and GADD45, and augmented the DNA damage response to C5b-9-induced injury [92]. Induction of DNA damage by C5b-9 may, in part, explain why GEC proliferation is limited after injury. Mechanisms for ERK activation by C5b-9 have been characterized in GECs in culture and PHN in vivo. Expression of a dominant-inhibitory mutant of Ras or of a constitutively-active RhoA mutant, or disassembly of F-actin, reduced complement-mediated activation of ERK [93], indicating that the mechanisms involve the Ras pathway and are facilitated by an intact actin cytoskeleton.

C5b-9-Induced Changes to the Podocyte Cytoskeleton

Actin filaments are the core structural component of podocyte foot processes. Condensation of the actin cytoskeleton at the base of effaced podocyte foot processes together with alterations in filtration slits, displacement and disruption of slit diaphragms are features of both human and experimental MN [94, 95]. In cultured GECs, C5b-9 disassembled F-actin filaments and focal adhesion complexes [96]. The latter was associated with ATP depletion and dephosphorylation of paxillin, but was not prevented by tyrosine phosphatase inhibition. There was no evidence of proteolysis of the focal adhesion or cytoskeletal proteins, which favors dynamic disassembly and reassembly, perhaps mediated by Rho GTPases. $\alpha 3\beta 1$ integrins, which are among the molecules responsible for attachment of GECs to the underlying matrix, remained intact. This is consistent with the normal distribution of $\beta 1$ integrins on the basal aspect of effaced foot processes in kidney biopsies of human MN [97].

Exposure of cultured GECs to C5b-9 increased RhoA activity and decreased activities of Rac1 and Cdc42 [98]. Analogous changes in the activities of these Rho-GTPases were observed in glomeruli of rats with PHN. Cultured GECs transfected with constitutively active RhoA showed a smaller and round contour and prominent cortical F-actin, while those transfected with constitutively active Rac1 underwent morphological changes resembling process formation. In addition, expression of constitutively active RhoA attenuated complement-mediated cytotoxicity, whereas cytotoxicity was augmented by a dominant negative RhoA. These results indicate that complement alters the balance of RhoA, Rac1 and Cdc42 activities in GECs, and suggest that the activity of Rac1 may contribute to process formation, while activation of RhoA (e.g. in the setting of complement attack) may have an opposite effect, i.e. contribute to foot process effacement. Activation of RhoA increased the resistance of GECs to complement-mediated injury. In keeping with these results, it was demonstrated that expression of constitutively active Ras protected GECs from complement cytotoxicity [99]. The protective effect of Ras was dependent on remodeling of the actin cytoskeleton, and was associated with a reduction in Rac activity, thereby altering the equilibrium in the activities of Rho GTPases.

Transient receptor potential channel (TRPC)-6 belongs to the TRP superfamily of ion channel-forming proteins, and is functionally connected to the podocyte actin cytoskeleton. Mutations in the TRPC6 channel are associated with hereditary proteinuric disease. Induced expression of wild-type TRPC6 was found in human proteinuric kidney diseases, including MN, and cultured podocytes that were exposed to complement upregulated TRPC6 protein [100]. This study suggests the involvement of TRPC6 and actin cytoskeleton in the pathology of MN. TRPC6 behaves as a nonselective cation channel, and given that the C5b-9 complex may itself function as a nonselective channel, it is presently unclear how TRPC6 channels may induce podocyte abnormalities in addition to the effects of C5b-9.

Complement-Induced Changes to the Filtration Slit Diaphragm

Nephrin was the first identified component of the filtration slit diaphragm, and several studies have addressed the integrity of the slit diaphragm in MN using nephrin expression, redistribution or post-translational modification as markers. Nephrin expression was reduced and its distribution altered in human MN [101–103]. Glomeruli of proteinuric rats with PHN exhibited a progressive decline in nephrin content after disease induction, and this decline was prevented by angiotensin II-converting enzyme blockade [104]. By analogy, the angiotensin II receptor blocker, candesartan, as well as the peroxisome proliferator-activated receptor-γ agonist, pioglitazone, reduced proteinuria and limited renal functional and structural changes in PHN [105]. Kidneys from rats with PHN showed lower nephrin mRNA and protein than normal rats, and both were restored by pioglitazone. The effect of pioglitazone may be related to its ability to interact with the nephrin promoter and increase nephrin transcription. An examination of rats with PHN prior to the onset of proteinuria showed complement-dependent disruption or dislocation of slit diaphragms and formation of occluding-type junctions, as well as a reduction in nephrin expression, an alteration in the distribution of nephrin, and reduction in the fraction of nephrin that was bound to actin [106, 107]. It was concluded that the onset of proteinuria in PHN is due to complement-dependent dissociation of nephrin from the actin cytoskeleton and loss of slit diaphragm integrity.

Podocin localizes in a lipid raft membrane domain at the insertion of the slit diaphragm and associates with the cytoplasmic tail of nephrin, providing a scaffolding function [108]. In PHN, the location of both nephrin and podocin was altered prior to the onset of proteinuria, and the two molecules dissociated from each other [109]. Furthermore, the lipid raft- or actin-associated portion of nephrin was reduced, and a fraction of nephrin (but not podocin) was apparently shed into the urine, as nephrin was found in the urine of rats with PHN. Shedding of nephrin from cultured cells injured by complement [102] may explain the findings that total nephrin is reduced in experimental and human MN; however, shedding of intact podocytes into the urine may also account for the loss of nephrin in PHN [110].

The situation is probably considerably more complex, as suggested by studies of nephrin translation, as well as post-translational regulation (tyrosine phosphorylation). PHN is associated with ER stress (discussed above) and there is evidence to indicate that translation of nephrin may be facilitated during ER stress [68]. Phosphorylation of eIF2α by PERK leads to a general reduction in protein translation during ER stress, but translation of a green fluorescent protein-reporter regulated by the 5′ end of mouse nephrin mRNA was paradoxically maintained. This effect was most likely due to the presence of short open reading frames in this 5′-flanking mRNA segment, and is in keeping with the observations that translation of mRNAs containing short open reading frames is more efficient after phosphorylation of eIF2α by PERK [66]. Tyrosine phosphorylation of nephrin may regulate interactions of nephrin with various binding partners [13]. The intracellular domain of nephrin contains six conserved tyrosine residues, and several of these residues can be phosphorylated by nonreceptor tyrosine kinases of the Src family. Phosphorylation of nephrin appears to augment its interaction with podocin [111]. In glomeruli of rats with PHN, there was activation of Src, and an increase in tyrosine phosphorylation of nephrin, which would be in keeping with an enhanced interaction of nephrin with podocin. In this study, there were no apparent differences in nephrin expression. These results in PHN are distinct from other proteinuric conditions, where nephrin was hypophosphorylated (Fyn-deficient mice) [112]. Further studies will be required to better define the role of nephrin phosphorylation and interactions with other proteins in MN. It should also be noted that additional binding partners of nephrin continue to be discovered (e.g. partitioning defective 3, partitioning defective 6, atypical protein kinase C), and that the components of the slit diaphragm may include various proteins previously shown to be associated with adherens and tight junctions [113, 114]. Furthermore, complex changes in expression and/or association of these proteins were demonstrated in experimental nephrotic syndrome (puromycin aminonucleoside nephrosis) [114]. Therefore, proper interactions of multiple proteins are required for maintenance of a normal slit diaphragm structure.

Extracellular Matrix Turnover and GBM Expansion: Implications for Chronic MN

Proteinuric kidneys transplanted from rats with active HN into normal syngeneic rats showed persistent low-grade proteinuria despite absence of new glomerular antibody deposition and resolution of complement deposition [115]. This result suggests that once MN enters a chronic phase, extensive remodeling of the GBM may prevent reconstitution of the normal podocyte-GBM interface and foot process architecture. In human MN, expansion of the GBM is due to increased deposition of type IV collagen isoforms $\alpha3$, $\alpha4$ and $\alpha5$, as well as laminin, nidogen and fibronectin, with subsequent accumulation of collagen IV isoforms $\alpha1$ and $\alpha2$ [116]. Studies in PHN suggest that GBM expansion is, at least in part, due to altered synthesis of extracellular matrix components by the injured podocytes [117]. Changes in GBM components or enzymes involved in extracellular matrix metabolism have been reported in active HN and PHN (fig. 1), although some results are variable among studies. In active HN, there was altered synthesis of laminin, fibronectin, entactin and heparan sulfate proteoglycans [118]. mRNA levels of $\alpha3$, $\alpha4$ and $\alpha5$ type IV collagen chains were increased in glomeruli of rats with PHN, and there was de novo synthesis of type I collagen [117]. Expression of agrin, the principal heparan sulfate proteoglycan of the GBM, was decreased in active HN, and this reduction correlated inversely with the amount of proteinuria [119]. Matrix metalloproteinase-9, an enzyme with proteolytic activity against type IV collagen, was upregulated in GECs of rats with PHN [120]. GEC heparanase expression was increased in PHN in a complement-dependent manner, and administration of anti-heparanase antibody reduced proteinuria [121, 122]. Transforming growth factor (TGF)-β family members may play a role in the metabolism of extracellular matrix components. Expression of TGF-β2 and TGF-β3 (but not TGF-β1), as well as TGF-β type I and type II receptors, were increased in PHN in a complement-dependent manner [123]. Therefore, in PHN, assembly of C5b-9 in GECs may promote formation of an expanded and disorganized GBM, while secreted proteolytic enzymes may be upregulated in GECs, alter breakdown or turnover of glomerular capillary wall components, and contribute to proteinuria.

Conclusion

Treatment of human MN has often relied on nonspecific immunosuppressive agents and tends to be dissociated from the understanding of disease mechanisms. Better insights into mechanisms will evolve into more specific concept-driven therapies (fig. 4). The identification of NEP as a human podocyte antigen has encouraged research into discovery of additional antigens and circulating autoantibodies. Conclusive proof for a pathogenic role of anti-PLA_2R autoantibodies is still pending; however, the recent identification of PLA_2R as a major MN antigen [22] represents a major breakthrough in the understanding of human MN. As a result, it may soon be possible to abandon the term 'idiopathic' when referring to primary human MN. Characterization of nephritogenic antigens will eventually allow development of diagnostic immunoassays to quantify autoantibodies, and perhaps establish biomarkers of disease activity. Such assays may facilitate rational therapy with anti-B cell (CD20) antibodies, e.g. in those cases with high titers of circulating nephritogenic antibodies [124]. Furthermore, definition of an antibody subclass (e.g. antibodies belonging to a complement-fixing subclass) may allow for targeted application of complement inhibitors and regulators [27].

Deposition of antibody and formation of immune deposits in glomeruli does not necessarily alter podocyte morphology or function [28]. Based on the need for C5b-9 in the pathogenesis of experimental MN, a multicenter double-blind controlled study of an inhibitory humanized anti-C5 monoclonal antibody (Eculizumab) was performed in 200 patients with idiopathic MN. There was no effect on proteinuria or renal function after 16 weeks of therapy; however, it

Steps in the pathogenesis of MN	Potential treatment options
Generation of immune response	Corticosteroids Cytotoxic drugs Calcineurin inhibitors Anti-CD20 antibody
Complement activation	Soluble CR1 or targeted CR1 Anti-C5 antibody Intravenous immunoglobulin
C5b-9-induced podocyte injury	Cyclooxygenase inhibitors Fish oil Antioxidants Modulators of stress pathways Protein kinase inhibitors/activators Corticosteroids Calcineurin inhibitors
Proteinuria	Angiotensin-converting enzyme inhibitors Angiotensin receptor blockers

Fig. 4. A mechanistic approach to the therapy of MN. Potential treatment options for the key steps in the pathogenesis of MN are outlined. The treatments include those based on cell culture or in vivo models of experimental MN, and those in clinical practice.

appeared that the dose regimens of Eculizumab were too low because inhibition of C5 was inadequate in most patients [124]. The patients who received Eculizumab in an open-label extension apparently had impressive responses [35]. Another area that may have therapeutic potential is modulation of complement activation via complement regulatory proteins [35, 125].

Identification of the signaling pathways activated by C5b-9 in podocytes provides other opportunities for therapeutic intervention. These may include drugs to alter glomerular arachidonate metabolism, oxidant and protease inhibitors, use of agents to modulate ER stress (e.g. salubrinal) [126] or ER protein quality control (e.g. chemical chaperones) [127], or selective modulators of protein serine-threonine kinases and tyrosine kinases. Antioxidant treatment of resistant idiopathic MN achieved a modest result [128]. Glucocorticoids improve renal structure and function in experimental renal diseases characterized by podocyte injury, and may target specific pathways in podocytes (e.g. p53) [129]. A better understanding of the mechanisms of cytoskeleton and slit diaphragm disassembly may provide opportunities for use of small molecule protein kinase inhibitors or activators to reverse podocyte injury. A recent study provides rationale for using calcineurin inhibitors to modulate the podocyte cytoskeleton and thereby reduce proteinuria [130]. Rapamycin reduced proteinuria in established PHN, suggesting that rapamycin may represent a viable option for treatment of MN in humans [131]. Effective therapy may require combinations of approaches to block antibody production, inhibit complement activation, modulate signaling pathways and induce cytoprotection. Finally, it should be noted that therapies in human MN are most likely to be effective when there are high titers of autoantibody, and active antibody deposition and complement activation, before extensive GBM remodeling has occurred.

Acknowledgments

This work was supported by Research Grants from the Canadian Institutes of Health Research (MOP-53264, MOP-84213), the Kidney Foundation of Canada and the Catherine McLaughlin Hakim Chair.

References

1 Nangaku M, Pippin J, Couser WG: Complement membrane attack complex (C5b-9) mediates interstitial disease in experimental nephrotic syndrome. J Am Soc Nephrol 1999;10:2323–2331.

2 Hsu SI, Couser WG: Chronic progression of tubulointerstitial damage in proteinuric renal disease is mediated by complement activation: a therapeutic role for complement inhibitors? J Am Soc Nephrol 2003;14: S186–S191.

3 Pippin JW, Brinkkoetter PT, Cormack-Aboud FC, Durvasula RV, Hauser PV, Kowalewska J, Krofft RD, Logar CM, Marshall CB, Ohse T, Shankland SJ: Inducible rodent models of acquired podocyte diseases. Am J Physiol Renal Physiol 2009;296:F213–F229.

4 Salant DJ, Quigg RJ, Cybulsky AV: Heymann nephritis: mechanisms of renal injury. Kidney Int 1989;35:976–984.

5 Cybulsky AV, Quigg RJ, Salant DJ: Experimental membranous nephropathy redux. Am J Physiol Renal Physiol 2005;289:F660–F671.

6 Nangaku M, Shankland SJ, Couser WG: Cellular response to injury in membranous nephropathy. J Am Soc Nephrol 2005;16:1195–1204.

7 Ronco P, Debiec H: Molecular pathomechanisms of membranous nephropathy: from Heymann nephritis to alloimmunization. J Am Soc Nephrol 2005;16:1205–1213.

8 Chen JS, Chen A, Chang LC, Chang WS, Lee HS, Lin SH, Lin YF: Mouse model of membranous nephropathy induced by cationic bovine serum albumin: antigen dose-response relations and strain differences. Nephrol Dial Transplant 2004;19:2721–2728.

9 Meyer TN, Schwesinger C, Wahlefeld J, Dehde S, Kerjaschki D, Becker JU, Stahl RA, Thaiss F: A new mouse model of immune-mediated podocyte injury. Kidney Int 2007; 72:841–852.

10 Groggel GC, Adler S, Rennke HG, Couser WG, Salant DJ: Role of the terminal complement pathway in experimental membranous nephropathy in the rabbit. J Clin Invest 1983; 72:1948–1957.

11 Pavenstadt H, Kriz W, Kretzler M: Cell biology of the glomerular podocyte. Physiol Rev 2003;83:253–307.

12 Ly J, Alexander M, Quaggin SE: A podocentric view of nephrology. Curr Opin Nephrol Hypertens 2004;13:299–305.

13 Tryggvason K, Patrakka J, Wartiovaara J: Hereditary proteinuria syndromes and mechanisms of proteinuria. N Engl J Med 2006;354:1387–1401.

14 Shankland SJ: The podocyte's response to injury: role in proteinuria and glomerulosclerosis. Kidney Int 2006;69:2131–2147.

15 Camussi G, Brentjens JR, Noble B, Kerjaschki D, Malavasi F, Roholt OA, Farquhar MG, Andres G: Antibody-induced redistribution of Heymann antigen on the surface of cultured glomerular visceral epithelial cells: possible role in the pathogenesis of Heymann glomerulonephritis. J Immunol 1985;135:2409–2416.

16 Kerjaschki D, Miettinen A, Farquhar MG: Initial events in the formation of immune deposits in passive Heymann nephritis. gp330-anti-gp330 immune complexes form in epithelial coated pits and rapidly become attached to the glomerular basement membrane. J Exp Med 1987;166:109–128.

17 Farquhar MG, Saito A, Kerjaschki D, Orlando RA: The Heymann nephritis antigenic complex: megalin (gp330) and RAP. J Am Soc Nephrol 1995;6:35–47.

18 Ronco P, Debiec H: Target antigens and nephritogenic antibodies in membranous nephropathy: of rats and men. Semin Immunopathol 2007;29:445–458.

19 Shah P, Tramontano A, Makker SP: Intramolecular epitope spreading in Heymann nephritis. J Am Soc Nephrol 2007;18:3060–3066.

20 Debiec H, Guigonis V, Mougenot B, Decobert F, Haymann JP, Bensman A, Deschenes G, Ronco PM: Antenatal membranous glomerulonephritis due to anti-neutral endopeptidase antibodies. N Engl J Med 2002;346:2053–2060.

21 Debiec H, Nauta J, Coulet F, van der Burg M, Guigonis V, Schurmans T, de Heer E, Soubrier F, Janssen F, Ronco P: Role of truncating mutations in MME gene in fetomaternal alloimmunisation and antenatal glomerulopathies. Lancet 2004;364:1252–1259.

22 Beck LH Jr, Bonegio RG, Lambeau G, Beck DM, Powell DW, Cummins TD, Klein JB, Salant DJ: M-type phospholipase A2 receptor as target antigen in idiopathic membranous nephropathy. N Engl J Med 2009;361:11–21.

23 Glassock RJ: Human idiopathic membranous nephropathy – a mystery solved? N Engl J Med 2009;361:81–83.

24 Ronco P, Debiec H: Target antigens and nephritogenic antibodies in membranous nephropathy. Arch Med Sci 2009;5:S442–S450.

25 Wakui H, Imai H, Komatsuda A, Miura AB: Circulating antibodies against alpha-enolase in patients with primary membranous nephropathy. Clin Exp Immunol 1999;118:445–450.

26 Walport MJ: Complement. First of two parts. N Engl J Med 2001;344:1058–1066.

27 Brown KM, Sacks SH, Sheerin NS: Mechanisms of disease: the complement system in renal injury – new ways of looking at an old foe. Nat Clin Pract Nephrol 2007;3:277–286.

28 Cybulsky AV, Rennke HG, Feintzeig ID, Salant DJ: Complement-induced glomerular epithelial cell injury. Role of the membrane attack complex in rat membranous nephropathy. J Clin Invest 1986;77:1096–1107.

29 Cybulsky AV, Quigg RJ, Salant DJ: The membrane attack complex in complement-mediated glomerular epithelial cell injury: formation and stability of C5b-9 and C5b-7 in rat membranous nephropathy. J Immunol 1986;137:1511–1516.

30 Baker PJ, Ochi RF, Schulze M, Johnson RJ, Campbell C, Couser WG: Depletion of C6 prevents development of proteinuria in experimental membranous nephropathy in rats. Am J Pathol 1989;135:185–194.

31 Couser WG, Johnson RJ, Young BA, Yeh CG, Toth CA, Rudolph AR: The effects of soluble recombinant complement receptor 1 on complement-mediated experimental glomerulonephritis. J Am Soc Nephrol 1995;5:1888–1894.

32 Kerjaschki D, Schulze M, Binder S, Kain R, Ojha PP, Susani M, Horvat R, Baker PJ, Couser WG: Transcellular transport and membrane insertion of the C5b-9 membrane attack complex of complement by glomerular epithelial cells in experimental membranous nephropathy. J Immunol 1989;143:546–552.

33 Nangaku M: Complement regulatory proteins in glomerular diseases. Kidney Int 1998;54:1419–1428.

34 Turnberg D, Botto M: The regulation of the complement system: insights from genetically-engineered mice. Mol Immunol 2003;40:145–153.

35 Cunningham PN, Quigg RJ: Contrasting roles of complement activation and its regulation in membranous nephropathy. J Am Soc Nephrol 2005;16:1214–1222.

36 Quigg RJ, Holers VM, Morgan BP, Sneed AE 3rd: Crry and CD59 regulate complement in rat glomerular epithelial cells and are inhibited by the nephritogenic antibody of passive Heymann nephritis. J Immunol 1995; 154:3437–3443.

37 Schiller B, He C, Salant DJ, Lim A, Alexander JJ, Quigg RJ: Inhibition of complement regulation is key to the pathogenesis of active Heymann nephritis. J Exp Med 1998;188:1353–1358.

38 Nangaku M, Pippin J, Richardson CA, Schulze M, Young BA, Alpers CE, Gordon KL, Johnson RJ, Couser WG: Beneficial effects of systemic immunoglobulin in experimental membranous nephropathy. Kidney Int 1996;50:2054–2062.

39 Spicer ST, Tran GT, Killingsworth MC, Carter N, Power DA, Paizis K, Boyd R, Hodgkinson SJ, Hall BM: Induction of passive Heymann nephritis in complement component 6-deficient PVG rats. J Immunol 2007;179:172–178.

40 Hinglais N, Kazatchkine MD, Bhakdi S, Appay MD, Mandet C, Grossetete J, Bariety J: Immunohistochemical study of the C5b-9 complex of complement in human kidneys. Kidney Int 1986;30:399–410.

41 Papagianni AA, Alexopoulos E, Leontsini M, Papadimitriou M: C5b-9 and adhesion molecules in human idiopathic membranous nephropathy. Nephrol Dial Transplant 2002;17:57–63.

42 Doi T, Kanatsu K, Nagai H, Suehiro F, Kuwahara T, Hamashima Y: Demonstration of C3d deposits in membranous nephropathy. Nephron 1984;37:232–235.

43 Moll S, Miot S, Sadallah S, Gudat F, Mihatsch MJ, Schifferli JA: No complement receptor 1 stumps on podocytes in human glomerulopathies. Kidney Int 2001;59:160–168.

44 Moll S, Lange S, Mihatsch MJ, Dragic Z, Schifferli JA, Inal JM: CRIT is expressed on podocytes in normal human kidney and upregulated in membranous nephropathy. Kidney Int 2006;69:1961–1968.

45 Morgan BP: Effects of the membrane attack complex of complement on nucleated cells. Curr Top Microbiol Immunol 1992;178:115–140.

46 Nicholson-Weller A, Halperin JA: Membrane signaling by complement C5b-9, the membrane attack complex. Immunol Res 1993;12:244–257.

47 Rus HG, Niculescu FI, Shin ML: Role of the C5b-9 complement complex in cell cycle and apoptosis. Immunol Rev 2001;180:49–55.

48 Cummings BS, McHowat J, Schnellmann RG: Phospholipase A(2) s in cell injury and death. J Pharmacol Exp Ther 2000;294:793–799.

49 Murakami M, Kudo I: Phospholipase A2. J Biochem 2002;131:285–292.

50 Cybulsky AV, Monge JC, Papillon J, McTavish AJ: Complement C5b-9 activates cytosolic phospholipase A2 in glomerular epithelial cells. Am J Physiol 1995;269:F739–F749.

51 Panesar M, Papillon J, McTavish AJ, Cybulsky AV: Activation of phospholipase A2 by complement C5b-9 in glomerular epithelial cells. J Immunol 1997;159:3584–3594.

52 Liu J, Takano T, Papillon J, Khadir A, Cybulsky AV: Cytosolic phospholipase A2-alpha associates with plasma membrane, endoplasmic reticulum and nuclear membrane in glomerular epithelial cells. Biochem J 2001;353:79–90.

53 Cybulsky AV, Salant DJ, Quigg RJ, Badalamenti J, Bonventre JV: Complement C5b-9 complex activates phospholipases in glomerular epithelial cells. Am J Physiol 1989;257:F826–F836.

54 Cohen D, Papillon J, Aoudjit L, Li H, Cybulsky AV, Takano T: Role of calcium-independent phospholipase A2 in complement-mediated glomerular epithelial cell injury. Am J Physiol Renal Physiol 2008;294:F469–F479.

55 Smith WL, Garavito RM, DeWitt DL: Prostaglandin endoperoxide H synthases (cyclooxygenases)-1 and -2. J Biol Chem 1996;271:33157–33160.

56 Takano T, Cybulsky AV: Complement C5b-9-mediated arachidonic acid metabolism in glomerular epithelial cells – role of cyclooxygenase-1 and -2. Am J Pathol 2000;156:2091–2101.

57 Takano T, Cybulsky AV, Yang X, Aoudjit L: Complement C5b-9 induces cyclooxygenase-2 gene transcription in glomerular epithelial cells. Am J Physiol Renal Physiol 2001;281:F841–F850.

58 Takano T, Cybulsky AV, Cupples WA, Ajikobi DO, Papillon J, Aoudjit L: Inhibition of cyclooxygenases reduces complement-induced glomerular epithelial cell injury and proteinuria in passive Heymann nephritis. J Pharmacol Exp Ther 2003;305:240–249.

59 Cybulsky AV, Lieberthal W, Quigg RJ, Rennke HG, Salant DJ: A role for thromboxane in complement-mediated glomerular injury. Am J Pathol 1987;128:45–51.

60 Zoja C, Benigni A, Verroust P, Ronco P, Bertani T, Remuzzi G: Indomethacin reduces proteinuria in passive Heymann nephritis in rats. Kidney Int 1987;31:1335–1343.

61 Blume C, Heise G, Muhlfeld A, Bach D, Schror K, Gerhardz CD, Grabensee B, Heering P: Effect of flosulide, a selective cyclooxygenase 2 inhibitor, on passive Heymann nephritis in the rat. Kidney Int 1999;56:1770–1778.

62 Nagao T, Nagamatsu T, Suzuki Y: Effect of DP-1904, a thromboxane A2 synthase inhibitor, on passive Heymann nephritis in rats. Eur J Pharmacol 1996;316:73–80.

63 Weise WJ, Natori Y, Levine JS, O'Meara YM, Minto AW, Manning EC, Goldstein DJ, Abrahamson DR, Salant DJ: Fish oil has protective and therapeutic effects on proteinuria in passive Heymann nephritis. Kidney Int 1993;43:359–368.

64 McCarthy ET, Sharma M: Indomethacin protects permeability barrier from focal segmental glomerulosclerosis serum. Kidney Int 2002;61:534–541.
65 Ron D, Walter P: Signal integration in the endoplasmic reticulum unfolded protein response. Nat Rev Mol Cell Biol 2007;8:519–529.
66 Yoshida H: ER stress and diseases. FEBS J 2007;274:630–658.
67 Cybulsky AV, Takano T, Papillon J, Khadir A, Liu J, Peng H: Complement C5b-9 membrane attack complex increases expression of endoplasmic reticulum stress proteins in glomerular epithelial cells. J Biol Chem 2002; 277:41342–41351.
68 Cybulsky AV, Takano T, Papillon J, Bijian K: Role of the endoplasmic reticulum unfolded protein response in glomerular epithelial cell injury. J Biol Chem 2005;280:24396–24403.
69 Inagi R, Kumagai T, Nishi H, Kawakami T, Miyata T, Fujita T, Nangaku M: Preconditioning with endoplasmic reticulum stress ameliorates mesangioproliferative glomerulonephritis. J Am Soc Nephrol 2008;19:915–922.
70 Thannickal VJ, Fanburg BL: Reactive oxygen species in cell signaling. Am J Physiol Lung Cell Mol Physiol 2000;279:L1005–L1028.
71 Greiber S, Munzel T, Kastner S, Muller B, Schollmeyer P, Pavenstadt H: NAD(P)H oxidase activity in cultured human podocytes: effects of adenosine triphosphate. Kidney Int 1998;53:654–663.
72 Neale TJ, Ullrich R, Ojha P, Poczewski H, Verhoeven AJ, Kerjaschki D: Reactive oxygen species and neutrophil respiratory burst cytochrome b558 are produced by kidney glomerular cells in passive Heymann nephritis. Proc Natl Acad Sci USA 1993;90:3645–3649.
73 Peng H, Takano T, Papillon J, Bijian K, Khadir A, Cybulsky AV: Complement activates the c-Jun N-terminal kinase/stress-activated protein kinase in glomerular epithelial cells. J Immunol 2002;169:2594–2601.
74 Neale TJ, Ojha PP, Exner M, Poczewski H, Ruger B, Witztum JL, Davis P, Kerjaschki D: Proteinuria in passive Heymann nephritis is associated with lipid peroxidation and formation of adducts on type IV collagen. J Clin Invest 1994;94:1577–1584.
75 Cybulsky AV, Takano T, Papillon J, McTavish AJ: Complement C5b-9 induces receptor tyrosine kinase transactivation in glomerular epithelial cells. Am J Pathol 1999;155:1701–1711.
76 Cybulsky AV, Papillon J, McTavish AJ: Complement activates phospholipases and protein kinases in glomerular epithelial cells. Kidney Int 1998;54:360–372.
77 Aoudjit L, Stanciu M, Li H, Lemay S, Takano T: p38 mitogen-activated protein kinase protects glomerular epithelial cells from complement-mediated cell injury. Am J Physiol Renal Physiol 2003;285:F765–F774.
78 Koshikawa M, Mukoyama M, Mori K, Suganami T, Sawai K, Yoshioka T, Nagae T, Yokoi H, Kawachi H, Shimizu F, Sugawara A, Nakao K: Role of p38 mitogen-activated protein kinase activation in podocyte injury and proteinuria in experimental nephrotic syndrome. J Am Soc Nephrol 2005;16:2690–2701.
79 Kyriakis JM, Avruch J: Mammalian mitogen-activated protein kinase signal transduction pathways activated by stress and inflammation. Physiol Rev 2001;81:807–869.
80 Ren G, Huynh C, Bijian K, Cybulsky AV: Role of apoptosis signal-regulating kinase 1 in complement-mediated glomerular epithelial cell injury. Mol Immunol 2008;45:2236–2246.
81 Shankland SJ, Floege J, Thomas SE, Nangaku M, Hugo C, Pippin J, Henne K, Hockenberry DM, Johnson RJ, Couser WG: Cyclin kinase inhibitors are increased during experimental membranous nephropathy: potential role in limiting glomerular epithelial cell proliferation in vivo. Kidney Int 1997;52:404–413.
82 Floege J, Johnson RJ, Alpers CE, Fatemi-Nainie S, Richardson CA, Gordon K, Couser WG: Visceral glomerular epithelial cells can proliferate in vivo and synthesize platelet-derived growth factor B-chain. Am J Pathol 1993;142:637–650.
83 Eitner F, Ostendorf T, Van Roeyen C, Kitahara M, Li X, Aase K, Grone HJ, Eriksson U, Floege J: Expression of a novel PDGF isoform, PDGF-C, in normal and diseased rat kidney. J Am Soc Nephrol 2002;13:910–917.
84 Floege J, Eitner F, Alpers CE: A new look at platelet-derived growth factor in renal disease. J Am Soc Nephrol 2008;19:12–23.
85 Paizis K, Kirkland G, Polihronis M, Katerelos M, Kanellis J, Power DA: Heparin-binding epidermal growth factor-like growth factor in experimental models of membranous and minimal change nephropathy. Kidney Int 1998;53:1162–1171.
86 Tsui CC, Shankland SJ, Pierchala BA: Glial cell line-derived neurotrophic factor and its receptor ret is a novel ligand-receptor complex critical for survival response during podocyte injury. J Am Soc Nephrol 2006;17: 1543–1552.
87 Hiromura K, Haseley LA, Zhang P, Monkawa T, Durvasula R, Petermann AT, Alpers CE, Mundel P, Shankland SJ: Podocyte expression of the CDK-inhibitor p57 during development and disease. Kidney Int 2001;60:2235–2246.
88 Petermann AT, Pippin J, Hiromura K, Monkawa T, Durvasula R, Couser WG, Kopp J, Shankland SJ: Mitotic cell cycle proteins increase in podocytes despite lack of proliferation. Kidney Int 2003;63:113–122.
89 Petermann A, Hiromura K, Pippin J, Blonski M, Couser WG, Kopp J, Mundel P, Shankland SJ: Differential expression of d-type cyclins in podocytes in vitro and in vivo. Am J Pathol 2004;164:1417–1424.
90 Mudge SJ, Paizis K, Auwardt RB, Thomas RJ, Power DA: Activation of nuclear factor-kappa B by podocytes in the autologous phase of passive Heymann nephritis. Kidney Int 2001;59:923–931.
91 Hussain S, Romio L, Saleem M, Mathieson P, Serrano M, Moscat J, Diaz-Meco M, Scambler P, Koziell A: Nephrin deficiency activates NF-kappaB and promotes glomerular injury. J Am Soc Nephrol 2009;20:1733–1743.
92 Pippin JW, Durvasula R, Petermann A, Hiromura K, Couser WG, Shankland SJ: DNA damage is a novel response to sublytic complement C5b-9-induced injury in podocytes. J Clin Invest 2003;111:877–885.
93 Cybulsky AV, Takano T, Papillon J, Bijian K, Guillemette J: Activation of the extracellular signal-regulated kinase by complement C5b-9. Am J Physiol Renal Physiol 2005;289: F593–F603.

94 Schneeberger EE, Grupe WE: The ultrastructure of the glomerular slit diaphragm in autologous immune complex nephritis. Lab Invest 1976;34:298–305.
95 Drumond MC, Kristal B, Myers BD, Deen WM: Structural basis for reduced glomerular filtration capacity in nephrotic humans. J Clin Invest 1994;94:1187–1195.
96 Topham PS, Haydar SA, Kuphal R, Lightfoot JD, Salant DJ: Complement-mediated injury reversibly disrupts glomerular epithelial cell actin microfilaments and focal adhesions. Kidney Int 1999;55:1763–1775.
97 Kerjaschki D, Ojha PP, Susani M, Horvat R, Binder S, Hovorka A, Hillemanns P, Pytela R: A beta 1-integrin receptor for fibronectin in human kidney glomeruli. Am J Pathol 1989;134:481–489.
98 Zhang H, Cybulsky AV, Aoudjit L, Zhu J, Li H, Lamarche-Vane N, Takano T: Role of Rho-GTPases in complement-mediated glomerular epithelial cell injury. Am J Physiol Renal Physiol 2007;293:F148–F156.
99 Huynh C, Ren G, Papillon J, Guillemette J, Takano T, Cybulsky AV: The cytoprotective role of Ras in complement-mediated glomerular epithelial cell injury. Clin Immunol 2009;131:343–353.
100 Moller CC, Wei C, Altintas MM, Li J, Greka A, Ohse T, Pippin JW, Rastaldi MP, Wawersik S, Schiavi S, Henger A, Kretzler M, Shankland SJ, Reiser J: Induction of TRPC6 channel in acquired forms of proteinuric kidney disease. J Am Soc Nephrol 2007;18:29–36.
101 Furness PN, Hall LL, Shaw JA, Pringle JH: Glomerular expression of nephrin is decreased in acquired human nephrotic syndrome. Nephrol Dial Transplant 1999;14:1234–1237.
102 Doublier S, Ruotsalainen V, Salvidio G, Lupia E, Biancone L, Conaldi PG, Reponen P, Tryggvason K, Camussi G: Nephrin redistribution on podocytes is a potential mechanism for proteinuria in patients with primary acquired nephrotic syndrome. Am J Pathol 2001;158:1723–1731.
103 Wang SX, Rastaldi MP, Patari A, Ahola H, Heikkila E, Holthofer H: Patterns of nephrin and a new proteinuria-associated protein expression in human renal diseases. Kidney Int 2002;61:141–147.
104 Benigni A, Tomasoni S, Gagliardini E, Zoja C, Grunkemeyer JA, Kalluri R, Remuzzi G: Blocking angiotensin II synthesis/activity preserves glomerular nephrin in rats with severe nephrosis. J Am Soc Nephrol 2001;12: 941–948.
105 Benigni A, Zoja C, Tomasoni S, Campana M, Corna D, Zanchi C, Gagliardini E, Garofano E, Rottoli D, Ito T, Remuzzi G: Transcriptional regulation of nephrin gene by peroxisome proliferator-activated receptor-gamma agonist: molecular mechanism of the antiproteinuric effect of pioglitazone. J Am Soc Nephrol 2006;17:1624–1632.
106 Yuan H, Takeuchi E, Taylor GA, McLaughlin M, Brown D, Salant DJ: Nephrin dissociates from actin, and its expression is reduced in early experimental membranous nephropathy. J Am Soc Nephrol 2002;13:946–956.
107 Saran AM, Yuan H, Takeuchi E, McLaughlin M, Salant DJ: Complement mediates nephrin redistribution and actin dissociation in experimental membranous nephropathy. Kidney Int 2003;64:2072–2078.
108 Schwarz K, Simons M, Reiser J, Saleem MA, Faul C, Kriz W, Shaw AS, Holzman LB, Mundel P: Podocin, a raft-associated component of the glomerular slit diaphragm, interacts with CD2AP and nephrin. J Clin Invest 2001;108:1621–1629.
109 Nakatsue T, Koike H, Han GD, Suzuki K, Miyauchi N, Yuan H, Salant DJ, Gejyo F, Shimizu F, Kawachi H: Nephrin and podocin dissociate at the onset of proteinuria in experimental membranous nephropathy. Kidney Int 2005;67:2239–2253.
110 Petermann AT, Krofft R, Blonski M, Hiromura K, Vaughn M, Pichler R, Griffin S, Wada T, Pippin J, Durvasula R, Shankland SJ: Podocytes that detach in experimental membranous nephropathy are viable. Kidney Int 2003;64:1222–1231.
111 Li H, Lemay S, Aoudjit L, Kawachi H, Takano T: Src-family kinase Fyn phosphorylates the cytoplasmic domain of nephrin and modulates its interaction with podocin. J Am Soc Nephrol 2004;15:3006–3015.
112 Verma R, Wharram B, Kovari I, Kunkel R, Nihalani D, Wary KK, Wiggins RC, Killen P, Holzman LB: Fyn binds to and phosphorylates the kidney slit diaphragm component nephrin. J Biol Chem 2003;278:20716–20723.
113 Simons M, Hartleben B, Huber TB: Podocyte polarity signalling. Curr Opin Nephrol Hypertens 2009;18:324–330.
114 Fukasawa H, Bornheimer S, Kudlicka K, Farquhar MG: Slit diaphragms contain tight junction proteins. J Am Soc Nephrol 2009;20:1491–1503.
115 Makker SP, Kanalas JJ: Course of transplanted Heymann nephritis kidney in normal host. Implications for mechanism of proteinuria in membranous glomerulonephropathy. J Immunol 1989;142:3406–3410.
116 Kim Y, Butkowski R, Burke B, Kleppel MM, Crosson J, Katz A, Michael AF: Differential expression of basement membrane collagen in membranous nephropathy. Am J Pathol 1991;139:1381–1388.
117 Minto AW, Kalluri R, Togawa M, Bergijk EC, Killen PD, Salant DJ: Augmented expression of glomerular basement membrane specific type IV collagen isoforms (alpha3-alpha5) in experimental membranous nephropathy. Proc Assoc Am Physicians 1998;110:207–217.
118 Floege J, Johnson RJ, Gordon K, Yoshimura A, Campbell C, Iruela-Arispe L, Alpers CE, Couser WG: Altered glomerular extracellular matrix synthesis in experimental membranous nephropathy. Kidney Int 1992;42:573–585.
119 Raats CJ, Luca ME, Bakker MA, Van Der Wal A, Heeringa P, Van Goor H, Van Den Born J, De Heer E, Berden JH: Reduction in glomerular heparan sulfate correlates with complement deposition and albuminuria in active Heymann nephritis. J Am Soc Nephrol 1999;10:1689–1699.
120 McMillan JI, Riordan JW, Couser WG, Pollock AS, Lovett DH: Characterization of a glomerular epithelial cell metalloproteinase as matrix metalloproteinase-9 with enhanced expression in a model of membranous nephropathy. J Clin Invest 1996;97:1094–1101.

121 Levidiotis V, Freeman C, Tikellis C, Cooper ME, Power DA: Heparanase is involved in the pathogenesis of proteinuria as a result of glomerulonephritis. J Am Soc Nephrol 2004;15:68–78.
122 Levidiotis V, Freeman C, Punler M, Martinello P, Creese B, Ferro V, van der Vlag J, Berden JH, Parish CR, Power DA: A synthetic heparanase inhibitor reduces proteinuria in passive Heymann nephritis. J Am Soc Nephrol 2004;15:2882–2892.
123 Shankland SJ, Pippin J, Pichler RH, Gordon KL, Friedman S, Gold LI, Johnson RJ, Couser WG: Differential expression of transforming growth factor-beta isoforms and receptors in experimental membranous nephropathy. Kidney Int 1996;50:116–124.
124 Cattran D: Management of membranous nephropathy: when and what for treatment. J Am Soc Nephrol 2005;16:1188–1194.
125 Atkinson C, Qiao F, Song H, Gilkeson GS, Tomlinson S: Low-dose targeted complement inhibition protects against renal disease and other manifestations of autoimmune disease in MRL/lpr mice. J Immunol 2008;180:1231–1238.
126 Boyce M, Bryant KF, Jousse C, Long K, Harding HP, Scheuner D, Kaufman RJ, Ma D, Coen DM, Ron D, Yuan J: A selective inhibitor of eIF2alpha dephosphorylation protects cells from ER stress. Science 2005;307:935–939.
127 Balch WE, Morimoto RI, Dillin A, Kelly JW: Adapting proteostasis for disease intervention. Science 2008;319:916–919.
128 Haas M, Mayer G, Wirnsberger G, Holzer H, Ratschek M, Neyer U, Neuweiler J, Kramar R, Schneider B, Breiteneder-Geleff S, Regele HM, Horl WH, Kerjaschki D: Antioxidant treatment of therapy-resistant idiopathic membranous nephropathy with probucol: a pilot study. Wien Klin Wochenschr 2002;114:143–147.
129 Wada T, Pippin JW, Marshall CB, Griffin SV, Shankland SJ: Dexamethasone prevents podocyte apoptosis induced by puromycin aminonucleoside: role of p53 and Bcl-2-related family proteins. J Am Soc Nephrol 2005;16:2615–2625.
130 Faul C, Donnelly M, Merscher-Gomez S, Chang YH, Franz S, Delfgaauw J, Chang JM, Choi HY, Campbell KN, Kim K, Reiser J, Mundel P: The actin cytoskeleton of kidney podocytes is a direct target of the antiproteinuric effect of cyclosporine A. Nat Med 2008;14:931–938.
131 Bonegio RG, Fuhro R, Wang Z, Valeri CR, Andry C, Salant DJ, Lieberthal W: Rapamycin ameliorates proteinuria associated tubulointerstitial inflammation and fibrosis in experimental membranous nephropathy. J Am Soc Nephrol 2005;16:2063–2072.
132 Kerjaschki D: Pathomechanisms and molecular basis of membranous glomerulopathy. Lancet 2004;364:1194–1196.

Andrey V. Cybulsky, MD
Division of Nephrology, Royal Victoria Hospital
687 Pine Avenue West
Montreal, QC H3A 1A1 (Canada)
Tel. +1 514 398 8148, Fax +1 514 843 2815, E-Mail andrey.cybulsky@mcgill.ca

Herrera GA (ed): Experimental Models for Renal Diseases: Pathogenesis and Diagnosis.
Contrib Nephrol. Basel, Karger, 2011, vol 169, pp 126–152

Prospects and Perspectives on IgA Nephropathy from Animal Models

Steven N. Emancipator

Louis Stokes Cleveland DVA Medical Center, and Case Western Reserve University, Department of Pathology, Cleveland, Ohio, USA

Abstract

The focus of this chapter is a systematic and critical review of the insights that experimental models have contributed to our understanding of IgA nephropathy (IgAN). We consider the generation of IgA subject to glomerular deposition, the partitioning of IgA between the circulation and glomeruli, the clearance of IgA and complexes containing IgA from the circulation, the 'upstream' effectors of glomerular pathophysiology, the inflammatory mediators that connect intraglomerular stimuli to functional and morphologic effects, and the contribution of animal models to our current understanding of the role that glycosylation of IgA plays in the genesis of IgAN. In each of these subsections, the evidence in favor of each principle or hypothesis is weighed in consideration of other potentially contradictory evidence and relevant issues related to IgAN in patients. The key limitations of each model system are presented, and where possible, reconciliation of discrepant observations are proposed. Subsequently, a synopsis of spontaneous models that do not offer particular mechanistic insights, and a compilation of experimental therapeutic initiatives, are reviewed. In all respects, the discussion of observations in patients or cells in vitro is limited to points where these data impinge upon interpretation of the data derived from the animal models.

Basic Model Systems

IgA nephropathy (IgAN) is a common but complex disease that, 40 years after its recognition, still evades clear insight to its genesis despite intensive multinational efforts. Many animal models of IgAN have accrued as a result of the attempt to delineate the mechanisms operative spontaneously in patients. These models, as variable and multifaceted as IgAN in patients, are codified in this chapter.

Passive models induced by parenteral injection represent the first, and technically easiest, category of experimental IgAN. Nearly 30 years ago, Rifai et al. [1] injected immune complexes of murine myeloma IgA into mice to establish the principle that circulating IgA immune complexes can deposit within the glomerular mesangium to elicit microhematuria. They demonstrated that the dose of complexes is related to glomerular response and propensity for IgA deposition. Shortly thereafter, Isaacs and Miller [2] used the same approach with a different myeloma protein specific for the glycosidic linkage in dextran, verifying that the experimental system is not dependent upon an abnormality unique to a particular IgA paraprotein or a particular antigen or hapten carrier. This model expanded on the role of the size of immune complexes, and introduced net electrostatic charge as a factor in the site of deposition and the glomerular response to such deposition of IgA immune complexes.

Schemes designed to elicit a predominantly IgA immune response using parenteral [3, 4] or oral routes [5] produced active serum sickness models of IgAN. These active systems probed the role that immunoregulation plays in the genesis of IgAN.

Models of secondary IgAN, consequent to severe hepatocellular injury elicited by carbon tetrachloride leading to hepatic cirrhosis [6, 7], experimental surgical cholestasis [8, 9] or ethanol-induced liver disease [10] were established. These models focused awareness on the role that the rate of clearance of IgA complexes from the circulation plays in glomerular deposition and instigated recognition of unique clearance pathways applicable to IgA and perhaps IgM, but not IgG.

Technical details for establishing these early experimentally inducible models of IgAN are provided in a methods-oriented protocol handbook that is updated periodically [11]. The models of IgAN induced by dietary mycotoxins [12, 13] are cited extensively herein; these should be readily reproducible by careful reading of the original reports. Other models arise as sporadic primary glomerulonephritis or spontaneously from mutation, infection or neoplasm in species ranging from rodents and lagomorphs to higher mammals, including primates. Methodologic detail is not presented for these spontaneous forms of IgAN because experimentation requires access to the appropriate livestock. In addition to the extensively studied models of spontaneous IgAN in ddY mice [14] and the variant of these mice (HIGA) selectively inbred for high IgA production [15, 16], a few recently established transgenic mouse models raise provocative issues about the pathogenesis of IgAN [17–19]. Most spontaneous models offer little mechanistic insight, and are cited simply to chronicle their existence.

Source of IgA Depositing in Glomeruli

Immune deposits in glomeruli result from a dynamic equilibrium between accretion of Ig, either complexed with antigen or other macromolecules or as free soluble protein, and removal by catabolism or excretion into the urine (or efferent arteriole). In turn, the rate of deposition or formation of complexes within glomeruli itself depends on the production of IgA and/or IgA complexes, entry of these components into the circulation, and partitioning of the immune reactants between the glomeruli and the various clearance mechanisms. IgA is the most abundant Ig class in mammalian hosts, whether assessed by mass content or synthetic rate [reviewed in 20, 21]. However, the concentration of IgA in plasma is far less than that of IgG. These superficially contradictory facts were reconciled by recognition of a mucosal immune system, characterized by distinct lymphocyte traffic patterns and regulatory pathways. Both the synthesis and concentration of IgA are centered on mucous membranes, and their associated secretions, but extramucosal lymphoid organs also contribute (especially to IgA in the primate circulation). The various means whereby IgA in nephritogenic form are generated, as complexes or macromolecular deposits, are presented in figure 1 (left and lower center portion). Active animal models of IgAN rely on both mucosal and extramucosal cellular sources of the deposited IgA, which will be considered separately. Models characterized by overproduction of IgA not evoked by vaccination are less clear, and will be introduced within the limits of current comprehension.

Generation of IgA by Mucosal Immunization

The proclivity for mucosal immunization to drive IgA production led first to oral and subsequently nasal vaccination of mice [5, 22–38] and rats [39–47] as a means to induce active models of IgAN. Inert proteins or infectious pathogens or their products served as antigens. This diversity lends credibility to the principle that mucosal immunization by a wide range of antigens favors an IgA response in mammals generally. In other rodent systems, parenteral vaccination was supplemented by use of mucosal boosting, polarizing immunization or stimulation by superantigens [36, 38, 48], and a few models leveraged genetic predisposition to favor Th2 responses and/or IgA production [33, 49–51].

The importance of the mucosal immune response for IgA antibody production, and for

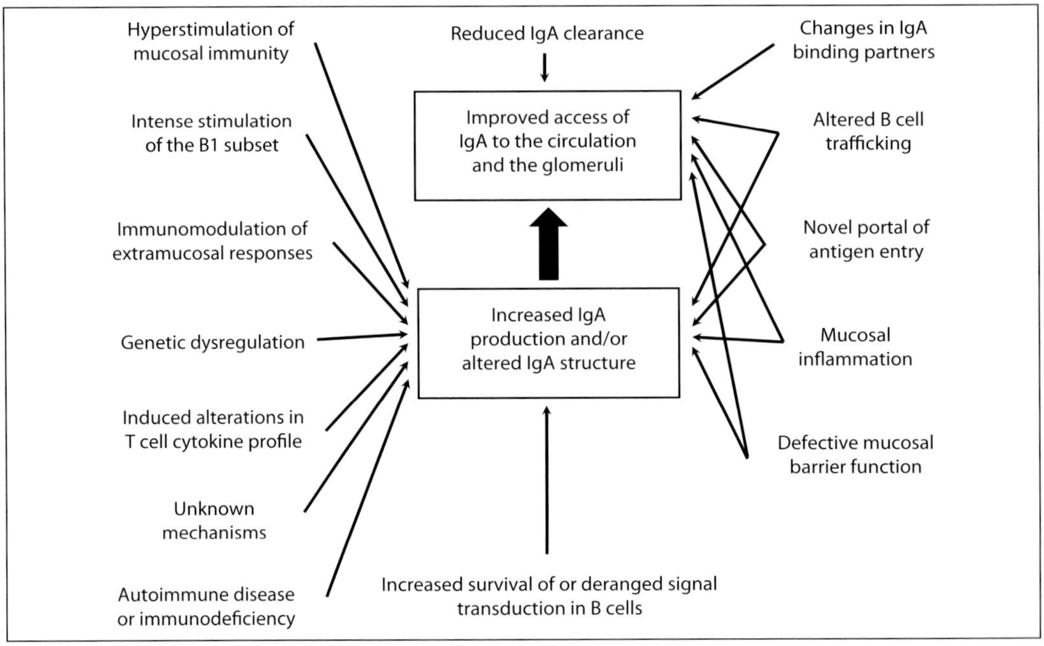

Fig. 1. Factors that promote increased IgA production and/or altered IgA structure (left and lower center) can synergize with factors that improve access of IgA and its complexes to the glomeruli (right and upper center). Increased IgA production can be instigated by physiologic hyperstimulation of the mucosal immune system [5, 22–47], by intense stimulation of the B1 subset of (extramucosal) lymphocytes [3, 4, 62–66] or by immunomodulation of extramucosal responses. Modulators include mucosal boosting with antigen [33, 38, 48], superantigens [36, 47, 49, 85] or stimuli of innate immunity [31, 32, 34, 49, 68]. Genetic or transgenic proclivity to dysregulation of B cells by T cells [50–52, 75] and induced alterations in T cell cytokine profiles [33, 38, 77] are also recognized. Often, immune dysregulation arises from unknown mechanisms ascribed to T cells largely by intuitive guesswork [10, 30, 35–37, 44–46, 53–56, 59–61, 69–72, 81]. Autoimmune and immunodeficiency diseases also frequently exhibit increased IgA synthesis and glomerular IgA deposits [58, 76–79, 88, 89] due to a failure to regulate adaptive responses correctly. Finally, prolonged B cell survival [73, 74] or aberrant intracellular signal transduction in B cells [79] leads to symptoms evocative of autoimmune disease, with heavy or predominant production and glomerular deposition of IgA. Irrespective of the mechanism(s) leading to increased IgA production, the resultant IgA may be wholly or partially antigen-specific, but might also be nonspecific [23, 24, 65, 66]. Although heightened IgA synthesis itself can promote glomerular IgA deposition (heavy arrow), changes in the polymerization status [96–99] and/or glycosylation pattern of the IgA [104, 105] appear to be more cogent, given the rarity of IgAN complicating myeloma in animals [90, 91] or in patients with myeloma or HIV infection [reviewed in 21]. Reduced rates of clearance of IgA or IgA complexes from the circulation via any of a number of active receptor systems on phagocytes and epithelial cells [6–9, 50, 52, 59, 61, 64, 117–130] enhances opportunity for accumulation of IgA in glomeruli. Changes in the levels of partners that bind IgA through mechanisms that are not immunologically specific can shift the partitioning of IgA and complexes between circulating and glomerular compartments; this likely represents the mechanism leading to IgAN in uteroglobin-deficient mice [113, 114]. Such partners might shield IgA from clearance and/or bridge IgA to intraglomerular ligands [106, 107]. Either misrouting of IgA-committed B cells and their plasma cell progeny from the site of physiological induction [78, reviewed in 92] or exposure of an immune host to antigen via a novel portal of entry [notably 31, 33, 35, 38] can contribute to defects in the usual barrier to antigen penetration of mucosal surfaces. Mucosal inflammation [52, 58, 75, 78] and defects in mucosal barrier function [10, 53–58, 61] also improve entry of antigen from mucosal lumens into the body proper. Each of these four mechanisms simultaneously drive production of structurally altered IgA by modulating and/or hyperstimulating the immune system while favoring formation of IgA complexes with antigen, which are more nephritogenic [96–99]. Penetration of macromolecules that bind IgA by nonimmune specificity (e.g. lectins) can also promote formation of nephritogenic complexes [106, 107]. If B cell homing shifts to the marrow or spleen, or antigen exposure is parenteral, further synergy is anticipated because extramucosal IgA complexes cannot be excreted efficiently from the body via pIgR [128].

genesis of IgAN, is underscored by abnormalities in mucosal structure and/or mucosal immunity in several other models of IgAN that are spontaneous, consequent to genetic manipulation, or induced by treatments distinct from vaccination, mucosal or otherwise. Specifically, transgenic overexpression of the T cell immunomodulator LIGHT leads to mucosal inflammation and diminished expression of pIgR by mucosal epithelial cells, in addition to enhanced B cell survival and IgAN [52]. The intramucosal inflammation and the reduction in pIgR-mediated IgA secretion are likely contributors to the evolution of IgAN, although this principle has not been established experimentally. Nivalenol-induced IgAN in mice [53, 54] and spontaneous IgAN in primates [55, 56] are associated with defective gut function and increased circulating antibodies specific for food antigens, especially gliadin and casein. Nivalenol-induced disease is also associated with increased antibodies specific for gut commensals and self-antigens [54], and both the specific antibodies as well as total circulating Ig pools are enriched in the IgA class, with reduced IgG and IgM. Furthermore, passive administration of monoclonal antibodies derived from mice given nivalenol induces IgAN in conventionally housed animals fed with normal chow [57].

In mice that develop a lupus-like autoimmune disease, defective mucosal immunity and increased gut permeability to dietary macromolecules are reported [58]. There is a progressive increase in a plasmacytic subset in the intestinal lamina propria of HIGA mice, relative to healthy control mice [59]. An herbal medicine that reduces the number of such cells and the spontaneous IgA secretion that arises from them also ameliorates glomerular IgA and IgG deposition, mesangial proliferation and proteinuria in this model [60]. Finally, heavy ingestion of ethanol by rats alters mucosal immunity in the gut [10, 61]. Thus, although glomerular injury and IgA deposition after administration of ethanol has focused on hepatocellular injury (see 'Models of Secondary IgAN' subsection', below), damage to mucous membranes and the mucosal immune system may be equally, or even more, culprit.

Generation of IgA by Extramucosal (Systemic) Immunization

Another approach to an active model employed intraperitoneal priming with the glucose polymer dextran, followed by repetitive intravenous challenges [3, 4, 62–66]. This approach exploits the capacity of lymphoid tissues associated with serous cavities to generate strong humoral responses that tend to be rich in IgA (and IgM), especially in response to antigens like dextran that are repeating polymers. The B cells stimulated by this immunization protocol are likely nontraditional lymphocytes that are relatively independent of T cell help. Recently, these variant cells, sometimes referred to as a 'B1' subset, were recognized as operative through pathways distinct from those whereby IgM positive B cells undergo gene rearrangement and somatic hypermutation in organized lymphoid tissues under T cell control [67]. As noted above, mucosal boosting after extramucosal priming can also promote strong IgA responses that lead to IgAN [32–34, 38, 48].

Other parenteral routes of vaccination can also lead to predominantly IgA immune responses and IgAN. In many of these systems, genetic proclivity for altered systemic immunity is recognized and likely explains the predominance of IgAN [49–51]. However, in some instances, the known parameters of the immune response would not necessarily predict predilection for a strong IgA response or alterations in immunoregulation; nonetheless, strong serum IgA responses and IgA predominance or co-dominance in a subsequent glomerulonephritis is observed [68–72]. The antigens in these systems are bacterial or viral pathogens that might invoke innate immunity, which modulates the adaptive antibody response through mechanisms that are currently being elucidated (see 'T Cell Regulation and Cytokine Polarity', below).

Alterations in B Cell Turnover

Many of the stimuli that drive B lymphocytes to clonal expansion, Ig class switching, affinity maturation and differentiation to plasmacytes can also promote apoptosis, autophagy and anergy. The region encoding the α-chain lies near the distal end of the Igh gene segments in both rodents and primates. Therefore, expression of IgA requires more intense and/or prolonged stimulation, with attendant heightened risk for cell cycle arrest and/or apoptosis. The balance between propagation and extinction can be fragile, and small displacements can precipitate large net effects. Generally, these features are under T cell control, but under some circumstances, B cell function becomes more autonomous of T cells.

The antiapoptotic factor bcl2 was first recognized in B lymphocytes and potently improves B cell survival; bcl2 also delays progression of the cell cycle from G1 to S. Therefore, it is not surprising that transgenic mice overexpressing bcl2 in B cells develop hypergammaglobulinemia and a humorally mediated multiorgan system autoimmune syndrome. Exaggerated IgA production progresses to glomerulonephritis with predominantly IgA deposition that is ultimately fatal [73]. Intensified IgA immune responses to antigenic stimulation and increases in IgA immune complexes and circulating autoantibodies are documented as well. Mice with a homozygous targeted deletion of the transcription factor E2F1, which controls the cell cycle and induces apoptosis, exhibit defective negative selection of T cells in the thymic medulla. Peripheral (T cell) tolerance, however, is preserved, and $E2F1^{-/-}$ mice do not develop autoimmunity. Superimposition of homozygous E2F1 deletion upon the B cell-specific overexpression of bcl2 elicits a mild autoimmune syndrome, without accelerated mortality, and the evolution of a mesangial proliferative glomerulonephritis with predominantly IgA deposits, with co-deposits of IgG and IgM [74].

The lymphotoxin-β receptor (LTβR) agonist LIGHT is expressed on activated T cells. Stimulation of this receptor is required for the generation of mucosa-associated lymphoid tissue, genesis of IgA-committed precursors and the homing of IgA-committed B cells to the gut. Whereas $LT\beta R^{-/-}$ mice are deficient in IgA and lack IgA or IgA+ cells in the mucosal secretions and gut, transgenic mice that overexpress LIGHT produce excess IgA and develop IgAN [52]. These mice also develop signs of inflammatory bowel disease and other abnormalities in mucosal immunity. The B cell activator of the tumor necrosis factor (TNF) family (BAFF) is required for B cell survival and differentiation, but overexpression causes B cell hyperplasia and autoimmunity. Transgenic mice that overexpress BAFF develop increased IgA levels, especially in the intestinal lamina propria, and glomerular deposits of IgA associated with mild defects in glomerular function [75]. In BAFF transgenic mice, LIGHT expression is not required, and co-transfection with LIGHT has no discernable effect.

Polyclonal Activation and Dysregulated Ig Production

Stimulation of lymphocytes can bypass the typical antigen-specific signal transduction pathways invoked during cognate and regulated immune responses. Indeed, co-stimulation with 'superantigen', vaccination of selected strains of rodents predisposed towards an immune response favoring IgA production and/or Th2 cytokine profiles, or immunomodulation by innate responses can also lead to IgAN [33, 36, 49–51, 68–72]. In addition, very intense cognate responses can drive immune responses with specificity for elements that are not a component of the intense stimulus, which is especially true for simpler antigens, those that resemble commensals or pathogens, or those that are prevalent in the environment in which the host survives.

Several different groups have reported that immunization of mice or rats with dextran does not inevitably lead to specific immune responses or demonstrable complexes of antibody and antigen in the circulation or in the glomeruli, notwithstanding the development of glomerulonephritis or IgAN. Although only 30% of dextran-immunized mice had detectable IgA immune complexes, 88% developed mesangial deposits of IgA, and the presence of IgA anti-dextran antibodies within immune complexes in the serum correlated well with mesangial IgA

[62]. Adaptation of the dextran-induced model of IgAN originally developed in mice for use in rats led to glomerulonephritis characterized by either no IgA or no Ig deposits of any class in glomeruli, and questionable antidextran-specific antibody responses [65, 66]. Few dextran deposits could be identified in these mouse or rat models. In some systems, oral immunization with several antigens led to glomerular IgA deposits and perhaps light microscopic abnormalities in glomeruli, with no antigen deposition and no functional evidence of glomerular disease [23, 24]. In another system, no glomerular IgA or antigen was detected, nor was an antibody response present in serum [26].

Overall, in addition to provoking IgAN with specific antibody and glomerular antigen deposits, a minority of reports suggest that at least some of the glomerular IgA could represent deposition of circulating IgA complexes and/or IgA polymers with diverse specificities that arise from intense stimulation with a single antigen. Based on refinement from initially unsuccessful attempts to establish an immunologically specific model of IgAN, one group underscored the importance of removing the putative antigen from the food supply, and presumably from other environmental sources [28]. Immune deposits of antigen, and antibody responses resulting in specific immune complexes, were restricted to immunized subjects if the antigen was withdrawn from the animals' chow.

An autoimmune lupus-like disease characterized by spontaneous glomerulonephritis that progresses to sclerosis develops in RF/J mice. These mice exhibit high levels of serum IgA and IgG, a variety of autoantibodies and hypocomplementemia that becomes more severe as the mice age. Early in the course of disease, mesangial proliferative glomerulonephritis with massive IgG and IgA deposits parallels increased IgG and IgA immune complexes; as the mice age, the cellular proliferation is eclipsed by mesangial matrix expansion, and IgA becomes more prominent in glomeruli and circulating immune complexes, as IgG wanes [76]. Other putatively autoimmune syndromes, associated with polyclonal lymphocyte activation, can also result in glomerular disease. Under some circumstances, the glomerulonephritis has features more or less akin to IgAN. In a chronic graft versus host disease similar to lupus, the administration of recombinant human keratinocyte growth factor alters the pattern of the immune response, with a collateral increase in glomerular involvement that closely resembles IgAN [77].

Immunodeficiency States with Autoimmunity and Hypergammaglobulinemia

Many immunodeficiency disease states are associated with paradoxical increases in Ig and/or immune complexes. Targeted deletion of a protein (WIP) that interacts with the protein (WASP) that undergoes mutation in Wiskott-Aldrich syndrome leads to inflammation of mucosal sites including colitis, pneumonitis, a lymphopenia that is more pronounced among B cells, elevated levels of autoantibodies and arthritis. The presumably autoimmune disease is accompanied by IgAN and underdevelopment of Peyer's patches [78]. Apparently, some combination of dysregulation of residual B cells and deranged mucosal immunity underlie the IgAN.

As anticipated, lyn$^{-/-}$ mice have humoral immunodeficiency derived from decreased numbers of B cells that exhibit defective responses to antigen and T cell help, among other stimuli. Paradoxically, serum IgA concentrations and the levels of autoantibodies are remarkably elevated. In addition to increased plasma cells, a population of unusual lymphoblast-like cells expressing Mac1 and cytoplasmic IgM spontaneously secrete a large amount of IgM. Many of these mice develop glomerulonephritis, poorly characterized at present but apparently autoimmune in nature [79]. Apparently, lyn supports elimination of autoreactive B cells, in addition to clonal expansion and differentiation.

T Cell Regulation and Cytokine Polarity

Alterations of T cell function, whether primary or induced by immunomodulators, can also lead to abnormally high IgA synthesis. Moreover, T cell responses regulate humoral immunity and control the features of oral or nasal tolerance [30, 35]. Mesangial proliferative glomerulonephritis and glomerular dysfunction, indicated by hematuria, proteinuria and/or increased serum creatinine is more pronounced in association with a Th2 response to vaccination with bacterial or viral pathogens or products. Despite similar mesangial deposits of IgA and C3 in Th1 responders, these latter mice had no or far less morphologic or functional evidence of glomerular disease [33, 36, 38, 49]. In the model instigated by immunization with *S. aureus*, less antigen was detected in the glomeruli of Th1-oriented mice, compared to Th2 responders. In HIGA mice, there is an inversion from Th1 to Th2 dominant cytokine production by splenic CD4+ T cells as the mice age, when synthesis of IgA accelerates and progressive IgAN ensues [80]. Furthermore, a shift from Th1 to Th2 predominance aggravates glomerulonephritis in chronic graft versus host disease with heavy glomerular IgA deposition [77].

Ingestion of vomitoxin provokes IgAN in mice. When cultures of lymphocytes or purified B cells from Peyer's patches or spleen are incubated in the presence of vomitoxin, synthesis of DNA and proteins (including Igs) is reduced. Nonetheless, vomitoxin potentiates a selective IgA response to the mitogen concanavalin A, via CD4+ T cell activation. This selective T cell help is associated with increased production of IL-6 and cytokines associated with a Th2 response in the presence of co-stimulation with vomitoxin, as opposed to stimulation with concanavalin A alone [81].

Genetically determined orientation of cytokine polarity apparently resides with hematopoietic stem cells. Upon engraftment of human bone marrow cells into immunodeficient mice, the preconditioning of the human donor cells in vitro shapes cytokine polarity, Ig synthesis and glomerular IgA deposition in the chimeric animals [82]. The clinical behavior and severity of IgAN that develops after bone marrow transplantation from ddY mice into other strains, or after transplantation of bone marrow from other strains into ddY recipients, is dictated by the background of the graft, and not by the recipient extramedullary tissues [83, 84].

Immunomodulators act through known pathogen-associated molecular patterns such as Toll-like receptors, or as 'superantigens' that bind antigen receptors and/or co-stimulatory molecules on lymphocytes by specificity for protein domains that lie outside of the physiological site for cognate interaction. One such modulator, *Staphylococcus* enterotoxin B, is a powerful adjuvant for oral immunization that promotes development of IgAN [36, 47]. Especially when combined with a phospholipid immunomodulator, such vaccination strategy impacts upon the Th2/Th1 ratio. The toxic shock syndrome toxin-1 superantigen drives Ig synthesis especially towards IgA in vitro [85]. If applicable to in vivo responses, such exposure might also elicit IgAN upon protracted antigen exposure. Expression of a constitutively activated mutant Toll-like receptor 4 leads to increased numbers of B cells in the intestinal lamina propria and heightened mucosal IgA synthesis independent of T cells. In rats, endotoxin drives IgA production and presumably contributes to the development of IgAN after oral immunization via activation of Toll-like receptor 4 [45, 46]. Herbal medicines that ameliorate IgA synthesis in HIGA mice act through unknown mechanisms that likely derive from effects on T cells [60].

The physiologic role of T cell-mediated regulation of B cell response has also been considered in the genesis of IgAN. Neonatal thymectomy of ddY mice that develop IgAN spontaneously ameliorates the disease [86]. This effect is predicted from the importance of T cell help for Ig class switching and for B cell differentiation. However, dextran is an antigen traditionally recognized as one that evokes a B cell response independent of T cells. Furthermore, the model of IgAN induced by dextran appears to rely heavily on the B1 subset of B cells that function independently of T cells (see above). Immunization of athymic mice with dextran leads to IgAN indistinguishable from that in euthymic mice. Therefore, it appears that the participation of T cells is not required for IgAN beyond the general need for promoting antigen driven B cell development [87].

The Akita mouse has a mutation in the insulin 2 gene, and serves as a model for diabetes. Males, characterized by more severe diabetes, exhibit increased serum IgA and collateral mesangial deposits of IgA at advanced age [88]. Glomerulosclerosis considered to be due to diabetes also develops in elderly mice; females, however, do not develop these changes. Another group reported that rats rendered diabetic by injection of streptozotocin, but not treated with insulin, exhibit markedly increased levels of IgA in plasma, with collateral increases in plasma secretory IgA and immune complexes containing C3 and IgA; insulin treatment decreased the levels of all these analytes in plasma, and withdrawal of insulin led to recrudescence. Late in life, these untreated diabetic rats showed mesangial deposits of IgA, C3 and secretory component [89]. Although the mechanism remains unknown, it is likely that the IgAN in diabetic rodents arises from a poorly characterized T cell defect, as type I diabetes in humans is a T cell-mediated autoimmune disease.

Partitioning of IgA between the Circulation and Glomeruli

IgA synthesis, whether driven by specific mucosal or extramucosal antigen or consequent to polyclonal activation or neoplastic transformation, is not the only requirement for the genesis of IgAN. IgA must eventually enter into and remain within glomeruli to form durable deposits. Despite the fact that increased IgA synthesis is observed in most patients and most animal models, even the extremely high levels of circulating IgA that result from advanced multiple myeloma or plasmacytoma elicit IgAN only rarely, and only under propitious circumstances [90, 91]. In the experimental systems where information is available, this transit of IgA from the site of synthesis to the glomeruli is by way of the bloodstream; this is the likely route in spontaneous disease as well [92].

The partitioning of IgA, whether free or bound within macromolecular complexes, is governed by physicochemical properties, affinity for intrinsic or extrinsic intraglomerular components and any affinity for circulating extraglomerular components that might interfere with or perhaps facilitate transit from the bloodstream to the glomeruli. Figure 1 (right and upper center portion) summarizes the various factors that contribute to access of nephritogenic IgA to the glomeruli. In addition, the rate of clearance of IgA or complexes containing IgA might also have an impact. Although clearance is discussed in the following section, some circulating molecules might serve at once to favor glomerular deposition and shield circulating macromolecular IgA from clearance mechanisms. These molecules are considered in this section.

Physicochemical Properties of IgA and Complexes
The physicochemical properties of IgA exert a powerful impact on the partitioning between the circulation and glomeruli. Particularly, the electrostatic charge and the hydrodynamic size of Ig molecules determine not only their half-time in the circulation (discussed in the next section), but also their proclivity to deposit and persist within glomeruli. Although these principles are best established with respect to IgG antibodies in immune complexes [93–95], they also apply to IgA and aggregates containing IgA. Most critically, IgA deposits in glomeruli only when present in complexes of approximately 1–2 MDa [96–98]. In the passive model of IgAN, addition of monomeric IgA specific for dinitrophenol hapten prolonged the circulation time of haptenated carrier complexed with polymeric IgA and reduced glomerular deposition of IgA, C3 and antigen, as well as the hematuria that follows such glomerular deposition [99]. Presumably, the monomeric IgA that is less capable of cross-linking antigen disrupts the lattice formation of the nephritogenic immune complexes, favoring small complexes and mitigating glomerular deposition.

Affinity for Glomerular Components
Circulating antibody can bind to glomerular deposits by immunologically specific or immunologically nonspecific means; reciprocally,

deposited antibody can capture circulating macromolecular antigens [100–103]. Administration of anti-idiotype antibody to mice with glomerular IgA immune complexes significantly and rapidly reduced persistence of the IgA, presumably by redistributing the immune lattices in situ [101]. Therefore, intraglomerular IgA lattices can undergo rearrangement in situ, promoting either accretion or dissolution of the complex, just as IgG complexes can [93–95].

Human IgAN is characterized by accumulation within glomeruli of polymeric IgA1 that is abnormally glycosylated. Enzymatic removal of sialyl and galactosyl residues from normal human IgA promotes binding of the IgA to human mesangial cells in vitro, relative to undigested IgA. Fractions of serum IgA1 purified from IgAN patients that are enriched in macromolecular IgA1 bearing underglycosylated hinge glycopeptides deposit well in rat glomeruli after injection, whereas other fractions from the same IgA1 pool localize poorly if at all [104]. In subsequent studies, rats given injections of human IgA1 pretreated in vitro sequentially with neuraminidase and β-galactosidase formed glomerular deposits of IgA, followed by influx of neutrophils into glomeruli, whereas rats given untreated human IgA1 did not show these alterations [105]. In summary, the abnormally glycosylated IgA1 found in patients exhibits proclivity for glomerular accumulation that can be ascribed at least in part to improved binding to glomerular mesangial cells and/or matrix.

Some macromolecules bind to glomerular cells or extracellular matrix. Such structures can thereby recruit Ig of any class to intraglomerular sites if the Ig binds to a macromolecule acting either as an antigen or as an immunologically nonspecific ligand. For example, in passive and active serum sickness models of IgAN, the size and charge of the antigen, dextran, influenced not only the extent, but also the site, of glomerular IgA antibody deposition [2–4]. The effects of charge in this system is based on binding to glomerular heparan sulfate and other polyanions. Another passive system exploits the binding of IgA to concanavalin A to promote glomerular deposition of IgA; both the binding of IgA to concanavalin A and the deposition of the resulting complexes in glomeruli arise by virtue of lectin affinity, unrelated to the specificity of the Ig [106]. The dietary lectin gliadin binds to rat mesangial cells in vitro, and to free soluble murine IgA [107]. Like the binding to concanavalin A, binding of IgA to gliadin occurs on the basis of lectin properties; the gliadin then serves as a bridge to trigger deposition of both specific and nonspecific IgA into glomeruli [108]. Modifications of the anti-Thy 1.1 model of glomerulonephritis have exploited the affinity of antigen or antibody within glomerular deposits to promote IgA deposits in rats [109–112]. Again, these models demonstrate the plasticity of established glomerular deposits subject to the influence of circulating macromolecules, similar to the previous observations with IgG immune deposits [93–95].

Affinity for Circulating Macromolecules
In healthy individuals as well as patients with IgAN, binary complexes of IgA with plasma fibronectin and ternary complexes with IgA, fibronectin and uteroglobin, are present in the circulation. Plasma fibronectin binds well to fibronectin in the extracellular matrix, and has ready access to the glomerular matrix because the glomerular endothelium is fenestrated. In turn, circulating fibronectin can bind both uteroglobin and IgA, especially the polymeric IgA1 bearing truncated *O*-linked glycans that may be critical for the genesis of IgAN.

Mice rendered deficient in uteroglobin production by targeted gene deletion or transgenic expression of an antisense vector develop IgAN, associated with glomerular fibronectin accumulation as well as functional defects [113, 114]. Apparently, uteroglobin competitively disrupts binding of IgA to fibronectin. In uteroglobin deficient mice, abnormally high levels of circulating IgA-fibronectin complexes develop, as in IgAN patients. Reconstitution of deficient mice by injecting exogenous uteroglobin diminished the formation of IgA-fibronectin complexes, their binding to glomerular cells and the development of IgAN. In contrast to mice, patients with adenosine at position 38 within the transcription initiation site of uteroglobin exon 1 did not have increased risk for IgAN, despite a 74% reduction in luciferase expression in vitro. However, progressive disease was more likely with an increasing number of A alleles, and patients with the AA genotype were approximately 5 times more likely to have progressive disease than those bearing the GG genotype [115]. In another study, patients with IgAN had levels of circulating complexes of uteroglobin with fibronectin similar to controls, whereas complexes of IgA with fibronectin, with or without intercalated uteroglobin, were higher than in controls. Increased uteroglobin content within IgA-fibronectin complexes was associated with lower proteinuria, which in turn correlates to reduced risk for progression of IgAN [116]. Patient IgA, bearing aberrant *O*-linked glycosylation, is more likely to bind extracellular matrix proteins such as fibronectin than control IgA, and the incorporation of uteroglobin into IgA complexes with fibronectin in humans may represent an interaction in the presence of abnormally glycosylated IgA that is more complex than in mice that lack *O*-glycans on circulating IgA. In addition to differences owing to the glycosylation of IgA, the clearance of IgA from the circulation is more robust in humans than mice, especially for polymeric IgA (see following). Accordingly, uteroglobin might exert less influence on the partitioning of IgA between circulating and glomerular compartments in patients, whereas in mice, uteroglobin might serve as an effective bridge to glomerular deposition and/or as a shield from clearance. In patients, although the risk for IgAN might be relatively independent of uteroglobin, the intraglomerular effects of IgA might be more pronounced than in mice, making progressive disease more likely if modulation by uteroglobin is diminished.

Clearance of IgA and Complexes of IgA from the Circulation

Circulating IgA, whether bound in an immune complex or not, is subject to clearance not only by the system of circulating and fixed tissue mononuclear phagocytes, but also by several other ligands. Similar to immune complexes containing IgG, immune complexes of murine IgA are removed from the circulation after injection into mice principally by nonparenchymal cells in the liver, presumably fixed phagocytes. Splenic uptake is less prominent. The molecular weight is the critical determinant for both the rates of clearance and of renal localization [96–98]. Very large complexes approximate particulates, and are rapidly removed. Intermediate complexes are cleared more slowly, and deposit in the kidney, whereas small complexes persist in the circulation for long periods and do not form durable glomerular deposits. The presence of J chain, or the polymeric nature of the IgA itself, is not relevant to either the clearance rate or glomerular deposition because complexes formed with monomeric IgA behave almost identically to the same-sized lattices prepared with polymeric IgA. The importance of J chain for glomerular deposition derives from its potential to favor formation of nephritogenic intermediately sized complexes. The distribution of uptake of radioactive IgA and IgG immune complexes into liver and spleen was similar in rats [117] and baboons [118], although the rate constants appear to differ among the species.

Models of Secondary IgAN

Given the importance of the liver in clearance of IgA immune complexes from the circulation and in cognizance of the occurrence of secondary IgAN in patients with cirrhosis and other hepatobiliary disease, several models of IgAN secondary to liver injury and/or cholestasis were developed in rats [6, 7, 9, 61, 117, 119–123] and mice [8, 50]. Severe hepatocellular injury induced by carbon tetrachloride progressed to cirrhosis; affected rats developed IgAN characterized by proteinuria and hematuria, and associated with increased levels of circulating immune complexes and serum IgA [6, 7]. Similarly, cholestasis instigated by ligation of the common bile duct produced a secondary IgAN [8, 9, 50]; cholestasis can synergize with oral immunization to intensify the glomerular changes [8, 45–47, 50]. Feeding ethanol or a lipotrope diet, and especially the combination of both insults also provoked secondary IgAN [61, 117, 120, 121]. A mutant rat strain with autosomal recessive hyperbilirubinuria similar to the Dubin-Johnson syndrome also develops mesangial proliferative glomerulonephritis that resembles IgAN [119]. In yet another model, two hepatoxins induced glomerular disease with features of secondary IgAN [122]; galactosamine was the more effective agent for inciting the IgAN.

Mononuclear Phagocytes

In each of these reports, the observed glomerular deposits were ascribed to defective clearance of circulating IgA immune complexes, presumed to result from deranged mononuclear phagocyte function. Indeed, blockade of the mononuclear phagocyte system by injecting colloidal carbon into mice resulted in glomerular IgA deposits after a protracted time, and synergized powerfully with oral immunization [22]. However, in the mutant hyperbilirubinemia model, biliary excretion rather than hepatic uptake seems culpable, casting some doubt on the exclusive contribution by Kupffer cells. Cholestasis itself, whether by surgical ligation or consequent to mutation, does not directly impact on the function of fixed phagocytes; rather, altered hepatic structure and function would likely diminish phagocytosis by Kupffer cells only secondarily. Moreover, in rats fed ethanol and a lipotrope diet, there was a selective defect in the clearance of IgA immune complexes, but the removal of IgG complexes from the circulation was intact [117]. Finally, in the active model of IgAN induced in mice by repetitive injection of dextran, the clearance of radiolabled IgA immune aggregates from the circulation was decreased, whereas the clearance of IgG aggregates was normal [64]. Binding of IgA to hepatocytes was lower than normal, whereas binding to Kupffer cells was increased. These investigators explicitly suggested that reduced uptake by hepatocytes, instead of or in addition to fixed phagocytes, underlies impaired liver clearance of IgA complexes, promoting their glomerular deposition. These observations suggest that defective mononuclear phagocyte function is not the exclusive cause of secondary IgAN.

Asialoglycoprotein Receptors

Beyond mononuclear phagocytes, other systems can impact upon the removal of IgA and/or IgA complexes from the circulation. Notably, as a glycoprotein, IgA is subject to interaction with the asialoglycoprotein receptor expressed by hepatocytes, among other cell lineages. Blockade of asialoglycoprotein receptors or Fc receptors had differential effects on the clearance of immune complexes from the mouse circulation, depending on the size and antigen content of the immune complexes, but each contributed significantly to overall clearance of the complexes [124, 125]. Localization studies identified hepatocytes as asialoglycoprotein receptor positive, whereas Fc receptors are expressed on the Kupffer cells. There are questions as to whether these studies, performed with IgG antibodies, are applicable to IgAN. Recently, the clearance of preparations of normal or abnormally glycosylated human IgA was examined in mice [126]; asialoglycoprotein receptors are implicated especially in clearing the higher molecular weight, abnormally glycosylated isoforms implicated in pathogenesis of IgAN. However, mouse phagocytes lack the receptor for IgA Fc regions. Therefore, experiments to explore the relative contributions of Fc receptor and asialoglycoprotein receptor to the clearance of IgA complexes have become feasible only recently with the development of CD89 transgenic mice [127]. To date, such experiments have not been reported.

Polymeric Ig Receptors

Transepithelial transport of IgA (and IgM) via polymeric Ig receptors (pIgR) is responsible for the secretory Ig that lines mucosal surfaces. In addition to supporting the secretion of free antibody, pIgR mediates the excretion of IgA immune complexes from extracellular fluid in the lamina propria into gut, airway and biliary lumens. In rodents, pIgR is expressed on hepatocytes as well as epithelial cells lining secretory mucosal surfaces, whereas in humans (and most primates), pIgR is not present on hepatocytes. IgA antibody in the intestinal lamina propria can bind antigen from the systemic circulation; the resultant complexes are excreted through mucosal epithelial cells into the gut lumen, via pIgR [128]. This excretory system may forestall deposition of the IgA complexes into glomeruli, whereas defects in this system might promote IgAN. Of interest, reduced pIgR expression is reported in the LIGHT transgenic mouse model of IgAN [52] and in aging ddY mice at risk for spontaneous IgAN [129]. Moreover, the reduced IgA excretion in the HIGA (selectively inbred) variant of ddY mice [59] is reasonably ascribed to reduced pIgR expression.

Receptors for the Fc Region of IgA

The receptor for IgA Fc regions, CD89, is expressed on a variety of primate cells, including neutrophils and platelets as well as circulating and fixed mononuclear phagocytes. The type I glycoprotein is shed from cell membranes, both spontaneously and upon stimulation of the cells. One such stimulus is the binding of IgA to CD89. The homologous receptor is not expressed by mouse cells; therefore, clearance of IgA and IgA complexes from the murine circulation differs from that in humans. Furthermore, circulating complexes of CD89 and IgA are increased in patients with IgAN. Transgenic mice that express human CD89 on macrophage/monocytes spontaneously develop mesangial IgA deposition associated with the morphologic and functional changes typical of IgAN [127]. Passive model systems indicate that IgA in glomerular deposits comes from circulating IgA-CD89 complexes, independent of the λ-chain (signal-transducing) common to Ig Fc receptors. Presumably, binding of soluble CD89 shields IgA from clearance mechanisms which are operative in the mouse. If so, CD89 could then serve as a 'Trojan horse' delivering IgA to a higher-affinity intraglomerular receptor or acceptor. On the other hand, injection of recombinant human soluble CD89 or a chimeric CD89-Fc protein into mice did not result in significant IgA or CD89 deposition in the renal mesangium, and no binding of mouse IgA to CD89 was detected [130]. Perhaps multipoint binding at the cell surface is required to support binding of murine IgA to CD89, whereas the higher affinity for human IgA promotes binding to free soluble CD89. At present, this mechanism has not been substantiated by any experiments. Moreover, the binding of CD89 would reduce clearance of IgA from the circulation mainly by competition with the Fc receptor on phagocytes; such competition would operate in humans or transgenic mice, but not in the wild-type mice used to perform the experiment that repudiates a role for CD89.

Effectors of Glomerular Pathophysiology

Despite considerable efforts, the stimuli that instigate changes in glomerular structure and function in IgAN remain enshrouded. As with glomerulonephritis in general, the traditional assumption has been that immune complexes deposit or form within the glomeruli to evoke direct receptor-mediated responses from indigenous glomerular cells, activate the complement cascade to promote indirect glomerular responses, and/or recruit and engage circulating leukocytes via interaction with receptors for Ig Fc regions, complement or perhaps other ligands. Attempts to elucidate which of these conceptually discrete paths apply have generated as much confusion as resolution. Traditional and novel sequences and schemes that may underlie glomerular IgA deposition and pathophysiology are summarized in figure 2. In part, the divergent views likely derive from the reality that parallel mechanisms operate, simultaneously or sequentially, even in a single model system. Concordance among experimental systems, and between any animal model and naturally occurring disease, is further stymied because each clinical case likely represents a narrow, if not unique, portfolio of mechanisms.

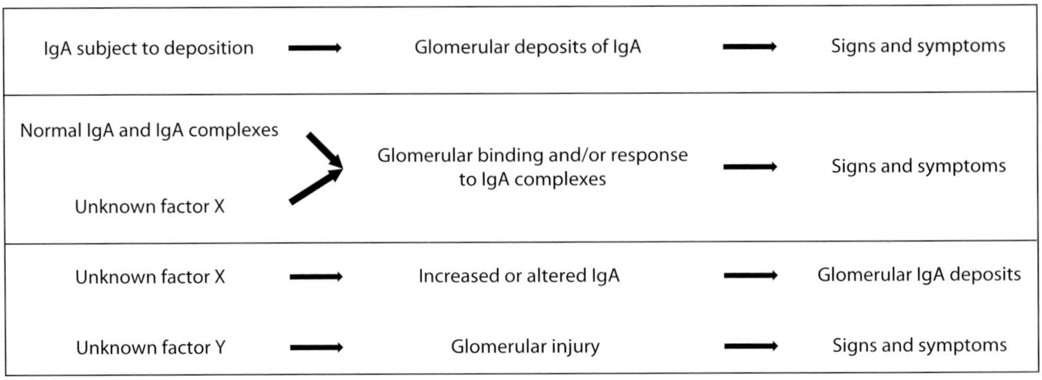

Fig. 2. Hypothetical sequences of development of IgAN. The most traditional scheme (top panel) posits that IgA, presumably as immune complexes but possibly including nonimmune binding partners, deposits within glomeruli and evokes glomerular dysfunction by interacting with mesangial cells or other elements. Co-deposits of IgG, complement, antigen or other components of the complexes might elicit physiological responses, instead of or in addition to the IgA [25, 27, 51, 98, 102, 103, 107–112, 131]. A newer scenario (middle panel) introduces the principle that one or more unknown or unspecified factors promote binding of IgA complexes to glomeruli and/or to favor or intensify the glomerular response to the bound IgA complexes [65, 66, 68–72, 106–108, 113–116, 131–137]. This sequence does not require abnormal amounts of IgA complexes or aberrant IgA structure, but could explain why only a small minority of HIV patients with very high levels of circulating IgA complexes develop IgAN. Increased production or access of complexes containing normal or abnormal IgA might be instigated by the same or a different factor (not shown). Finally, a third model (bottom panel) predicts that glomerular IgA deposits are serendipitous; the glomerular injury and consequent signs and symptoms arise in response to one (or more) factor(s) Y, whereas glomerular IgA deposition that is not causal of and potentially totally unrelated to the glomerular dysfunction, is a consequence of factor(s) X [19, 76, 131, 132, 139, 140, 147–154]. Even if X and Y are the same, distinct properties independently evoke the genesis or deposition of IgA complexes and the signs and symptoms of glomerulopathy. In fact, glomerular IgA deposition might be consequential to dysfunction, rather than its cause [88, 89, 138].

Co-Deposits of Other Ig Classes and Complement

Considerable focus has been devoted to the role of IgG, and to a lesser extent IgM, co-deposits and the influence of complement activation. Although IgA itself can bind to mesangial cells and perhaps visceral glomerular epithelial cells (podocytes) in vitro to elicit cellular responses, IgA is relatively indolent [30, 35]. Notably, modifications of the well established IgG- (and complement-) mediated anti-Thy1 glomerulonephritis supports the view that IgG and complement co-deposition is important. One approach reported that an IgA switch variant of a monoclonal antibody specific for Thy 1.1 elicited hematuria comparable to that evoked by injection of the original IgG2a antibody, but only the latter promoted intraglomerular complement activation, glomerular leukocyte influx or proteinuria [110]. Furthermore, admixture of the IgA switch variant with a subnephritogenic dose of the IgG2a antibody provoked albuminuria that was not observed with either class of antibody alone [111]. Hematuria developed in all groups, but mesangial matrix increase was significantly more pronounced in rats given both antibodies.

In a different approach, this same group reported that injection of subnephritogenic quantities of haptenated F(ab')2 IgG anti-Thy1 antibody into rats, followed by injection of antibody specific for the hapten, leads to IgAN. If small amounts of IgG anti-hapten antibody were combined with larger quantities of IgA, much more glomerular complement deposition and proteinuria was evoked, compared to rats given either IgA or IgG anti-hapten antibody alone [112]. Again, hematuria was equal among the three groups that developed glomerular immune complex deposition, but morphologic evidence of injury was more pronounced in rats given the combined IgA and IgG antibody preparation. In the model of IgAN instigated by injection of haptenated F(ab')2 anti-Thy1 antibody followed by rat IgA specific for the hapten, proteinuria and glomerular C3 deposition were correlated and developed after only injection of polymeric (as opposed to monomeric) IgA antibody, as did glomerular influx of monocytes [109]. Upon deposition of pIgA in the mesangium, rats pre-treated with cobra venom factor did not develop the proteinuria, glomerular deposits of C3 or C9 or glomerular macrophage influx observed in normal rats, despite an identical pattern of glomerular IgA [131].

Another approach employed controlled admixtures of Ig classes in a passive serum sickness model in mice. Again, more glomerular injury ensued when complement-fixing IgG co-deposited with the (predominant) IgA, whereas a noncomplement-fixing switch variant of this IgG did not potentiate glomerular injury beyond that observed with IgA alone [25, 27]. These groups also employed modification of an active serum sickness model in the same strain of mice to demonstrate that the IgG component of the humoral immune response that operates in the absence of mucosal (oral) tolerance is important for glomerular dysfunction, and that depletion or congenital absence of complement (particularly of C3 rather than C5) diminishes the phlogistic effects of IgG co-deposits.

In yet another system that employs immunization with apoferritin, mice deficient in C4 produced much less IgG antibody than did wild-type mice. Upon antigenic challenge, the C4-deficient mice developed an IgA-predominant mesangial proliferative glomerulonephritis with more mesangial hypercellularity than the IgG-predominant mesangial and capillary glomerulonephritis that evolved in the wild-type mice [51]. Deficiency of C3, alone or with C4 deficiency, leads to mesangial deposits rich in IgA, but without the mesangial proliferation, glomerular neutrophil infiltration and, of course, the glomerular C3 deposition that develop in C4-deficient mice replete with C3. In several distinct systems, injection of IgA immune complexes that contain an antigenic component that promotes complement activation into mice results in glomerular C3 deposition and altered glomerular function, whereas structurally similar complexes without this antigenic component localize to glomeruli to a comparable degree, without concomitant complement deposits or urinary abnormalities [98, 102, 103]. Overall, these data in rats and mice implicate complement activation as a linchpin for physiologic or morphologic response, at least in these models of IgAN.

Other Co-Deposits

As already noted (see 'Affinity for Glomerular Components' subsection, above) a variety of macromolecules can bind to glomerular cells or extracellular matrix. In addition to recruiting components to establish or enlarge glomerular deposits, molecular structures with such affinity might activate as well as bind to glomerular cells. Moreover, some macromolecules can activate the complement cascade, and thus indirectly influence glomerular cell function. Gliadin not only binds to rat mesangial cells in vitro on the basis of lectinic affinity, but also stimulates these cells to produce tumor necrosis factor, and shifts cellular metabolism from the normally predominant vasodilatory prostaglandins (E2) towards the contractile eicosanoid thromboxane (B2) [107]. Other agents including infectious pathogens, dietary/environmental agents and endogenous macromolecules also bind to glycoconjugates or other acceptors on the surface of mesangial cells and activate them in vitro. The same effects can alter glomerular function in vivo [108]. These direct functional effects of potential antigens depend on binding to the cells.

Adaptation of immunization with cationic dextran to rats has generated confusion. The first such attempt reported disease with properties very similar to those of the original mouse model [132]. However, two independent subsequent reports indicate that although rats develop hematuria, proteinuria and variable mesangial proliferation during the period in which dextran is administered (intravenously), all these signs subside subsequent to cessation of the antigen injections. In one report, there was no deposition of IgA, IgM, IgG or C3 [66], whereas another group reported no glomerular IgA (after concerted effort), but did detect IgG, IgM, C3 and small C5b-9 deposits in glomeruli of nephritic rats [65]. One of these groups also detected circulating IgG, but not IgA, antidextran antibodies in immunized rats [65]. When added to rat mesangial cells in culture, polycationic macromolecular polypeptide precipitated cell injury, whereas similar polycations with lower molecular weight did not; the effect was inhibited by pre- or co- incubation with anions [133]. Cell injury in this system appears consequent to cross-linking of cell surface molecules bound on a charge basis to an extrinsic macromolecule similar to the polycationic dextran preparations employed in vivo [65, 66]. If so, this is another example whereby co-deposits within a macromolecular aggregate can lead to glomerular injury.

Intraglomerular antigen may also intensify damage through activation of innate immunity. In a passive IgAN model, in situ glomerular deposition of a microbial polysaccharide containing the relevant hapten (phosphorylcholine) evoked much more severe morphologic and urinary abnormalities than a similar phosphorylcholine preparation devoid of a pathogen-associated molecular pattern, despite comparable C3 deposition [103]. Similarly, in an active system, mice primed with a protein carrier and then boosted with the carrier haptenated with phosphorylcholine developed an anamnestic response to the hapten. Subsequent challenge with hapten contained within a macromolecular polysaccharide derived from bacterial cell wall instigated mesangial proliferative glomerulonephritis with predominant glomerular deposition of IgA, whereas a mild glomerulonephritis with only weak IgM deposits ensued after challenge with hapten on the inert carrier used for priming [134]. These experiments cannot discriminate indirect effects of innate responses of circulating leukocytes from direct effects due to innate responses of indigenous glomerular cells. Contributions of both extrarenal effects introduced by exogenous cytokines [135] and intrarenal effects of innate responses [136] are recognized, and may underlie exacerbation of IgAN by pathogens [68–72]. Although innate responses can shape adaptive immunity, the recently reported effects of pathogen-derived DNA on the ddY model of IgAN appear to represent intrarenal effects that do not stem from immunomodulation [137].

Glomerular Responses and Hemodynamic Factors
Whether instigated by a separate process or incidental to nascent glomerulonephritis, increased glomerular permeability to macromolecules, which manifests clinically by proteinuria, may also enhance glomerular deposition of IgA immune complexes [138]. Mice made proteinuric by pretreatment with adriamycin or bovine serum albumin developed markedly enhanced mesangial IgA, antigen and C3 deposition after injection of preformed IgA complexes, compared to nonproteinuric mice. Neither pretreatment influenced the distribution of glomerular deposits or the rate of clearance of the IgA complexes. Therefore, proteinuric IgAN could lead to positive feedback and progressive intensification of glomerular dysfunction. This effect might contribute to IgAN in diabetic rodents [88, 89], but cannot explain the changes in serum IgA or cellular immunity.

Mediators

Several mediators have been implicated in connecting the operative stimuli to glomerular dysfunction in many distinct model systems. Implication arises from detection of elevated mediator (or surrogate) concentration in or production by tissue from diseased kidney relative to control tissue, effect(s) instigated by administration of the exogenous mediator and/or observation of salutary effects of known or putative inhibitors of the mediator or its signal transduction.

Reactive Oxygen and Nitrogen Species
Malondialdehyde, a product of adduction of reactive oxygen species to membrane phospholipids, is increased in extracts of glomeruli from rats with an orally induced active serum sickness model of IgAN [41, 42]. Levels of the oxidation product are reduced after treatment with the antioxidant or radical scavenger agents vitamin E and/or fish oil. In this and a mouse model induced by feeding nivalenol, the amelioration of glomerular function by vitamin E or fish oil lends further credence to the principle that oxygen-based free radicals contribute to deranged glomerular function [41–43, 139, 140]. In HIGA mice with IgAN, increases in intrarenal reactive oxygen species are correlated with increased activation of the renin-angiotensin system, and angiotensin receptor antagonists reduce oxidant stress within glomeruli [141].

In a rat model induced by immunization with ovalbumin including oral challenge to drive IgA production, increased inducible nitric oxide synthetase expression correlated with proteinuria levels and co-localized with activated NF-kB in indigenous glomerular cells and infiltrating leukocytes [48]. Rats with higher levels of glomerular IgA (and milder proteinuria) expressed significantly higher levels of inducible nitric oxide synthetase and produced more nitrite than controls, but less than the levels in (more severely afflicted) rats with only IgG immune complexes. In vitro, IgA immune complexes increased generation of nitric oxide and inducible nitric oxide synthetase expression by mesangial cells and monocytes, albeit to a much lower degree than IgG immune complexes. Inhibitors of NF-kB activation reduced these effects. In sum, these observations support a role for nitric oxide in immune complex-mediated glomerular injury.

Eicosanoids
Elevated levels of eicosanoids produced in vitro by glomeruli isolated from diseased rats were associated with altered glomerular hemodynamics in a model induced by oral immunization with an inert protein, followed by parenteral challenge [39]. In the same model, other investigators demonstrated the beneficial effects of fish oil [42, 43], rich in ω-3 PUFA that shunt eicosanoid metabolism towards less phlogogenic isoforms. Collaterally, similar effects of eicosapentaneoic or docosahexaenoic acids (also ω-3 PUFAs) have proven salutary in the mouse model of IgAN-induced by vomitoxin [139, 140, 142]. Although genetic ablation or pharmacologic inhibition of cyclooxygenase 2 further exaggerated IgA production in this model, reduced activity of this enzyme did not influence disease parameters [143]. Cyclooxygenase 2 preferentially promotes synthesis of vasodilatory eicosanoids such as PGE2 and prostacyclin, with little effect on vasoconstrictors such as thromboxane. Diminished production or effect of thromboxane in rats [39] reduced hematuria associated with IgAN; the lack of effect of cyclooxygenase 2 reduction on glomerular function in the mouse model [143] does not exculpate eicosanoids as a factor in the glomerular response to deposition of IgA immune complexes because other isoforms of prostaglandin and other isomers of cyclooxygenase are minimally affected or left intact. Overall, these experiments in rats and mice corroborate the view that arachidonate metabolism contributes to altered glomerular function.

Growth Factors and Cytokines
In the mouse model of IgAN instigated by immunization and repetitive challenge with dextran, markedly increased platelet-derived growth factor (PDGF) was detected in glomeruli (immunohistochemically) at the protein level as well as at the level of mRNA in comparison to control mice [144]. Kidney cDNA derived from diseased HIGA mice is enriched in RNA coding for PDGF and endothelial differentiation gene 5, a receptor that is upregulated in response to PDGF [145]. In the same model, a putative drug mixture that

inhibits the receptor tyrosine kinase associated with PDGF ameliorated the glomerular disease [146]; further analysis revealed that the active components were specific to the PDGF-B receptor, without influencing other growth factor receptor tyrosine kinases. In a lupus-like spontaneous model of glomerulonephritis in the RF/J strain of mice that is very rich in glomerular IgA deposits, PDGF is overexpressed in kidney tissue at earlier phases of disease characterized by mesangial proliferation, but is overshadowed by TGF-β at later times when glomerulosclerosis supervenes [76].

Interleukin (IL)-6 also represents an intriguing focal point for evaluation of mediators that promote IgAN. Beyond the copious data derived from studies of patient urine and biopsies, several animal models are noteworthy. Transgenic mice that overexpress IL-6 in B cells (under the control of an Igh promoter sequence) develop massive plasmacytosis, hypergammaglobulinemia (including a three-fold elevation in serum IgA levels) and a mesangial proliferative glomerulonephritis [147]; no glomerular immunofluorescence was reported. Subsequent studies of mice bearing an IL-6 expression vector under the promoter for major histocompatibility class I developed similar plasmacytosis and mesangial proliferative glomerulonephritis associated with proteinuria and hematuria; blocking antibody specific for the IL-6 receptor or all-transretinoic acid (known to downregulate both IL-6 receptor and its major signal-transducing adaptor protein) ameliorated the morphologic and functional indicators of glomerular disease as well as the plasmacytosis [148, 149]. These observations are in keeping with the principle that IL-6 is a mesangial cell growth factor [150, 151], a concept that is at least somewhat controversial at present. In addition to effects on glomerular mesangial cells, IL-6 is a potent driver of B cell differentiation and Ig secretion; in some systems, IgA production is stimulated preferentially. The mouse model of IgAN elicited by feeding trichothecene mycotoxins to mice appears particularly dependent upon IL-6; genetic ablation of IL-6 resulted in significant reductions in serum IgA concentration, glomerular IgA deposition and hematuria compared to syngeneic wild-type mice [152]. IL-6 deficiency did not affect the metabolic wasting induced by vomitoxin. Diminished IL-6 production attendant upon ω-3 PUFA dietary supplementation was associated with marked reduction of the signs of IgAN in this model [139, 140]. Suspensions of mouse peritoneal macrophages in vitro increase IL-6 expression in response to addition of mycotoxin through the effect of RNA-activated protein kinase phosphorylation; these kinases then act upon the cAMP response element-binding protein (CREB) [142]. In each of these systems, overproduction of IgA and glomerular deposition of IgA are roughly correlated, and it is difficult to determine whether the effects of IL-6 reside principally with increased IgA production, with the glomerular response to IgA deposition, or from synergy between these two possibilities. Indeed, as discussed in a recent review, the mesangial proliferative glomerulonephritis and urinary abnormalities might develop independently of the IgA (or other Ig) deposition in the glomeruli, and instead represent direct effects of IL-6 and/or other mediators upon glomerular cells [65, 66, 153, 154]. This view is consistent with the transgenic mouse models [147–150], that do not report specifically upon glomerular Ig deposition and document IgG1 overproduction and plasmacytosis far in excess of the IgA synthesis and differentiation. In a passive model characterized by a fixed and reproducible level of glomerular IgA immune complex deposition, infusion of exogenous IL-6 alone had no measurable effect; however, IL-6 exhibited strong synergy with collateral IL-1 administration [135].

Mice with IgAN induced by parenteral injection of myeloma IgA and the relevant specific antigen express increased transforming growth factor (TGF)-β in the glomerular mesangium [155]; reduction of TGF-β is collateral with diminished mesangial proliferation and hematuria in this system, suggesting a contribution of TGF-β to glomerular pathophysiology. Mice that develop IgAN from appropriate polarizing immunization and challenge with antigen derived from

S. aureus also exhibit increased glomerular TGF-β mRNA [49]. Three different groups also reported increased RNA encoding TGF-β in renal tissue of rats with IgAN induced by oral immunization and challenge with inert proteins [41–44, 46], and another group reported similar findings in the HIGA mouse model of spontaneous IgAN [16]. In the HIGA mice, immunohistochemical staining disclosed TGF-β protein in the glomerular mesangium. Finally, increased TGF-β expression in glomeruli is also reported in a different mouse model of spontaneous proteinuric glomerulonephritis, the RF/J mouse [76]. Although IgG deposition is initially more intense in this latter model, there is very intense glomerular IgA that increases with age, and levels of serum IgA and circulating IgA immune complexes are elevated and progress with age to eclipse the increases in IgG. Therefore, this model may represent a form of immune complex glomerulonephritis evolving towards an IgA-dominant pattern. In all these systems, TGF-β is correlated with urinary abnormalities, mesangial matrix expansion and/or expression of increased matrix proteins in the glomeruli. Extraglomerular TGF-β may also exert intraglomerular effects; splenic T cells from HIGA mice exhibit upregulation of TGF-β, which may promote both increased IgA production and progressive glomerulosclerosis in HIGA mice [80].

TNF-α is a pluripotent cytokine, especially powerful at evoking IL-6 production, NF-kB activation and other cellular responses implicated in the pathogenesis of IgAN, both in patients and animal models. In the mouse model of IgAN evoked by dietary nivalenol, TNF-α acts through its receptor TNFR1 to increase concentrations of IgA and IgA immune complexes in the circulation, and increase mesangial IgA deposits [156]. No effects of TNF-α on anorexia or growth were detected. It is unclear from this study whether the effects of TNF-α on the experimental IgAN relate to intensified glomerular response to IgA complexes, or derive entirely from increased glomerular deposition of IgA. In favor of the former view, studies of patient renal biopsies and human cells in vitro may contribute. Specifically, chemokines of the pentraxin (PTX) family, especially PTX3, amplify a variety of inflammatory reactions. PTX3 is expressed in patients with IgAN, by cells that are probably infiltrating leukocytes. Either TNF-α or IgA stimulate PTX3 production by human mesangial cells in vitro, but no attempt to determine (a) synergistic response(s) has been reported [157]. In turn, PTX3 binds to human mesangial cells, and stimulates production of platelet-activating factor as well as contractile changes in the cells. These autocrine effects have not, to date, been explored in animal models (or in naturally occurring disease in patients), but do provoke the principle that TNF-α, PTX3 and platelet-activating factor may contribute to the glomerular response to IgA deposits. Of course, TNF-α is a potent stimulator of IL-6 synthesis, which in turn can have significant effects on glomerular function (see above).

Other Mediators

Macrophage migration inhibitory factor (MIF) is implicated in connecting glomerular immune complex deposition to physiological response in the passive serum sickness model of IgAN in mice; specifically, exogenous anti-MIF reduces glomerular TGF-β expression as well as hematuria and mesangial hypercellularity [155]. This report also documents that exogenous MIF itself stimulates TGF-β production in vitro, an effect that is blocked by neutralizing antibody specific for MIF. Monocyte chemoattractant protein-1 expression in glomeruli is elevated in rodent models of IgAN instigated by synergy between vaccination and hepatocellular injury [45, 46] or cholestasis [50]. In mice genetically selected for low acute inflammatory responses and a monocyte-predominant inflammatory response to injury, the IgA predominant responses relative to mice with a neutrophil-predominant high acute inflammatory response evoke less intense inflammation, but appear especially associated with monocyte chemoattractant protein despite a lack of glomerular influx of monocytes or macrophages [50]. IL-12 may contribute to glomerular inflammation, acting through IFN-γ or by other means. If recombinant murine IL-12 is injected (i.p.) repetitively into the high-IgA inbred variant (HIGA) of ddY mice, glomerular damage is intensified, with crescent formation, macrophage accumulation in glomeruli and heightened interstitial cell infiltration [80]. Platelet activating factor synthesis by mesangial cells is induced by pentraxin 3, in turn stimulated by TNF-α or IgA (see above, [157]).

IgA Glycosylation in the Genesis of IgAN

Abnormal glycosylation of IgA is an important feature in patients with IgAN, and is perhaps causal [21]. Recent animal models offer refinements to our appreciation of the role of IgA glycosylation in the genesis of IgAN. As already mentioned, natural or enzymatically generated human IgA1 bearing abnormal glycans deposits more readily into rat glomeruli [104, 105] and activates mesangial cells in culture more effectively than control preparations of normal human IgA1 [158]. In a transgenic model in which the mouse Ig genes are replaced by the corresponding human genes, IgAN induced by injection of abnormally glycosylated human IgA1 is prevented by prior immunization with a hinge peptide bearing the abnormal glycans [17]. Glomerular deposition of injected IgA1 is inversely correlated with the titer of the antibody. Thus, antibodies specific for the aberrantly glycosylated hinge, which may cause IgAN, can also confer protection from IgAN. Presumably, opsonic effects of the antibody accelerate clearance of the IgA1 from the circulation and/or alter the lattice structure of glomerular IgA deposits (see above and [93–95, 99–103]). However, these transgenic mice do not express the CD89 IgA Fc receptor that is present on human phagocytes, and measurement of the clearance or deposition of human or mouse IgA has limited value for extrapolation to disease in patients.

Another important difference between people and mice relates to the fact that human IgA1 bears *O*-linked as well as *N*-linked glycans, whereas mouse IgA has only the latter. To date, focus has been on abnormal *O*-linked glycosylation as underlying IgAN in patients [21]. Nonetheless, when transgenically altered human IgA is injected into mice, the *N*-linked glycans appear more important than the *O*-linked glycans for shaping the partitioning of IgA complexes among the circulation, hepatic clearance and glomerular deposition [126]. The important role for abnormalities in *N*-glycosylation in mice is further corroborated by other approaches. Mice that bear targeted deletion of the specific galactosyl transferase that supports *N*-linked glycosylation develop IgAN spontaneously [18]. Other murine models of IgAN in (bcl1) transgenic mice and mucosally immunized mice report defects in *N*-glycans on IgA in association with the glomerular disease [33, 38, 73]. Finally, the spontaneous development of progressive glomerulonephritis in mice with an induced mutation is very provocative [19]. A point mutation only in the core 1 galactosyltransferase gene results in severe loss of function of the enzyme that is critical for normal assembly of the *O*-linked glycan in the human IgA1 hinge peptide. However, mouse IgA does not contain the relevant peptide, and is likely normally glycosylated in these mice. Mesangial proliferative glomerulonephritis, with features evocative of IgAN but devoid of glomerular Ig deposits, develops in these mutant mice. Perhaps the defective glycosylation of podocalyxin and/or other substrates observed in these mice [19] also underlies spontaneous human IgAN. Therefore, glomerular deposition of IgA1, favored by the increased affinity of defectively glycosylated IgA1 for glomerular substituents, may be a serendipitous feature of the human disease rather than its cause, at least in some cases [153].

IgAN in Other Species

In addition to the aforementioned spontaneous IgAN that arises in primates [55, 56], macaques infected with simian immunodeficiency virus (with or without complicating infection with *M. bovis*) also develop IgAN [159] or an IgAN-like glomerulonephritis [160]. In pigs, spontaneous IgAN is associated with mycoplasma infection [161], and feeding nivalenol induces IgAN [162] similar to that provoked in mice by this dietary toxin [53]. In dogs, some cases of sporadic glomerulonephritis have prominent IgA deposits [163–166]; a few cases of infection-related canine glomerulonephritis also have a pattern that could represent IgAN [167, 168]. A single case of myeloma associated IgAN is reported in dogs [91]. In one large autopsy series, an IgAN pattern was more common among dogs with liver disease or enteritis [165]. However, other series including one focused on glomerulopathy associated with congenital portosystemic shunts detected no glomerular IgA [169–172]. Active serum sickness induced in rabbits with a bacterial vaccine elicits glomerulonephritis, and although deposits are frequent, IgAN is an unusual pattern [173, 174]. Equine glomerulonephritis associated with purpura and IgA immune complex deposition, reminiscent of Henoch-Schonlein purpura, develops after streptococcal infection [175], and leishmanial infection of hamsters can lead to an IgAN-like nephritis [176]. A spontaneous IgAN was also reported in association with Aleutian mink disease decades ago [177]. A model of spontaneous progressive glomerulosclerosis in mice also has predominantly IgA deposits, but has not been investigated further [178].

Therapeutic Initiatives in Experimental IgAN

The most extensive experimental therapeutics relate to altering eicosanoids by use of ω-3 polyunsaturated fatty acids [41, 43, 139, 140, 142] and diminution of oxidative stress owing to these agents or, more pointedly, to use of vitamin E [41–43], all with favorable results to at least some extent. Antagonists of cytokines [46], growth factors [146] or chemokines [155] have also ameliorated varying signs of experimental IgAN, as have treatment with probiotics [37] or traditional herbal remedies with unknown mechanisms [36, 44, 45, 60]. One herbal was combined with the antiplatelet agent dipyridamole [36], which has been used to treat patients with some success. Therapy aimed at disrupting the immune lattice by use of anti-idiotype (pseudoantigen) antibody [101], monomeric IgA [99] or specific [179] or nonspecific hydrolytic enzymes [63] are also reportedly effective, whereas pharmacologic modulation of the response to vaccination (to favor Th2) aggravated experimental IgAN [36]. Sodium cromoglycate, used to treat asthma in patients and known to have effects on the specific immune response, prevents development of experimental IgAN, but has no influence on established disease [29].

References

1 Rifai A, Small PA Jr, Teague PO, Ayoub EM: Experimental IgA nephropathy. J Exp Med 1979;150:1161–1173.

2 Isaacs KL, Miller F: Antigen size and charge in immune complex glomerulonephritis. II. Passive induction of immune deposits with dextran-anti-dextran immune complexes. Am J Pathol 1983;111:298–306.

3 Isaacs K, Miller F, Lane B: Experimental model for IgA nephropathy. Clin Immunol Immunopathol 1981;20:419–426.

4 Isaacs KL, Miller F: Role of antigen size and charge in immune complex glomerulonephritis. Lab Invest 1982;47:198–205.

5 Emancipator SN, Gallo GR, Lamm ME: Experimental IgA nephropathy induced by oral immunization. J Exp Med 1983;157:572–582.

6 Gormly AA, Smith PS, Seymour AE, Clarkson AR, Woodroffe AJ: IgA glomerular deposits in experimental cirrhosis. Am J Pathol 1981;104:50–54.

7 Iida H, Izumino K, Matsumoto M, Takata M, Mizumura Y, Sugimoto T: Glomerular deposition of IgA in experimental hepatic cirrhosis. Acta Pathol Jpn 1985;35:561–567.

8 Emancipator SN, Gallo GR, Razaboni R, Lamm ME: Experimental cholestasis promotes the deposition of glomerular IgA immune complexes. Am J Pathol 1983;113: 19–26.

9 Melvin T, Burke B, Michael AF, Kim Y: Experimental IgA nephropathy in bile duct ligated rats. Clin Immunol Immunopathol 1983;27:369–377.

10 Smith SM, Tokuda S, Tsukamoto H: Mucosal immune dysfunction associated with alcoholic IgA nephropathy. Clin Immunol Immunopathol 1992;64:205–209.

11 Emancipator SN: Animal models of IgA nephropathy. Curr Protoc Immunol 2001;Chapter 15: Unit 15.11.

12 Pestka JJ, Dong W, Warner RL, Rasooly L, Bondy GS: Effect of dietary administration of the trichothecene vomitoxin (deoxynivalenol) on IgA and IgG secretion by Peyer's patch and splenic lymphocytes. Food Chem Toxicol 1990;28:693–699.

13 Dong W, Sell JE, Pestka JJ: Quantitative assessment of mesangial immunoglobulin A (IgA) accumulation, elevated circulating IgA immune complexes, and hematuria during vomitoxin-induced IgA nephropathy. Fundam Appl Toxicol 1991;17:197–207.

14 Imai H, Nakamoto Y, Asakura K, Miki K, Yasuda T, Miura AB: Spontaneous glomerular IgA deposition in ddY mice: an animal model of IgA nephritis. Kidney Int 1985;27: 756–761.

15 Miyawaki S, Muso E, Takeuchi E, Matsushima H, Shibata Y, Sasayama S, Yoshida H: Selective breeding for high serum IgA levels from noninbred ddY mice: isolation of a strain with an early onset of glomerular IgA deposition. Nephron 1997;76: 201–207.

16 Muso E, Yoshida H, Takeuchi E, Yashiro M, Matsushima H, Oyama A, Suyama K, Kawamura T, Kamata T, Miyawaki S, Izui S, Sasayama S: Enhanced production of glomerular extracellular matrix in a new mouse strain of high serum IgA ddY mice. Kidney Int 1996;50:1946–1957.

17 Hiki Y, Takahashi K, Shimozato S, Odani H, Yamamoto K, Tomita M, Hasegawa M, Murakami K, Nabeshima K, Nakai S, Fujita Y, Ishida I, Iwase H, Sugiyama S: Protective role of anti-synthetic hinge peptide antibody for glomerular deposition of hypoglycosylated IgA1. Clin Exp Nephrol 2008;12: 20–27.

18 Nishie T, Miyaishi O, Azuma H, Kameyama A, Naruse C, Hashimoto N, Yokoyama H, Narimatsu H, Wada T, Asano M: Development of immunoglobulin A nephropathy-like disease in beta-1,4-galactosyltransferase-I-deficient mice. Am J Pathol 2007;170:447–456.

19 Alexander WS, Viney EM, Zhang JG, Metcalf D, Kauppi M, Hyland CD, Carpinelli MR, Stevenson W, Croker BA, Hilton AA, Ellis S, Selan C, Nandurkar HH, Goodnow CC, Kile BT, Nicola NA, Roberts AW, Hilton DJ: Thrombocytopenia and kidney disease in mice with a mutation in the C1galt1 gene. Proc Natl Acad Sci USA 2006;103:16442–16447.

20 Strober W, McGhee JR: Inductive and effector tissues and cells of the mucosal immune system: an overview; in Mestecky J, Lamm ME, Strober W, Bienenstock J, McGhee JR and Mayer L (eds): Mucosal Immunology, ed 3. Amsterdam, Elsevier Academic Press, 2005, pp 371–374.

21 Emancipator SN, Mestecky J, Lamm ME: IgA nephropathy and related diseases; in Mestecky J, Lamm ME, Strober W, Bienenstock J, McGhee JR and Mayer L (eds): Mucosal Immunology, ed 3. Amsterdam, Elsevier Academic Press, 2005, pp 1579–1600, ISBN: 0-12-491543-4.

22 Sato M, Ideura T, Koshikawa S: Experimental IgA nephropathy in mice. Lab Invest 1986;54: 377–384.

23 Genin C, Laurent B, Sabatier JC, Colon S, Berthoux FC: IgA mesangial deposits in C3H/HeJ mice after oral immunization with ferritin or bovine serum albumin. Clin Exp Immunol 1986;63:385–394.

24 Genin C, Sabatier JC, Berthoux FC: IgA mesangial deposits in C3H/HeJ mice after oral immunisation. Proc Eur Dial Transplant Assoc Eur Ren Assoc 1985;21:703–708.

25 Emancipator SN, Ovary Z, Lamm ME: The role of mesangial complement in the hematuria of experimental IgA nephropathy. Lab Invest 1987;57:269–276.

26 Biewenga J, Noordhoek GT, Donker AJ, Beukhof JR: IgA nephropathy is not induced in mice by oral administration of TNP-conjugated ovalbumin. Nephron 1987;47: 295–298.

27 Emancipator SN, Lamm ME: The role of IgG, IgM, and C3 in experimental murine IgA nephropathy. Semin Nephrol 1987;7:286–288.

28 Coppo R, Mazzucco G, Martina G, Roccatello D, Amore A, Novara R, Bargoni A, Piccoli G, Sena LM: Gluten-induced experimental IgA glomerulopathy. Lab Invest 1989;60:499–506.

29 Jin SY, Choi IJ: The effect of sodium cromoglycate on the induction of experimental IgA nephropathy. Yonsei Med J 1990;31:33–48.

30 Gesualdo L, Lamm ME, Emancipator SN: Defective oral tolerance promotes nephritogenesis in experimental IgA nephropathy induced by oral immunization. J Immunol 1990;145:3684–3691.

31 Jessen RH, Emancipator SN, Jacobs GH, Nedrud JG: Experimental IgA-IgG nephropathy induced by a viral respiratory pathogen. Dependence on antigen form and immune status. Lab Invest 1992;67:379–386.

32 Endo Y, Kanbayashi H, Hara M: Experimental immunoglobulin A nephropathy induced by Gram-negative bacteria. Nephron 1993;65:196–205.

33 Chintalacharuvu SR, Nagy NU, Sigmund N, Nedrud JG, Amm ME, Emancipator SN: T cell cytokines determine the severity of experimental IgA nephropathy by regulating IgA glycosylation. Clin Exp Immunol 2001; 126:326–333.

34 Yamamoto C, Suzuki S, Kimura H, Yoshida H, Gejyo F: Experimental nephropathy induced by *Haemophilus parainfluenzae* antigens. Nephron 2002;90:320–327.

35 Amore A, Coppo R, Nedrud JG, Sigmund N, Lamm ME, Emancipator SN: The role of nasal tolerance in a model of IgA nephropathy induced in mice by Sendai virus. Clin Immunol 2004;113:101–108.

36 Jia L, Wang C, Kong H, Yang J, Li F, Lv S, Xu G: Effect of PA-MSHA vaccine on plasma phospholipids metabolic profiling and the ratio of Th2/Th1 cells within immune organ of mouse IgA nephropathy. J Pharm Biomed Anal 2007;43:646–654.

37 Soylu A, Berktaş S, Sarioğlu S, Erbil G, Yilmaz O, Demir BK, Tufan Y, Yeşüilirmak D, Türkmen M, Kavukçu S. *Saccharomyces boulardii* prevents oral-poliovirus vaccine-induced IgA nephropathy in mice. Pediatr Nephrol 2008;23:1287–1291.

38 Chintalacharuvu SR, Yamashita M, Bagheri N, Blanchard TG, Nedrud JG, Lamm ME, Tomino Y, Emancipator SN: T cell cytokine polarity as a determinant of immunoglobulin A (IgA) glycosylation and the severity of experimental IgA nephropathy. Clin Exp Immunol 2008;153:456–462.

39 Gesualdo L, Emancipator SN, Kesselheim C, Lamm ME: Glomerular hemodynamics and eicosanoid synthesis in a rat model of IgA nephropathy. Kidney Int 1992;42:106–114.

40 Yi ZW, Rodriguez GE, Krieg RJ Jr, Tokieda K, Chan JC: Rat macrophages in experimental IgA nephropathy. Biochem Mol Med 1996; 57:152–155.

41 Trachtman H, Chan JC, Chan W, Valderrama E, Brandt R, Wakely P, Futterweit S, Maesaka J, Ma C: Vitamin E ameliorates renal injury in an experimental model of immunoglobulin A nephropathy. Pediatr Res 1996;40:620–626.

42 Chan W, Krieg RJ Jr, Norkus EP, Chan JC: alpha-Tocopherol reduces proteinuria, oxidative stress, and expression of transforming growth factor beta 1 in IgA nephropathy in the rat. Mol Genet Metab 1998;63:224–229.

43 Kuemmerle NB, Chan W, Krieg RJ Jr, Norkus EP, Trachtman H, Chan JC: Effects of fish oil and alpha-tocopherol in immunoglobulin A nephropathy in the rat. Pediatr Res 1998;43: 791–797.

44 Zhong Q, Leung JC, Chan LY, Tsang AW, Chen X, Lai KN: The study of Chinese medicinal herbal formula Shen San Fang in the treatment of experimental IgA nephropathy. Am J Chin Med 2005;33:613–626.

45 Zhang GZ, Wu XC, Peng XJ, Yi ZW, Mo SH, Tan H: Effects of astragalus on renal tubulointerestitial lesions and expression of NF-kappaB and MCP-1 in renal tissues in rat experimental IgA nephropathy (in Chinese). Zhongguo Dang Dai Er Ke Za Zhi 2008;10: 173–178.

46 Tang Y, Lou TQ, Wang C, Peng H, Liu X, Tang H: Study on protective effects and its mechanism of leflunomide on renal tissue in rat IgA nephropathy model (in Chinese). Zhongguo Wei Zhong Bing Ji Jiu Yi Xue 2008;20:283–285.

47 Peng W, Liu ZR: Comparison of two rat models of IgA nephropathy (in Chinese). Nan Fang Yi Ke Da Xue Xue Bao 2008;28: 1842–1845.

48 Gómez-Guerrero C, López-Franco O, Suzuki Y, Sanjuán G, Hernández-Vargas P, Blanco J, Egido J: Nitric oxide production in renal cells by immune complexes: role of kinases and nuclear factor-kappaB. Kidney Int 2002;62: 2022–2034.

49 Sharmin S, Shimizu Y, Hagiwara M, Hirayama K, Koyama A: *Staphylococcus aureus* antigens induce IgA-type glomerulonephritis in Balb/c mice. J Nephrol 2004;17: 504–511.

50 Kurihara RS, Yokoo M, Domingues WV, Cabrera WH, Ribeiro OG, Ibanez OM, Malheiros DA, Barros RT, de Almeida Prado EB: Genetic potential for an acute inflammatory response in IgA glomerulonephritis in mice. Braz J Med Biol Res 2005;38:1807–1815.

51 Quigg RJ, Lim A, Haas M, Alexander JJ, He C, Carroll MC: Immune complex glomerulonephritis in C4- and C3-deficient mice. Kidney Int 1998;53:320–330.

52 Wang J, Anders RA, Wu Q, Peng D, Cho JH, Sun Y, Karaliukas R, Kang HS, Turner JR, Fu YX: Dysregulated LIGHT expression on T cells mediates intestinal inflammation and contributes to IgA nephropathy. J Clin Invest 2004;113:826–835.

53 Pestka JJ: Deoxynivalenol-induced IgA production and IgA nephropathy-aberrant mucosal immune response with systemic repercussions. Toxicol Lett 2003;140–141: 287–295.

54 Rasooly L, Pestka JJ: Vomitoxin-induced dysregulation of serum IgA, IgM and IgG reactive with gut bacterial and self antigens. Food Chem Toxicol 1992;30:499–504.

55 Schroeder C, Osman AA, Roggenbuck D, Mothes T: IgA-gliadin antibodies, IgA-containing circulating immune complexes, and IgA glomerular deposits in wasting marmoset syndrome. Nephrol Dial Transplant 1999;14:1875–1880.

56 Brack M, Schroeder C, Fooke M, Schlumberger W: IgM/IgA nephropathy in callitrichids: antigen studies. Nephron 1999; 82:221–231.

57 Yan D, Rumbeiha WK, Pestka JJ: Experimental murine IgA nephropathy following passive administration of vomitoxin-induced IgA monoclonal antibodies. Food Chem Toxicol 1998;36:1095–1106.

58 Akadegawa K, Ishikawa S, Sato T, Suzuki J, Yurino H, Kitabatake M, Ito T, Kuriyama T, Matsushima K: Breakdown of mucosal immunity in the gut and resultant systemic sensitization by oral antigens in a murine model for systemic lupus erythematosus. J Immunol 2005;174:5499–5506.

59 Kamata T, Nogaki F, Fagarasan S, Sakiyama T, Kobayashi I, Miyawaki S, Ikuta K, Muso E, Yoshida H, Sasayama S, Honjo T: Increased frequency of surface IgA-positive plasma cells in the intestinal lamina propria and decreased IgA excretion in hyper IgA (HIGA) mice, a murine model of IgA nephropathy with hyperserum IgA. J Immunol 2000;165:1387–1394.

60 Makino T, Ono T, Matsuyama K, Nogaki F, Miyawaki S, Honda G, Muso E: Suppressive effects of *Perilla frutescens* on IgA nephropathy in HIGA mice. Nephrol Dial Transplant 2003;18:484–490.

61 Lopez MC, Watzl B, Hebert AR, Murphy L, Huang DS, Watson RR: Ethanol-induced changes in the intestinal mucosa-associated immune system of young and mature mice. Alcohol 1993;10:555–557.

62 González-Cabrero J, Egido J, Barat A, González E: Detection and characterization of circulating and glomerular immune complexes in experimental IgA nephropathy. Immunology 1990;70:296–302.

63 Gesualdo L, Ricanati S, Hassan MO, Emancipator SN, Lamm ME: Enzymolysis of glomerular immune deposits in vivo with dextranase/protease ameliorates proteinuria, hematuria, and mesangial proliferation in murine experimental IgA nephropathy. J Clin Invest 1990;86:715–722.

64 González E, González-Cabrero J, Egido J: Defective hepatic handling of IgA immune aggregates by mice with experimental IgA nephropathy. Immunology 1989;67:308–313.

65 Burg M, Gröne HJ, Koch KM, Floege J: Cationic dextran induces mesangioproliferative nephritis in rats independent of glomerular IgA deposition. Nephrol Dial Transplant 1997;12:1856–1862.
66 Pasi A, Dendorfer U, Holthöfer H, Nelson PJ, Tazzari S, Armelloni S, Fornasieri A, D'Amico G, Schlöndorff D: Characterization of nephropathy induced by immunization with high molecular weight dextran. Nephrol Dial Transplant 1997;12:1849–1855.
67 Macpherson AJ, Lamarre A, McCoy K, Harriman GR, Odermatt B, Dougan G, Hengartner H, Zinkernagel RM: IgA production without mu or delta chain expression in developing B cells. Nat Immunol 2001;2:625–631.
68 Kavukçu S, Soylu A, Sarioğlu S, Türkmen M, Küpelioğlu A, Pekçetin C, Güre A: IgA nephropathy in mice following repeated administration of conjugated *Haemophilus influenzae* type B vaccine (PRP-T). Tokai J Exp Clin Med 1997;22:167–174.
69 Kamiyama S: Experimental glomerulonephritis induced by Coxsackie B4 virus in mice–glomerular changes associated with intermittent viral inoculations (in Japanese). Nippon Jinzo Gakkai Shi 1990;32:939–948.
70 Yoshida K, Suzuki J, Suzuki S, Kume K, Mutoh S, Kato K, Suzuki H: Experimental IgA nephropathy induced by coxsackie B4 virus in mice. Am J Nephrol 1997;17:81–88.
71 Isome M, Yoshida K, Suzuki S, Kume K, Suzuki J, Kato K, Suzuki H: Experimental glomerulonephritis following successive inoculation of five different serotypes of group B coxsackieviruses in mice. Nephron 1997;77:93–99.
72 Kawasaki Y, Mitsuaki H, Isome M, Nozawa R, Suzuki H: Renal effects of Coxsackie B4 virus in hyper-IgA mice. J Am Soc Nephrol 2006;17:2760–2769.
73 Marquina R, Díez MA, López-Hoyos M, Buelta L, Kuroki A, Kikuchi S, Villegas J, Pihlgren M, Siegrist CA, Arias M, Izui S, Merino J, Merino R: Inhibition of B cell death causes the development of an IgA nephropathy in (New Zealand white x C57BL/6) F(1)-bcl-2 transgenic mice. J Immunol 2004;172:7177–7185.
74 Marín MJ, García I, Peña M, Bolívar A, Zubiaga A, Lóez-Hoyos M: E2F1$^{-/-}$ C57BL/6 mice overexpressing a human Bcl-2 transgene in B cells develop a mild autoimmune syndrome. Ann NY Acad Sci 2005;1051:156–165.
75 McCarthy DD, Chiu S, Gao Y, Summers-deLuca LE, Gommerman JL: BAFF induces a hyper-IgA syndrome in the intestinal lamina propria concomitant with IgA deposition in the kidney independent of LIGHT. Cell Immunol 2006;241:85–94.
76 Kamata T, Muso E, Yashiro M, Kawamura T, Oyama A, Matsushima H, Takeuchi E, Yoshida H, Sasayama S: Up-regulation of glomerular extracellular matrix and transforming growth factor-beta expression in RF/J mice. Kidney Int 1999;55:864–876.
77 Ellison CA, Gibson IW, Hayglass KT, Gartner JG: Effect of palifermin in a murine model of graft-versus-host disease (GVHD) associated with Th2 cytokine production, autoantibody production, and glomerulonephritis. J Clin Immunol 2006;26:485–494.
78 Curcio C, Pannellini T, Lanzardo S, Forni G, Musiani P, Antón IM: WIP null mice display a progressive immunological disorder that resembles Wiskott-Aldrich syndrome. J Pathol 2007;211:67–75.
79 Nishizumi H, Taniuchi I, Yamanashi Y, Kitamura D, Ilic D, Mori S, Watanabe T, Yamamoto T: Impaired proliferation of peripheral B cells and indication of autoimmune disease in lyn-deficient mice. Immunity 1995;3:549–560.
80 Nogaki F, Muso E, Kobayashi I, Kusano H, Shirakawa K, Kamata T, Oyama A, Ono T, Miyawaki S, Yoshida H, Sasayama S: Interleukin 12 induces crescentic glomerular lesions in a high IgA strain of ddY mice, independently of changes in IgA deposition. Nephrol Dial Transplant 2000;15:1146–1154.
81 Warner RL, Brooks K, Pestka JJ: In vitro effects of vomitoxin (deoxynivalenol) on T-cell interleukin production and IgA secretion. Food Chem Toxicol 1994;32:617–625.
82 Senpuku H, Asano T, Matin K, Salam MA, Tsuha Y, Horibata S, Shimazu Y, Soeno Y, Aoba T, Sata T, Hanada N, Honda M: Effects of human interleukin-18 and interleukin-12 treatment on human lymphocyte engraftment in NOD-scid mouse. Immunology 2002;107:232–242.
83 Imasawa T, Nagasawa R, Utsunomiya Y, Kawamura T, Zhong Y, Makita N, Muso E, Miyawaki S, Maruyama N, Hosoya T, Sakai O, Ohno T: Bone marrow transplantation attenuates murine IgA nephropathy: role of a stem cell disorder. Kidney Int 1999;56:1809–1817.
84 Suzuki H, Suzuki Y, Aizawa M, Yamanaka T, Kihara M, Pang H, Horikoshi S, Tomino Y: Th1 polarization in murine IgA nephropathy directed by bone marrow-derived cells. Kidney Int 2007;72:319–327.
85 Shimizu Y, Sakurai H, Hirayama K, Seki M, Yoh K, Yamagata K, Koyama A: Staphylococcal cell membrane antigen, a possible antigen in post-methicillin resistant *Staphylococcus aureus* (MRSA) infection nephritis and IgA nephropathy, exhibits high immunogenic activity that is enhanced by superantigen. J Nephrol 2005;18:249–256.
86 Nagasawa R, Mitarai T, Utsunomiya Y, Yoshida H, Kitamura M, Yamura W, Maruyama N, Isoda K, Sakai O: Neonatal thymectomy diminishes renal IgA deposition in IgA nephropathy-prone ddY mice. Nephron 1994;66:326–332.
87 Bagheri N, Pepple DA, Hassan MO, Harding CV, Emancipator SN: Development of immune-complex glomerulonephritis in athymic mice: T cells are not required for the genesis of glomerular injury. Lab Invest 2005;85:354–363.
88 Haseyama T, Fujita T, Hirasawa F, Tsukada M, Wakui H, Komatsuda A, Ohtani H, Miura AB, Imai H, Koizumi A: Complications of IgA nephropathy in a non-insulin-dependent diabetes model, the Akita mouse. Tohoku J Exp Med 2002;198:233–244.
89 Miller LL, Izzo MJ, Wemett D, Panner BJ, Schenk EA: Increased plasma IgA, sIgA, and C3- and IgA-containing immune complexes with renal glomerular deposits in diabetic rats. Diabetes 1988;37:185–193.

90 Kennel-De March A, Prin-Mathieu C, Kohler CH, Kolopp-Sarda MN, Faure GC, Béné MV: Back-pack mice as a model of renal mesangial IgA dimers deposition. Int J Immunopathol Pharmacol 2005;18:701–708.
91 Ramaiah SK, Seguin MA, Carwile HF, Raskin RE: Biclonal gammopathy associated with immunoglobulin A in a dog with multiple myeloma. Vet Clin Pathol 2002;31:83–89.
92 Suzuki Y, Tomino Y: Potential immunopathogenic role of the mucosa-bone marrow axis in IgA nephropathy: insights from animal models. Semin Nephrol 2008;28:66–77.
93 Agodoa LY, Gauthier VJ, Mannik M: Precipitating antigen-antibody systems are required for the formation of subepithelial electron-dense immune deposits in rat glomeruli. J Exp Med 1983;158:1259–1271.
94 Mannik M, Agodoa LY, David KA. Rearrangement of immune complexes in glomeruli leads to persistence and development of electron-dense deposits. J Exp Med. 1983 May 1;157(5):1516–28.
95 Agodoa LY, Gauthier VJ, Mannik M: Antibody localization in the glomerular basement membrane may precede in situ immune deposit formation in rat glomeruli. J Immunol 1985;134:880–884.
96 Rifai A, Mannik M. Clearance kinetics and fate of mouse IgA immune complexes prepared with monomeric or dimeric IgA. J Immunol. 1983 Apr;130(4):1826–32.
97 Rifai A, Millard K: Glomerular deposition of immune complexes prepared with monomeric or polymeric IgA. Clin Exp Immunol 1985;60:363–368.
98 Rifai A: Characteristics of nephritogenic IgA immune complexes. Am J Kidney Dis 1988; 12:402–405.
99 Chen A, Sheu LF, Ding SL, Lee WH: Monomeric immunoglobulin A protects glomerulus against polymeric immunoglobulin A immune complex in experimental immunoglobulin A nephropathy. Lab Invest 1996;74:737–746.
100 Chen A, Wong SS, Rifai A: Glomerular immune deposits in experimental IgA nephropathy. A continuum of circulating and in situ formed immune complexes. Am J Pathol 1988;130:216–222.
101 Chen A, Sheu LF, Shaio MF, Huang TH, Ding SL, Lee WH: Anti-idiotype antibody directly interferes with glomerular IgA immune complex deposition. Nephron 1998;78:464–473.
102 Rifai A, Chen A, Imai H: Complement activation in experimental IgA nephropathy: an antigen-mediated process. Kidney Int 1987; 32:838–844.
103 Montinaro V, Esparza AR, Cavallo T, Rifai A: Antigen as mediator of glomerular injury in experimental IgA nephropathy. Lab Invest 1991;64:508–519.
104 Hiki Y, Kokubo T, Iwase H, Masaki Y, Sano T, Tanaka A, Toma K, Hotta K, Kobayashi Y: Underglycosylation of IgA1 hinge plays a certain role for its glomerular deposition in IgA nephropathy. J Am Soc Nephrol 1999;10: 760–769.
105 Sano T, Hiki Y, Kokubo T, Iwase H, Shigematsu H, Kobayashi Y: Enzymatically deglycosylated human IgA1 molecules accumulate and induce inflammatory cell reaction in rat glomeruli. Nephrol Dial Transplant 2002;17:50–56.
106 Davin JC, Dechenne C, Lombet J, Rentier B, Foidart JB, Mahieu PR: Acute experimental glomerulonephritis induced by the glomerular deposition of circulating polymeric IgA-concanavalin A complexes. Virchows Arch A Pathol Anat Histopathol 1989;415:7–20.
107 Amore A, Emancipator SN, Roccatello D, Gianoglio B, Peruzzi L, Porcellini MG, Piccoli G, Coppo R: Functional consequences of the binding of gliadin to cultured rat mesangial cells: bridging immunoglobulin A to cells and modulation of eicosanoid synthesis and altered cytokine production. Am J Kidney Dis 1994;23:290–301.
108 Emancipator SN, Rao CS, Amore A, Coppo R, Nedrud JG: Macromolecular properties that promote mesangial binding and mesangiopathic nephritis. J Am Soc Nephrol 1992;2(Suppl 10):S149–S158.
109 Stad RK, Bruijn JA, van Gijlswijk-Janssen DJ, van Es LA, Daha MR: An acute model for IgA-mediated glomerular inflammation in rats induced by monoclonal polymeric rat IgA antibodies. Clin Exp Immunol 1993; 92:514–521.
110 van Dixhoorn MG, Gorter A, Sato T, van der Wal AM, van Eendenburg JD, Rozing J, Daha MR, de Heer E: Induction of microhematuria by an IgA isotype switch variant of a monoclonal anti-Thy-1.1 antibody in the rat. Kidney Int 1996;50:1612–1623.
111 Van Dixhoorn MG, Sato T, Muizert Y, Van Gijlswijk-Janssen DJ, de Heer E, Daha MR: Combined administration of IgA and IgG anti-Thy-1 antibodies enhances renal inflammation in rats. Kidney Int 1999;55:2299–2309.
112 van Dixhoorn MG, Sato T, Muizert Y, van Gijlswijk-Janssen DJ, De Heer E, Daha MR: Combined glomerular deposition of polymeric rat IgA and IgG aggravates renal inflammation. Kidney Int 2000;58:90–99.
113 Zheng F, Kundu GC, Zhang Z, Ward J, DeMayo F, Mukherjee AB: Uteroglobin is essential in preventing immunoglobulin A nephropathy in mice. Nat Med 1999;5:1018–1025.
114 Zhang Z, Kundu GC, Zheng F, Yuan CJ, Lee E, Westphal H, Ward J, DeMayo F, Mukherjee AB: Insight into the physiological function(s) of uteroglobin by gene-knockout and antisense-transgenic approaches. Ann NY Acad Sci 2000;923:210–233.
115 Kim YS, Kang D, Kwon DY, Park WY, Kim H, Lee DS, Lim CS, Han JS, Kim S, Lee JS: Uteroglobin gene polymorphisms affect the progression of immunoglobulin A nephropathy by modulating the level of uteroglobin expression. Pharmacogenetics 2001;11:299–305.
116 Coppo R, Chiesa M, Cirina P, Peruzzi L, Amore A, European IgACE Study Group: In human IgA nephropathy uteroglobin does not play the role inferred from transgenic mice. Am J Kidney Dis 2002;40:495–503.
117 Amore A, Roccatello D, Picciotto G, Emancipator SN, Ropolo R, Cacace G, Suriani A, Gianoglio B, Sena LM, Cirina P, Mazzucco G, Alfieri V, Piccoli G, Coppo R, De Filippi PG: Processing of IgA aggregates in a rat model of chronic liver disease. Clin Immunol Immunopathol 1997;84:107–114.

118 Waxman FJ, Hebert LA, Cosio FG, Smead WL, VanAman ME, Taguiam JM, Birmingham DJ: Differential binding of immunoglobulin A and immunoglobulin G1 immune complexes to primate erythrocytes in vivo. Immunoglobulin A immune complexes bind less well to erythrocytes and are preferentially deposited in glomeruli. J Clin Invest 1986;77:82–89.
119 Hosokawa S, Tagaya O, Mikami T, Nozaki Y, Kawaguchi A, Yamatsu K, Shamoto M: A new rat mutant with chronic conjugated hyperbilirubinemia and renal glomerular lesions. Lab Anim Sci 1992;42:27–34.
120 Smith SM, Tsukamoto H: Time dependency of IgA nephropathy induction in alcohol ingestion. Alcohol Clin Exp Res 1992;16: 471–473.
121 Amore A, Coppo R, Roccatello D, Piccoli G, Mazzucco G, Gomez-Chiarri M, Lamm ME, Emancipator SN: Experimental IgA nephropathy secondary to hepatocellular injury induced by dietary deficiencies and heavy alcohol intake. Lab Invest 1994;70:68–77.
122 Stad RK, Bogers WM, Muizert Y, van Es LA, Daha MR: Deposition of IgA is associated with macrophage influx in the kidney of rats. Scand J Immunol 1991;34:81–89.
123 Smith SM, Yu GS, Tsukamoto H: IgA nephropathy in alcohol abuse. An animal model. Lab Invest 1990;62:179–184.
124 Finbloom DS, Magilavy DB, Harford JB, Rifai A, Plotz PH: Influence of antigen on immune complex behavior in mice. J Clin Invest 1981;68:214–224.
125 Rifai A, Finbloom DS, Magilavy DB, Plotz PH: Modulation of the circulation and hepatic uptake of immune complexes by carbohydrate recognition systems. J Immunol 1982;128:2269–2275.
126 Rifai A, Fadden K, Morrison SL, Chintalacharuvu KR: The N-glycans determine the differential blood clearance and hepatic uptake of human immunoglobulin (Ig) A1 and IgA2 isotypes. J Exp Med 2000;191:2171–2182.
127 Launay P, Grossetête B, Arcos-Fajardo M, Gaudin E, Torres SP, Beaudoin L, Patey-Mariaud de Serre N, Lehuen A, Monteiro RC: Fcalpha receptor (CD89) mediates the development of immunoglobulin A (IgA) nephropathy (Berger's disease). Evidence for pathogenic soluble receptor-Iga complexes in patients and CD89 transgenic mice. J Exp Med 2000;191:1999–2009. Erratum in: J Exp Med 2000;192:following 309.
128 Robinson JK, Blanchard TG, Levine AD, Emancipator SN, Lamm ME: A mucosal IgA-mediated excretory immune system in vivo. J Immunol 2001;166:3688–3692.
129 Yanagihara T, Kumagai Y, Norose Y, Moro I, Nanno M, Murakami M, Takahashi H: Age-dependent decrease of polymeric Ig receptor expression and IgA elevation in ddY mice: a possible cause of IgA nephropathy. Lab Invest 2004;84:63–70.
130 van der Boog PJ, van Kooten C, van Zandbergen G, Klar-Mohamad N, Oortwijn B, Bos NA, van Remoortere A, Hokke CH, de Fijter JW, Daha MR: Injection of recombinant FcalphaRI/CD89 in mice does not induce mesangial IgA deposition. Nephrol Dial Transplant 2004;19:2729–2736.
131 Stad RK, van Gijlswijk-Janssen DJ, van Es LA, Daha MR: Complement depletion abolishes IgA-mediated glomerular inflammation in rats. Exp Nephrol 1994;2:182–189.
132 Fornasieri A, D'Amico G: Experimental IgA mesangial nephropathy: the role of antigen and antibody. Contrib Nephrol 1995;111: 149–154.
133 Broestl JA, Emancipator SN: Rat mesangial cell lysis in vitro is induced by cationic polypeptides. Am J Pathol 1993;142:529–539.
134 Imai H, Yasuda T, Wakui H, Miura AB: *Streptococcal pneumoniae* polysaccharide increases IgA-class antibody activity under the immunological memory of a protein antigen: two signals on experimental IgA nephropathy. Nippon Jinzo Gakkai Shi 1995; 37:323–329.
135 Montinaro V, Hevey K, Aventaggiato L, Fadden K, Esparza A, Chen A, Finbloom DS, Rifai A: Extrarenal cytokines modulate the glomerular response to IgA immune complexes. Kidney Int 1992;42:341–353.
136 Kobayashi N, Bagheri N, Nedrud JG, Strieter RM, Tomino Y, Lamm ME, Emancipator SN: Differential effects of Sendai virus infection on mediator synthesis by mesangial cells from two mouse strains. Kidney Int 2003;64: 1675–1684.
137 Suzuki H, Suzuki Y, Narita I, Aizawa M, Kihara M, Yamanaka T, Kanou T, Tsukaguchi H, Novak J, Horikoshi S, Tomino Y: Toll-like receptor 9 affects severity of IgA nephropathy. J Am Soc Nephrol 2008;19:2384–2395.
138 Chen A, Ding SL, Sheu LF, Song YB, Shieh SD, Shaio MF, Chou WY, Ho YS: Experimental IgA nephropathy. Enhanced deposition of glomerular IgA immune complex in proteinuric states. Lab Invest 1994;70: 639–647.
139 Shi Y, Pestka JJ: Attenuation of mycotoxin-induced IgA nephropathy by eicosapentaenoic acid in the mouse: dose response and relation to IL-6 expression. J Nutr Biochem 2006;17:697–706.
140 Jia Q, Zhou HR, Bennink M, Pestka JJ: Docosahexaenoic acid attenuates mycotoxin-induced immunoglobulin a nephropathy, interleukin-6 transcription, and mitogen-activated protein kinase phosphorylation in mice. J Nutr 2004;134:3343–3349.
141 Ohashi N, Katsurada A, Miyata K, Satou R, Saito T, Urushihara M, Kobori H: Role of activated intrarenal reactive oxygen species and renin-angiotensin system in IgA nephropathy model mice. Clin Exp Pharmacol Physiol 2009;36:750–755.
142 Shi Y, Pestka JJ: Mechanisms for suppression of interleukin-6 expression in peritoneal macrophages from docosahexaenoic acid-fed mice. J Nutr Biochem 2009;20:358–368.
143 Jia Q, Pestka JJ: Role of cyclooxygenase-2 in deoxynivalenol-induced immunoglobulin a nephropathy. Food Chem Toxicol 2005;43: 721–728.
144 Gesualdo L, Pinzani M, Floriano JJ, Hassan MO, Nagy NU, Schena FP, Emancipator SN, Abboud HE: Platelet-derived growth factor expression in mesangial proliferative glomerulonephritis. Lab Invest 1991;65:160–167.

145 Katsuma S, Shiojima S, Hirasawa A, Suzuki Y, Takagaki K, Murai M, Kaminishi Y, Hada Y, Koba M, Muso E, Miyawaki S, Ohgi T, Yano J, Tsujimoto G: Genomic analysis of a mouse model of immunoglobulin A nephropathy reveals an enhanced PDGF-EDG5 cascade. Pharmacogenomics J 2001;1:211–217.
146 Hattori T, Sadakane C, Koseki J, Kase Y, Takeda S: Saireito probably prevents mesangial cell proliferation in HIGA mice via PDGF-BB tyrosine kinase inhibition. Clin Exp Nephrol 2007;11:275–282.
147 Suematsu S, Matsuda T, Aozasa K, Akira S, Nakano N, Ohno S, Miyazaki J, Yamamura K, Hirano T, Kishimoto T: IgG1 plasmacytosis in interleukin 6 transgenic mice. Proc Natl Acad Sci USA 1989;86:7547–7551.
148 Katsume A, Saito H, Yamada Y, Yorozu K, Ueda O, Akamatsu K, Nishimoto N, Kishimoto T, Yoshizaki K, Ohsugi Y: Anti-interleukin 6 (IL-6) receptor antibody suppresses Castleman's disease like symptoms emerged in IL-6 transgenic mice. Cytokine 2002;20:304–311.
149 Shima Y, Iwano M, Yoshizaki K, Tanaka T, Kawase I, Nishimoto N: All-trans-retinoic acid inhibits the development of mesangial proliferative glomerulonephritis in interleukin-6 transgenic mice. Nephron Exp Nephrol 2005;100:e54–62.
150 Horii Y, Iwano M, Hirata E, Shiiki M, Fujii Y, Dohi K, Ishikawa H: Role of interleukin-6 in the progression of mesangial proliferative glomerulonephritis. Kidney Int Suppl 1993; 39:S71–S75.
151 Ruef C, Budde K, Lacy J, Northemann W, Baumann M, Sterzel RB, Coleman DL: Interleukin 6 is an autocrine growth factor for mesangial cells. Kidney Int 1990;38:249–257.
152 Pestka JJ, Zhou HR: Interleukin-6-deficient mice refractory to IgA dysregulation but not anorexia induction by vomitoxin (deoxynivalenol) ingestion. Food Chem Toxicol 2000; 38:565–575.
153 Floege J, Eitner F, Barratt J, Smith AC, Feehally J: Mutant mice provide new insight into the role of (mis-)glycation in IgA nephropathy and other glomerular diseases. Nephrol Dial Transplant 2007;22:1518–1520.
154 Tsushima Y, Tomino Y, Shimizu M, Ohmuro H, Takeuchi K, Kuramoto T, Shirato I, Koide H: IgA deposits might not influence the production of extracellular matrix in glomeruli of ddY mice, a spontaneous animal model for IgA nephropathy. Nephron 1994;66:81–86.
155 Leung JC, Chan LY, Tsang AW, Liu EW, Lam MF, Tang SC, Lai KN: Anti-macrophage migration inhibitory factor reduces transforming growth factor-beta 1 expression in experimental IgA nephropathy. Nephrol Dial Transplant: 2004;19:1976–1985.
156 Pestka JJ, Zhou HR: Effects of tumor necrosis factor type 1 and 2 receptor deficiencies on anorexia, growth and IgA dysregulation in mice exposed to the trichothecene vomitoxin. Food Chem Toxicol 2002;40:1623–1631.
157 Bussolati B, Peri G, Salvidio G, Verzola D, Mantovani A, Camussi G: The long pentraxin PTX3 is synthesized in IgA glomerulonephritis and activates mesangial cells. J Immunol 2003;170:1466–1472.
158 Zhang JJ, Xu LX, Zhang Y, Zhao MH: Binding capacity of in vitro deglycosylated IgA1 to human mesangial cells. Clin Immunol 2006;119:103–109.
159 Borda JT, Pauley DR, MacKey JJ, Alvarez X, Simon MA, Klumpp SA: Immunoglobulin-A nephropathy with crescentic glomerulonephritis in a pigtailed macaque (*Macaca nemestrina*). Vet Pathol 2004;41:44–49.
160 Stephens EB, Tian C, Li Z, Narayan O, Gattone VH 2nd: Rhesus macaques infected with macrophage-tropic simian immunodeficiency virus (SIVmacR71/17E) exhibit extensive focal segmental and global glomerulosclerosis. J Virol 1998;72:8820–8832.
161 Yoshie T: Study on the swine glomerulopathy resembling human IgA nephropathy (in Japanese). Nippon Jinzo Gakkai Shi 1991;33: 179–189.
162 Hedman R, Thuvander A, Gadhasson I, Reverter M, Pettersson H: Influence of dietary nivalenol exposure on gross pathology and selected immunological parameters in young pigs. Nat Toxins 1997;5:238–246.
163 Koeman JP, Biewenga WJ, Gruys E: Proteinuria in the dog: a pathomorphological study of 51 proteinuric dogs. Res Vet Sci: 1987;43:367–378.
164 Center SA, Smith CA, Wilkinson E, Erb HN, Lewis RM: Clinicopathologic, renal immunofluorescent, and light microscopic features of glomerulonephritis in the dog: 41 cases (1975–1985). J Am Vet Med Assoc 1987;190: 81–90.
165 Miyauchi Y, Nakayama H, Uchida K, Uetsuka K, Hasegawa A, Goto N: Glomerulopathy with IgA deposition in the dog. J Vet Med Sci 1992;54:969–975.
166 Harris CH, Krawiec DR, Gelberg HB, Shapiro SZ: Canine IgA glomerulonephropathy. Vet Immunol Immunopathol 1993;36:1–16.
167 Nieto CG, Navarrete I, Habela MA, Serrano F, Redondo E: Pathological changes in kidneys of dogs with natural Leishmania infection. Vet Parasitol 1992;45:33–47.
168 Zaragoza C, Barrera R, Centeno F, Tapia JA, Mañe MC: Canine pyometra: a study of the urinary proteins by SDS-PAGE and Western blot. Theriogenology 2004;61:1259–1272.
169 Tisdall PL, Rothwell TL, Hunt GB, Malik R: Glomerulopathy in dogs with congenital portosystemic shunts. Aust Vet J 1996;73:52–54.
170 Dambach DM, Smith CA, Lewis RM, Van Winkle TJ: Morphologic, immunohistochemical, and ultrastructural characterization of a distinctive renal lesion in dogs putatively associated with *Borrelia burgdorferi* infection: 49 cases (1987–1992). Vet Pathol 1997;34:85–96.
171 Elsinghorst TA: First cases of animal diseases published since 2000. 1. Dogs. Vet Q 2003;25: 112–123.
172 Aresu L, Pregel P, Bollo E, Palmerini D, Sereno A, Valenza F: Immunofluorescence staining for the detection of immunoglobulins and complement (C3) in dogs with renal disease. Vet Rec 2008;163:679–682.
173 Stinson MW, Nisengard RJ, Neiders ME, Albini B: Serology and tissue lesions in rabbits immunized with *Streptococcus mutans*. J Immunol 1983;131:3021–3027.
174 Albini B, Nisengard RJ, Glurich I, Neiders ME, Stinson MW: *Streptococcus mutans*-induced nephritis in rabbits. Am J Pathol 1985;118:408–418.
175 Galan JE, Timoney JF: Immune complexes in purpura hemorrhagica of the horse contain IgA and M antigen of *Streptococcus equi*. J Immunol 1985;135:3134–3137.

176 Sartori A, Roque-Barreira MC, Coe J, Campos-Neto A: Immune complex glomerulonephritis in experimental kala-azar. II: detection and characterization of parasite antigens and antibodies eluted from kidneys of *Leishmania donovani*-infected hamsters. Clin Exp Immunol 1992;87:386–392.
177 Portis JL, Coe JE: Deposition of IgA in renal glomeruli of mink affected with Aleutian disease. Am J Pathol 1979;96:227–236.
178 Yoshida F, Matsuo S, Fujishima H, Kim HK, Tomita T: Renal lesions of the FGS strain of mice: a spontaneous animal model of progressive glomerulosclerosis. Nephron 1994;66:317–325.
179 Lamm ME, Emancipator SN, Robinson JK, Yamashita M, Fujioka H, Qiu J, Plaut AG: Microbial IgA protease removes IgA immune complexes from mouse glomeruli in vivo: potential therapy for IgA nephropathy. Am J Pathol 2008;172:31–36.

Steven N. Emancipator, MD
Chief, Pathology and Laboratory Medicine Service
113(W), 10701 East Boulevard
Cleveland, OH 44106 (USA)
Tel. +1 216 791 3800 ext. 4039, E-Mail sne@case.edu

Herrera GA (ed): Experimental Models for Renal Diseases: Pathogenesis and Diagnosis.
Contrib Nephrol. Basel, Karger, 2011, vol 169, pp 153–160

Infection-Related Glomerulonephritis

Sergey V. Brodsky · Tibor Nadasdy

Department of Pathology, The Ohio State University, Columbus, Ohio, USA

Abstract

The epidemiology of infection related glomerulonephritis has been changing during recent decades. With the successful treatment of streptococcal infections, the incident of classic post-streptococcal glomerulonephritis is decreasing in most developed countries. In contrast, we see staphylococcus infection related glomerulonephritis more frequently. Animal models reproducing these glomerulonephritides are scant, particularly because it is difficult to develop clinically relevant animal models of infection related glomerulonephritis. Species specific immune responses as well as differences in susceptibility between species and even between animal strains make the interpretation of animal models difficult. Still, these models significantly contributed to our understanding of infection related glomerulonephritis. This is a short review focusing on the existing animal models of infection associated glomerulonephritis.

Acute Post-Streptococcal Glomerulonephritis, Glomerulonephritis Associated with Infective Endocarditis, Deep-Seated Abscesses, and Infected Ventriculoatrial Shunts

Acute glomerulonephritis has long been known to follow certain acute infections. Although Hippocrates, described the occurrence of back pain and gross hematuria leading to oliguria or anuria more than two millennia ago [1], it was Wells [2] who noted bloody urine in patients with scarlet fever and post-scarlatinal anasarca. Bright [3], too, noted the association with scarlatina and described the finding of blood in the urine and swelling of the face in what were probably attacks of post-infectious glomerulonephritis (PIGN). Bright observed that, with or without treatment, the patient's urine lost its 'red particles' and that the swelling subsided, although in some instances he recorded that the patient returned at a later time with what we would now regard as chronic renal failure.

With the introduction of microscopic examination of the kidney over the ensuing decades, it became apparent that changes took place in the glomeruli, and Langhans [4] described a category of Bright's disease with glomerular inflammation. Schick [5] commented on the similarity of the latent period in serum sickness to that of acute glomerulonephritis.

Scarlet fever was noted to be associated with acute glomerulonephritis in the late 1800s and early 1900s, and it soon became apparent that there was a common association with a previous infection of group A streptococcus [6–8]. Varying attack rates of acute glomerulonephritis were noted in patients with scarlet fever, ranging from 3% [9] to 18% [10] of patients. Caiger [9] found a greater attack rate of nephritis in children who were younger than 15 years of age compared with adults. These varying rates were probably largely dependent on the strain of streptococcus causing the infection, and it is now well recognized that only a relatively limited number of certain types of the organism are capable of inducing acute post-infectious glomerulonephritis. Ophüls [11], as early as 1917, suggested that only certain strains of streptococci appeared to have a nephritogenic capacity.

Most cases of acute glomerulonephritis (caused by the streptococci) seen today do not have the clinical signs and symptoms of scarlet fever [12]. In colder northern climates, most cases of acute glomerulonephritis follow upper-respiratory tract infections, such as pharyngitis or tonsillitis [13], but the disease can also follow skin infections, especially in warmer climates [14]. Most cases of PIGN are caused by group A streptococci, which are also responsible for rheumatic fever.

Whitworth et al. [15] and Beaufils et al. [16] first described the association of glomerulonephritis and visceral suppuration in the absence of infective endocarditis. The temporal association between deep-seated abscesses and immune complex glomerulonephritis suggests an etiologic relationship between these two conditions. Deep suppurative infections are caused either by Gram-positive or Gram-negative organisms; bacterial endocarditis also can be present. The majority of cases are due to intrathoracic abscesses, but wound infections and abscesses in other locations have also been reported [17].

Renal complications of bacterial endocarditis include glomerulonephritis, which was first named as 'embolic nonsuppurative focal nephritis' by Löhlein [18] at the turn of the 20th century. Now it is apparent that the mechanism of the disease is virtually always immunologic. The most common organisms leading to glomerulonephritis as a complication of bacterial endocarditis are the streptococcus viridans group and coagulase-negative staphylococci, including *Staphylococcus epidermidis*, although other microorganisms have been reported.

Infection of ventriculoatrial shunts (inserted for the relief of hydrocephalus of congenital or acquired origin) may lead to glomerulonephritis [19–21]. Renal symptoms in patients with ventriculoperitoneal shunts are uncommon, but they have been reported and are similar to those seen with ventriculoatrial shunts [21]. The infecting organism is usually a coagulase-negative staphylococcus *(S. epidermidis)*, which accounts for approximately 75% of shunt infections [20]. To our knowledge, there is no adequate experimental model of shunt glomerulonephritis reported.

Despite the fact that the disease has long been recognized in humans, experimental models of PIGN are scant. To study the pathophysiological changes in the kidney, several models of glomerulonephritis were used. It is now widely accepted that acute post-streptococcal and other forms of PGIN stem from an immunologic phenomenon. Much of the support for immune complex pathogenesis comes from the analogy to acute 'one shot' serum sickness in the rabbit, experimentally induced by a single injection of bovine serum albumin (BSA).

When human acute PIGN is compared to acute serum sickness, there are obvious similarities. Von Pirquet [22], in 1911, thought that the similarities between PIGN and acute serum sickness syndrome supported his concept of allergy. Longcope [23] showed that parenteral administration of foreign protein in experimental animal models could induce glomerulonephritis. There is a latent interval between the injection of a foreign protein and development of acute glomerulonephritis that is quite reminiscent of the latent interval between streptococcal infection and the clinical onset of human renal disease. Morphologic changes in experimental animals were similar to the lesions described in human acute PIGN (fig. 1), namely: glomerular endocapillary hypercellularity with many inflammatory cells, increased thickness of the glomerular basement membrane and podocyte foot process effacement [24].

By immunofluorescence techniques, the presence of antigen, antibody and complement in the glomeruli was demonstrated. This finding correlates with the formation of a variety of circulating immune complexes. Ultrastructural studies in both humans and experimental animal models show similar glomerular subepithelial electron-dense 'immune type' deposits [25]. Low levels of serum complement are found in both instances. As with human PIGN, experimental acute serum sickness

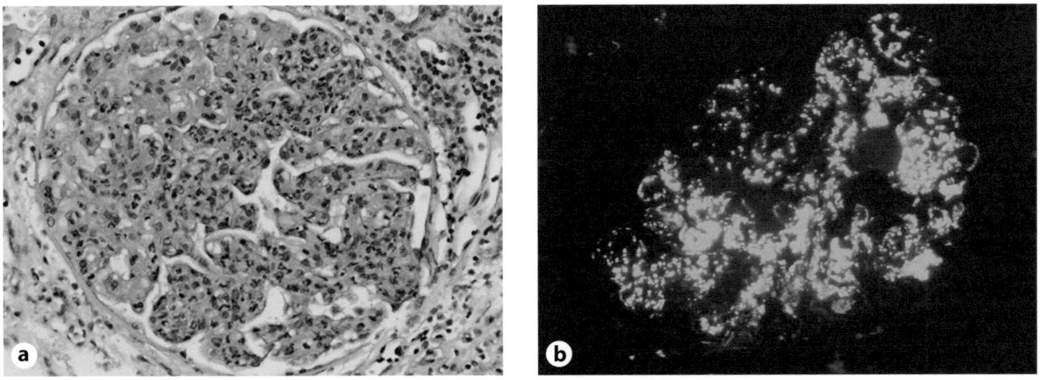

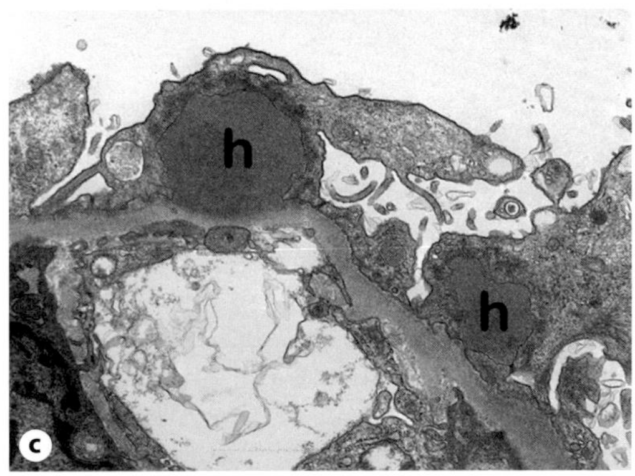

Fig. 1. Morphologic findings in post-streptococcal/postinfectious glomerulonephritis. **a** prominent intracapillary hypercellularity in this enlarged glomerulus. Many of the intracapillary cells are polymorphonuclear leukocytes. Hematoxylin and eosin. **b** The coarsely granular glomerular immune complex deposits are primarily positive for C3. Direct immunofluorescence. **c** Electron microscopy reveals large subepithelial electron-dense deposits, called humps (h). Uranyl acetate-lead citrate.

is a self-limited disease that generally resolves over a period of weeks. A simplified scheme on the pathogenesis of infection-associated glomerulonephritis is provided in figure 2.

Morphologic changes, similar to the rabbit model of serum sickness, were found in rats after serial injections of BSA (fig. 3) [26]. However, in this experimental model, the numerous immune complex deposits were not confined to the kidney and were seen in other organs as well [27]. Interestingly, injection of human IgG to different rat strains resulted in different renal pathologic changes. Thus, pre-immunization with IgG and (8 weeks later) daily IgG injections for 4 weeks resulted in endocapillary proliferative glomerulonephritis with heavy proteinuria in spontaneously hypertensive rats, whereas the same treatment in Wistar-King-Aptekman rats resulted in mesangial immune complex depositions without severe proliferative lesions [28].

Experimental studies confirmed that mononuclear phagocytes appear in the glomerular tuft in various forms of experimentally induced glomerulonephritis, resembling PIGN [28–32]. Monocytes were found to be present in the serum sickness model of acute glomerulonephritis by Hunsicker et al. [31] and by Holdsworth et al. [32], using electron microscopy and staining for nonspecific esterase. Several investigators have suggested that infiltrating mononuclear cells and their soluble mediators are responsible for inducing the growth and proliferation of endothelial and mesangial cells [33]. On the other hand, mesangial cells may regulate the physiology of the infiltrating mononuclear cells and modify the local immune response [34]. Reactive oxygen species, produced by glomerular macrophages, have been shown to contribute to glomerular basement membrane injury in a rabbit model of diffuse proliferative glomerulonephritis [35], but macrophages can also contribute

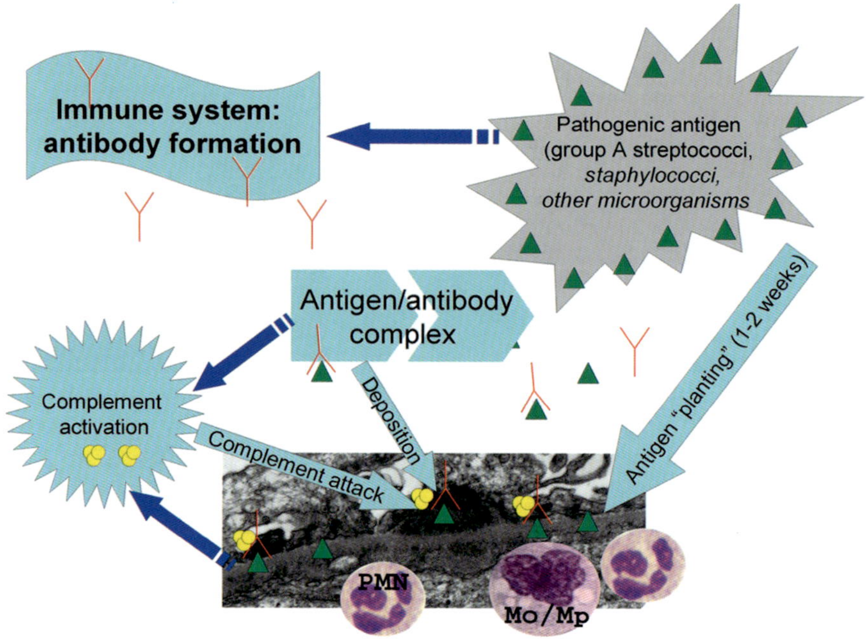

Fig. 2. Pathophysiologic mechanisms of infection-associated glomerulonephritis. Pathogenic antigens, presented on microorganisms (e.g., group A streptococci, staphylococci) initiate immune response and antibodies against these antigens are produced. Destruction of the microorganism results in appearance of circulating antigens and antigen-antibody complexes are formed in the circulation. These immune complexes may be entrapped along the glomerular basement membrane (GBM) and in the mesangium. Alternately, direct antigen 'planting' in the GBM may occur. This may result in in situ formation of antigen-antibody complexes along the GBM. These glomerular immune complexes (deposited or formed in situ) will activate the complement system. Components of the complement cascade (in particular C5a) will attract inflammatory cells to the site of the immune complex deposition. These inflammatory cells will accumulate in the glomerular capillaries and cause glomerular capillary injury leading to hematuria and proteinuria and impaired renal function. PMN = Polymorphonuclear leukocyte; Mo/Mp = monocyte/macrophage.

to the glomerular handling of circulating immune complexes [36].

Many attempts have been made to induce an animal model of PIGN by the injection of intact streptococci [37, 38], crude culture supernatants [39] or specific components of streptococci [39, 40]. Although some of these experimental manipulations produced histologic lesions somewhat similar to the disease pattern in humans, most called for the periodic administration of the putative factors thought to be involved; this, of course, does not precisely mimic the gradual release of streptococcal products that probably occurs at the site of infection in humans [39]. Also, many of the experimental studies were performed at a time when electron and immunofluorescence microscopy and other biochemical determinations were not available, making an adequate comparison difficult [39].

Holm [39] and Nordstrand et al. [40] developed animal models in rabbits and mice that permitted the establishment of a focal infection using subcutaneous tissue cages and continuous monitoring of the infection and the release of streptococcal factors. According to authors, these models simulate PIGN clinically, histologically and immunologically; however, not

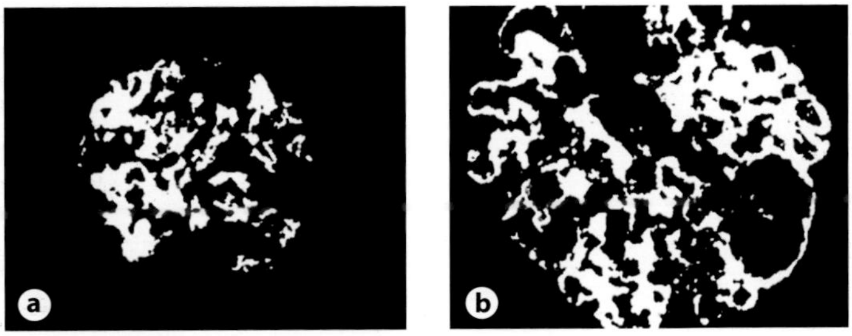

Fig. 3. Glomerular immune complex deposits in a rat model of immune complex glomerulonephritis, induced by daily administration of bovine serum albumin (BSA) for 4 weeks. At 10 weeks, the immune complexes are localized primarily to the mesangium (**a**). Later, at 12 weeks, the deposits are also present along the glomerular capillary walls. Immunofluorescence using an anti-BSA antibody (with permission from [27]).

all investigators agree that this is a solid model of the human disease. Yoshizawa et al. [41, 42] developed a rabbit model with the infusion of 43-kDa protein from nephritogenic streptococci, called pre-absorbing antigen. After infusing 18 mg of this antigen to rabbits for 8 days, the animals developed proliferative immune complex glomerulonephritis with glomerular deposition of C3 without deposition of immunoglobulins. Electron microscopy revealed hump-like subepithelial deposits. Burova et al. [43] induced proliferative glomerulonephritis in the rabbit following repeat infusions of group A streptococcal strains bearing IgG-binding M family proteins. The animals had glomerular IgG and C3 deposition. Infusion of mutant bacterial strains, lacking the IgG-binding proteins, did not result in glomerulonephritis. The authors postulated that streptococcal IgG-binding proteins had an important role in triggering PIGN in their model [43].

Experimental studies by Cronin and Lange [44], using Wistar-Furth rats, showed deposition of endostreptosin along the glomerular basement membrane one day after injection of immunoaffinity-isolated endostreptosin. Rats showed increasing deposition of IgG and C3 with diminished staining for endostreptosin by day 8 and 12. No anti-endostreptosin antibodies were detected in the sera in the first 3 days, whereas rats from day 4 onward had low levels of these antibodies. According to these authors, endostreptosin does not appear to be immunologically related to streptococcal exoenzymes or the streptococcal cell wall [44].

Holm et al. [45], in a steel net tissue cage model of 'slow release' in a rabbit model of acute poststreptococcal glomerulonephritis, showed loss of nephritogenic potential by deletion of a streptokinase gene (in a nephritogenic type 49 strain of streptococcus) in a molecular construct prepared by electrotransformation. The same group demonstrated an important role for streptokinase in the pathogenesis of PIGN in their mouse model. Nephritogenic group A streptococci could cause glomerulonephritis only if they contained the nephritis-associated streptokinase gene (SKA1). Strains with a deleted SKA1 gene did not cause glomerulonephritis in their mouse model [46]. These data indicate that variance of streptokinase may be associated with different affinities to glomerular structures and may, thereby, affect the nephrogenetic potential of the various streptococcal strains in this experimental model.

Thörig et al. [47] induced nonbacterial thrombotic endocarditis in rabbits, followed by injections of live streptococci. The animals developed bacterial endocarditis and renal disease. Before inducing the bacterial endocarditis, one group of animals was immunized by heat-killed streptococci for 12 weeks. Although, both immunized

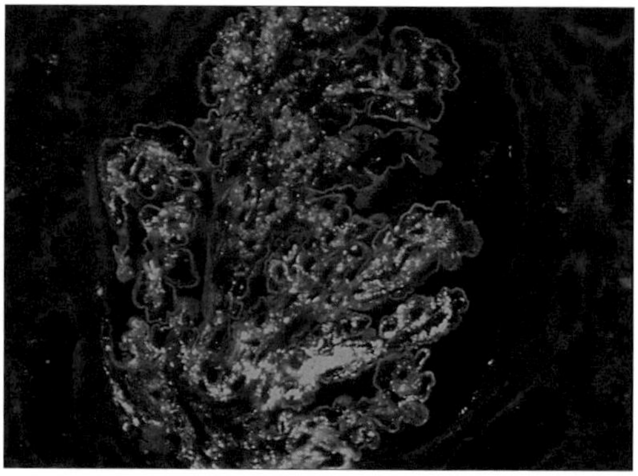

Fig. 4. Granular IgA containing glomerular immune complex deposits in a patient with proliferative glomerulonephritis following MRSA infection. Direct immunofluorescence.

and non-immunized animals had septic renal micro-abscesses, only the immunized animals with high circulating immune complexes were found to have mesangial immune complex deposits [47]. Although the glomerular immune complex deposition was mild, this model resembled human endocarditis-associated glomerulonephritis, where the glomerular damage is primarily related to immune complex deposits.

Staphylococcal Infection and Glomerulonephritis

In the past, most glomerulonephritides associated with staphylococcus infections were either secondary to endocarditis, deep-seated visceral abscesses or ventriculoatrial shunts [48]. Interestingly, in more recent years, a number of publications have drawn attention to the association of staphylococcus infection and glomerulonephritis not related to endocarditis or ventriculoatrial shunts [49]. Most of these staphylococcal organisms are coagulase-positive *Staphylococcus aureus*. The site of infection includes osteomyelitis, septic arthritis, discitis, pneumonia, infected leg ulcers and a variety of other infections, including rectal abscess, other deep-seated abscesses, peritonitis and pancreatitis, as well as an undisclosed primary site of infection with positive blood cultures [49–54]. Many of these infections were associated with methicillin-resistant *S. aureus*. Cases of glomerulonephritis, associated with other staphylococci, including methicillin-resistant *S. epidermidis*, have been reported [51]. The increase in the incidence of staphylococcus infection-associated glomerulonephritis may be related to the increasing number of elderly patients and probably to the increased incidence of nosocomial infections with methicillin-resistant *S. aureus* and other staphylococci [54]. These patients usually have an underlying debilitating condition, such as diabetes, malignancy or chronic inflammatory diseases.

The association between methicillin-resistant *S. aureus* and glomerulonephritis has been well documented [52–54]. Based on the clinical reports, renal biopsy findings are characterized by glomerular IgA deposits, resembling idiopathic IgA nephropathy [52]. Indeed, *S. aureus* cell envelope antigen was co-localized with IgA deposits in renal biopsies [55]. Moreover, a form of IgA glomerulonephritis in mice can be induced by immunization with antigens, derived from *S. aureus* [56]. Thus, immunization of Balb/c and C57BL/6 mice for 4 months with antigens derived from *S. aureus* admixed with Freund's incomplete adjuvant resulted in mesangial proliferative glomerulonephritis with IgA, IgG and C3 depositions in Balb/c mice. Interestingly, C57BL/6 mice had similar morphologic findings, but with C3 mesangial deposits only. In Balb/c mice, but not in C57BL/6 mice, *S. aureus* antigens were detected in the glomeruli. Also, Balb/c mice had an increased TGF-β messenger RNA expression, as compared to C57BL/6 mice

[56]. These findings suggest that *S. aureus* antigens may play an important role in the pathogenesis of immune complex glomerulonephritis, but other genetic (host) factors are also important.

In conclusion, the face and epidemiology of infection-related glomerulonephritis is changing [49, 57]. With the successful treatment of streptococcal infections, the incidence of the classic post-streptococcal glomerulonephritis is decreasing in most developed countries. In contrast, we see staphylococcal infection-related glomerulonephritides more and more frequently. Reproducing these infection-related glomerulonephritides in experimental animals is difficult because of species-specific immune responses and differences in susceptibility even between different animal strains. Unfortunately, only few animal models exist, but the results of these could be interpreted as relevant in studying human infection-related glomerulonephritis.

References

1 Hippocrates. Coan prognosis; in Chadwick J, Mann WN (trans): The medical works of Hippocrates. Oxford, Blackwell & Thomas, 1950, p 219.
2 Wells WC: Transactions of a Society for the Improvement of Medical and Chirurgical Knowledge. London, Society for the Improvement of Medical and Chirurgical Knowledge, 1812, vol 3, p 3.
3 Bright R: Cases and observations illustrative of renal disease accompanied with the secretion of albuminous urine. Guys Hosp Rep 1836;1:338.
4 Langhans T: Über die Veränderung der Glomeruli bei der Nephritis nebst einigen Bemerkungen über die Entstehung der Fibrincylinder. Virchows Arch A Pathol Anat Histopathol 1879;76:85.
5 Schick B: Die Nachkrankheiten des Scharlach: Pathogenese der Nachkrankheiten; in Escherich T, Shick B (eds): Scharlach. Leipzig, Holder, 1912, p 151
6 Hansborg H: Untersuchungen über die Prognose der Scharlachnephritis. Acta Med Scand 1924;61:570.
7 Hayman JM, Martin JW: Acute nephritis: review of 77 cases. Am J Med Sci 1940;200:505.
8 Osman AA, Close HG, Carter H: Studies in Bright's disease. VIII. Observations on the aetiology of scarlatinal nephritis. Guys Hosp Rep 1933;83:360.
9 Caiger FF: Scarlet fever; in Allbutt TC (ed): A System of Medicine. New York, Macmillan, 1898, vol 3, p 122.
10 Wahrer CF: An epidemic of hemorrhagic nephritis following scarlet fever. JAMA 1908; 51:1410.
11 Ophüls W: The etiology and development of nephritis. JAMA 1917;69:1223.
12 Earle DP: Poststreptococcal acute glomerulonephritis. Hosp Pract 1985;20:84.
13 Rammelkamp CH: Acute hemorrhagic glomerulonephritis; in McCarty M (ed): *Streptococcal infections*. New York, Columbia University Press, 1954.
14 Dillon HC, Reeves MS, Maxted WR: Acute glomerulonephritis following skin infection due to streptococci of M-type 2. Lancet 1968;1:543.
15 Whitworth JA, Morel-Maroger L, Mignon F, Richet G: The significance of extracapillary proliferation: clinicopathological review of 60 patients. Nephron 1976;16:1.
16 Beaufils M, Morel-Maroger L, Sraer JD: Acute renal failure of glomerular origin during visceral abscesses. N Engl J Med 1976; 295:185.
17 Beaufils M: Glomerular disease complicating abdominal sepsis. Kidney Int 1981;19:609–618.
18 Löhlein M: Über hämorrhagische Nierenaffektionen bei chronischer ulzeröser Endokarditis (Embolische nichteiterige Herdnephritis). Med Klin 1910;10:375.
19 Black JA, Challacombe DN, Ockenden BG: Nephrotic syndrome associated with bacteremia after shunt operations for hydrocephalus. Lancet 1965;2:291.
20 Narchi H, Taylor R, Azmy AF, Murphy AV, Beattie TJ: Shunt nephritis. J Pediatr Surg 1988;23:839.
21 Haffner D, Schindera F, Aschoff A, et al: The clinical spectrum of shunt nephritis. Nephrol Dial Transplant 1997;12:1143.
22 Von Pirquet CE: Allergy. Arch Intern Med 1911;7:383.
23 Longcope WT: The production of experimental nephritis by repeated protein intoxication. J Exp Med 1913;18:678.
24 Feldman JD: Electron microscopy of serum sickness nephritis. J Exp Med 1958;108:957–962.
25 Germuth FG Jr: A comparative histologic and immunologic study in rabbits of induced hypersensitivity of the serum sickness type. J Exp Med 1953;97:257.
26 Arisz L, Noble B, Milgrom M, Brentjens JR, Andres GA: Experimental chronic serum sickness in rats. A model of immune complex glomerulonephritis and systemic immune complex deposition. Int Arch Allergy Appl Immunol 1979;60:80–88.
27 Miyazaki S, Kawasaki K, Yaoita E, Yamamoto T, Kihara I: Bovine serum albumin (BSA) nephritis in rats. III. Antigen distribution in various organs. Clin Exp Immunol 1985;59:293–299.
28 Yamamoto K, Oite T, Kihara I, Shimizu F: Experimental glomerulonephritis induced by human IgG in rats. Clin Exp Immunol 1984; 57:575–582.
29 Rodriguez-Iturbe B: Epidemic poststreptococcal glomerulonephritis. Kidney Int 1984;25:129.

30 Schreiner GF, Kiely J, Cotran RS, Unanue ER: Characterization of resident glomerular cells in the rat expressing Ia determinants and manifesting genetically restricted interactions with lymphocytes. J Clin Invest 1981; 68:920.

31 Hunsicker LG, Shearer TP, Plattner SB, Weisenburger D: The role of monocytes in serum sickness nephritis. J Exp Med 1979; 150:413.

32 Holdsworth SR, Neale TJ, Wilson CB: The participation of macrophages and monocytes in experimental immune complex glomerulonephritis. Clin Immunol Immunopathol 1980;15:510.

33 Dubois CH, Foidart JB, Hautier MB, Dechenne CA, Lemaire MJ, Mahieu PR: Proliferative glomerulonephritis in rats: evidence that mononuclear phagocytes infiltrating the glomeruli stimulate the proliferation of endothelial and mesangial cells. Eur J Clin Invest 1981;11:91.

34 MacCarthy EP, Hsu A, Ooi YM, Ooi BS: Evidence for a mouse mesangial cell derived factor that stimulates lymphocyte proliferation. J Clin Invest 1985;76:426.

35 Boyce NW, Tipping PG, Holdsworth SR: Glomerular macrophages produce reactive oxygen species in experimental glomerulonephritis. Kidney Int 1989;35:778.

36 Boyce NW, Holdsworth SR: Quantitation of the intrarenal uptake of immunoglobulin aggregates by macrophages in diffuse proliferative glomerulonephritis. Clin Exp Immunol 1986;64:638.

37 Becker CG, Murphy GE: The experimental induction of glomerulonephritis like that in man by infection with group A streptococci. J Exp Med 1968;127:1.

38 Miyamoto Y, Kotake I, Ochiai S, Kodama T: Experimental nephritis produced in rabbits by inoculation of group A hemolytic streptococci isolated from epidemic nephritis patients: a focal infection experiment in the paranasal sinuses. Jpn J Microbiol 1960; 4:105.

39 Holm SE: The pathogenesis of acute poststreptococcal glomerulonephritis in new lights. APMIS 1988;96:189.

40 Nordstrand A, Norgren M, Holm S: An experimental model for acute poststreptococcal glomerulonephritis in mice. APMIS 1996;104:805.

41 Yoshizawa N, Oshima S, Sagel I, Shimizu J, Treser G: Role of a streptococcal antigen in the pathogenesis of acute poststreptococcal glomerulonephritis: characterization of the antigen and a proposed mechanism for the disease. J Immunol 1992;148:3110.

42 Yoshizawa N, Oshima S, Takeuchi A: Experimental acute glomerulonephritis induced in the rabbit with a specific streptococcal antigen. Clin Exp Immunol 1997; 107:61.

43 Burova L, Thern A, Pigarevsky P: Role of group A streptococcal IgG binding proteins in triggering experimental glomerulonephritis in the rabbit. APMIS 2003;111:955.

44 Cronin WJ, Lange K: Immunologic evidence for the in situ deposition of a cytoplasmic Streptococcal antigen (endostreptosin) on the glomerular basement membrane in rats. Clin Nephrol 1990;34:143.

45 Holm SE, Ferretti JJ, Simon D, Johnston KH: Deletion of a streptokinase gene eliminates the nephritogenic capacity of a type 49 strain; in Orefici G (ed): New Perspectives on Streptococci and Streptococcal Infections. Stuttgart, Gustav Fischer, 1992, p 261.

46 Nordstrand A, McShan WM, Ferretti JJ: Allele substitution of the streptokinase gene reduces the nephritogenic capacity of group A streptococcal strain NZ131. Infect Immun 2000;68:1019.

47 Thörig L, Daha MR, Eulderink F, Kooy-Bauer WC, Thompson J: Experimental *Streptococcus sanguis* endocarditis: immune complexes and renal involvement. Clin Exp Immunol 1980;40:469.

48 Maher ER, Hamilton DV, Thiru S, Wheatley T: Acute renal failure due to glomerulonephritis associated with staphylococcal infection. Postgrad Med J 1984;60:433.

49 Nasr SH, Markowitz GS, Stokes MB, Said SM, Valeri AM, D'Agati VD: Acute postinfectious glomerulonephritis in the modern era: experience with 86 adults and review of the literature. Medicine (Baltimore) 2008;87:21–32.

50 Griffin M, Bjornsson J, Erickson S: Diffuse proliferative glomerulonephritis and acute renal failure associated with acute staphylococcal osteomyelitis. J Am Soc Nephrol 1997; 8:1633.

51 Peel R, Sellars L, Long ED, Bhandari S: A man with a backache and renal failure. Am J Kidney Dis 2003;41:1.

52 Satoskar AA, Nadasdy G, Plaza JA, Sedmak D, Shidham G, Hebert L, Nadasdy T: Staphylococcus infection-associated glomerulonephritis mimicking IgA nephropathy. Clin J Am Soc Nephrol 2006;1:1179–1186.

53 Handa T, Ono T, Watanabe H, et al: Glomerulonephritis induced by methicillin-sensitive *Staphylococcus aureus* infection. Clin Exp Nephrol 2003;7:247.

54 Kobayashi M, Koyoma A: Methicillin-resistant *Staphylococcus aureus* (MRSA) infection in glomerulonephritis – a novel hazard emerging on the horizon. Nephrol Dial Transplant 1998;13:2999–3001.

55 Koyama A, Sharmin S, Sakurai H, Shimizu Y, Hirayama K, Usui J, Nagata M, Yoh K, Yamagata K, Muro K, Kobayashi M, Ohtani K, Shimizu T, Shimizu T: *Staphylococcus aureus* cell envelope antigen is a new candidate for the induction of IgA nephropathy. Kidney Int 2004;66:121–132.

56 Sharmin S, Shimizu Y, Hagiwara M, et al: *Staphylococcus aureus* antigens induce IgA type glomerulonephritis in Balb/c mice. J Nephrol 2004;17:504.

57 Wen YK: Clinicopathological study of infection-associated glomerulonephritis in adults. Int Urol Nephrol 2009, E-pub ahead of print.

Tibor Nadasdy, MD
Department of Pathology, Renal and Transplant Pathology Laboratory
Ohio State University, M-015 Starling Loving Hall
320 W. 10th Ave Columbus, OH 43210 (USA)
Tel. +1 614 293 9563, Fax +1 614 293 4255, E-Mail Tibor.Nadasdy@osumc.edu

Herrera GA (ed): Experimental Models for Renal Diseases: Pathogenesis and Diagnosis.
Contrib Nephrol. Basel, Karger, 2011, vol 169, pp 161–174

Discovery of Genes Related to Diabetic Nephropathy in Various Animal Models by Current Techniques

Jun Wada[a] · Lin Sun[b] · Yashpal S. Kanwar[b]

[a]Department of Medicine and Clinical Science, Okayama University Graduate School of Medicine, Okayama, Japan;
[b]Department of Pathology, Northwestern University Medical School, Chicago, Ill., USA

Abstract

One of the major problems facing clinical nephrology currently throughout the world is an exponential increase in patients with end-stage renal disease (ESRD), which is largely related to a high incidence of diabetic nephropathy. The latter is characterized by a multitude of metabolic and signaling events following excessive channeling of glucose, which leads to an increased synthesis of extracellular matrix (ECM) glycoproteins resulting in glomerulosclerosis, interstitial fibrosis and ultimately ESRD. With the incidence of nephropathy at pandemic levels and a high rate of ESRD, physicians around the world must treat a disproportionately large number of diabetic patients with up-to-date innovative measures. In this regard, identification of genes that are crucially involved in the progression of diabetic nephropathy would enhance the discovery of new biomarkers and could also promote the development of novel therapeutic strategies. Over the last decade, we focused on the recent methodologies of high-throughput and genome-wide screening for identification of relevant genes in various animal models, which included the following: (1) single nucleotide polymorphism-based genome-wide screening; (2) the transcriptome approach, such as differential display reverse transcription polymerase chain reaction (DDRT-PCR), representational difference analysis of cDNA (cDNA-RDA)/suppressive subtractive hybridization, SAGE (serial analysis of gene expression) and DNA Microarray; and (3) the proteomic approach and *2*-dimensional polyacrylamide gel *e*lectrophoresis (2D-PAGE) coupled with mass spectroscopic analysis. Several genes, such as Tim44 (translocase of inner mitochondrial membrane-44), RSOR/MIOX (renal specific oxidoreductase/*myo*-inositol oxygenase), UbA52, Rap1b (Ras-related GTPase), gremlin, osteopontin, hydroxysteroid dehydrogenase-3β isotype 4 and those of the Wnt signaling pathway, were identified as differentially expressed genes in kidneys of diabetic rodents. Functional analysis of these genes and the subsequent translational research in the clinical settings would be very valuable in the prevention and treatment of diabetic nephropathy. Future trends for identification of the biomarkers and therapeutic target genes should also include genome scale DNA/histone-methylation profiling, metabolomic approaches (e.g. metabolic phenotyping by 1H spectroscopy) and lectin microarray for glycan profiling along with the development of robust data-mining strategies.

Background of Diabetic Nephropathy

The most problematic issue in clinical nephrology is the progressive worldwide increase in patients with end-stage renal disease (ESRD) that is commonly associated with diabetic nephropathy. Over the last 30 years, extensive investigatory efforts have been made to delineate the mechanisms involved in the pathogenesis of diabetic nephropathy, resulting in the identification of three major signaling pathways (fig. 1)

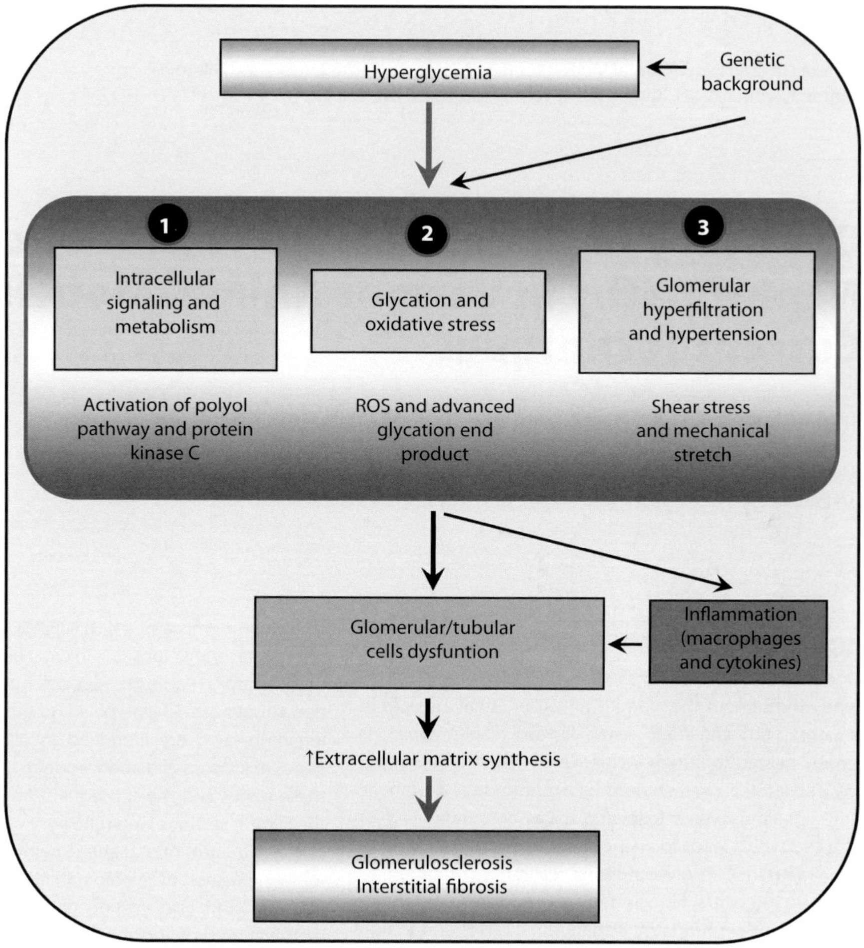

Fig. 1. An overview of the signaling events induced by exposure of renal cells to high glucose ambience leading to altered expression of various genes and cellular abnormalities culminating in diabetic nephropathy. The schematic highlights three sets of major signaling events which converge to induce dysfunction of glomerular and tubular cells leading to glomerulosclerosis and interstitial fibrosis and ultimately ESRD. ROS = Reactive oxygen species.

[1]. Upstream of these pathways, hyperglycemia seems to be the major driving force for initiating various signaling events. The downstream events include setting in of microinflammation and increased synthesis of extracellular matrix (ECM) glycoproteins by the renal glomerular, tubular and interstitial cells. This leads to the amassing of the ECM in the glomerulus resulting in thickening of the glomerular basement membranes, mesangial expansion and nodule formation, a characteristic of Kimmelstiel-Wilson lesion and a hallmark of diabetic nephropathy [2] (fig. 2). At the same time, the tubulointerstitial compartment is also affected. One can observe a thickening of the tubular basement membranes and increased interstitial fibrosis, adding further to the glomerular injury and, consequentially, chronic irreversible progressive renal failure.

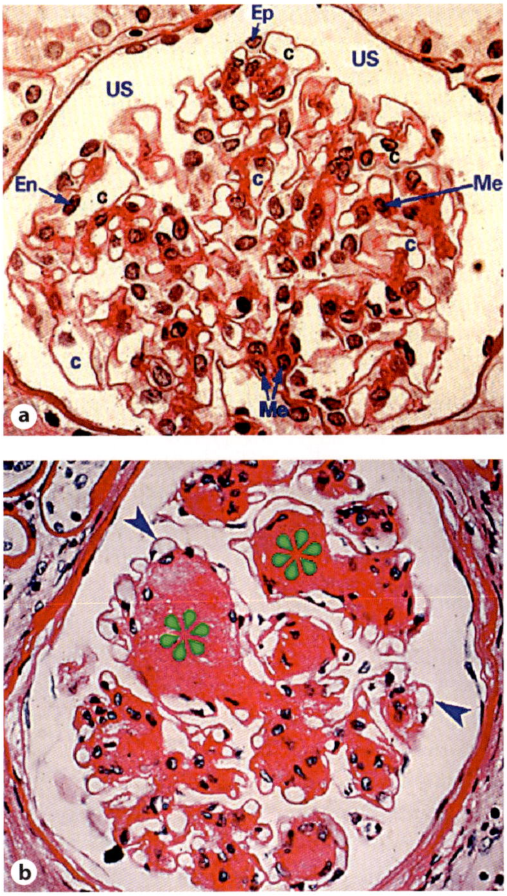

Fig. 2. Light photomicrographs illustrating glomerular lesions in diabetic nephropathy. **a** Normal glomerulus. **b** Thickened basement membranes (arrowheads) and nodular appearance of the mesangial regions characteristic of Kimmelstiel-Wilson lesions (asterisks). Me = Mesangium; Ep = visceral epithelial cell (podocyte); En = endothelial cell; C = capillary lumen; US = urinary space.

Conceivably, it is likely that the lesions in both compartments are the result of common signaling pathways initiated by excessive intracellular channeling of glucose.

The first major pathway relates to metabolic abnormalities in which there is a heightened activity of the polyol pathway and activation of protein kinase C-β (PKC- β). The events related to this pathway have been described in an animal model of STZ-induced diabetic rats (model of type-1 diabetes) and *db/db* mice (model of type 2 diabetes) [3]. Interestingly, an ameliorative effect was observed when these animals were treated with inhibitors of PKC-β or of the polyol pathway, i.e. aldose reductase [4–6]. These agents, especially the inhibitor of PKC-β, could possibly serve as a therapeutic measure to dampen the progression of diabetic nephropathy, a subject matter that would be of attractive clinical value in future investigations.

The characteristics of the second pathway include the formation of advanced glycation end-products (AGEs) from excessive imbibing of glucose [7]. The AGEs via interaction with their receptor, RAGE, transduce a complex series of signaling events that result in cellular dysfunctions, thus generating an inflammatory response and reactive oxygen species (ROS), which in turn cause oxidative stress [7]. Both in vitro and in vivo studies support the relevance of this pathway in the pathogenesis of diabetic nephropathy [7]. The fact that several inhibitors of AGEs, such as pyridoxamine, LR-90 and KIOM-79, have been demonstrated to be beneficial in various murine models of diabetes emphasizes the role of AGE:RAGE interactions [8–10]. Although these inhibitors may be effective in murine models, their efficacy certainly needs to be evaluated in diabetic nephropathy in humans.

The oxidative stress due to ROS also induces cellular dysfunction, which compounds the glomerular and tubular injury further, thus worsening the outcome of diabetic nephropathy [11, 12]. The ROS are mainly generated via the NADPH oxidase system or at the level of the mitochondria. In this regard, the NADPH oxidase inhibitors in the amelioration of diabetic nephropathy have been reported in various animal model systems [11, 12]. During the last 10 years, glomerular hyperfiltration leading to intraglomerular hypertension has emerged as a third pathway, and this pathway adversely influences the course of diabetic nephropathy [13]. Intriguingly, reduction of glomerular hyperfiltration by renin-angiotensin system inhibitors apparently has remarkable beneficial effects in the amelioration of proteinuria and progression of nephropathy, both in animal experiments and various clinical studies in patients with diabetes mellitus [13].

These three major pathways converge to induce dysfunction of glomerular, tubular and interstitial cells. There is good evidence that cellular injury affects the endothelial cells of renal vasculature, accompanied with increased expression of adhesion molecules and chemokines, resulting in a macrophage influx into the renal parenchyma and subjecting it to low-grade inflammation [14]. There are a few reports from the literature supporting this prolonged microinflammation as a common pathway for the progression of diabetic nephropathy [14, 15]; therefore, it is likely that anti-inflammatory agents serving as supplemental agents to treat diabetic nephropathy would be another fertile area of investigation in coming years. Counterintuitive to the notion of microinflammation in diabetic nephropathy is the upregulation of a profibrogenic cytokine, transforming growth factor-β (TGF-β). TGF-β is not only a potent downregulator of an immune/inflammatory response, but it is also a well-characterized critical molecule for the increased synthesis ECM glycoproteins leading to glomerulosclerosis and interstitial fibrosis [6, 16]. As expected, the administration of neutralizing anti-TGF-β antibody has been reported to prevent the renal insufficiency in *db/db* mice [16]. Along these lines, the efficacy of administration of such neutralizing antibodies to other relevant growth factors, such as connective tissue growth factor and inflammatory cytokines, or the genetic deletion of adhesion molecules, such as ICAM-1 and macrophage scavenger receptor-A, on the amelioration of diabetic nephropathy and prevention of ESRD need to be aggressively evaluated since there has been a precipitous increase in the incidence of this dire renovascular complication of diabetes in recent years [14, 15].

The incidence reported for ESRD is highest in Taiwan, at $404/10^6$ persons in 2005, followed by the United States, Jalisco (Mexico), Shanghai (China) and Japan (in 2004) with the respective rates being 351, 302, 275 and 267 for these 4 geographic regions. Most countries with a high rate of ESRD have a disproportionately large proportion of patients with diabetes mellitus. For example, in Jalisco 60% of new ESRD patients have a primary diagnosis of diabetes, while approx. 12% in Russia and Norway have such a diagnosis and only 5% in Iceland (http://www.usrds.org/adr.htm).

The impact of diabetic nephropathy on the incidence of chronic kidney disease and ESRD is enormous. To combat this problem, intensified multi-prong approaches have been performed in patients with type 2 diabetes mellitus, resulting in reduced risk of microangiopathy, cardiovascular complications and mortality in randomized studies carried out at the Steno Memorial Hospital, Denmark, by Gaede et al. [17–19]. However, the incidence of ESRD is still progressively increasing worldwide. In such a dilemma, identification of the genes that are critically involved in the progression of diabetic nephropathy is warranted, the discovery of which should eventually lead to the development of biomarkers and new treatment strategies. In view of the above considerations, we devoted our efforts to identifying genes relevant to diabetic nephropathy in various animal models using various current biological approaches, including high-throughput and genome-wide screening methods.

Single Nucleotide Polymorphism-Based Genome-Wide Screening

Attempts have been made to identify the susceptibility gene variants for diabetic nephropathy by single nucleotide polymorphism (SNP)-based case-control association screening in patients with type 1 and 2 diabetes mellitus. The advent of high-throughput and low-cost genotyping technologies, discovery of sufficiently dense SNP markers, and vast improvements in collection methods for DNA analysis has made genome-wide screening readily feasible. For example, a combination of multiplex PCR and Invader assay has enabled the genotyping of one SNP with as low as 0.4 ng of genomic DNA, and theoretically more than 200,000 SNPs can be genotyped in one cycle with100 μg [20, 21]. In addition, microarray systems, such as Genechip Arrays (Affymetrix) and Infinium beads array (Illumina) can genotype hundreds of thousands of SNPs using 250–750 ng of genomic DNA.

In line with the development of genotyping technology, the International HapMap Project (IHMP) has further promoted the identification of SNPs [22]. The goal of IHMP is to compare the gene sequences and to identify chromosomal regions where the genetic variants are shared among different individuals. By making this information freely accessible, the IHMP will enable investigators to search for genes involved in various diseases and will also aid in delineating the responsiveness to differential therapies.

Utilizing genotyping technologies and SNP methods in the Genetics of Kidneys in Diabetes (GoKinD) study, a large number of individuals with type 1 diabetes were screened to identify biomarkers in two subsets of patients: one with well-documented kidney disease and the other with normal kidney functions despite long-term diabetes [23]. A single marker, rs11886047 (located upstream of PLEKHH2 promoter), was found to be associated with diabetic nephropathy by transmission disequilibrium testing in the case trios ($p = 0.0307$). Also, there was a increase of heterozygous genotypes in cases, relative to controls, from the 601 case and 577 control GoKinD singleton case/control population ($p = 0.00256$) [24]. The results of the GoKinD study suggest that the PLEKHH2, which is exclusively expressed in the kidney glomerulus, may be a genetic risk factor for susceptibility to diabetic nephropathy. In another study by Maeda et al. [25], patients with type 2 diabetes were divided into two groups: one having retinopathy with overt nephropathy, and the other control group having diabetic retinopathy with no renal dysfunctions. They genotyped over 80,000 gene-based SNPs and found that substitution of Arg913 to Gln in the gene, belonging to solute carrier family 12 member 3 (SLC12A3), seems to reduce the risk in the development of diabetic nephropathy [26]. They also noted that SNP is also prevalent in the engulfment and cell motility 1 gene (ELMO1), and thus concluded that it is a likely candidate in conferring susceptibility for diabetic nephropathy [27].

Transcriptome Approach Using Cultured Cells, Animal Models and Human Samples

Gene expression profiling has been used to identify genes that are up- or downregulated in cultured cells subjected to various forms of pathophysiological stresses. The differentially expressed genes that are induced under such stresses may serve as good candidates to develop potential novel biomarkers and targets for new therapies. Such gene profiling information has also been forthcoming from various animal models, such as streptozotocin (STZ)-induced diabetes in C57B6/J and ICR mice [28, 29] and Wistar rats [30], as well as rodents developing diabetes spontaneously, i.e. non-obese diabetic, *db/db* and KK/TA mice [28, 31–34], and OLETF rats [35].

The STZ-induced diabetic animals are a model of type 1 diabetes which have rapid onset of hyperglycemia within a couple of days. Since these mice develop renal changes several weeks or months later, the differential gene expression can be readily assessed during the early phases of diabetic nephropathy. However, one cannot absolutely rule out the possibility that the altered gene expression may also be related to the ROS that are induced upon STZ administration during the early phases of diabetic nephropathy. On the other hand, in spontaneous type 1 and 2 rodent models, one needs to take into account the time course-dependent changes in the gene expression that would be correlative of the histological lesions seen during various phases of renal hypertrophy versus sclerosis or fibrosis.

The insurmountable problem in rodent animal studies is that there is really no good model which mimics lesions in human diabetic nephropathy, such as glomerular hypertrophy, basement membrane thickening, increase in mesangial matrix and nodular sclerosis. In this regard, the triple transgenic mice overexpressing megsin, RAGE (receptor for advanced glycation end-products) and iNOS (inducible nitric oxide synthase) exhibit severe albuminuria and renal changes comparable to human diabetic nephropathy in addition to inflammatory cell infiltration and interstitial fibrosis [36]. Conceivably, these mice may be the most suitable model for identifying genes that may be critical in the pathogenesis of diabetic nephropathy.

Another critical problem in gene expression profiling studies is that there are many cell types in the kidney, each contributing to varying degrees in the pathogenesis of diabetic nephropathy; thus, it seems that analyses of each of the cells under high glucose ambient cultures would be the most appropriate measure for gene profiling relevant to diabetic nephropathy. Along this notion, many investigators have indeed utilized mesangial cells [37], conditionally immortalized mouse podocytes [38], human proximal tubular cells (HK-2 cells) [39, 40] and rat-immortalized renal proximal tubular cells to study the pathogenesis of diabetic nephropathy [30]. For human samples, peripheral blood mononuclear cells of diabetic patients [41], renal biopsy tissues [42], laser capture microdissection and analysis of the glomeruli in these biopsy specimens [32], and examination of glomerular podocytes in the urinary sediment [43] may serve as the next group of potential targets of investigation for gene profiling studies.

DDRT-PCR and RDA-cDNA/SSH

The elegant method of DDRT (differential display reverse transcription)-PCR to study the differential gene regulation between two populations of RNA was originally described by Liang and Pardee [44] in the early 90s. In this technique, two sets of oligonucleotide primers are used: one anchored to the polyadenylate tail of a subset of mRNAs, and the other being short and arbitrary in sequence so that it can anneal at different positions relative to the first primer. The mRNA subpopulations are then amplified following reverse transcription. The cDNAs are then resolved by sequencing gel and the bands from two sets of populations are compared side by side. With the application of DDRT-PCR for cDNAs isolated from mesangial cells grown in 5.5 mM (low) and 25 mM (high) D-glucose, Song et al. [45] discovered that the munc13s gene was differentially regulated.

Although DDRT-PCR was considered a powerful method to identify the differentially regulated genes, it had certain technical problems such as false-positivity for certain irrelevant genes. To overcome this problem, Hubank and Schatz [47] devised the PCR-coupled subtractive process of representational difference analysis (RDA) and adapted it for the use with mammalian cDNA [46]. Along the principles of cDNA-RDA, the suppressive subtractive hybridization (SSH) technique was developed later [47]. By applying RDA/SSH for the cDNAs prepared from STZ-induced diabetic mice kidneys, we identified a number of genes, such as Tim44 (translocase of inner mitochondrial membrane-44) [48–50], RSOR/MIOX (renal specific oxido-reductase/*myo*-inositol oxygenase) [51, 52], UbA52 [53] and Ras-like GTPase, Rap1b [54, 55]. All of these were differentially regulated in hyperglycemic state. Similarly, McMahon et al. [56] and Murphy et al. [57] reported gremlin to be upregulated in mesangial cells in culture subjected to high glucose ambience. The above approaches are certainly very useful in identifying the differentially expressed genes; however, they lack the capacity for gene profiling of global mRNA expressions. Therefore, microarray methods were developed over the past decade.

Serial Analysis of Gene Expression and DNA Microarray

Serial analysis of gene expression (SAGE) allows the quantitative and simultaneous analysis of a large number of transcripts. The method is based on the isolation of unique sequence tags (9- to 13-bp in length) from individual mRNAs and concatenation of tags serially into long DNA molecules for lump-sum sequencing [58]. So far, the usage of SAGE has been restricted to the expression profiling of kidney transcriptomes [59], in particular the renal collecting duct cells of mice subjected to potassium depletion [60], and comparative gene analyses between various cell types, such as glomerular versus aortic endothelia [61]. In this regard, the advent of high-throughput next generation sequencers, such as the 1G Genome Analyzer (Illumina/Saleza) and SOLID System (Applied Biosystems) have greatly facilitated gene expression profiling by SAGE tag with higher efficiency and lower costs. The application of these sequencers should enhance the capabilities of profiling mRNA expressions as well as discovery of miRNAs relevant to a given disease process.

In recent years, cDNA microarray has been the most commonly used method for profiling mRNA expression compared to SAGE [62]. The microarray molecular profiling of human kidney [42] and rodents [29] has resulted in the generation of a large amount of data, and in such a scenario it may be difficult to pinpoint the gene(s) that are relevant to the pathogenesis of diabetic nephropathy. These difficulties can be overcome when microarray is combined with reliable analytical procedures that can correlate with a given phenotype.

Susztak et al. [28] employed weighted vote-based analytical methods to delineate the genes whose expression can be correlated with the presence or absence of mesangial matrix expansion. They identified hydroxysteroid dehydrogenase-3β isotype 4 and osteopontin as the lead classifier genes that are correlative of mesangial matrix expansion phenotype. Along these lines Cohen et al. [63] performed Affymetrix oligonucleotide microarrays (HG-U133A) by standard robust multi-array analysis (RMA) on microdissected human renal tissues and compared the gene profiles of patients with and without diabetic nephropathy.

These studies indicated limited detection of certain biological processes that are also relevant to the pathogenesis of diabetic nephropathy. These included genes pertinent to inflammation and angiogenesis. The limited detection was thought to be attributed to the apparent lack of sensitivity that was associated with the gene-oriented averaging probe signals. This shortcoming was rectified by the use of ChipInspector, which is based on single probe analysis and de novo gene annotation that bypasses the probe set definition based on the out-of-date genomic data. In doing so, the single probe-based analysis yielded reduced background noise with enhanced sensitivity and fewer false positives. It also successfully identified the Wnt signaling pathway activated in diabetic nephropathy [63].

Another approach to identify the biologically relevant genes is to investigate pathway analysis of the gene expression. Naito et al. [64] compared gene expression of glomerular cells harvested from kidneys of *db/db* and *db/m* mice by laser capture microdissection. They then performed pathway analysis of 550 upregulated probes using the Ingenuity signal analysis. The results pointed out that the mitochondrial oxidative phosphorylation pathway is the most significantly affected canonical pathway in diabetic nephropathy [64]. The above studies emphasize that the efficient use of microarray data for identification of the relevant genes would require further improvements for the specificity and sensitivity of the mRNA detection and specifically the development of new software for the analysis of the vast amount of the available current data sets.

Proteomic Approaches

Proteomic analysis using two-dimensional polyacrylamide gel electrophoresis (2D-PAGE) coupled with Mass spectroscopic analysis has been used to survey the changes in protein expression in diabetic nephropathy. Thongboonkerd et al. [65] compared the renal proteome of 120 day-old OVE26 transgenic mice manifesting hypoinsulinemia, hyperglycemia, hyperlipidemia and proteinuria with background FVB non-diabetic mice. Proteins derived from whole-kidney lysate were separated by 2D-PAGE and identified by matrix-assisted laser desorption ionization time-of-flight (MALDI-TOF) mass spectrometry. They demonstrated marked accumulation of elastin in the macula densa, collecting ducts and pelvicalyceal epithelia of diabetic kidneys [65]. Tilton et al. [66] analyzed the renal cortical proteome of *db/db* mice using 2D-PAGE combined with MALDI-TOF, MALDI-TOF/TOF and LC-MS/MS. The top-ranked network described by Ingenuity Pathway Analysis indicated that PPAR-α may be the common nodal point of interactions for several enzymes whose expression levels are influenced by diabetic state [66].

Besides renal tissues, urine proteome analysis in diabetic patients may also allow early detection of diabetic nephropathy and could thus provide certain prognostic information. Along these lines, an increase in α1-anti-trypsin, UbA52 and type I collagen expression has been reported in patients with diabetic nephropathy [67–69]. In view of the above observations, it is conceivable that the panel of biomarkers identified in patients with diabetic nephropathy with their signatures in the serum and urinary proteins/peptides along with their unique pattern of expression and profiling could serve as a novel clinical diagnostic approach for the detection of early diabetic nephropathy [69, 70].

Newly Identified Genes Relevant in the Progression of Diabetic Nephropathy

The cellular events such as increased flux of polyols and hexosamines; generation of AGEs; increased activity of PKC, transforming growth factor-β-Smad-MAPK (mitogen-activated protein kinase) pathway and GTP-binding proteins; G1 cell cycle arrest associated with altered expression of cyclin kinases and their inhibitors; and generation of ROS are responsible for a final outcome of increased synthesis and deposition of ECM. The ROS, whether mitochondrial or cell membrane-derived, are also responsible for the activation of the renin-angiotensin system that eventually contributes to glomerular hyperfiltration and subsequent renal fibrosis (fig. 1) [71]. In addition to these macromolecules, newly identified genes, such as RSOR/MIOX, Tim44 and Rap1b, may also be an integral part of the hyperglycemia-induced cytosolic and mitochondrial processes that culminate in the development of diabetic nephropathy [48–55].

Under high glucose ambience, there is an increased activity of accessory cytosolic polyol pathway; as a result, glucose is reduced to sorbitol by a NADPH-dependent enzyme, aldose reductase. The sorbitol is oxidized to fructose by sorbitol dehydrogenease utilizing NAD+ as a cofactor. Although aldose reductase inhibitors have been used for diabetic neuropathy in Japan and other countries, the relevance of the polyol pathway is somewhat unclear in diabetic nephropathy. Another enzyme that has come to light in the context of glucose metabolism is RSOR/MIOX [52, 72]. It is expressed in the renal proximal tubules, and is upregulated in diabetic mouse kidneys. It is responsible for the oxidation of *myo*-inositol generated from glucose-6-phosphate. Phosphatidyl-inositol, a metabolic product of *myo*-inositol is involved in cellular signaling and osmoregulatory functions of the kidney. MIOX expression is upregulated in the experimental diabetes, while renal concentration of *myo*-inositol decreases and its supplementation normalizes glucose-induced proliferation and collagen synthesis in tubular cells, thus implicating the role of the cytosolic MIOX in diabetic nephropathy [73].

In the mitochondria, during oxidative phosphorylation, the electron donors generate a high membrane potential by pumping protons across the mitochondrial inner membrane. As a consequence, the electron transport is inhibited and the half-life of free-radical intermediates of ubiquinone increases, which reduces O_2 to O_2^- with ensuing oxidant stress in high glucose ambience. Rotenone, an inhibitor of electron transport, blocks the DCF-sensitive ROS generation and the fact that a similar effect is observed by overexpression of mitochondrial manganese superoxide dismutase (Mn-SOD) or uncoupling protein-1 (UCP-1) supports the notion that mitochondria are the major source of ROS following increased glucose flux into the cell. The ROS produced can thus modulate various cellular processes and expression of proteins, including those involved

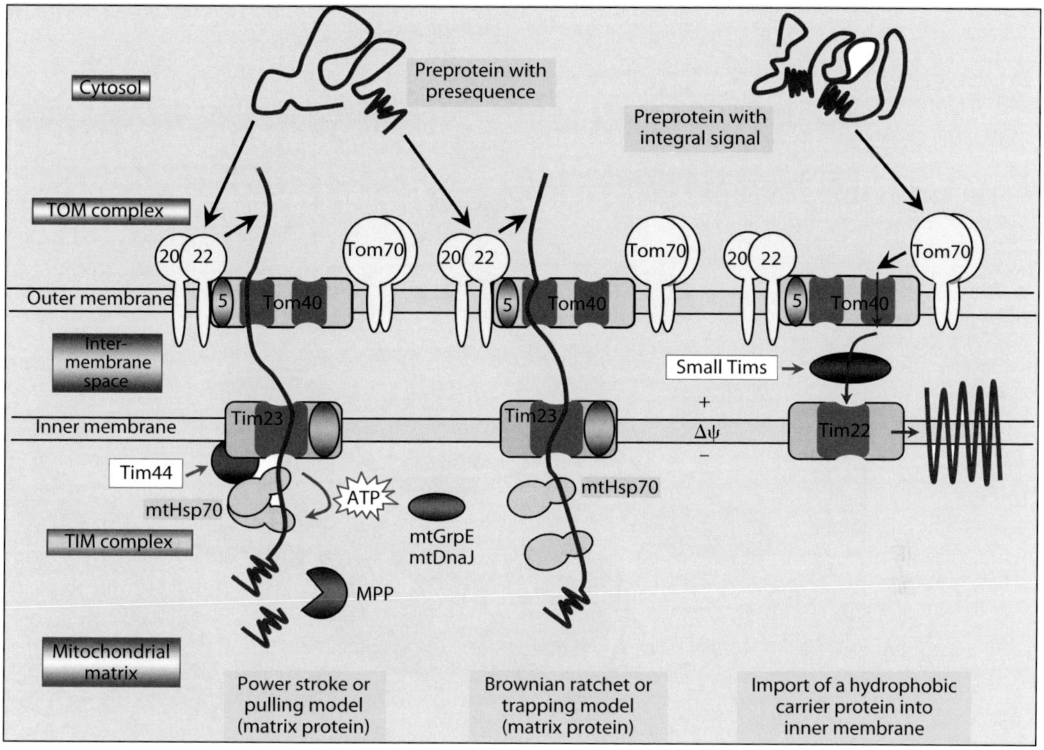

Fig. 3. Main pathways of protein import into mitochondrial matrix and inner membrane. Mitochondria matrix preproteins with an amino-terminal presequence as well as proteins with integral target sequence translocate through the translocase outer membrane (TOM) complex, made up of 5-, 20-, 22-, 40- and 70-kDa proteins. Preproteins with presequence are translocated across the inner membrane by the Tim23 complex. Two models are proposed: power stroke and Brownian ratchet. The process requires the inner membrane potential (Δψ) and ATP-dependent action of mtHsp70. In the power stroke model, Tim44 seems to function as a membrane anchor of mtHsp70 to the TIM23 complex. The proproteins with integral signals are guided by small Tims proteins across the intermembrane space to the Tim22 complex of the inner membrane and are then inserted into the membrane in a Δψ-dependent step. Tim44 is upregulated in hyperglycemia and its overexpression reduces proteinuria, renal hypertrophy, cell proliferation and apoptosis, and superoxide production in STZ-induced diabetes in CD-1 mice [48–50].

in the mitochondrial transport. One of the transport proteins, known as Tim44, is upregulated in hyperglycemic state [50]. It functions as a membrane anchor of mtHsp70 to the TIM23 complex and is involved in the import of preproteins with mitochondrial-targeted pre-sequences into the mitochondrial matrix (fig. 3). Its importance in the pathogenesis of diabetic nephropathy is underscored by the fact that gene delivery of Tim44 by the HVJ-envelope vector inhibits proteinuria, renal hypertrophy, cell proliferation and apoptosis, and superoxide production in STZ-induced diabetes in CD-1 mice [48]. Also, gene delivery of Tim44 in human proximal tubule cell line HK-2 facilitates the import of antioxidative enzymes such as SOD and glutathione peroxidase along with the inhibition of the ROS production, reduction of ATP contents and normalization of alterations in inner mitochondrial membrane potential [48].

In recent years, the role of small GTP-binding proteins in the pathobiology of hyperglycemia-

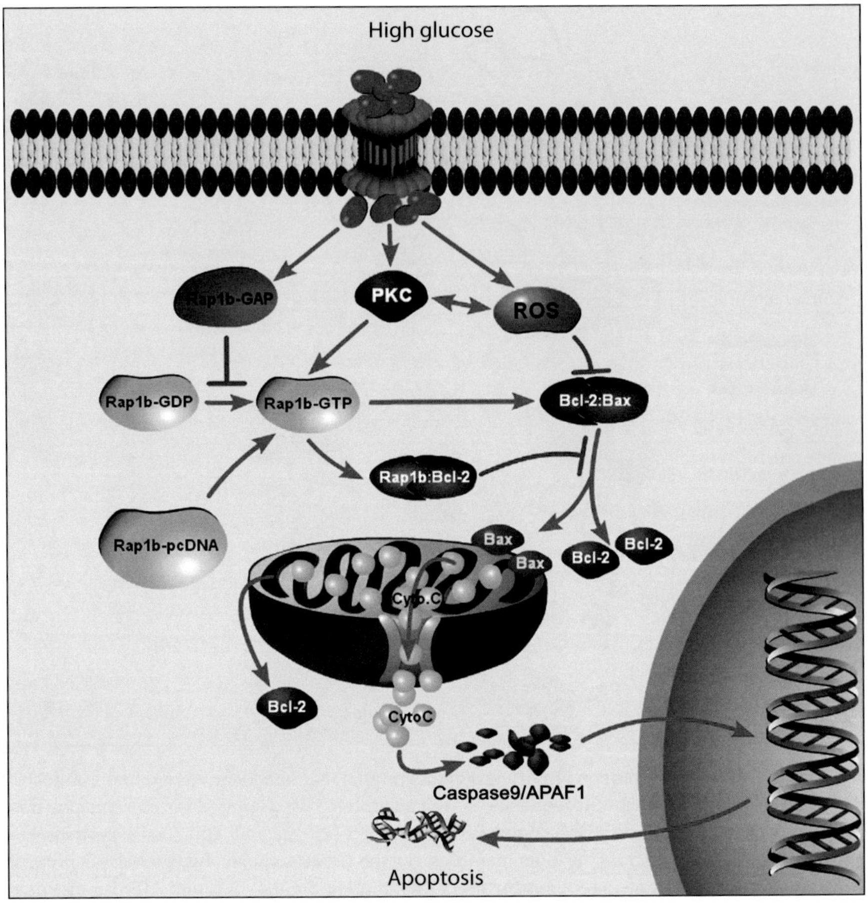

Fig. 4. Schematic drawing of various events following exposure to high glucose ambience and overexpression of Rap1b-GTPase in kidney cells. High glucose ambience leads to activation of PKC and generation of ROS. The latter perturb Bcl2:Bax interactions that leads to the release of mitochondrial cytochrome C, activation of caspases, fragmentation of DNA and consequential apoptosis. Transfection of GTPase Rap1b conceivably stabilizes Bcl2:Bax and Rap1b:Bcl2 interactions and inhibit the release of cytochrome C, oxidant stress, DNA fragmentation and apoptosis [54, 74].

induced renovascular injury has attracted much attention of many investigators. Major small GTP-binding proteins activated in hyperglycemic state belong to the Ras and Rho family of small GTPases. They exert pleiotropic effects and regulate growth, morphogenesis, cell motility, axonal guidance, cytokinesis and intracelullar trafficking by cycling between an inactive (GDP-bound) and active (GTP-bound) state. Rap1b is a homolog of Ras-related family members that are endowed with anti-apoptotic properties mediated via their effectors (e.g. Raf) [54]. Rap1b is known to stimulate B-Raf activity, thus serving as a modulator of cell proliferation and apoptosis [74]. In the kidney of diabetic mice, apoptosis of tubular cells and dysmorphic mitochondria are observed to be associated with reduction of Bcl-2 and increased Bax expression [54]. Interestingly, the GTPase activity seems to be reduced significantly, although Rap1b expression is increased.

Overexpression of Rap1b in HK-2 cells partially reverses these high glucose-induced abnormalities, such as increased DNA fragmentation, altered mitochondrial morphology and disrupted Bcl-2-Bax and Bcl-2-Rap1b interactions (fig. 4) [54]. Overall, it seems that these newly identified genes, RSOR/MIOX, Tim44 and Rap1b, are indeed functionally linked to various intracellular events in kidney tissues subjected to high glucose ambience. The discovery of these new genes relevant to diabetic nephropathy in various animal models should give some impetus to future translational research and to the development of pharmaceutical interventions where these can serve as molecular targets.

Conclusions and Future Perspectives

Current approaches such as transcriptome and proteome profiling, as well as molecular genetics, using various cell lines, animal models and human samples have greatly facilitated the understanding of the mechanism(s) relevant to the progression of diabetic nephropathy. Based on the data generated by using these techniques, the newly discovered biomarkers could serve as therapeutic targets for the amelioration of diabetic nephropathy, which certainly contribute to the reduction in mortality and morbidity in chronic kidney disease patients that progress to ESRD. In addition to transcriptome and proteome approaches, the future trends for the identification of the biomarkers and therapeutic target genes could include genome-scale DNA methylation profiling [75]. The emerging role of epigenome control of the cancer cells, germ cells and pluripotent stem cells has been emphasized in the transcriptional regulation of various genes that receive sustained long-term injury for years and decades. Intensive long-term versus conventional short-interval symptomatic therapy seems to have remarkable beneficial effects on the risk of cardiovascular disease in patients with type 1 diabetes and this suggests that there may be alterations in the genomic DNA- or histone-methylation pattern which may be linked to the long-term 'metabolic memory' for the progression of vascular complications of diabetes [76]. Such a methylation-related profiling would certainly advance the field, especially with respect to development of new biomarkers and various therapeutic strategies. In addition to the delineation of epigenome control of the genes, metabolic phenotyping using 1H spectroscopy [77] and lectin microarray [78] for the glycan profiling would also promote the identification of the new biomarkers of diabetic nephropathy. Finally, integration of the information from different sources using system biology approaches would be an important step in data-mining for the identification of relevant genes that are pertinent to the diagnosis and therapy for diabetic nephropathy.

Acknowledgements

Supported by NIH grants DK28492 and DK60635.

References

1 Balakumar P, Arora MK, Reddy J, Anand-Srivastava MB: Pathophysiology of diabetic nephropathy: involvement of multifaceted signaling mechanism. J Cardiovasc Pharmacol 2009;54:129–138.

2 Reddy AS: Diabetic Nephropathy: Theory and Practice. East Hanover, College Book Publishers, 2004, pp 1–563, 2004.

3 Kikkawa R, Koya D, Haneda M: Progression of diabetic nephropathy. Am J Kidney Dis 2003;41(3 Suppl 1):S19–S21.

4 Sugimoto H, Shikata K, Hirata K, Akiyama K, Matsuda M, Kushiro M, Shikata Y, Miyatake N, Miyasaka M, Makino H: Increased expression of intercellular adhesion molecule-1 (ICAM-1) in diabetic rat glomeruli: glomerular hyperfiltration is a potential mechanism of ICAM-1 upregulation. Diabetes 1997;46:2075–2081.

5 Koya D, Haneda M, Nakagawa H, Isshiki K, Sato H, Maeda S, Sugimoto T, Yasuda H, Kashiwagi A, Ways DK, King GL, Kikkawa R: Amelioration of accelerated diabetic mesangial expansion by treatment with a PKC beta inhibitor in diabetic db/db mice, a rodent model for type 2 diabetes. FASEB J 2000;14:439–447.

6 Sharma K, McCue P, Dunn SR: Diabetic kidney disease in the *db/db* mouse. Am J Physiol Renal Physiol 2003;284:F1138–F1144.

7 Chung AC, Zhang H, Kong YZ, Tan JJ, Huang XR, Kopp JB, Lan HY: Advanced glycation end-products induce tubular CTGF via TGF-beta-Independent Smad3 Signaling. J Am Soc Nephrol 2010;21:249–260.
8 Alderson NL, Chachich ME, Youssef NN, Beattie RJ, Nachtigal M, Thorpe SR, Baynes JW: The AGE inhibitor pyridoxamine inhibits lipemia and development of renal and vascular disease in Zucker obese rats. Kidney Int 2003;63:2123–2133.
9 Figarola JL, Loera S, Weng Y, Shanmugam N, Natarajan R, Rahbar S: LR-90 prevents dyslipidaemia and diabetic nephropathy in the Zucker diabetic fatty rat. Diabetologia 2008; 51:882–891.
10 Kim YS, Kim J, Kim CS, Sohn EJ, Lee YM, Jeong IH, Kim H, Jang DS, Kim JS: KIOM-79, an inhibitor of AGEs-protein cross-linking, prevents progression of nephropathy in Zucker diabetic fatty rats. Evid Based Complement Alternat Med 2009, E-pub ahead of print.
11 Satoh M, Fujimoto S, Haruna Y, Arakawa S, Horike H, Komai N, Sasaki T, Tsujioka K, Makino H, Kashihara N: NAD(P)H oxidase and uncoupled nitric oxide synthase are major sources of glomerular superoxide in rats with experimental diabetic nephropathy. Am J Physiol Renal Physiol 2005;288: F1144–F1152.
12 Asaba K, Tojo A, Onozato ML, Goto A, Quinn MT, Fujita T, Wilcox CS: Effects of NADPH oxidase inhibitor in diabetic nephropathy. Kidney Int 2005;67:1890–1898.
13 Wolf G: New insights into the pathophysiology of diabetic nephropathy: From hemodynamics to molecular pathology. Eur J Clin Invest 2004;34:785–796.
14 Tesch GH: Role of macrophages in complications of type 2 diabetes. Clin Exp Pharmacol Physiol 2007;34:1016–1019.
15 Usui HK, Shikata K, Sasaki M, Okada S, Matsuda M, Shikata Y, Ogawa D, Kido Y, Nagase R, Yozai K, Ohga S, Tone A, Wada J, Takeya M, Horiuchi S, Kodama T, Makino H: Macrophage scavenger receptor-A-deficient mice are resistant against diabetic nephropathy through amelioration of microinflammation. Diabetes 2007;56:363–372.
16 Ziyadeh FN, Hoffman BB, Han DC, Iglesias-De La Cruz MC, Hong SW, Isono M, Chen S, McGowan TA, Sharma K: Long-term prevention of renal insufficiency, excess matrix gene expression, and glomerular mesangial matrix expansion by treatment with monoclonal antitransforming growth factor-beta antibody in db/db diabetic mice. Proc Natl Acad Sci USA 2000;97:8015–8020.
17 Gaede P, Vedel P, Parving HH, Pedersen O: Intensified multifactorial intervention in patients with type 2 diabetes mellitus and microalbuminuria: the Steno type 2 randomized study. Lancet 1999;353:617–622.
18 Gaede P, Vedel P, Larsen N, Jensen GV, Parving HH, Pedersen O: Multifactorial intervention and cardiovascular disease in patients with type 2 diabetes. N Engl J Med 2003;348:383–393.
19 Gaede P, Lund-Andersen H, Parving HH, Pedersen O: Effect of a multifactorial intervention on mortality in type 2 diabetes. N Engl J Med 2008;358:580–591.
20 Hosono N, Kubo M, Tsuchiya Y, Sato H, Kitamoto T, Saito S, Ohnishi Y, Nakamura Y: Multiplex PCR-based real-time invader assay (mPCR-RETINA): a novel SNP-based method for detecting allelic asymmetries within copy number variation regions. Hum Mutat 2008;29:182–189.
21 Ohnishi Y, Tanaka T, Ozaki K, Yamada R, Suzuki H, Nakamura Y: A high-throughput SNP typing system for genome-wide association studies. J Hum Genet 2001;46:471–477.
22 The international HapMap project. Nature 2003;426:789–796.
23 Mueller PW, Rogus JJ, Cleary PA, Zhao Y, Smiles AM, Steffes MW, Bucksa J, Gibson TB, Cordovado SK, Krolewski AS, Nierras CR, Warram JH: Genetics of Kidneys in Diabetes (GoKinD) study: a genetics collection available for identifying genetic susceptibility factors for diabetic nephropathy in type 1 diabetes. J Am Soc Nephrol 2006;17:1782–1790.
24 Greene CN, Keong LM, Cordovado SK, Mueller PW: Sequence variants in the PLEKHH2 region are associated with diabetic nephropathy in the GoKinD study population. Hum Genet 2008;124:255–262.
25 Maeda S: Genome-wide search for susceptibility gene to diabetic nephropathy by gene-based SNP. Diabetes Res Clin Pract 2004;66 (Suppl 1):S45–S47.
26 Tanaka N, Babazono T, Saito S, Sekine A, Tsunoda T, Haneda M, Tanaka Y, Fujioka T, Kaku K, Kawamori R, Kikkawa R, Iwamoto Y, Nakamura Y, Maeda S: Association of solute carrier family 12 (sodium/chloride) member 3 with diabetic nephropathy, identified by genome-wide analyses of single nucleotide polymorphisms. Diabetes 2003;52:2848–2853.
27 Shimazaki A, Kawamura Y, Kanazawa A, Sekine A, Saito S, Tsunoda T, Koya D, Babazono T, Tanaka Y, Matsuda M, Kawai K, Iizumi T, Imanishi M, Shinosaki T, Yanagimoto T, Ikeda M, Omachi S, Kashiwagi A, Kaku K, Iwamoto Y, Kawamori R, Kikkawa R, Nakajima M, Nakamura Y, Maeda S: Genetic variations in the gene encoding ELMO1 are associated with susceptibility to diabetic nephropathy. Diabetes 2005;54:1171–1178.
28 Susztak K, Bottinger E, Novetsky A, Liang D, Zhu Y, Ciccone E, Wu D, Dunn S, McCue P, Sharma K: Molecular profiling of diabetic mouse kidney reveals novel genes linked to glomerular disease. Diabetes 2004;53:784–794.
29 Wada J, Zhang H, Tsuchiyama Y, Hiragushi K, Hida K, Shikata K, Kanwar YS, Makino H. Gene expression profile in streptozotocin-induced diabetic mice kidneys undergoing glomerulosclerosis. Kidney Int 2001;59:1363–1373.
30 Hsieh TJ, Chen R, Zhang SL, Liu F, Brezniceanu ML, Whiteside CI, Fantus IG, Ingelfinger JR, Hamet P, Chan JS: Upregulation of osteopontin gene expression in diabetic rat proximal tubular cells revealed by microarray profiling. Kidney Int 2006;69: 1005–1015.
31 Wilson KH, McIndoe RA, Eckenrode S, Morel L, Agarwal A, Croker BP, She JX: Alterations of renal phenotype and gene expression profiles due to protein overload in NOD-related mouse strains. BMC Nephrol 2005;6:17.

32 Naito, Uchiyama K, Kuroda M, Mizushima K, Aoi W, Kokura S, Ichikawa H, Yoshida N, Yoshikawa T: Laser capture microdissection/GeneChip analysis of gene expression in glomerular cells in diabetic db/db mice. Redox Rep 2004;9:307–312.
33 Makino H, Miyamoto Y, Sawai K, Mori K, Mukoyama M, Nakao K, Yoshimasa Y, Suga S. Altered gene expression related to glomerulogenesis and podocyte structure in early diabetic nephropathy of db/db mice and its restoration by pioglitazone. Diabetes 2006;55:2747–2756.
34 Fan Q, Shike T, Shigihara T, Tanimoto M, Gohda T, Makita Y, Wang LN, Horikoshi S, Tomino Y: Gene expression profile in diabetic KK/Ta mice. Kidney Int 2003;64:1978–1985.
35 Ko GJ, Kang YS, Han SY, Lee MH, Song HK, Han KH, Kim HK, Han JY, Cha DR: Pioglitazone attenuates diabetic nephropathy through an anti-inflammatory mechanism in type 2 diabetic rats. Nephrol Dial Transplant 2008;23:2750–2760.
36 Inagi R, Yamamoto Y, Nangaku M, Usuda N, Okamato H, Kurokawa K, van Ypersele de Strihou C, Yamamoto H, Miyata T: A severe diabetic nephropathy model with early development of nodule-like lesions induced by megsin overexpression in RAGE/iNOS transgenic mice. Diabetes 2006;55:356–366.
37 Cheng DW, Jiang Y, Shalev A, Kowluru R, Crook ED, Singh LP: An analysis of high glucose and glucosamine-induced gene expression and oxidative stress in renal mesangial cells. Arch Physiol Biochem 2006;112:189–218.
38 Han SH, Yang S, Jung DS, Li JJ, Kim JJ, Kwak SJ, Kim DK, Moon SJ, Lee JE, Han DS, Kang SW: Gene expression patterns in glucose-stimulated podocytes. Biochem Biophys Res Commun 2008;370:514–518.
39 Qi W, Chen X, Zhang Y, Holian J, Mreich E, Gilbert RE, Kelly DJ, Pollock CA: High glucose induces macrophage inflammatory protein-3 alpha in renal proximal tubule cells via a transforming growth factor-beta 1 dependent mechanism. Nephrol Dial Transplant 2007;22:3147–3153.
40 Qi W, Chen X, Gilbert RE, Zhang Y, Waltham M, Schache M, Kelly DJ, Pollock CA: High glucose-induced thioredoxin-interacting protein in renal proximal tubule cells is independent of transforming growth factor-beta1. Am J Pathol 2007;171:744–754.
41 Moczulski DK, Fojcik H, Wielgorecki A, Trautsolt W, Gawlik B, Kosiorz-Gorczynska S, Oczko-Wojciechowska M, Wiench M, Strojek K, Zukowska-Szczechowska E, Grzeszczak W: Expression pattern of genes in peripheral blood mononuclear cells in diabetic nephropathy. Diabet Med 2007;24:266–271.
42 Baelde HJ, Eikmans M, Doran PP, Lappin DW, de Heer E, Bruijn JA: Gene expression profiling in glomeruli from human kidneys with diabetic nephropathy. Am J Kidney Dis 2004;43:636–650.
43 Szeto CC, Lai KB, Chow KM, Szeto CY, Yip TW, Woo KS, Li PK, Lai FM: Messenger RNA expression of glomerular podocyte markers in the urinary sediment of acquired proteinuric diseases. Clin Chim Acta 2005;361:182–190.
44 Liang P, Pardee AB: Differential display of eukaryotic messenger RNA by means of the polymerase chain reaction. Science 1992;257:967–971.
45 Song Y, Ailenberg M, Silverman M: Cloning of a novel gene in the human kidney homologous to rat munc13s: its potential role in diabetic nephropathy. Kidney Int 1998;53:1689–1695.
46 Lisitsyn N, Wigler M: Cloning the differences between two complex genomes. Science 1993;259:946–951.
47 Hubank M, Schatz DG: Identifying differences in mRNA expression by representational difference analysis of cDNA. Nucleic Acids Res 1994;22:5640–5648.
48 Zhang Y, Wada J, Hashimoto I, Eguchi J, Yasuhara A, Kanwar YS, Shikata K, Makino H: Therapeutic approach for diabetic nephropathy using gene delivery of translocase of inner mitochondrial membrane 44 by reducing mitochondrial superoxide production. J Am Soc Nephrol 2006;17:1090–1101.
49 Matsuoka T, Wada J, Hashimoto I, Zhang Y, Eguchi J, Ogawa N, Shikata K, Kanwar YS, Makino H: Gene delivery of Tim44 reduces mitochondrial superoxide production and ameliorates neointimal proliferation of injured carotid artery in diabetic rats. Diabetes 2005;54:2882–2890.
50 Wada J, Kanwar YS: Characterization of mammalian translocase of inner mito-chondrial membrane (Tim44) isolated from diabetic newborn mouse kidney. Proc Natl Acad Sci USA 1998;95:144–149.
51 Kanwar YS, Akagi S, Nayak B, Sun L, Wada J, Xie P, Thakur A, Chugh SS, Danesh FR: Renal-specific oxidoreductase biphasic expression under high glucose ambience during fetal versus neonatal development. Kidney Int 2005;68:1670–1683.
52 Yang Q, Dixit B, Wada J, Tian Y, Wallner EI, Srivastva SK, Kanwar YS: Identification of a renal-specific oxido-reductase in newborn diabetic mice. Proc Natl Acad Sci USA 2000;97:9896–9901.
53 Sun L, Pan X, Wada J, Haas CS, Wuthrich RP, Danesh FR, Chugh SS, Kanwar YS: Isolation and functional analysis of mouse UbA52 gene and its relevance to diabetic nephropathy. J Biol Chem 2002;277:29953–29962.
54 Sun L, Xie P, Wada J, Kashihara N, Liu FY, Zhao Y, Kumar D, Chugh SS, Danesh FR, Kanwar YS: Rap1b GTPase ameliorates glucose-induced mitochondrial dysfunction. J Am Soc Nephrol 2008;19:2293–2301.
55 Lin S, Chugh S, Pan X, Wallner EI, Wada J, Kanwar YS: Identification of up-regulated Ras-like GTPase, Rap1b, by suppression subtractive hybridization. Kidney Int 2001;60:2129–2141.
56 McMahon R, Murphy M, Clarkson M, Taal M, Mackenzie HS, Godson C, Martin F, Brady HR: IHG-2, a mesangial cell gene induced by high glucose, is human gremlin. Regulation by extracellular glucose concentration, cyclic mechanical strain, and transforming growth factor-beta1. J Biol Chem 2000;275:9901–9904.
57 Murphy M, Godson C, Cannon S, Kato S, Mackenzie HS, Martin F, Brady HR: Suppression subtractive hybridization identifies high glucose levels as a stimulus for expression of connective tissue growth factor and other genes in human mesangial cells. J Biol Chem 1999;274:5830–5834.

58 Velculescu VE, Zhang L, Vogelstein B, Kinzler KW: Serial analysis of gene expression. Science 1995;270:484–487.

59 Schelling JR, El-Meanawy MA, Barathan S, Dodig T, Iyengar SK, Sedor JR: Generation of kidney transcriptomes using serial analysis of gene expression. Exp Nephrol 2002;10:82–92.

60 Cheval L, Duong Van Huyen JP, Bruneval P, Verbavatz JM, Elalouf JM, Doucet A: Plasticity of mouse renal collecting duct in response to potassium depletion. Physiol Genomics 2004;19:61–73.

61 Sengoelge G, Luo W, Fine D, Perschl AM, Fierlbeck W, Haririan A, Sorensson J, Rehman TU, Hauser P, Trevick JS, Kulak SC, Wegner B, Ballermann BJ: A SAGE-based comparison between glomerular and aortic endothelial cells. Am J Physiol Renal Physiol 2005;288:F1290–1300.

62 Schena M, Shalon D, Davis R, Brown PO: Quantitative monitoring of gene expression patterns with a complementary DNA microarray. Science 1995;270:467–470.

63 Cohen CD, Lindenmeyer MT, Eichinger F, Hahn A, Seifert M, Moll AG, Schmid H, Kiss E, Grone E, Grone HJ, Kretzler M, Werner T, Nelson PJ: Improved elucidation of biological processes linked to diabetic nephropathy by single probe-based microarray data analysis. *PLoS One* 2008;3:e2937.

64 Naito Y, Uchiyama K, Mizushima K, Kuroda M, Akagiri S, Takagi T, Handa O, Kokura S, Yoshida N, Ichikawa H, Takahashi J, Yoshikawa T: Microarray profiling of gene expression patterns in glomerular cells of astaxanthin-treated diabetic mice: a nutrigenomic approach. Int J Mol Med 2006;18:685–695.

65 Thongboonkerd V, Barati MT, McLeish KR, Benarafa C, Remold-O'Donnell E, Zheng S, Rovin BH, Pierce WM, Epstein PN, Klein JB: Alterations in the renal elastin-elastase system in type 1 diabetic nephropathy identified by proteomic analysis. J Am Soc Nephrol 2004;15:650–662.

66 Tilton RG, Haidacher S, Lejeune W, Zhang X, Zhao Y, Kurosky A, Brasier A, Denner L: Diabetes-induced changes in the renal cortical proteome assessed with 2-dimensional gel electrophoresis and mass spectrometry. Proteomics 2007;7:1729–1742.

67 Sharma K, Lee S, Han S, Francos B, McCue P, Wassell R, Shaw MA, RamachandraRao SP: Two-dimensional fluorescence difference gel electrophoresis analysis of the urine proteome in human diabetic nephropathy. Proteomics 2005;5:2648–2655.

68 Dihazi H, Muller GA, Lindner S, Meyer M, Asif AR, Oellerich M, Strutz F: Characterization of diabetic nephropathy by urinary proteomic analysis: identification of a processed ubiquitin form as a differentially excreted protein in diabetic nephropathy patients. Clin Chem 2007;53:1636–1645.

69 Rossing K, Mischak H, Dakna M, Zurbig P, Novak J, Julian BA, Good DM, Coon JJ, Tarnow L, Rossing P: Urinary proteomics in diabetes and CKD. J Am Soc Nephrol 2008; 19:1283–1290.

70 Yang YH, Zhang S, Cui JF, Lu B, Dong XH, Song XY, Liu YK, Zhu XX, Hu RM: Diagnostic potential of serum protein pattern in Type 2 diabetic nephropathy. Diabet Med 2007;24:1386–1392.

71 Kanwar YS, Wada J, Sun L, Xie P, Wallner EI, Chen S, Chugh S, Danesh FR: Diabetic nephropathy: mechanisms of renal disease progression. Exp Biol Med (Maywood) 2008; 233:4–11.

72 Nayak B, Xie P, Akagi S, Yang Q, Sun L, Wada J, Thakur A, Danesh FR, Chugh SS, Kanwar YS: Modulation of renal-specific oxidoreductase/myo-inositol oxygenase by high-glucose ambience. Proc Natl Acad Sci USA 2005;102:17952–17957.

73 Ziyadeh FN, Simmons DA, Snipes ER, Goldfarb S: Effect of myo-inositol on cell proliferation and collagen transcription and secretion in proximal tubule cells cultured in elevated glucose. J Am Soc Nephrol 1991;1: 1220–1229.

74 Lin S, Sahai A, Chugh SS, Pan X, Wallner EI, Danesh FR, Lomasney JW, Kanwar YS: High glucose stimulates synthesis of fibronectin via a novel protein kinase C, Rap1b, and B-Raf signaling pathway. J Biol Chem 2002;277:41725–41735.

75 Meissner A, Mikkelsen TS, Gu H, Wernig M, Hanna J, Sivachenko A, Zhang X, Bernstein B, Nusbaum C, Jaffe D, Gnirke A, Jaenisch R, Lander E: Genome-scale DNA methylation maps of pluripotent and differentiated cells. Nature 2008;454:766–770.

76 Nathan DM, Cleary PA, Backlund JY, Genuth SM, Lachin JM, Orchard TJ, Raskin P, Zinman B: Intensive diabetes treatment and cardiovascular disease in patients with type 1 diabetes. N Engl J Med 2005;353:2643–2653.

77 Holmes E, Loo RL, Stamler J, Bictash M, Yap IK, Chan Q, Ebbels T, De Iorio M, Brown IJ, Veselkov KA, Daviglus ML, Kesteloot H, Ueshima H, Zhao L, Nicholson JK, Elliott P: Human metabolic phenotype diversity and its association with diet and blood pressure. Nature 2008;453:396–400.

78 Ebe Y, Kuno A, Uchiyama N, Koseki-Kuno S, Yamada M, Sato T, Narimatsu H, Hirabayashi J: Application of lectin microarray to crude samples: differential glycan profiling of lec mutants. J Biochem 2006;139:323–327.

Yashpal S. Kanwar, MD, PhD
Department of Pathology, Northwestern University Medical School
303 East Chicago Avenue
Chicago, IL 60611 (USA)
Tel. +1 312 503 0004, Fax +1 312 503 0627, E-Mail y-kanwar@northwestern.edu

Herrera GA (ed): Experimental Models for Renal Diseases: Pathogenesis and Diagnosis.
Contrib Nephrol. Basel, Karger, 2011, vol 169, pp 175–182

Glomerular Basement Membrane Disorders in Experimental Models for Renal Diseases: Impact on Understanding Pathogenesis and Improving Diagnosis

Clifford E. Kashtan[a,b] · Yoav Segal[c,d]

[a]Department of Pediatrics, University of Minnesota Medical School, University of Minnesota Amplatz Children's Hospital, [b]Alport Syndrome Treatments and Outcomes Registry (ASTOR), [c]Division of Renal Diseases and Hypertension, University of Minnesota Medical School, and [d]Veterans Administration Medical Center, Minneapolis, Minn., USA

Abstract

Animal models have provided important insights into human renal diseases that arise from mutations in genes that encode or regulate the synthesis of glomerular basement membrane proteins. This chapter describes several well-characterized animal models of type IV collagen disorders (Alport syndrome, HANAC syndrome), a laminin disorder (Pierson syndrome), nail-patella syndrome and HERNS syndrome. These models can be exploited in studies of the pathogenesis and treatment of such disorders.

The glomerular basement membrane (GBM) is a critical component of the glomerular filtration barrier, providing anchorage for glomerular endothelial cells and glomerular visceral epithelial cells (podocytes). Alterations in the normal protein composition of the GBM have various consequences, including weakening of the glomerular capillary wall, leading to hematuria; induction of podocyte foot process fusion, associated with proteinuria; and initiation of cellular pathways that eventuate in glomerulosclerosis. Studies of animal models have provided important insights into the roles of several GBM proteins in maintaining normal glomerular structure and function. These models have also helped illuminate the pathophysiology of genetic disorders that affect GBM proteins.

The major protein constituents of GBM are type IV collagen, laminin, nidogen and perlecan, a proteoglycan. While the GBM of the developing glomerulus contains solely the α1 and α2 chains of type IV collagen [α1(IV) and α2(IV)], the mature GBM is rich in α3(IV), α4(IV) and α5(IV) chains. The predominant laminin in the normal, mature GBM is laminin-521, formerly called laminin-11, made up of the laminin α5, β2 and γ1 chains. Mutations in type IV collagen and GBM laminin cause renal disease in both animals and humans, but no renal phenotype has been associated with perlecan or nidogen mutations in humans or in mice thus far [1–4]. Therefore this discussion will focus on type IV collagen and GBM laminin.

Type IV Collagen

Type IV Collagen Genes and Proteins

The type IV collagen family consists of six genetically related but distinct chains [α1(IV) – α6(IV)] encoded by six different genes (COL4A1 – COL4A6). The type IV collagen genes are arranged in pairs on three chromosomes in humans: COL4A1-COL4A2 on chromosome 13, COL4A3-COL4A4 on chromosome 2 and COL4A5-COL4A6 on the X chromosome. The 5′ ends of each gene pair are adjacent to each other, separated by sequences of various lengths containing regulatory motifs.

Each type IV collagen α chain has a major collagenous domain of about 1,400 residues consisting of the repetitive triplet sequence (Gly)-X-Y, a C-terminal noncollagenous (NC1) domain of about 230 residues and a noncollagenous N-terminal sequence of 15–20 residues (NC2). Type IV collagen α chains associate with each other to form a limited array of heterotrimers, directed by recognition sequences located within the NC1 domains. Three type IV collagen heterotrimers have been found in mammalian basement membranes: α1α1α2, α3α4α5 and α5α5α6. Like heterotrimers form open nonfibrillar networks through C-terminal and N-terminal interactions that are further reinforced through lateral interactions. Basement membranes form through nidogen-mediated association of type IV collagen and laminin assemblies.

Renal Disease Due to Type IV Collagen Mutations

Alport Syndrome

Alport syndrome is a progressive hereditary nephropathy due to mutations in COL4A3, COL4A4 or COL4A5 that interfere with the formation, deposition and/or function of the type IV collagen α3α4α5 network. Sensorineural deafness and ocular anomalies are important features of human Alport syndrome, but since these features of the disorder have yet to be described in any animal models, this discussion will focus on the renal aspects of Alport syndrome in humans and animal models.

Glomerulogenesis proceeds normally in developing Alport kidneys, but absence of the α3α4α5 network results in biochemical, structural and functional changes in GBM. The α1α1α2 network, type V collagen and type VI collagen (all normally confined to a subendothelial location in mature GBM), are deposited throughout the GBM. Laminin expression is also altered in Alport GBM, with the laminin α2 chain deposited as laminin-2 (α2β1γ1) or laminin-4 (α2β2γ1); laminin α1 and β1 chains are also abnormally deposited. Hematuria is present from very early in life, perhaps birth, and is followed by the development of proteinuria and progressive deterioration in glomerular filtration, ultimately leading to uremia. By electron microscopy, the GBM is initially attenuated and podocyte foot processes are well-preserved, but over time the GBM undergoes progressive thickening with disorganization of the lamina densa, associated with diffuse foot process effacement. These glomerular events are accompanied by tubular atrophy and interstitial fibrosis.

Alport syndrome is primarily an X-linked disorder (XLAS), with approximately 80% of the disease burden arising from COL4A5 mutations. Reported COL4A5 mutations number in the hundreds, include nonsense, missense and splicing mutations in addition to deletions and other major rearrangements, and are spread throughout the COL4A5 gene. Mutations in both alleles of COL4A3 or COL4A4 cause autosomal recessive Alport syndrome (ARAS) and account for about 15% of patients, while about 5% of affected individuals have autosomal dominant Alport syndrome due to heterozygous COL4A3 or COL4A4 mutations. However, most heterozygous carriers of mutations at these loci are asymptomatic or have isolated microscopic hematuria associated with thin glomerular basement membranes (thin basement membrane nephropathy).

Alport syndrome occurs spontaneously in dogs. The earliest described canine form of XLAS occurred in a Samoyed breed and resulted from a stop codon in exon 35 of COL4A5 [5]. XLAS in a mixed breed kindred was traced to a 10-bp deletion in exon 9 of COL4A5 that produces a frameshift and premature stop codon in exon 10 [6]. In both models, the GBM of affected males completely lacks the α3α4α5 network, while heterozygous females exhibit mosaic loss of expression of the network [7]. ARAS occurs in English Cocker Spaniels and arises from a premature stop codon in exon 3 of COL4A4 [8]. In this model, α3(IV) and α4(IV) chains are missing from the GBM, yet the α5(IV) chain is present, albeit at reduced levels compared to normal littermates [9]. The α6(IV) chain, which in normal kidneys is found only in the basement membranes of Bowman's capsules and collecting ducts, is ectopically present in the GBM of these dogs, indicating deposition of α5α5α6 heterotrimers [9].

Several murine models of Alport syndrome have been described. Two ARAS models were developed by gene targeting at the *Col4a3* locus [10, 11] and one by insertional mutagenesis at the *Col4a3-Col4a4* locus [12]. All were initially characterized in mixed genetic backgrounds and displayed similar pathological and functional abnormalities and outcomes. The influence of genetic background on the natural history of murine ARAS was studied by back-crossing onto both the 129/SvJ and the C57BL/6J backgrounds [13]. Compared to mice on the 129 background, C57BL/6 *Col4a3*$^{-/-}$ mice exhibited later onset of measurable proteinuria (100 vs. 35 days) and end-stage renal failure (194 vs. 66 days). Two as-yet unidentified quantitative trait loci on chromosomes 9 and 6 were found to influence the rate of progression to uremia. A murine model of XLAS was generated on the C57BL/6 background by introducing a G5X mutation in the *Col4a5* gene [14].

Recently, strain-dependent differences in GBM composition that may affect the rate of progression to renal failure were identified in *Col4a3*$^{-/-}$ mice [15]. Slow-progressing C57BL/6 *Col4a3*$^{-/-}$ mice exhibited robust GBM deposition of α5(IV) chains accompanied by ectopically expressed α6(IV) chains. The two chains were biochemically associated with each other and with α1 and α2(IV) chains, indicating co-assembly as a α1α2/α5α6(IV) collagen network. Much lower amounts of the α5(IV) and α6(IV) chains were detected in the GBM of 129/Sv *Col4a3*$^{-/-}$ mice, due to less synthesis or defective assembly of α6(IV) chain-containing networks in tissues from 129/Sv mice, independent of the Alport mutation. In reciprocal F1(B6x129) crosses of *Col4a3*$^{-/-}$ mice, it was found that strong GBM deposition of α5/α6(IV) chains was inherited in a pattern consistent with X-linked transmission, and correlated with a 46% increase in the longevity observed in C57BL/6- versus 129/Sv-affected mice. By comparing GBM composition and renal survival in *Col4a3*$^{-/-}$ mice on pure and hybrid backgrounds, it was estimated that ectopic α5α6(IV) collagen in the Alport GBM contributed about 28% of the added lifespan observed in C57BL/6- versus 129/Sv-affected mice.

Several arguments against an important role for GBM α5α6(IV) expression as a modifier of Alport syndrome outcome can be made. GBM α5(IV) expression does not occur in human ARAS unless the α3(IV) and α4(IV) chains are also present, and α6(IV) expression in GBM of human subjects with ARAS has never been described [16]. A survival difference similar to that described in murine ARAS is observed when XLAS mice on the 129/Sv and C57BL/6 backgrounds are compared (Y. Segal, unpubl. observations), despite the fact that these XLAS mice cannot generate α5α5α6 heterotrimers. Finally, GBM α5α6(IV) expression has yet to be reported in human XLAS, although some patients exhibit preservation of α3α4α5(IV) deposition in GBM [17, 18].

Multiple therapeutic interventions have been shown to have beneficial effects in 129/Sv ARAS mice, including angiotensin antagonism [19, 20], TGFβ-1 inhibition [21], chemokine receptor 1 suppression [22], BMP-7 administration [23], blockade of matrix metalloproteinases [24] and HMG-CoA reductase inhibition [25]. Amelioration of murine ARAS by bone marrow transplantation has been described [26–28], but data from a subsequent study suggested that the therapeutic effect was attributable to irradiation [29].

Hereditary Angiopathy with Nephropathy, Aneurysms and Cramps

Missense mutations in the COL4A1 gene have been described in three families displaying an autosomal dominant hereditary angiopathy associated with nephropathy, aneurysms and muscle cramps (HANAC syndrome) [30]. The reported mutations affect three highly conserved glycine residues in the collagenous domain of the α1(IV) chain. Retinal arteriolar tortuosity and retinal hemorrhages were common to affected individuals in all three families, as were intracranial aneurysms, leukoencephalopathy and elevated creatine kinase levels. In two families, affected individuals had muscle cramps. Renal findings in affected individuals included microscopic and gross hematuria in one family, mild renal insufficiency in two families and renal cysts in all three families.

Renal biopsy in affected individuals with hematuria showed no abnormalities of GBM structure or type IV collagen expression. However, basement membranes of Bowman's capsules, tubules and interstitial capillaries exhibited irregular thickening, splitting into multiple layers and focal interruptions.

COL4A1 mutations produce heterogeneous phenotypes with variable renal penetrance. Indeed, childhood porencephaly, adult stroke and retinal tortuosity without clear evidence of renal involvement have been described in several reports of COL4A1 mutations [31–33]. Combined with the findings of normal GBM ultrastructure in HANAC and mouse *Col4a1* and *Col4a2* mutants (see below), this variability in renal penetrance suggests that, when intact, the GBM α3α4α5 network can compensate for defects in the α1α1α2 network.

The involvement of COL4A1 mutations in human disease was first suggested by studies of a mouse mutant in which focal disruptions of vascular basement membranes were associated with porencephaly [34]. The vascular defects were shown to result from a semi-dominant mutation in *Col4a1,* and COL4A1 mutations were found in human families with dominant transmission of porencephaly. Affected mice do not exhibit ultrastructural abnormalities of GBM [34]. Several additional *Col4A1* and *Col4a2* mutant strains have emerged from large-scale mutagenesis screens for eye defects. *Bru* mice harboring a *Col4a1* missense mutation have normal GBM ultrastructure, but thickening and splitting of Bowman's capsule BM [35]. Several *Col4a1* and *Col4a2* mutant strains have mild proteinuria without demonstrable pathological abnormalities of the kidneys [36]. Targeted deletion at the *Col4a1-Col4a2* locus is homozygous-lethal, with embryogenesis proceeding through early organ development, but terminating with widespread evidence of basement membrane failure [37]. Heterozygous-null mutants appear normal, further suggesting that dominant negative effects, and not haploinsufficiency, govern the phenotypes of induced missense mutations.

Renal Disease Due to Laminin Mutations: Pierson Syndrome

The association of congenital nephrotic syndrome with eye abnormalities in siblings was first described by Pierson in 1963 [38]. Affected infants died within the first year of life with end-stage renal disease. Subsequent reports of additional patients described variable abnormalities on fetal ultrasound, including kidney enlargement and hyperechogenicity, oligohydramnios, placental enlargement and/or pulmonary hypoplasia [39, 40]. Infants surviving to term exhibited congenital nephrotic syndrome, with renal biopsy findings of mesangial sclerosis and diffuse GBM abnormalities, accompanied by ocular globe enlargement (buphthalmos) with reduction in the size and reactivity of pupils (microcoria). Additional ocular findings in some patients included cataract, posterior rupture of the lens capsule and retinal abnormalities. Some affected children were also found to have muscular hypotonia, central nervous system hemangioma and genital abnormalities [41, 42].

The gene for Pierson syndrome has been mapped to chromosome 3p14-p22 and identified as LAMB2, the gene encoding the laminin β2 chain [39]. Subsequent reports indicate that LAMB2 mutations may be associated with isolated congenital nephrotic syndrome and with mild variants of Pierson syndrome [43–46]. Absence of the laminin β2 chain in transgenic mice results in massive proteinuria along with retinal and neuromuscular abnormalities [47]. In these mice, proteinuria precedes the appearance of podocyte abnormalities, suggesting that the GBM contributes to the barrier function of the glomerular capillary wall.

Renal Disease Due to Other Mutations

Nail-Patella Syndrome

Nail-patella syndrome (NPS) is a rare autosomal dominant disorder characterized by dystrophic nails, hypoplasia or absence of the patellae, dysplasia of the elbows and iliac horns, and renal disease [48, 49]. The severity of renal involvement, which occurs in 30–40% of patients, determines prognosis.

Targeted disruption of the LIM homeodomain transcription factor gene Lmx1b in mice resulted in hypoplastic nails and absent patellae as well as renal disease, suggesting a possible locus for human NPS [50]. LMX1B, the human homologue of Lmx1b, and the NPS locus were found to map to chromosome 9q34, and heterozygous mutations in LMX1B were identified in NPS patients [51]. The results of in vitro experiments in which the transcriptional effects of mixing wild-type and mutant LMX1B protein are measured, along with the variety of observed mutation types, suggest that the NPS phenotype results from haploinsufficiency rather than a dominant-negative mechanism [52]. Mutations in the homeodomain of LMX1B protein and female gender may confer an increased risk of renal involvement [53].

LMX1B protein is specifically expressed in glomerular podocytes, beginning at the S-shaped body stage of glomerular development [50]. Lmx1b-null mice exhibit marked reduction in GBM expression of the α3 and α4 chains of type IV collagen, and the slit diaphragm proteins podocin and CD2AP [54–56]. However, changes in expression of these putative targets of LMX1B were not observed in kidneys of patients with NPS nephropathy [57], perhaps because patients with NPS are heterozygous for LMX1B mutations, leaving the mechanisms for the renal effects of LMX1B mutations uncertain.

Hereditary Endotheliopathy with Retinopathy, Nephropathy and Stroke (HERNS)

HERNS was described as a disorder of recurrent stroke and collected eye abnormalities transmitted as an autosomal dominant trait [58]. The renal manifestations include proteinuria, hematuria and chronic kidney disease. Kidney biopsy findings are marked by glomerular capillary wall thickening with lamellation of the GBM confined to the subendothelial space and evidence of injury to overlying endothelial cells. HERNS has been shown to be allelic with the related disorders cerebroretinal vasculopathy and hereditary retinal vasculopathy, and to result from truncating mutations in the DNA exonuclease *TREX1* on chromosome 3p21 [59]. While not yet reported, animal models are expected to be important in understanding the mechanisms disturbing glomerular and other vascular basement membranes in these disorders.

References

1 Dong L, Chen Y, Lewis M, Hsieh JC, Reing J, Chaillet JR, Howell CY, Melhem M, Inoue S, Kuszak JR, DeGeest K, Chung AE: Neurologic defects and selective disruption of basement membranes in mice lacking entactin-1/nidogen-1. Lab Invest 2002;82:1617–1630.

2 Murshed M, Smyth N, Miosge N, Karolat J, Krieg T, Paulsson M, Nischt R: The absence of nidogen 1 does not affect murine basement membrane formation. Mol Cell Biol 2000;20:7007–7012.

3 Rossi M, Morita H, Sormunen R, Airenne S, Kreivi M, Wang L, Fukai N, Olsen BR, Tryggvason K, Soininen R: Heparan sulfate chains of perlecan are indispensable in the lens capsule but not in the kidney. EMBO J 2003;22:236–245.

4 Schymeinsky J, Nedbal S, Miosge N, Poschl E, Rao C, Beier DR, Skarnes WC, Timpl R, Bader BL: Gene structure and functional analysis of the mouse nidogen-2 gene: nidogen-2 is not essential for basement membrane formation in mice. Mol Cell Biol 2002;22:6820–6830.

5 Zheng K, Thorner PS, Marrano P, Baumal R, McInnes RR: Canine X chromosome-linked hereditary nephritis: a genetic model for human X-linked hereditary nephritis resulting from a single base mutation in the gene encoding the α5 chain of collagentype IV. Proc Natl Acad Sci USA 1994;91:3989–3993.

6 Cox M, Lees GE, Kashtan CE, Murphy KE: Genetic cause of X-linked Alport syndrome in a family of domestic dogs. Mamm Genome 2003;14:396–403.

7 Lees G, Helman RG, Kashtan CE, Michael AF, Homco LD, Millichamp NJ, Camacho ZT, Templeton JW, Ninomiya Y, Sado Y, Kim Y: A new form of X-linked dominant hereditary nephritis in dogs. Am J Vet Res 1999;60:373–383.

8 Davidson AG, Bell RJ, Lees GE, Kashtan CE, Davidson GS, Murphy KE: Genetic cause of autosomal recessive hereditary nephropathy in the English Cocker Spaniel. J Vet Intern Med 2007;21:394–401.

9 Lees G, Helman RG, Kashtan CE, Michael AF, Homco LD, Millichamp NJ, Ninomiya Y, Sado Y, Kim Y: A model of autosomal recessive Alport syndrome in English cocker spaniel dogs. Kidney Int 1998;54:706–719.

10 Miner JJ, Sanes JR: Molecular and functional defects in kidneys of mice lacking collagen α3(IV): implications for Alport syndrome. J Cell Biol 1996;135:1403–1413.

11 Cosgrove D, Meehan DT, Grunkemeyer JA, Kornak JM, Sayers R, Hunter WJ, Samuelson GC: Collagen COL4A3 knockout: a mouse model for autosomal Alport syndrome. Genes Dev 1996;10:2981–2992.

12 Lu W, Phillips CL, Killen PD, Hlaing T, Harrison WR, Elder FFB, Miner JH, Overbeek P, Meisler MH: Insertional mutation of the collagen genes Col4a3 and Col4a4 in a mouse model of Alport syndrome. Genomics 1999;61:113–124.

13 Andrews KL, Mudd JL, Li C, Miner JH: Quantitative trait loci influence renal disease progression in a mouse model of Alport syndrome. Am J Pathol 2002;160:721–730.

14 Rheault MN, Kren SM, Thielen BK, Mesa HA, Crosson JT, Thomas W, Sado Y, Kashtan CE, Segal Y: Mouse model of X-linked Alport syndrome. J Am Soc Nephrol 2004;15:1466–1474.

15 Kang JS, Wang XP, Miner JH, Morello R, Sado Y, Abrahamson DR, Borza DB: Loss of alpha3/alpha4(IV) collagen from the glomerular basement membrane induces a strain-dependent isoform switch to alpha5alpha6(IV) collagen associated with longer renal survival in Col4a3-/- Alport mice. J Am Soc Nephrol 2006;17:1962–1969.

16 Kashtan CE, Kleppel MM, Gubler MC: Immunohistologic findings in Alport syndrome. Contrib Nephrol 1996;117:142–153.

17 Nakanishi K, Yoshikawa N, Iijima K, Kitagawa K, Nakamura H, Ito H, Yoshioka K, Kagawa M, Sado Y: Immunohistochemical study of α1–5 chains of type IV collagen in hereditary nephritis. Kidney Int 1994;46:1413–1421.

18 Peissel B, Geng L, Kalluri R, Kashtan C, Rennke HG, Gallo GR, Yoshioka K, Sun MJ, Hudson BG, Neilson EG, Zhou J: Comparative distribution of the α1(IV), α5(IV) and α6(IV) collagen chains in normal human adult and fetal tissues and in kidneys from X-linked Alport syndrome patients. J Clin Invest 1995;96:1948–1957.

19 Gross O, Schulze-Lohoff E, Koepke ML, Beirowski B, Addicks K, Bloch W, Smyth N, Weber M: Antifibrotic, nephroprotective potential of ACE inhibitor vs AT1 antagonist in a murine model of renal fibrosis. Nephrol Dial Transplant 2004;19:1716–1723.

20 Gross O, Beirowski B, Koepke ML, Kuck J, Reiner M, Addicks K, Smyth N, Schulze-Lohoff E, Weber M: Preemptive ramipril therapy delays renal failure and reduces renal fibrosis in COL4A3-knockout mice with Alport syndrome. Kidney Int 2003;63:438–446.

21 Sayers R, Kalluri R, Rodgers KD, Shield CF, Meehan DT, Cosgrove D: Role for transforming growth factor-beta 1 in Alport renal disease progression. Kidney Int 1999;56:1662–1673.

22 Ninichuk V, Gross O, Reichel C, Kandoga A, Pawar RD, Ciubar R, Segerer S, Belemezova E, Radomska E, Lcukow B, Perez de Lema G, Murphy PM, Gao J, Henger A, Kretzler M, Horuk R, Weber M, Krombach F, Schlondorff D, Anders H: Delayed chemokine receptor 1 blockade prolongs survival in collagen 4A3-deficient mice with Alport disease. J Am Soc Nephrol 2005;16:977–985.

23 Zeisberg M, Bottiglio C, Kumar N, Maeshima Y, Strutz F, Muller GA, Kalluri R: Bone morphogenic protein-7 inhibits progression of chronic renal fibrosis associated with two genetic mouse models. Am J Physiol Renal Physiol 2003;285:F1060–F1067.

24 Zeisberg M, Khurana M, Rao VH, Cosgrove D, Rougier JP, Werner MC, Shield CF, 3rd, Werb Z, Kalluri R: Stage-specific action of matrix metalloproteinases influences progressive hereditary kidney disease. PLoS Med 2006;3:e100.

25 Koepke ML, Weber M, Schulze-Lohoff E, Beirowski B, Segerer S, Gross O: Nephroprotective effect of the HMG-CoA-reductase inhibitor cerivastatin in a mouse model of progressive renal fibrosis in Alport syndrome. Nephrol Dial Transplant 2007;22:1062–1069.
26 Sugimoto H, Mundel TM, Sund M, Xie L, Cosgrove D, Kalluri R: Bone-marrow-derived stem cells repair basement membrane collagen defects and reverse genetic kidney disease. Proc Natl Acad Sci USA 2006;103:7321–7326.
27 Prodromidi EI, Poulsom R, Jeffery R, Roufosse CA, Pollard PJ, Pusey CD, Cook HT: Bone marrow-derived cells contribute to podocyte regeneration and amelioration of renal disease in a mouse model of Alport syndrome. Stem Cells 2006;24:2448–2455.
28 Floege J, Kunter U, Weber M, Gross O: Bone marrow transplantation rescues Alport mice. Nephrol Dial Transplant 2006;21:2721–2723.
29 Katayama K, Kawano M, Naito I, Ishikawa H, Sado Y, Asakawa N, Murata T, Oosugi K, Kiyohara M, Ishikawa E, Ito M, Nomura S: Irradiation prolongs survival of Alport mice. J Am Soc Nephrol 2008;19:1692–1700.
30 Plaisier E, Gribouval O, Alamowitch S, Mougenot B, Prost C, Verpont MC, Marro B, Desmettre T, Cohen SY, Roullet E, Dracon M, Fardeau M, Van Agtmael T, Kerjaschki D, Antignac C, Ronco P: COL4A1 mutations and hereditary angiopathy, nephropathy, aneurysms, and muscle cramps. N Engl J Med 2007;357:2687–2695.
31 van der Knaap MS, Smit LM, Barkhof F, Pijnenburg YA, Zweegman S, Niessen HW, Imhof S, Heutink P: Neonatal porencephaly and adult stroke related to mutations in collagen IV A1. Ann Neurol 2006;59:504–511.
32 Breedveld G, de Coo IF, Lequin MH, Arts WF, Heutink P, Gould DB, John SW, Oostra B, Mancini GM: Novel mutations in three families confirm a major role of COL4A1 in hereditary porencephaly. J Med Genet 2006;43:490–495.
33 Vahedi K, Kubis N, Boukobza M, Arnoult M, Massin P, Tournier-Lasserve E, Bousser MG: COL4A1 mutation in a patient with sporadic, recurrent intracerebral hemorrhage. Stroke 2007;38:1461–1464.
34 Gould DB, Phalan FC, Breedveld GJ, van Mil SE, Smith RS, Schimenti JC, Aguglia U, van der Knaap MS, Heutink P, John SW: Mutations in Col4a1 cause perinatal cerebral hemorrhage and porencephaly. Science 2005;308:1167–1171.
35 Van Agtmael T, Schlotzer-Schrehardt U, McKie L, Brownstein DG, Lee AW, Cross SH, Sado Y, Mullins JJ, Poschl E, Jackson IJ: Dominant mutations of Col4a1 result in basement membrane defects which lead to anterior segment dysgenesis and glomerulopathy. Hum Mol Genet 2005;14:3161–3168.
36 Favor J, Gloeckner CJ, Janik D, Klempt M, Neuhauser-Klaus A, Pretsch W, Schmahl W, Quintanilla-Fend L: Type IV procollagen missense mutations associated with defects of the eye, vascular stability, the brain, kidney function and embryonic or postnatal viability in the mouse, *Mus musculus*: an extension of the Col4a1 allelic series and the identification of the first two Col4a2 mutant alleles. Genetics 2007;175:725–736.
37 Poschl E, Schlotzer-Schrehardt U, Brachvogel B, Saito K, Ninomiya Y, Mayer U: Collagen IV is essential for basement membrane stability but dispensable for initiation of its assembly during early development. Development 2004;131:1619–1628.
38 Pierson M, Cordier J, Hervouuet F, Rauber G: An unusual congenital and familial congenital malformative combination involving the eye and kidney (in French). J Genet Hum 1963;12:184–213.
39 Zenker M, Aigner T, Wendler O, Tralau T, Muntefering H, Fenski R, Pitz S, Schumacher V, Royer-Pokora B, Wuhl E, Cochat P, Bouvier R, Kraus C, Mark K, Madlon H, Dotsch J, Rascher W, Maruniak-Chudek I, Lennert T, Neumann LM, Reis A: Human laminin beta2 deficiency causes congenital nephrosis with mesangial sclerosis and distinct eye abnormalities. Hum Mol Genet 2004;13:2625–2632.
40 Hofstaetter C, Neumann I, Lennert T, Dudenhausen JW: Prenatal diagnosis of diffuse mesangial glomerulosclerosis by ultrasonography: a longitudinal study of a case in an affected family. Fetal Diagn Ther 1996;11:126–131.
41 Glastre C, Cochat P, Bouvier R, Colon S, Cottin X, Giffon D, Wright C, Dijoud F, David L: Familial infantile syndrome with ocular abnormalities. Pediatr Nephrol 1990;4:340–342.
42 Swietlinski J, Maruniak-Chudek I, Niemir ZI, Wozniak A, Wilinska M, Zacharzewska J: A case of atypical congenital nephrotic syndrome. Pediatr Nephrol 2004;19:349–352.
43 Choi HJ, Lee BH, Kang JH, Jeong HJ, Moon KC, Ha IS, Yu YS, Matejas V, Zenker M, Choi Y, Cheong HI: Variable phenotype of Pierson syndrome. Pediatr Nephrol 2008.
44 Kagan M, Cohen AH, Matejas V, Vlangos C, Zenker M: A milder variant of Pierson syndrome. Pediatr Nephrol 2008;23:323–327.
45 Wuhl E, Kogan J, Zurowska A, Matejas V, Vandevoorde RG, Aigner T, Wendler O, Lesniewska I, Bouvier R, Reis A, Weis J, Cochat P, Zenker M: Neurodevelopmental deficits in Pierson (microcoria-congenital nephrosis) syndrome. Am J Med Genet A 2007;143:311–319.
46 Matejas V, Al-Gazali L, Amirlak I, Zenker M: A syndrome comprising childhood-onset glomerular kidney disease and ocular abnormalities with progressive loss of vision is caused by mutated LAMB2. Nephrol Dial Transplant 2006;21:3283–3286.
47 Jarad G, Cunningham J, Shaw AS, Miner JH: Proteinuria precedes podocyte abnormalities inLamb2-/- mice, implicating the glomerular basement membrane as an albumin barrier. J Clin Invest 2006;116:2272–2279.
48 Hawkins CF, Smith OE: Renal dysplasia in a family with multiple hereditary abnormalities including iliac horns. Lancet 1950;I:803–808.
49 Sweeney E, Fryer A, Mountford R, Green A, McIntosh I: Nail patella syndrome: a review of the phenotype aided by developmental biology. J Med Genet 2003;40:153–162.
50 Chen H, Lun Y, Ovchinnikov D, Kokubo H, Oberg KC, Pepicelli CV, Gan L, Lee B, Johnson RL: Limb and kidney defects in Lmx1b mutant mice suggest an involvement of LMX1B in human nail patella syndrome. Nat Genet 1998;19:51–55.

51 Dreyer SD, Zhou G, Baldini A, Winterpacht A, Zabel B, Cole W, Johnson RL, Lee B: Mutations in LMX1B cause abnormal skeletal patterning and renal dysplasia in nail patella syndrome. Nat Genet 1998;19:47–50.
52 Dreyer SD, Morello R, German MS, Zabel B, Winterpacht A, Lunstrum GP, Horton WA, Oberg KC, Lee B: LMX1B transactivation and expression in nail-patella syndrome. Hum Mol Genet 2000;9:1067–1074.
53 Bongers EM, Huysmans FT, Levtchenko E, de Rooy JW, Blickman JG, Admiraal RJ, Huygen PL, Cruysberg JR, Toolens PA, Prins JB, Krabbe PF, Borm GF, Schoots J, van Bokhoven H, van Remortele AM, Hoefsloot LH, van Kampen A, Knoers NV: Genotype-phenotype studies in nail-patella syndrome show that LMX1B mutation location is involved in the risk of developing nephropathy. Eur J Hum Genet 2005;13:935–946.
54 Morello R, Zhou G, Dreyer SD, Harvey SJ, Ninomiya Y, Thorner PS, Miner JH, Cole W, Winterpacht A, Zabel B, Oberg KC, Lee B: Regulation of glomerular basement membrane collagen expression by LMX1B contributes to renal disease in nail patella syndrome. Nat Genet 2001;27:205–208.
55 Miner JH, Morello R, Andrews KL, Li C, Antignac C, Shaw AS, Lee B: Transcriptional induction of slit diaphragm genes by Lmx1b is required in podocyte differentiation. J Clin Invest 2002;109:1065–1072.
56 Rohr C, Prestel J, Heidet L, Hosser H, Kriz W, Johnson RL, Antignac C, Witzgall R: The LIM-homeodomain transcription factor Lmx1b plays a crucial role in podocytes. J Clin Invest 2002;109:1073–1082.
57 Heidet L, Bongers EM, Sich M, Zhang SY, Loirat C, Meyrier A, Broyer M, Landthaler G, Faller B, Sado Y, Knoers NV, Gubler MC: In vivo expression of putative LMX1B targets in nail-patella syndrome kidneys. Am J Pathol 2003;163:145–155.
58 Jen J, Cohen AH, Yue Q, Stout JT, Vinters HV, Nelson S, Baloh RW: Hereditary endotheliopathy with retinopathy, nephropathy, and stroke (HERNS). Neurology 1997;49:1322–1330.
59 Richards A, van den Maagdenberg AM, Jen JC, Kavanagh D, Bertram P, Spitzer D, Liszewski MK, Barilla-Labarca ML, Terwindt GM, Kasai Y, McLellan M, Grand MG, Vanmolkot KR, de Vries B, Wan J, Kane MJ, Mamsa H, Schäfer R, Stam AH, Haan J, de Jong PT, Storimans CW, van Schooneveld MJ, Oosterhuis JA, Gschwendter A, Dichgans M, Kotschet KE, Hodgkinson S, Hardy TA, Delatycki MB, Hajj-Ali RA, Kothari PH, Nelson SF, Frants RR, Baloh RW, Ferrari MD, Atkinson JP: C-terminal truncations in human 3′-5′ DNA exonuclease TREX1 cause autosomal dominant retinal vasculopathy with cerebral leukodystrophy. Nat Genet 2007;39:1068–1070.

Clifford E. Kashtan, MD
University of Minnesota Medical School, Department of Pediatrics
Division of Pediatric Nephrology
420 Delaware Street SE
MMC 491
Minneapolis, MN 55455 (USA)
Tel. +1 612 626 2922, Fax +1 612 626 2791, E-Mail kasht001@umn.edu

Herrera GA (ed): Experimental Models for Renal Diseases: Pathogenesis and Diagnosis.
Contrib Nephrol. Basel, Karger, 2011, vol 169, pp 183–197

Experimental Models of Lupus Nephritis

Joseph P. Grande

Mayo Clinic, Rochester, Minn., USA

Abstract

A number of murine models recapitulate many of the features of systemic lupus. Spontaneous models of lupus have provided the basis for genetic linkage studies to define loci that confer an increased risk of nephritis in susceptible mice. Functional relevance of genes within susceptibility loci have been verified through the use of transgenic mice bearing targeted deletions or overexpression of candidate genes. Critical gene targets are involved in the production of autoantigens, stimulation of B cells to produce autoantibodies, stimulation of T cells to provide B cell help and cytokine-mediated damage following tissue deposition of immune complexes. In this overview, features of the commonly employed models of human lupus nephritis are reviewed. Strategies to identify candidate genes involved in the initiation and progression of lupus nephritis in these models are discussed. Finally, key studies to validate target genes as pathophysiologically relevant mediators of lupus nephritis and studies to identify potential targets for therapeutic intervention are summarized.

Systemic lupus is a multisystem disease characterized by the development of autoantibodies against a number of cell surface antigens. Although systemic lupus has a waxing and waning clinical course, end-organ damage to the kidney and cardiovascular system are leading causes of morbidity and mortality. Murine models of lupus have played a critical role in helping clinicians understand the pathophysiology of lupus nephritis and have provided the basis for better, more targeted therapeutic interventions. This chapter provides an overview of the most commonly-used models of lupus nephritis and how they have been used to identify specific pathogenic mechanisms, develop prognostic and diagnostic markers and provide the basis for new therapeutic interventions to prevent the development and/or progression of lupus nephritis.

What Is Lupus Nephritis?

Lupus is a chronic remitting and relapsing multisystem autoimmune disease that can affect skin, kidneys, lungs, nervous system, serous membranes and other organs [1]. The prevalence of lupus ranges from 40 cases per 100,000 persons among North Europeans to more than 200 per 100,000 persons among blacks [2]. A wide variety of autoantibodies are characteristically produced in lupus, and can recognize diverse nuclear and cytoplasmic components of the cell in a fashion that is neither organ- nor tissue-specific. Antibodies to double-stranded DNA and the Smith antigen are diagnostic of lupus (the Smith antibody is directed against small nuclear ribonucleoprotein particles and core proteins).

Table 1. Pathogenic mechanisms underlying the development of lupus nephritis

Development of autoantigens
Development of autoantibodies by B cells
T helper cell-mediated expansion of B cell repertoire
Autoantibodies and immune complexes induce renal damage through cytokine production

A variety of drugs can induce lupus nephritis, including hydralazine, procainamide and quinidine. Skin and joint manifestations of drug-induced lupus are common, whereas renal and neurologic manifestations are rare. Genetic, immunologic, hormonal, infectious and environmental factors have been associated with the development of lupus. Glomerulonephritis is seen in 50–80% of patients with lupus, and end-stage renal disease develops in 10–20% of patients. Renal involvement is characterized by glomerular immunoglobulin, and complement deposition and the recruitment of mononuclear inflammatory cells contribute to damage, inflammation and loss of renal function. A variety of histopathologic lesions are observed in patients with lupus nephritis [3, 4], ranging from minimal alterations, mesangial proliferative glomerulonephritis, focal-segmental or diffuse proliferative glomerulonephritis with capillary wall and immune complex deposition, membranous nephropathy, or advanced sclerosing glomerulonephritis. Immune complex deposition leads to leukocyte influx and tissue inflammation [5]. Acute lupus glomerulonephritis is characterized by proliferation of glomerular cells, leukocyte exudation, karyorrhexis, fibrinoid necrosis, crescent formation, hyaline deposits and interstitial inflammation. Activation of both circulating leukocytes and parenchymal cells, generation of cytokines and reactive oxygen species, activation of complement, and the coagulation/fibrinolytic cascades are characteristic features of acute lupus nephritis.

In all too many cases, patients with lupus nephritis develop chronic histopathologic alterations, characterized by extracellular matrix deposition and loss of tissue architecture. There have been intensive research efforts directed towards understanding the basic mechanisms of how autoantigens are generated, which in turn leads to the production of a number of antibodies by B cells. These studies, largely using murine models of lupus glomerulonephritis, have led to the development of a number of potential therapeutic strategies.

A Brief Overview of the Pathogenesis of Lupus Nephritis

Key pathogenic factors responsible for the development of lupus nephritis are summarized in table 1 [6, 7].

A characteristic feature of lupus is the development of multiple autoantibodies that have specificity towards DNA. Nephritogenic antigens are likely to be DNA complexed with chromatin or nucleosomes [8, 9]. DNA containing apoptotic bodies are normally cleared by macrophages without triggering an inflammatory or specific immunologic response. However, situations characterized by defects in apoptosis or reduced clearance of apoptotic bodies are associated with the development of autoantibodies to nucleosomes and DNA. Autoantibody production is also facilitated by polyclonal B cell hyperactivity, which has been observed in several models of lupus nephritis, including the MRL/lpr and BXSB mouse [10–12].

Once formed, autoantibodies stimulate production of CD4-positive helper T cells to assist in the pathogenic production of additional anti-DNA antibodies by B cells [13]. The anti-DNA antibodies localize in the kidney by binding glomerular cell surfaces, matrix antigen or by in situ complex formation of autoantibodies and circulating autoantigens that deposit within glomeruli [14, 15]. Nucleosomes bind to glomerular structures that target circulating anti-DNA antibodies to the kidney [8, 16]. Nephritogenic antibodies isolated from patients with systemic lupus tend to be highly cross-reactive with multiple autoantigens, including DNA basement membrane complements, proteins and phospholipids [17]. Localization of immune complexes in the kidney depends upon many factors, including antibody class, charge, idiotype and antigenic specificity [18]. The severity of lupus is dictated in part by distribution of immune complexes [19]; subepithelial deposits produce epithelial cell injury and proteinuria with little leukocyte infiltration or proliferation. Large subendothelial deposits promote proliferative and necrotizing lesions with inflammation. Once deposited in the kidney, immune complexes activate inflammatory cascades leading to complement activation, coagulation/fibrinolysis, chemotactic factor production, cytokine generation and reactive oxygen species generation [20]. Both circulating cells and parenchymal cells are activated upon exposure to immune complexes and contribute to inflammation. Circulating anti-DNA antibodies amplify this response.

There is evidence that serum containing immune complexes can promote inflammatory responses. For example, serum from systemic lupus erythematosus (SLE) patients increases neutrophil adherence to cultured endothelial cells [21]. Anti-double-stranded DNA antibodies induce endothelial cell expression of IL-8 and TGF-β [22]. Anti-double-stranded DNA antibodies isolated from patients with lupus bind to cultured mesangial cells and trigger apoptosis [23].

What Are the Most Relevant Animal Models of Lupus Nephritis?

Animal models of lupus nephritis have been employed to define basic pathways leading from autoantigen production to the development of autoantibodies and to the progression of acute renal injury to failure. These models have provided mechanistic insights related to the pathogenesis of lupus nephritis. A summary of the most commonly employed models of lupus is provided in table 2.

Spontaneous Models of Lupus Nephritis

Several inbred strains of mice have been found to spontaneously develop lupus nephritis. In general, these animals have a number of genetic susceptibility loci which have provided the basis for linkage studies to define critical genes responsible for the development and progression of lupus nephritis.

NZB Hybrids

The NZB/NZW model was among the first murine models described for lupus nephritis. NZB/NZW strain, also termed the BW strain, was formed as an F1 cross between New Zealand Black Mice (NZB) and New Zealand White Mice (NZW) [24, 25]. There are contributing autoimmunity genes from both parent strains. The NZB parent strain develops antithymocyte and anti-red blood cell antibodies, the latter of which leads to severe hemolytic anemia that is fatal by 15–18 months [26]. NZB mice have several immunologic defects, including intrinsic B cell proliferation and immunoglobulin production, loss of suppressor cell function, and defective apoptosis of resting B cells [reviewed in 27].

The NZW parent strain also has an autoimmune background with a low level of autoantibodies, frequent nephritis and a shortened lifespan. NZW mice develop mesangial glomerulonephritis; however, this rarely leads to renal failure [28]. NZW mice also develop a low level of autoantibodies [29].

The NZB/NZW cross develops high-titer autoantibody levels by 4–5 months of age. [26]. NZB/NZW mice develop immune complex-mediated glomerulonephritis that resembles human diffuse proliferative lupus nephritis

Table 2. Murine models of lupus nephritis

Spontaneous models	(NZWxNZB)F1MRL/lpr BXSB
Transgenic models	Bcl-2 IFN-γ FLICE inhibitory protein CD19 BAFF BCMA (BAFF receptor) IEX-1 CSF-1
Gene-targeted deletions	DNAseI PCD-1 PTEN Lyn CD22 TGF-β1 Complement (C1q, C2) CR2 PD-1 SHP-1 Coronin 1A TSAd CCR2 (protective) CXCR3 (protective) Fc-γ receptor (protective) Fn14 (protective)
Miscellaneous	GVHD mice

[24, 30–34]. Female NZB/NZW mice are more severely affected than males. There is a 50% mortality rate at 8 months in female mice and at 15 months in male mice.

There are many genetic alterations in both parental strains that independently or possibly additively contribute to the risk of lupus nephritis. Susceptibility genes interact in a complex manner to produce the full-blown lupus glomerulonephritis phenotype. In general, susceptibility genes identified to date contain polymorphic alleles that confer differences in function rather than true mutations. However, studies employing transgenic animals bearing targeted deletion or overexpression of target genes have been used to modulate the risk of lupus glomerulonephritis in NZB/NZW mice.

MRL/lpr Mice

The MRL/lpr mouse is a generally accepted model of lupus [35]. The MRL/lpr mouse strain was derived from cross breeding LG/J mice with the AKR/J, C57BL/6 and C3HDi mouse strains [36]. These mice possess the lpr (lymphoproliferation) gene, which encodes a defective form of Fas, a cell surface protein that mediates rapid apoptosis when cross-linked by antibodies [37, 38]. The defect in Fas is due to insertion of an endogenous retrovirus in the Fas (CD95) or lpr gene [39], leading to inactivation of the Fas gene, and impaired Fas-mediated apoptosis [39, 40]. The Fas receptor is required for regulation of B and T cell lifespan. MRL lpr/lpr mice develop an autoimmune syndrome resembling that found in human SLE and other autoimmune diseases

[41]. A similar phenotype is observed in mice with generalized lymphoproliferative disease. These mice have a mutation on the extracellular domain of the Fas ligand, resulting in lack of binding to Fas.

MRL/lpr mice develop massive adenopathy due to accumulation of T cells, polyclonal B cell expansion and numerous autoantibodies. Most animals develop albuminuria and loss of renal function between 3 and 6 months of age. There is a 50% mortality rate at 6 months of age. Mice develop a diffuse proliferative glomerulonephritis. Immunofluorescence studies have revealed extensive glomerular immune complex deposition along capillary walls and mesangial regions. Although IgG2a is the predominant serum immunoglobulin, the primary IgG eluted from kidneys of MRL/lpr mice is IgG3 [42]. Microarray analysis of glomerular gene expression in MRL/lpr mice has shown a number of genes to be upregulated [43], including complement components, adhesion molecules, chemokines and their receptors, and molecules related to antigen presentation.

Although the MRL/lpr mouse provides a robust model of lupus glomerulonephritis, its direct relevance to human lupus is not clear [44]. Fas mutations in humans are associated with a rare fatal autoimmune lymphoproliferative syndrome [45, 46] rather than classical human systemic lupus.

BXSB Mice

BXSB mice were produced by crossing a C57GL/6 (B6) female with an SB/Le male. Male mice develop more severe disease, with a 50% mortality rate at 6 months versus 50% mortality at 20 months for females. Hormones are not responsible for the more severe phenotype in males, as castration does not affect the disease course. The Yaa gene located on the BXSB Y chromosome appears to be responsible for the lupus phenotype. The mutation involves a translocation from the telemetric end of the X chromosome containing the TLR7 gene onto the Y chromosome [47, 48], leading to overexpression of TLR7. As discussed below, Toll-like receptors (TLRs) play a major role in innate immunity directed against a variety of compounds, including DNA.

Graft-vs.-Host Disease Models

In this model, nonautoimmune BALB/c mice are neonatally injected with MHC class II incompatible splenocytes. These animals develop polyclonal B cell activation and autoimmune features with features of graft-vs.-host disease. Renal lesions in these animals include a membranous nephropathy with segmental sclerosing features [49].

Genetic Basis of SLE

The spontaneous models of SLE have provided insights related to genes that are responsible for the development and progression of lupus nephritis [50]. The spontaneous models have polygenetic factors that contribute to the SLE phenotype. For example, in the (NZBxNZW)F1 mouse strain, at least 20–30 loci are associated with the development of lupus [34, 50–52].

Congenic dissection has been used to analyze individual susceptibility loci on a resistant mouse background. Mapping studies indicate that loci on murine chromosomes 1 (Sle1), 4 (Sle2), 7 (Sle3) and 17 (Sle4) confer disease susceptibility in the NZM2410 lupus strain. For example, when NZM2410 lupus-prone mice possessing the Sle1 locus on chromosome 1 are crossed with the lupus-resistant B6 strain (B6.Sle1), the mice develop anti-H2A/H2B/DNA antibodies, with spontaneous activation of B and T cells. These mice do not develop significant nephritis [53]. Lupus-prone mice with the Sle2 locus on chromosome 4 crossed with lupus-resistant B6 mice have overproduction of polyclonal IgM, with increased numbers of splenic B cells. They do not develop antinuclear antibodies (ANA). However, T cells show increased expression of activation markers. These animals also do not develop significant nephritis [54]. Lupus-prone mice with the Sle3 locus on chromosome 7 crossed with lupus-resistant B6 mice are characterized by polyclonal B and T cell activation, increased

CD4/8 ratio and decreased activation-induced cell death of T cells. These mice develop moderate nephritis [55].

Analysis of susceptibility loci has led to the identification of several gene products involved in the clearance of apoptotic bodies and regulation of B cell function [56].

For example, the complement receptor 2 (CR2) gene is a possible candidate for Sle1c. Deletion of the CR2 gene in B6 mice, which are not prone to develop lupus, has only limited effects on autoantibody production. However, autoantibody production is higher in B6/lpr Cr2$^{-/-}$ mice than B6/lpr controls. These studies suggest that the CR2 may be linked to production of autoantibodies by modifying the effect of other lupus susceptibility genes [57].

The Sle1 locus has two Fc-γ receptors that are linked to the development of lupus nephritis. Fc-γIIa is a major receptor for IgG2. Mutations in the Fc-γIIa receptor lead to autoantibody production due to lack of clearance of dsDNA-IgG2 complexes. The lupus-prone NZB, BXSB and MRL strains share the NZB-type defective allele of the gene encoding Fc-γRIIB, an Fc receptor which negatively regulates B cell receptor function [58, 59].

PDCD1 has been shown to reside within the Sleb2 locus. PDCD1 is an inhibitory transmembrane receptor that downregulates lymphocyte proliferation and cytokine secretion. Loss of PDCD1 function leads to breakdown of peripheral tolerance and lymphocyte hyperactivity. PDCD1 knockout mice develop lupus nephritis and arthritis [56, 60].

Pathogenesis of Lupus Nephritis: Lessons from Experimental Studies and Transgenic Animals

Whereas the spontaneous models of lupus nephritis provided the basis for genetic studies to identify potential mechanistic targets, transgenic animals have been used to define the relevance of specific signaling pathway(s) to the development of lupus nephritis. A general approach has been to isolate the target gene of interest and put it into a mouse that is not otherwise susceptible to developing lupus nephritis. These studies have shown that defects in apoptosis, lymphocyte signaling, T cell selection, cell-to-cell adhesion and antigen presentation, macrophage function, cytokine expression, and negative regulation of lymphocyte activation are related to the pathogenesis of lupus nephritis [56]. In this section, a brief overview of studies to define the nature of autoantibodies that form in lupus nephritis is provided, followed by a discussion of transgenic models that have been employed to define critical signaling pathways involved in the pathogenesis of lupus nephritis.

Development of anti-DNA antibodies is a characteristic feature of human SLE, and is shared by virtually all murine models of lupus nephritis. The pathogenicity of anti-dsDNA antibodies has been demonstrated by administration of anti-dsDNA antibodies to normal or preautoimmune mice [61]. Many of the autoantigens in SLE can be found on the surface of apoptotic cells, indicating that apoptotic cells may be an important source of autoantigen in patients who develop lupus. However, direct injection of apoptotic cells does not produce lupus nephritis – animals develop IgM ANA, anticardiolipin and anti-ssDNA without developing anti-dsDNA or renal damage [62]. Immunization of mice with dendritic cells leads to the production of anti-IgG anti-dsDNA and glomerular immune complex deposits. However, lupus nephritis does not develop in these animals [63]. This may be due to the fact that apoptotic cells suppress the immune response through release of anti-inflammatory cytokines from phagocytic cells [64]. Full-blown lupus nephritis may be established in C57BL/6 or BALB/c mice immunized with human β2 glycoprotein I, a protein that binds apoptotic cells and is one of the first autoantigens that develops in humans with systemic lupus and treated with lipopolysaccharide, a potent proinflammatory cytokine that activates immune cells. These mice develop class II (mesangial proliferative) and class III (focal proliferative) lupus glomerulonephritis [65, 66]. Recent studies have indicated that exposure to chromatin dictates the nephritogenic potential of anti-dsDNA antibodies in (NZWxNZB)F1 mice [67, 68]. Finally, the methylation status of DNA has been shown to be important in induction of lupus nephritis. Hypomethylated DNA induces lupus-like autoimmune disease even in BALB/c mice not prone to lupus [69]. Of note, apoptotic DNA has lower methylation levels than necrotic DNA, which is not effective in inducing lupus nephritis. Methylation of apoptotic DNA decreases its nephritogenic potential, whereas demethylation of necrotic DNA increases its nephritogenic potential.

TLRs play an important role in clearance of nucleoproteins from apoptotic cells. TLRs are a family of receptors that recognize molecular motifs commonly found in microbial pathogens. Ligation of TLRs promotes a robust inflammatory response through induction of cytokine expression. Apoptotic blebs are consumed by dendritic cells and bind to TLR7 and TLR9 through exposed RNA structures or unmethylated CpG motifs, respectively [68]. Signaling through TLR7 and TLR9 promotes dendritic cell maturation, upregulation of co-stimulatory molecules and processing and presentation of antigens derived from chromatin [70]. These interactions stimulate the production of autoantibody-producing B cells. TLR7 and TLR9 are localized in the endosomal compartment of the cell, so their activation requires internalization of ligands by other receptors which recognize chromatin. CD32 (FC-γRIIa) may play such a role in transferring anti-dsDNA antibody-containing immune complexes from TLR9 to lysosomes. This interaction induces dendritic cell activation and IFN-α production. IFN-α drives maturation of monocytes into dendritic cells. The functional relevance of TLR7 and TLR9 was established in studies of lupus-prone mice (NZBxNXW)F1 mice treated with a dual inhibitor of TLR7 and TLR9. Autoantibody production was decreased and renal survival was improved in NZBxNZW F1 mice treated with a dual TLR7/TLR9 inhibitor [71]. Glomerulonephritis in lupus-prone mice was less severe in mice lacking TLR7 [72] or the adaptor molecule MyD88, which is involved in signaling through all TLRs except TLR3 [73]. On the other hand, activation of TLR9 with CpG oligonucleotide induces progression of renal disease in MRL/lpr mice [74]. Activation of TLR7 also exacerbates lupus nephritis. As previously discussed, translocation of TLR7 to the Y chromosome is largely responsible for the lupus nephritis phenotype in BXSB mice [75].

DNAseI has been identified as a lupus nephritis susceptibility gene. Patients with SLE have reduced serum DNAse activity. A nonsense mutation in the DNAseI gene has been reported in two patients with SLE [76]. DNAseI knockout mice have increased production of autoantibodies directed against a variety of nuclear antigens and develop glomerulonephritis at 6–8 months of age. DNAse deficiency is associated with impaired clearance of autoantigenic chromatin, which is liberated during cell death [77]. Renal lesions of DNAseI-deficient mice is similar to that of (NZBxNZW)F1 mice but tends to be less severe [78]. DNAseI-deficient mice develop predominantly anti-nucleosomal antibodies, whereas the range of antibody specificity in (NZBxNZW) F1 mice is broader [79, 80]. The fundamental importance of DNAseI deficiency is underscored by studies which show that progression of murine lupus nephritis is linked to acquired renal DNAseI deficiency and not directly to upregulated apoptosis [81]. In (NZBxNZW)F1 mice, anti-dsDNA antibodies appeared in serum and mesangial immune complexes were detected in the kidney before apoptosis was activated. Transformation of a mesangial lesion to a diffuse proliferative lesion was associated with increased caspase 3 activity and a dramatic decrease in DNAseI transcription [81].

A major cause of ANA in SLE is impaired clearance of cell debris [82, 83]. Further evidence for the importance of apoptotic DNA as an autoantigen that promotes lupus nephritis stems from observations that defective apoptosis is associated with the development of SLE. Transgenic mice that overexpress the cell survival protein bcl-2 gene develop lymphocytic infiltrates in arteries and solid organs, autoantibodies and immune complex-mediated renal disease

[84]. The autoantibodies appear to be generated through inappropriate survival of autoreactive B cell clones that are normally lost through negative selection and apoptosis [84, 85].

As discussed above, the MRL/lpr mice have a defect in Fas, a key apoptotic effector. Mice deficient for PTEN, or which overexpress, produce a phenotype similar to that of mice with Fas FLICE inhibitory protein deficiency [86, 87]. Interactions between TWEAK (TNF-like weak inducer of apoptosis) and its ligand Fn14 exacerbate lupus in the host-versus-graft model of lupus through induction of cytokine synthesis. Fn14 deficiency improves renal function and survival in association with decreased production of IL-6, MCP-1 and IL-10 [88, 89].

It is well established that patients with SLE have a defect in immune complex clearance. Murine models of lupus involving defects in immune complex clearance include knockout strains for complement proteins and Fc-γ receptors, two systems involved with immune complex processing [90].

C1q initiates the classical complement pathway and thereby plays an important role in clearance of immune complexes. Defects in C1q have been identified in patients with systemic lupus [91, 92]. Although rare (41 cases reported), 90% of patients with homozygous C1q deficiency develop a severe lupus-like disorder characterized by malar rash, production of ANA and glomerulonephritis [93, 94].

Ninety percent of homozygous C1q-deficient mice develop a systemic lupus-like syndrome, with high ANA titers and histone antibodies. Numerous apoptotic cell bodies are found in these mice, indicating a profound defect in the clearance of apoptotic cells. Twenty-five percent of mice have immune complex-mediated glomerulonephritis [95, 96]. However, the genetic background has a significant effect on the C1q knockout phenotype. For example, C1q deficiency has a lupus-like phenotype in MRL/Mp mice, but not C57BL/6 or 129 Sv mice [52]. Complete C4 deficiency is rare (28 cases) because humans have 2–8 copies of the C4d gene, each encoding a C4a or C4B molecule. However, the prevalence of SLE in patients with complete C4 deficiency is 75% [97]. The most common inherited complement deficiency in humans is homozygous C2 deficiency, which has a prevalence of 1:20,000, 33% of whom develop a mild form of SLE [98]. CRP (C-reactive protein) is also involved in clearance of chromatin. CRP knockout mice develop autoimmunity and glomerulonephritis due to lack of chromatin clearance [99].

Fc-γ receptors are critical in the clearance of antibodies and immune complexes. The murine lupus susceptibility loci Nba1 and Sle1 contain two Fc-γ receptors linked to lupus nephritis. Ligation of the Fc-γ receptor with antibodies leads to the production of proinflammatory cytokines, including IL-1b, which promotes renal damage [100] through induction of other proinflammatory cytokines such as MCP-1 and IL-8 [101]. Deletion of the Fc-γR chain in NZB/NZW mice prolongs survival and attenuates renal injury with less inflammation [102]. NZB/NZW mice crossed with Fc-γRI and Fc-γRIII knockout mice develop ANA and immune complex deposition in the glomeruli. However, inflammatory nephritis and renal failure does not occur [102].

B Cell Hyperactivity and Lupus Nephritis

Murine models have provided important insights related to mechanisms underlying B cell hyperactivity in lupus nephritis. Signals through the B cell antigen receptor (BCR) are critical to development and B cell responses to antigen. Defective B cell receptor signaling can promote immunodeficiency and autoimmunity. Antigen or antireceptor antibodies initiate signal transduction through the receptor CD79a (Ig-α) and CD79b (Ig-β) subunits expressed on B cells. The cytoplasmic domain of CD79a and CD79b contain sites for tyrosine phosphorylation and docking, termed immunoreceptor tyrosine-based activation motifs. An intracellular phosphorylation cascade is initiated, leading to B cell activation [103]. Key intracellular regulatory molecules include Lyn, Btk, Vav and a protein phosphatase, SHP1 [104, 105].

Co-receptors which positively or negatively regulate B cell signaling are involved in the pathogenesis of lupus nephritis. A co-receptor complex that positively regulates B cell signaling is the CD19-CD21 complex. Like the B cell receptor, the cytoplasmic domain of this co-receptor complex contains sites for tyrosine phosphorylation and docking, thereby propagating activation signals. Phosphorylation of CD19 increases the activity of a number of protein tyrosine kinases, thereby amplifying the signal [106].

Mice with altered CD19, CD21, CD22, Lyn or SHP1 expression produce autoantibodies and develop autoimmunity to varying degrees. Peripheral tolerance is disrupted in mice that overexpress CD19, resulting in hyperactive B cells and spontaneous production of IgG antibodies [107]. Mice with only a 15–30% increase in CD19 expression have systemic lupus-like symptoms [41]. Lyn-deficient mice develop immune complex glomerulonephritis with autoantibodies [108].

CD19 and CD21 expression on B cells isolated from patients with systemic sclerosis is 20% higher compared to healthy controls. Of note, overexpression of CD19 by only 20% induced autoantibody production in normally nonimmune strains of mice [109]. Viable mice with SHP1 mutations develop autoantibodies, hypergammaglobulinemia and tissue immune complex deposition [110].

Several critical inhibitory co-receptors have been identified which abrogate B cell signaling, including CD22. The cytoplasmic domain of CD22 contains an immunoreceptor tyrosine-based inhibitory motif, which acts through activation of the tyrosine phosphatase SHP1 [106]. The CD22 gene is located near a lupus susceptibility site identified in NZW mice. NZW mice contain the CD22a allele, which generates an aberrantly spliced mRNA [111], producing a protein that is not normally upregulated following lipopolysaccharide stimulation [112]. CD22 may antagonize B cell activation by raising the threshold necessary to initiate BCR-induced signals. Following BCR engagement, CD22 is phosphorylated within immunoreceptor tyrosine-based inhibitory motifs present on the cytoplasmic domain of CD22. Phosphorylation of CD22 initiates an intracellular phosphorylation cascade that leads to downregulation of CD19 phosphorylation and BCR-mediated signal transduction. CD22-deficient mice have hyperactive B cells, anti-DNA antibodies and manifestations of SLE [113–115]. Based on these considerations, CD19 and CD22 appear to regulate signaling thresholds needed for expansion of the peripheral B cell pool [115].

Once an antigen is presented to a B cell, co-stimulatory molecules facilitate activation of the B cell. These include B7-CD28 co-stimulation, which may be blocked by CTLA4-Ig (abatacept).

CTLA4-Ig is beneficial in murine lupus, when given before autoantibodies arise [116, 117]. Follicular Th cells that drive B cell help in germinal centers are completely dependent on CD28 co-stimulation [118]. However, TLR-mediated signals together with cytokines, including BAFF, IL-21 and IL-17, can substitute for B7-CD28 co-stimulation and drive reactivation of memory B cells.

BAFF (BlyS, TALL-1, THANK and zTNF4) is a member of the TNF family of cytokines produced by dendritic cells, monocytes and macrophages. BAFF is a co-stimulator for B cell survival when coupled with B cell antigen receptor ligation. BAFF ligation correlates with increased Bcl-2 expression and NF-κB activation, both of which increase B cell survival [119]. BAFF is responsible for the polyclonal activation of B cells typically seen in models of lupus nephritis. BAFF engages 3 receptors: BCMA (B cell maturation antigen), TACI (transmembrane activator, calcium modulator and cyclophilin ligand interactor) and BAFF receptor [120].

BAFF expression is elevated in both MRL/lpr mice and (NZWxNZB)F1 hybrid mice, and correlates with disease progression in both models [119]. Mice transgenic for BCMA or BAFF have SLE-like disease, with anti-DNA antibodies, increased IgM, vasculitis and glomerulonephritis [119, 121, 122]. Administration of BMCA-Ig and TACI-Ig fusion proteins as decoy BAFF receptors to NZWxNZB]F1 mice leads to increased survival and slower renal disease progression [119, 120, 123].

Elevated BAFF levels are found in a number of SLE patients, suggesting that this may provide a therapeutic target for patients with lupus [119]. Along those lines, belimumab, a humanized monoclonal anti-BAFF antibody has been developed as a selective inhibitor of BAFF.

T Cells and Lupus Nephritis

Although intrinsic B cell hyperactivity is a central feature of lupus nephritis, particularly early in the course of the disease, maturation of B cells into autoantibody secreting cells requires T cell help [6, 124]. T cells drive autoreactive B cells to generate ANA in several models of lupus, particularly through a CD40-gp39 interaction [125].

Depletion of CD4+ T cells has different effects in MRL/lpr mice and (NZB xNZW)F1 mice. In (NZBxNZW)F1 mice, development of increased polyclonal IgM antibody production was still observed following depletion of CD4+ T cells. However, in MRL/lpr mice, development of increased polyclonal IgM appears to require CD4+ T cells. In both MRL/lpr and (NZBxNZW) F1 mice, CD4+ T cells are required for increased production of IgG antibodies [126].

Studies with knockout mice have identified several proteins that play a major role in the regulation of T cell function and are involved in lupus nephritis. PD-1, a cell surface receptor belonging to the immunoglobulin superfamily inhibits the proliferation of activated T cells. Mice deficient in PD-1 develop-lupus like arthritis and glomerulonephritis with IgG deposition [127]. The Coronin-1A gene, which regulates cytoskeletal structure, is a lupus susceptibility gene in the MRL/lpr mouse. Mutation of Coronin-1A is associated with a number of abnormalities in T cell function, including activation, survival and migration [128]. T cell adaptor protein (TSAd) is a T cell-specific signaling adaptor involved in assembly of intracellular signaling complexes. TSAd-deficient mice have defective T cell death in vitro, accumulation of T and B cells in the spleen, hypergammaglomerulinemia, dsDNA and glomerulonephritis [129]. Overexpression of the immediate early response gene IEX-1 reduces T cell apoptosis, leading to accumulation of activated T cells, splenomegaly and adenopathy with development of anti-dsDNA antibodies and increased urine protein excretion [130].

Cytokines and Renal Disease Progression

Antigen-antibody complexes trigger renal injury through production of proinflammatory cytokines. As discussed above, IFN-γ is a major product of TLR7 and TLR9 signaling. IFNs are produced by resident renal cells and promote renal disease [131]. IFN-γ is detected in the serum of lupus patients. Mice transgenic for IFN-γ develop ANA and lupus nephritis [132], whereas mice with a deficiency of the type I IFN receptor are protected from development of experimental lupus nephritis [133], although immune complex deposits are identified within glomeruli [134].

Of note, IFN-α treatment of female (NZWxBXSB)F1 mice mimics some, but not all, features associated with the Yaa mutation. As discussed above, male BXSB mice develop antiphospholipid syndrome accelerated glomerulonephritis due to the Yaa locus encoding an extra copy of TLR7. A major function of TLR7 is induction of IFN-α [135].

A wide variety of chemokines are produced during the progression of lupus nephritis. MRL/lpr mice with chemokine receptor type 2 deficiency (CCR2, the receptor for MCP-1) have less adenopathy, less proteinuria, less glomerular inflammation and longer survival than MRL/lpr mice, despite similar levels of circulating immunoglobulins and immune complex deposition within glomeruli [136]. CXCR3 mediates renal Th1 and Th17 immune responses in lupus nephritis. CXCR3 is highly expressed on Th1 cells. In the MRL/lpr model, mice with deletion of CXCR3 have fewer IFN-producing Th1 and IL-17-producing Th17 cells and less severe glomerulonephritis [137].

IL-21 is produced by activated T cells and its interactions with IL-21R are required for isotype switching and differentiation of B cells into Ab-secreting cells. Blockade of IL-21 signaling with IL-21R.Fc reduces renal disease progression in the lupus-prone MRL/lpr model [138].

TGF-β is a pleiotropic cytokine that has a wide variety effects on growth, inflammation and matrix deposition. TGF-β knockout mice develop antibodies reactive with dsDNA, ssDNA and Sm ribonucleoprotein, resulting in glomerular immune complex deposits [139].

CSF-1 promotes the development of monocyte and macrophage phenotypes that enhance lupus nephritis. CSF-1 is a primary growth factor for macrophages; CSF-1 deficiency protects MRL-Fas/lpr mice from kidney disease. Increasing CSF-1 exacerbates lupus in MRL/lpr mice [140].

Some growth factors inhibit lupus nephritis. For example, hepatocyte growth factor inhibits Th2 responses, is antiapoptotic, and proregenerative. Increased HGF expression in the murine GVHD model of lupus nephritis prevents proteinuria and ameliorates glomerulonephritis [141].

Conclusions

Spontaneous models of lupus nephritis have helped genetic linkage studies to identify regions that confer an increased risk of lupus nephritis in susceptible animals. In many cases, these loci have homologs in humans with lupus nephritis. Functional relevance of target genes has been established through the use of transgenic mice that overexpress the gene of interest or through the use of knockout mice bearing deletion of the gene of interest. Critical genes are involved in autoantigen production, autoantibody production and cytokine-mediated tissue injury. These studies have identified a number of potential therapeutic targets for preventing the development or progression of lupus nephritis.

References

1 Peutz-Kootstra CJ, de Heer E, Hoedemaeker PJ, Abrass CK, Bruijn JA: Lupus nephritis: lessons from experimental animal models. J Lab Clin Med 2001;137:244–260.
2 Johnson AE, Gordon C, Palmer RG, Bacon PA: The prevalence and incidence of systemic lupus erythematosus in Birmingham, England. Relationship to ethnicity and country of birth. Arthritis Rheum 1995;38:551–558.
3 Grande JP, Balow JE: Renal biopsy in lupus nephritis. Lupus 1998;7:611–617.
4 Weening JJ, D'Agati VD, Schwartz MM, et al: The classification of glomerulonephritis in systemic lupus erythematosus revisited. J Am Soc Nephrol 2004 ;15:241–250.
5 Grande JP: Mechanisms of progression of renal damage in lupus nephritis: pathogenesis of renal scarring. Lupus 1998;7:604–610.
6 Mohan C, Datta SK: Lupus: key pathogenic mechanisms and contributing factors. Clin Immunol Immunopathol 1995;77:209–220.
7 Hahn BH: Antibodies to DNA. N Engl J Med 1998;338:1359–1368.
8 Berden J: Lupus nephritis. Kidney Int 1997;52:538–558.
9 Burlingame RW, Boey ML, Starkebaum G, Rubin RL: The central role of chromatin in autoimmune responses to histones and DNA in systemic lupus erythematosus. J Clin Invest 1994;94:184–192.
10 Perkins DL, Glaser RM, Mahon CA, Michaelson J, Marshak-Rothstein A: Evidence for an intrinsic B cell defect in lpr/lpr mice apparent in neonatal chimeras. J Immunol 1990;145:549–555.
11 Sobel ES, Katagiri T, Katagiri K, Morris SC, Cohen PL, Eisenberg RA: An intrinsic B cell defect is required for the production of autoantibodies in the lpr model of murine systemic autoimmunity. J Exp Med 1991;173: 1441–1449.
12 Nemazee D, Guiet C, Buerki K, Marshak-Rothstein A: B lymphocytes from the autoimmune-prone mouse strain MLR/lpr manifest an intrinsic defect in tetraparental MRL/lpr in equilibrium DBA/2 chimeras. J Immunol 1991;147:2536–2539.
13 Mohan C, Adams S, Stanik V, Datta SK: Nucleosome: a major immunogen for pathogenic autoantibody-inducing T cells of lupus. J Exp Med 1993;177:1367–1381.
14 Madaio MP, Shlomchik MJ: Emerging concepts regarding B cells and autoantibodies in murine lupus nephritis. B cells have multiple roles; all autoantibodies are not equal. J Am Soc Nephrol 1996;7:387–396.
15 Couser WG: Mediation of immune glomerular injury. J Am Soc Nephrol 1990;1:13–29.
16 van Bruggen MC, Kramers C, Walgreen B, Elema JD, Kallenberg CG, van den Born J, et al: Nucleosomes and histones are present in glomerular deposits in human lupus nephritis. Nephrol Dial Transplant 1997;12:57–66.

17 Sabbaga J, Pankewycz OG, Lufft V, Schwartz RS, Madaio MP: Cross-reactivity distinguishes serum and nephritogenic anti-DNA antibodies in human lupus from their natural counterparts in normal serum. J Autoimmunity 1990;3:215–235.
18 Foster MH, Cizman B, Madaio MP: Nephritogenic autoantibodies in systemic lupus erythematosus: Immunochemical properties, mechanisms of immune deposition, and genetic origins. Lab Invest 1993;69: 494–507.
19 Couser WG: Mediation of immune glomerular injury. J Am Soc Nephrol 1990;1:13–29.
20 Eddy AA, Michael AF. Immunopathogenic mechanisms of glomerular injury; in Tisher CC, Brenner BM (eds): Renal Pathology. Philadelphia, J. B. Lippincott Co., 1994, pp 162–221.
21 Hashimoto Y, Ziff M, Hurd ER: Increased endothelial cell adherence, aggregation, and superoxide generation by neutrophils incubated in systemic lupus erythematosus and Felty's syndrome sera. Arthritis Rheum 1982;25:1409–1418.
22 Lai KN, Leung JCK, Lai KB, Lai CKW: Effect of anti-DNA autoantibodies on the gene expression of interleukin 8, transforming growth factor-β, and nitric oxide synthase in cultured endothelial cells. Scand J Rheumatol 1997;26:461–467.
23 Tsai CY, Wu TH, Sun KH, Liao TS, Lin WM, Yu CL: Polyclonal IgG anti-dsDNA antibodies exert cytotoxic effect on cultured rat mesangial cells by binding to cell membrane and augmenting apoptosis. Scand J Rheumatol 1993;22:162–171.
24 Helyer BJ, Howie JB: Renal disease associated with positive lupus erythematosus tests in a cross-bred strain of mice. Nature 1963;197:197.
25 Bielschowsky M, Helyer B, Howie J: Spontaneous anemia in mice of the NZB/BL strain. Proc Univ Otago Med Sch 1959;37:9–11.
26 Andrews BS, Eisenberg RA, Theofilopoulos AN, et al: Spontaneous murine lupus-like syndromes. Clinical and immunopathological manifestations in several strains. J Exp Med 1978;148:1198–1215.
27 Stoll ML, Gavalchin J: Systemic lupus erythematosus-messages from experimental models. Rheumatology (Oxford) 2000;39:18–27.
28 Kelley VE, Winkelstein A: Age- and sex-related glomerulonephritis in New Zealand white mice. Clin Immunol Immunopathol 1980;16:142–150.
29 Hahn BH, Shulman LE: Autoantibodies and nephritis in the white strain (NZW) of New Zealand mice. Arthritis Rheum 1969;12:355–364.
30 Dubois EL, Horowitz RE, Demopoulos HB, Teplitz R: NZB/NZW mice as a model of systemic lupus erythematosus. JAMA 1966;195:285–289.
31 Channing AA, Kasuga T, Horowitz RE, Dubois EL, Demopoulos HB: An ultrastructural study of spontaneous lupus nephritis in the NZB-BL-NZW mouse. Am J Pathol 1965;47: 677–694.
32 Burnet FM, Holmes MC: The natural history of the NZB/NZW F1 hybrid mouse: a laboratory model of systemic lupus erythematosus. Australas Ann Med 1965;14:185–191.
33 Howie JB, Helyer BJ: The immunology and pathology of NZB mice. Adv Immunol 1968;9:215–266.
34 Morel L, Wakeland EK: Susceptibility to lupus nephritis in the NZB/W model system. Curr Opin Immunol 1998;10:718–725.
35 Theofilopoulos AN, Dixon FJ: Murine models of systemic lupus erythematosus. Adv Immunol 1985;37:269–390.
36 Murphy E, Roths J: A single gene for massive lymphoproliferation with immune complex disease in a new mouse strain. Proceedings of the 16th International Congress in Hematology Exerpta Medica, Amsterdam. 1976;69:69–80.
37 Itoh N, Yonehara S, Ishii A, et al: The polypeptide encoded by the cDNA for human cell surface antigen Fas can mediate apoptosis. Cell 1991;66:233–243.
38 Cohen PL, Eisenberg RA: Lpr and gld: single gene models of systemic autoimmunity and lymphoproliferative disease. Annu Rev Immunol 1991;9:243–269.
39 Watanabe-Fukunaga R, Brannan CI, Copeland NG, Jenkins NA, Nagata S: Lymphoproliferation disorder in mice explained by defects in Fas antigen that mediates apoptosis. Nature 1992;356:314–317.
40 Nagata S, Golstein P. The Fas death factor. Science. 1995 Mar 10;267(5203):1449–56.
41 Chan OT, Madaio MP, Shlomchik MJ: The central and multiple roles of B cells in lupus pathogenesis. Immunol Rev 1999;169:107–121.
42 Takahashi S, Nose M, Sasaki J, Yamamoto T, Kyogoku M: IgG3 production in MRL/lpr mice is responsible for development of lupus nephritis. J Immunol 1991;147:515–519.
43 Teramoto K, Negoro N, Kitamoto K, et al: Microarray analysis of glomerular gene expression in murine lupus nephritis. J Pharmacol Sci 2008;106:56–67.
44 Mysler E, Bini P, Drappa J, et al: The apoptosis-1/Fas protein in human systemic lupus erythematosus. J Clin Invest 1994;93:1029–1034.
45 Fisher GH, Rosenberg FJ, Straus SE, et al: Dominant interfering Fas gene mutations impair apoptosis in a human autoimmune lymphoproliferative syndrome. Cell 1995;81:935–946.
46 Rieux-Laucat F, Le Deist F, Hivroz C, et al: Mutations in Fas associated with human lymphoproliferative syndrome and autoimmunity. Science 1995;268:1347–1349.
47 Pisitkun P, Deane JA, Difilippantonio MJ, Tarasenko T, Satterthwaite AB, Bolland S: Autoreactive B cell responses to RNA-related antigens due to TLR7 gene duplication. Science 2006;312:1669–1672.
48 Subramanian S, Tus K, Li QZ, et al: A Tlr7 translocation accelerates systemic autoimmunity in murine lupus. Proc Natl Acad Sci USA 2006;103:9970–9975.
49 Florquin S, Abramowicz D, de Heer E, et al: Renal immunopathology in murine host-versus-graft disease. Kidney Int 1991;40:852–861.
50 Santiago-Raber ML, Laporte C, Reininger L, Izui S: Genetic basis of murine lupus. Autoimmun Rev 2004;3(1):33–39.

51 Fairhurst AM, Wandstrat AE, Wakeland EK: Systemic lupus erythematosus: multiple immunological phenotypes in a complex genetic disease. Adv Immunol 2006;92:1–69.

52 Jorgensen TN, Gubbels MR, Kotzin BL: New insights into disease pathogenesis from mouse lupus genetics. Curr Opin Immunol 2004;16:787–793.

53 Mohan C, Alas E, Morel L, Yang P, Wakeland EK: Genetic dissection of SLE pathogenesis. Sle1 on murine chromosome 1 leads to a selective loss of tolerance to H2A/H2B/DNA subnucleosomes. J Clin Invest 1998;101:1362–1372.

54 Mohan C, Morel L, Yang P, Wakeland EK: Genetic dissection of systemic lupus erythematosus pathogenesis: Sle2 on murine chromosome 4 leads to B cell hyperactivity. J Immunol 1997;159:454–465.

55 Mohan C, Yu Y, Morel L, Yang P, Wakeland EK: Genetic dissection of Sle pathogenesis: Sle3 on murine chromosome 7 impacts T cell activation, differentiation, and cell death. J Immunol 1999;162:6492–6502.

56 Hicks J, Bullard DC: Review of autoimmune (lupus-like) glomerulonephritis in murine models. Ultrastruct pathol 2006;30:345–359.

57 Wu X, Jiang N, Deppong C, et al: A role for the Cr2 gene in modifying autoantibody production in systemic lupus erythematosus. J Immunol 2002;169:1587–1592.

58 Jiang Y, Hirose S, Abe M, et al: Polymorphisms in IgG Fc receptor IIB regulatory regions associated with autoimmune susceptibility. Immunogenetics 2000;51:429–435.

59 Pritchard NR, Cutler AJ, Uribe S, Chadban SJ, Morley BJ, Smith KG: Autoimmune-prone mice share a promoter haplotype associated with reduced expression and function of the Fc receptor FcgammaRII. Curr Biol 2000;10:227–230.

60 Cunninghame Graham DS, Vyse TJ: The candidate gene approach: have murine models informed the study of human SLE? Clin Exp Immunol 2004;137:1–7.

61 Vlahakos D, Foster MH, Ucci AA, Barrett KJ, Datta SK, Madaio MP: Murine monoclonal anti-DNA antibodies penetrate cells, bind to nuclei, and induce glomerular proliferation and proteinuria in vivo. J Am Soc Nephrol 1992;2:1345–1354.

62 Mevorach D, Zhou JL, Song X, Elkon KB: Systemic exposure to irradiated apoptotic cells induces autoantibody production. J Exp Med 1998;188:387–392.

63 Georgiev M, Agle LM, Chu JL, Elkon KB, Ashany D: Mature dendritic cells readily break tolerance in normal mice but do not lead to disease expression. Arthritis Rheum 2005;52:225–238.

64 Voll RE, Herrmann M, Roth EA, Stach C, Kalden JR, Girkontaite I: Immunosuppressive effects of apoptotic cells. Nature 1997;390:350–351.

65 Arbuckle MR, McClain MT, Rubertone MV, et al: Development of autoantibodies before the clinical onset of systemic lupus erythematosus. N Engl J Med 2003;349:1526–1533.

66 Levine JS, Subang R, Nasr SH, et al: Immunization with an apoptotic cell-binding protein recapitulates the nephritis and sequential autoantibody emergence of systemic lupus erythematosus. J Immunol 2006; 177:6504–6516.

67 Mjelle JE, Kalaaji M, Rekvig OP: Exposure of chromatin and not high affinity for dsDNA determines the nephritogenic impact of anti-dsDNA antibodies in (NZBxNZW)F1 mice. Autoimmunity 2009;42:104–111.

68 Mortensen ES, Rekvig OP: Nephritogenic potential of anti-DNA antibodies against necrotic nucleosomes. J Am Soc Nephrol 2009;20:696–704.

69 Wen ZK, Xu W, Xu L, et al: DNA hypomethylation is crucial for apoptotic DNA to induce systemic lupus erythematosus-like autoimmune disease in SLE-non-susceptible mice. Rheumatology (Oxford) 2007;46:1796–1803.

70 Santiago-Raber ML, Baudino L, Izui S. Emerging roles of TLR7 and TLR9 in murine SLE. J Autoimmun. 2009 Nov-Dec;33(3–4):231–238.

71 Barrat FJ, Meeker T, Chan JH, Guiducci C, Coffman RL: Treatment of lupus-prone mice with a dual inhibitor of TLR7 and TLR9 leads to reduction of autoantibody production and amelioration of disease symptoms. Eur J Immunol 2007;37:3582–3586.

72 Savarese E, Steinberg C, Pawar RD, et al: Requirement of Toll-like receptor 7 for pristane-induced production of autoantibodies and development of murine lupus nephritis. Arthritis Rheum 2008;58:1107–1115.

73 Sadanaga A, Nakashima H, Akahoshi M, et al: Protection against autoimmune nephritis in MyD88-deficient MRL/lpr mice. Arthritis Rheum 2007;56:1618–1628.

74 Anders H-J, Vielhauer V, Eis V, Linde Y, Kretzler M, Perez de Lema G, et al: Activation of toll-like receptor-9 induces progression of renal disease in MRL-Fas(lpr) mice. FASEB J. 2004 January 20, 2004:03–0646fje.

75 Deane JA, Pisitkun P, Barrett RS, et al: Control of Toll-like receptor 7 expression is essential to restrict autoimmunity and dendritic cell proliferation. Immunity 2007;27:801–810.

76 Yasutomo K, Horiuchi T, Kagami S, et al: Mutation of DNASE1 in people with systemic lupus erythematosus. Nat Genet 2001;28:313–314.

77 Mannherz HG, Peitsch MC, Zanotti S, Paddenberg R, Polzar B: A new function for an old enzyme: the role of DNase I in apoptosis. Curr Top Microbiol Immunol 1995;198:161–174.

78 Jacob M, Napirei M, Ricken A, Dixkens C, Mannherz HG: Histopathology of lupus-like nephritis in Dnase1-deficient mice in comparison to NZB/W F1 mice. Lupus 2002; 11:514–527.

79 Bruns A, Blass S, Hausdorf G, Burmester GR, Hiepe F: Nucleosomes are major T and B cell autoantigens in systemic lupus erythematosus. Arthritis Rheum 2000;43:2307–2315.

80 Jacob L, Viard JP: Anti-DNA antibodies and their relationships with anti-histone and anti-nucleosome specificities. Eur J Med 1992;1:425–431.

81 Seredkina N, Zykova SN, Rekvig OP: Progression of murine lupus nephritis is linked to acquired renal Dnase1 deficiency and not to up-regulated apoptosis. Am J Pathol 2009;175:97–106.

82 Herrmann M, Winkler T, Gaipl U, Lorenz H, Geiler T, Kalden JR: Etiopathogenesis of systemic lupus erythematosus. Int Arch Allergy Immunol 2000;123:28–35.

83 Walport MJ: Lupus, DNase and defective disposal of cellular debris. Nat Genet 2000;25:135–136.

84 Strasser A, Whittingham S, Vaux DL, et al: Enforced BCL2 expression in B-lymphoid cells prolongs antibody responses and elicits autoimmune disease. Proc Natl Acad Sci USA 1991;88:8661–8665.

85 Kuo P, Bynoe MS, Wang C, Diamond B: Bcl-2 leads to expression of anti-DNA B cells but no nephritis: a model for a clinical subset. Eur J Immunol 1999;29:3168–3178.

86 Di Cristofano A, Kotsi P, Peng YF, Cordon-Cardo C, Elkon KB, Pandolfi PP: Impaired Fas response and autoimmunity in Pten+/- mice. Science 1999;285:2122–2125.

87 Van Parijs L, Refaeli Y, Abbas AK, Baltimore D: Autoimmunity as a consequence of retrovirus-mediated expression of C-FLIP in lymphocytes. Immunity 1999;11:763–770.

88 Zhao Z, Burkly LC, Campbell S, et al: TWEAK/Fn14 interactions are instrumental in the pathogenesis of nephritis in the chronic graft-versus-host model of systemic lupus erythematosus. J Immunol 2007;179: 7949–7958.

89 Molano A, Lakhani P, Aran A, Burkly LC, Michaelson JS, Putterman C: TWEAK stimulation of kidney resident cells in the pathogenesis of graft versus host induced lupus nephritis. Immunol Lett 2009;125:119–128.

90 Birmingham DJ, Rovin BH, Yu CY, Hebert LA: Of mice and men: the relevance of the mouse to the study of human SLE. Immunol Res 2001;24:211–224.

91 Slingsby JH, Norsworthy P, Pearce G, et al: Homozygous hereditary C1q deficiency and systemic lupus erythematosus. A new family and the molecular basis of C1q deficiency in three families. Arthritis Rheum 1996;39:663–670.

92 Casciola-Rosen LA, Anhalt G, Rosen A: Autoantigens targeted in systemic lupus erythematosus are clustered in two populations of surface structures on apoptotic keratinocytes. J Exp Med 1994;179:1317–1330.

93 Walport MJ: Complement and systemic lupus erythematosus. Arthritis Res 2002;4 (Suppl 3):S279–S293.

94 Walport MJ, Davies KA, Botto M: C1q and systemic lupus erythematosus. Immunobiology 1998;199:265–285.

95 Mitchell DA, Pickering MC, Warren J, et al: C1q deficiency and autoimmunity: the effects of genetic background on disease expression. J Immunol 2002;168:2538–2543.

96 Botto M, Dell'Agnola C, Bygrave AE, et al: Homozygous C1q deficiency causes glomerulonephritis associated with multiple apoptotic bodies. Nat Genet 1998;19:56–59.

97 Rupert KL, Moulds JM, Yang Y, et al: The molecular basis of complete complement C4A and C4B deficiencies in a systemic lupus erythematosus patient with homozygous C4A and C4B mutant genes. J Immunol 2002;169:1570–1578.

98 Sullivan KE, Petri MA, Schmeckpeper BJ, McLean RH, Winkelstein JA: Prevalence of a mutation causing C2 deficiency in systemic lupus erythematosus. J Rheumatol 1994;21:1128–1133.

99 Bao L, Haas M, Boackle SA, et al: Transgenic expression of a soluble complement inhibitor protects against renal disease and promotes survival in MRL/lpr mice. J Immunol 2002;168:3601–3607.

100 Marsh CB, Wewers MD, Tan LC, Rovin BH: Fc(gamma) receptor cross-linking induces peripheral blood mononuclear cell monocyte chemoattractant protein-1 expression: role of lymphocyte Fc(gamma)RIII. J Immunol 1997;158:1078–1084.

101 Marsh CB, Anderson CL, Lowe MP, Wewers MD: Monocyte IL-8 release is induced by two independent Fc gamma R- mediated pathways. J Immunol 1996;157:2632–2637.

102 Clynes R, Dumitru C, Ravetch JV: Uncoupling of immune complex formation and kidney damage in autoimmune glomerulonephritis. Science 1998;279:1052–1054.

103 Benschop RJ, Cambier JC: B cell development: signal transduction by antigen receptors and their surrogates. Curr Opin Immunol 1999;11:143–151.

104 Cornall RJ, Cyster JG, Hibbs ML, et al: Polygenic autoimmune traits: Lyn, CD22, and SHP-1 are limiting elements of a biochemical pathway regulating BCR signaling and selection. Immunity 1998;8:497–508.

105 Tedder TF: Introduction: response-regulators of B lymphocyte signaling thresholds provide a context for antigen receptor signal transduction. Sem Immunol 1998;10:259–265.

106 Kurosaki T: Regulation of B-cell signal transduction by adaptor proteins. Nature Reviews 2002;2:354–363.

107 Inaoki M, Sato S, Weintraub BC, Goodnow CC, Tedder TF: CD19-regulated signaling thresholds control peripheral tolerance and autoantibody production in B lymphocytes. J Exp Med 1997;186:1923–1931.

108 Nishizumi H, Taniuchi I, Yamanashi Y, et al: Impaired proliferation of peripheral B cells and indication of autoimmune disease in lyn-deficient mice. Immunity 1995;3:549–560.

109 Saito E, Fujimoto M, Hasegawa M, et al: CD19-dependent B lymphocyte signaling thresholds influence skin fibrosis and autoimmunity in the tight-skin mouse. J Clin Invest 2002;109:1453–1462.

110 Tsui HW, Siminovitch KA, de Souza L, Tsui FW: Motheaten and viable motheaten mice have mutations in the haematopoietic cell phosphatase gene. Nat Genet 1993;4:124–129.

111 Mary C, Laporte C, Parzy D, et al: Dysregulated expression of the Cd22 gene as a result of a short interspersed nucleotide element insertion in Cd22a lupus-prone mice. J Immunol 2000;165:2987–2996.

112 Nitschke L, Lajaunias F, Moll T, et al: Expression of aberrant forms of CD22 on B lymphocytes in Cd22a lupus-prone mice affects ligand binding. Int Immunol 2006;18:59–68.

113 Sato S, Miller AS, Inaoki M, et al: CD22 is both a positive and negative regulator of B lymphocyte antigen receptor signal transduction: altered signaling in CD22-deficient mice. Immunity 1996;5:551–562.

114 O'Keefe TL, Williams GT, Batista FD, Neuberger MS: Deficiency in CD22, a B cell-specific inhibitory receptor, is sufficient to predispose to development of high affinity autoantibodies. J Exp Med 1999;189:1307–1313.

115 Fujimoto M, Bradney AP, Poe JC, Steeber DA, Tedder TF: Modulation of B lymphocyte antigen receptor signal transduction by a CD19/CD22 regulatory loop. Immunity 1999;11:191–200.

116 Mihara M, Tan I, Chuzhin Y, et al: CTLA4Ig inhibits T cell-dependent B-cell maturation in murine systemic lupus erythematosus. J Clin Invest 2000;106:91–101.

117 Schiffer L, Sinha J, Wang X, et al: Short term administration of costimulatory blockade and cyclophosphamide induces remission of systemic lupus erythematosus nephritis in NZB/W F1 mice by a mechanism downstream of renal immune complex deposition. J Immunol 2003;171:489–497.
118 Linterman MA, Rigby RJ, Wong R, et al: Roquin differentiates the specialized functions of duplicated T cell costimulatory receptor genes CD28 and ICOS. Immunity 2009;30:228–241.
119 Mackay F, Mackay CR: The role of BAFF in B-cell maturation, T-cell activation and autoimmunity. Trends Immunol 2002;23:113–115.
120 Mackay F, Browning JL: BAFF: a fundamental survival factor for B cells. Nature Reviews 2002;2:465–475.
121 Gross JA, Johnston J, Mudri S, et al: TACI and BCMA are receptors for a TNF homologue implicated in B-cell autoimmune disease. Nature 2000;404:995–999.
122 Khare SD, Sarosi I, Xia XZ, et al: Severe B cell hyperplasia and autoimmune disease in TALL-1 transgenic mice. Proc Natl Acad Sci USA 2000;97:3370–3375.
123 Jacob CO, Pricop L, Putterman C, et al: Paucity of clinical disease despite serological autoimmunity and kidney pathology in lupus-prone New Zealand mixed 2328 mice deficient in BAFF. J Immunol 2006;177:2671–2680.
124 Klinman DM: Polyclonal B cell activation in lupus-prone mice precedes and predicts the development of autoimmune disease. J Clin Invest 1990;86:1249–1254.
125 Mohan C, Shi Y, Laman JD, Datta SK: Interaction between CD40 and its ligand gp39 in the development of murine lupus nephritis. J Immunol 1995;154:1470–1480.
126 Merino R, Iwamoto M, Fossati L, Izui S: Polyclonal B cell activation arises from different mechanisms in lupus-prone (NZB x NZW)F1 and MRL/MpJ-lpr/lpr mice. J Immunol 1993;151:6509–6516.
127 Nishimura H, Nose M, Hiai H, Minato N, Honjo T: Development of lupus-like autoimmune diseases by disruption of the PD-1 gene encoding an ITIM motif-carrying immunoreceptor. Immunity 1999;11:141–151.
128 Haraldsson MK, Louis-Dit-Sully CA, Lawson BR, et al: The lupus-related Lmb3 locus contains a disease-suppressing Coronin-1A gene mutation. Immunity 2008;28:40–51.
129 Drappa J, Kamen LA, Chan E, et al: Impaired T cell death and lupus-like autoimmunity in T cell-specific adapter protein-deficient mice. J Exp Med 2003;198:809–821.
130 Zhang Y, Schlossman SF, Edwards RA, Ou CN, Gu J, Wu MX: Impaired apoptosis, extended duration of immune responses, and a lupus-like autoimmune disease in IEX-1-transgenic mice. Proc Natl Acad Sci USA 2002;99:878–883.
131 Fairhurst AM, Xie C, Fu Y, et al: Type I interferons produced by resident renal cells may promote end-organ disease in autoantibody-mediated glomerulonephritis. J Immunol 2009;183:6831–6838.
132 Seery JP, Carroll JM, Cattell V, Watt FM: Antinuclear autoantibodies and lupus nephritis in transgenic mice expressing interferon gamma in the epidermis. J Exp Med 1997;186:1451–1459.
133 Nacionales DC, Kelly-Scumpia KM, Lee PY, et al: Deficiency of the type I interferon receptor protects mice from experimental lupus. Arthritis Rheum 2007;56:3770–3783.
134 Jorgensen TN, Roper E, Thurman JM, Marrack P, Kotzin BL: Type I interferon signaling is involved in the spontaneous development of lupus-like disease in B6.Nba2 and (B6.Nba2 x NZW)F(1) mice. Genes immun 2007;8:653–662.
135 Ramanujam M, Kahn P, Huang W, et al: Interferon-alpha treatment of female (NZW x BXSB)F(1) mice mimics some but not all features associated with the Yaa mutation. Arthritis Rheum 2009;60:1096–1101.
136 Perez de Lema G, Maier H, Franz TJ, et al: Chemokine receptor Ccr2 deficiency reduces renal disease and prolongs survival in MRL/lpr lupus-prone mice. J Am Soc Nephrol 2005;16:3592–3601.
137 Steinmetz OM, Turner JE, Paust HJ, et al: CXCR3 mediates renal Th1 and Th17 immune response in murine lupus nephritis. J Immunol 2009;183:4693–4704.
138 Herber D, Brown TP, Liang S, Young DA, Collins M, Dunussi-Joannopoulos K: IL-21 has a pathogenic role in a lupus-prone mouse model and its blockade with IL-21R.Fc reduces disease progression. J Immunol 2007;178:3822–3830.
139 Dang H, Geiser AG, Letterio JJ, et al: SLE-like autoantibodies and Sjogren's syndrome-like lymphoproliferation in TGF-beta knockout mice. J Immunol 1995;155:3205–3212.
140 Menke J, Rabacal WA, Byrne KT, et al: Circulating CSF-1 promotes monocyte and macrophage phenotypes that enhance lupus nephritis. J Am Soc Nephrol 2009;20:2581–2592.
141 Kuroiwa T, Iwasaki T, Imado T, Sekiguchi M, Fujimoto J, Sano H. Hepatocyte growth factor prevents lupus nephritis in a murine lupus model of chronic graft-versus-host disease. Arthritis Res Ther 2006;8:R123.

Joseph P. Grande, MD, PhD
Mayo Clinic
200 First Street SW
Rochester, MN 55905 (USA)
Tel. +1 507 266 9356, Fax +1 507 266 1163, E-Mail grande.joseph@mayo.edu

Herrera GA (ed): Experimental Models for Renal Diseases: Pathogenesis and Diagnosis.
Contrib Nephrol. Basel, Karger, 2011, vol 169, pp 198–210

Experimental Models of Membranoproliferative Glomerulonephritis, Including Dense Deposit Disease

Katherine A. Vernon · Matthew C. Pickering · H. Terence Cook

Centre for Complement and Inflammation Research, Department of Medicine, Imperial College, London, UK

Abstract

Membranoproliferative glomerulonephritis (MPGN) is characterised by mesangial expansion and hypercellularity and capillary wall thickening with capillary wall and mesangial deposits of immunoglobulin and/or complement. Two main forms are described in humans: MPGN type I with subendothelial and mesangial electron-dense deposits on electron microscopy, and MPGN type II, or dense deposit disease, with electron dense transformation of the glomerular capillary wall. Spontaneous MPGN type I has been described in dogs and sheep in association with C3 deficiency. Induced models of MPGN type I have been described in mice with cryoglobulinaemia. Glomerulonephritis resembling MPGN type II has occurred spontaneously in pigs that have a genetic deficiency of the complement control protein factor H. The animals develop capillary wall deposits of C3 before birth. Mice have been genetically engineered with a deficiency of factor H and similarly develop glomerular capillary wall C3 with MPGN. This model has been used to study both pathogenesis and therapeutic interventions. In particular, MPGN associated with factor H deficiency is absolutely dependent on both the ability to activate C3 and on the ability of factor I to cleave C3b. There is an important role for C5 activation in the development of glomerular inflammation in this model. Factor H dysfunction is associated with an increased susceptibility to complement-activating nephrotoxic insults and in these scenarios C5 activation appears to play a major role in mediating glomerular injury.

'Membranoproliferative' describes a pattern of glomerular injury characterised by mesangial expansion and hypercellularity, and thickening of glomerular capillary walls. It is a pattern that may be seen with immunoglobulin and/or complement deposits in glomeruli or in other conditions, such as chronic thrombotic microangiopathy. The term membranoproliferative glomerulonephritis (MPGN) is best reserved for those cases in which there is glomerular immunoglobulin and/or complement, and electron-dense deposits on electron microscopy. These deposits are typically subendothelial and mesangial in location and often associated with duplication of the glomerular basement membrane and mesangial cell interposition. The terms membranoproliferative glomerulonephritis and mesangiocapillary glomerulonephritis (MCGN) are synonymous.

At least three variants of MPGN have been described in humans. Type I MPGN is the most common. Light microscopy shows increased glomerular lobulation with mesangial expansion and hypercellularity, and capillary wall thickening with a double contour appearance on silver staining. Electron microscopy demonstrates mesangial and subendothelial electron-dense deposits and often scattered subepithelial deposits. Type I MPGN is a pattern of glomerular injury that may be associated with a number of chronic infections, including viral hepatitis. It may be seen in mixed cryoglobulinaemia, and lupus glomerulonephritis may also show a membranoproliferative appearance.

Type II MPGN shows a characteristic electron microscopic appearance with dense osmiophilic transformation of the glomerular basement membrane and this appearance is the reason for the alternative name of dense deposit disease (DDD). It is now clear that the appearance of MPGN is only one morphological appearance of DDD. DDD typically occurs in the setting of hypocomplementaemia resulting from uncontrolled activation of the alternative pathway of complement.

Type III MPGN has features of type I with, in addition, either conspicuous subepithelial deposits similar to membranous glomerulonephritis (Burkholder type) or disruption and layering of the glomerular basement membrane (Strife type). Many pathologists consider these appearances to be variants of MPGN type I rather than separate entities.

MPGN is a condition which continues to pose a challenge, not only because the pathogenesis remains unclear, but also due to uncertainty over optimal treatment regimes and its poor prognosis, with 50% of patients developing end-stage renal disease after 10 years. The need for experimental models of MPGN to enable further investigation into the mechanisms involved in the initiation and progression of the disease, and to enable targeted therapies to be designed, is therefore clear. Experimental models to date can be divided into those that have occurred spontaneously, as in the dog and pig, and those that have been engineered, as in the case of genetically manipulated mice.

Experimental Models of Type I MPGN

Spontaneous MPGN has been reported in the dog [1], Finnish Landrace cross lamb [2], horse [3] and pig [4]. The MPGN described in the pig is akin to human DDD and will therefore be discussed later. There are also engineered models of cryoglobulinaemic MPGN in the mouse.

Spontaneous Models of MPGN Type I

In the dog, an autosomal recessive deficiency of C3 [5, 6] leads to C3 levels of less than 0.001% of normal [7], and histological evidence of type I MPGN [1]. The histopathological changes seen in the dog were similar to those in humans, with increased mesangial cellularity and matrix, mesangial cell interposition in the glomerular capillary wall, and mesangial and subendothelial deposits staining for IgG and IgM. Interestingly, infusion of C3-containing plasma into these dogs led to worsening of both their proteinuria and histological lesions.

Type I MPGN has also occurred spontaneously in Finnish Landrace cross lambs [2]. The affected lambs were born hypocomplementaemic, and circulating levels of C3 were reported to be less than 5% of normal [8]. Histologically there was mesangial hypercellularity, thickening of the capillary wall, peri-glomerular fibrosis and subendothelial electron-dense deposits on electron microscopy. Immunofluorescent studies showed granular to linear deposits of C3, IgG, IgM and IgA in glomerular capillary walls. Affected lambs were under 4 months of age, and were found to have hypoalbuminaemia, elevated plasma urea levels and albuminuria [9]. Again, a genetic predisposition to the disease, recessive in nature, was thought to be responsible [10].

In both lambs and dogs, genetic deficiency of C3 has therefore been associated with MPGN type I. In humans with C3 deficiency, MPGN has also been described (reviewed in [11]). The mechanism is thought to relate to the role of complement in the physiological removal of immune complexes from the circulation. In the absence of effective opsonisation of immune complexes by complement, complexes accumulate in the circulation and deposit in tissues, such as the kidney, triggering inflammation. Deficiency of classical pathway components such as C2 have also been associated with MPGN type I. Interestingly, in contrast to the deleterious effects of plasma infusion in the C3-deficient dogs, plasma therapy in human C2 deficiency has been beneficial [12].

Engineered Models of Cryoglobulinaemic MPGN

Cryoglobulins and MPGN

Cryoglobulins are immunoglobulins (Ig) or complexes of immunoglobulins that precipitate below 37°C. They are responsible for systemic disease in humans in which renal involvement commonly manifests as MPGN. Cryoglobulins are divided into three groups according to the cryoprecipitate composition. Type I cryoglobulins consist of a monoclonal immunoglobulin or light chain, usually in association with lymphoproliferative disorders. Mixed cryoglobulins are complexes of two or more immunoglobulins, in which IgG is bound by immunoglobulin with anti-IgG (rheumatoid factor) activity. In type II cryoglobulins, the anti-globulin component is monoclonal (IgM), whilst in type III cryoglobulins the IgM component is polyclonal. Conditions associated with mixed cryoglobulins include autoimmune diseases such as systemic lupus erythematosus, and those with chronic antigenaemia such as hepatitis C. In fact hepatitis C is now thought to be responsible for the many cases of mixed cryoglobulinaemia and MPGN previously felt to be idiopathic.

Mixed cryoglobulinaemia is possibly the most common cause of type I MPGN, with approximately 30% of patients with mixed cryoglobulinaemia developing MPGN. A model to study the pathogenesis of the condition was therefore important for expanding understanding of the mechanisms responsible for the development of MPGN type I.

The Thymic Stromal Lymphopoietin Transgenic Mouse Line

A model of cryoglobulinaemic MPGN that has been extensively studied is induced in the mouse by transgenic overexpression of thymic stromal lymphopoietin (TSLP) under the control of the *lck* promoter [13–15]. TSLP is an interleukin-7-like cytokine previously isolated from conditioned medium of a thymic stromal cell line, and supports the differentiation of IgM-positive B cells [16, 17]. TSLP transgenic mice were derived from a single founder and backcrossed for more than eight generations to a C57BL/6 background.

TSLP transgenic mice develop a characteristic phenotype, notably more severe in females than males with mixed (type III) cryoglobulin formation, and a systemic inflammatory disease involving the liver, kidney, lungs, spleen and skin [15]. All mice developed a pattern of MPGN closely resembling cryoglobulinaemic glomerulonephritis in humans, with increased blood urea nitrogen levels and proteinuria. The ultrastructural appearance of the cryoglobulin deposits on electron microscopy resembled that seen in humans. This model has since been used to study both pathogenesis and the efficacy of therapeutic interventions.

Detailed studies of this model have been carried out using immunohistochemistry and in situ hybridisation. TSLP mice showed increased mesangial expression of α smooth muscle actin starting at 1 month associated with increased expression of type IV collagen and laminin, but not type I collagen [13]. These findings are consistent with human MPGN [18, 19]. There was increased platelet-derived growth factor (PDGF) B chain and PDGF β-receptor (β-R) mRNA in the mesangium and parietal epithelial cells. Modulators of the plasmin pathway such as tissue plasminogen activator (tPA), plasminogen activator inhibitor (PAI-1) and protease nexin-1 (PN-1) have been examined in the TSLP model of MPGN [14].

Plasminogen activators and inhibitors play a significant role in regulating renal fibrosis [20–22], with production of plasmin leading to degradation of extracellular matrix (ECM) components and activation of matrix metalloproteinases, thereby reducing fibrosis. tPa is the main plasminogen activator in glomeruli and is inhibited by PAI-1 and PN-1. PN-1 and tPA were significantly upregulated in TSLP mice compared to wild-type mice. Incubation of mesangial cells with cryoglobulins in vitro stimulated TGF-β1 secretion, and was associated with increased expression of PN-1 mRNA and its product PN-1, in addition to tPA and PAI-1.

The role of Fc receptors for IgG, Fcγ receptors, has been investigated in this model [23, 24]. Four murine Fcγ receptors (FcγRs) have been described. FcγRI, III and IV are activatory receptors that share the FcRγ chain, while FcγRIIb is an inhibitory receptor that downregulates B cell responsiveness and myeloid cell activation. FcγRIIb is a single subunit receptor containing a tyrosine-based inhibitory motif. To investigate the role of FcγRIIb, the receptor was deleted in the TSLP transgenic mouse [24]. TSLP/FcγRIIb$^{-/-}$ mice were found to have significantly impaired renal function, greater proteinuria, exacerbated glomerular histology with an increase in glomerular extracellular matrix and cellularity (due in part to a greater influx of macrophages), and reduced survival. It was concluded that FcγRIIb is involved in limiting immune complex-induced recruitment of macrophages, and in processes that lead to the production of ECM and cellular proliferation.

To investigate the role of activating FcγRs, FcRγ chain knockout (FcRγ$^{-/-}$) mice were crossed with TSLP transgenic mice [23]. Both TSLP and TSLP/FcRγ$^{-/-}$ mice developed MPGN, but the TSLP/FcRγ$^{-/-}$ mice had higher levels of circulating cryoglobulins and more glomerular immune complex deposits compared to the TSLP mice, although proteinuria was similar. The authors suggested that the absence of activating FcγRs on macrophages led to a reduced clearance of immune complexes and, thus, a greater concentration of circulating cryoglobulins, thereby resulting in increased cryoglobulin deposition in the kidney.

The effect of overexpressing complement receptor-1 related protein Y (Crry), an inhibitor of C3 convertase, has also been studied in this model [25]. Crry is a membrane-bound protein found in rodents which inhibits the cleavage of C3 into its active metabolites, thereby blocking both the classical and alternative pathways of complement activation. The overexpression of Crry in TSLP mice however did not ameliorate the renal pathology, with no reduction seen in macrophage influx, complement and immunoglobulin deposition, or glomerular C9 deposition [an indicator of the presence of the membrane attack complex (MAC)]. ECM and glomerular collagen type IV deposition were also increased similarly in the TSLP mice. It was suggested that Crry overexpression was insufficient to prevent renal damage in this model on account of its inhibitory capacity being overwhelmed by the degree of complement activation.

A number of potential therapies have been investigated in this transgenic mouse model, including interferon-α (IFN-α), imatinib and all-trans-retinoic acid (ATRA). Oral administration of IFN-α showed no significant effect in this model, which suggests that its beneficial effect in human MPGN is primarily due to its antiviral effects [26]. Imatinib is a selective tyrosine kinase inhibitor which blocks the effects of PDGF receptors (PDGFRs) amongst others. Given previous work reporting an association between glomerular cell proliferation and ECM accumulation with increased expression of PDGF-B and PDGFR-β in cryoglobulinaemic glomerulonephritis, the TSLP model has been used to assess the effect of imatinib in this setting [27]. Treatment with imatinib inhibited cryoglobulin production, resulting in attenuation of systemic disease and also renal injury with reduced immune complex (IC) deposition, both in early and established disease. Finally, the third agent tested in this model is all-transretinoic acid [28]. With retinoids having previously been shown to have a therapeutic effect in other animal models of renal disease, the effect of ATRA (a carboxylic acid form of vitamin A) on the cryoglobulinaemic MPGN phenotype was examined. ATRA was actually found to aggravate renal injury in TSLP mice, and circulating cryoglobulins were also increased, as were their systemic manifestations involving the lungs and liver, characterised by increased leucocyte infiltration.

Murine MPGN Induced by IgG3 Monoclonal Hybridoma

The second murine model of MPGN is also a cryoglobulinaemic form of glomerulonephritis, induced by an infusion of murine hybridoma cells [29] producing monoclonal IgG3, which behaves as a cryoglobulin. The model has since been advanced by the development of mice that overexpress the IgG3 (clone 6–19) which has both rheumatoid factor (anti-IgG2a) and cryoglobulin activity [30]. Histopathological assessment in these mice reveals confluent subendothelial deposits of cryoglobulins in glomerular capillary walls, and neutrophil-rich glomerular infiltrates.

The model has been used to answer a number of questions, including what aspect of the IgG3 molecule confers its renal pathogenicity. Mice implanted with hybridoma secreting 6–19 IgG3 anti-IgG2a rheumatoid factor (with cryoglobulin activity) develop cutaneous leucocytoclastic vasculitis and acute glomerulonephritis, typified by 'wire-loop'-like glomerular lesions [30]. However although rheumatoid factor and cryoglobulin activity are required for development of the skin lesions, the anti-IgG2a rheumatoid factor activity is not necessary for induction of glomerular lesions.

The injection of 6–19 hybridoma cells to induce cryoglobulinaemia has also been used to investigate the role of complement in cryoglobulin-induced IC glomerulonephritis [31]. Complement-deficient mice on both a C57BL/6 and BALB/c genetic background were compared with strain-matched wild-type (WT) controls. By using these two genetic backgrounds, type I and type II cryoglobulinaemia could be compared. In BALB/c mice, the IgG3 cryoglobulin produced has rheumatoid factor activity against IgG2a (of allotype α) which is not present in C57BL/6 mice. As a result, the cryoprecipitate in C57BL/6 mice was composed of the IgG3 cryoglobulin only (type I), whilst in BALB/c mice there were IgG3-IgG2a complexes (type II). Proliferative glomerulonephritis with endocapillary hypercellularity and neutrophil infiltration occurred after intraperitoneal injection of hybridoma cells, and this was followed by systemic manifestations.

The authors found no apparent difference in the role of complement between the two strains and, thus, in cryoglobulinaemia types I and II, and there was no survival advantage in complement-deficient mice. Neutrophil influx into glomeruli was less in C3-, factor B- and C5-deficient mice compared to WT and C1q-deficient mice, suggesting an important role for the alternative pathway. The lack of CD59a, the membrane inhibitor of the MAC, did not lead to an increase in neutrophil infiltration, leading the authors to suggest that it is C5a generation which accounts for neutrophil recruitment in this model, and that C5 blockade may therefore be a potential therapeutic strategy for attenuating renal involvement in cryoglobulinaemia.

Key Points Learnt from the Engineered Experimental Models of Type I MPGN

The engineered models of MPGN described are murine models of cryoglobulinaemic MPGN. The TSLP transgenic mouse in which cryoglobulinaemic MPGN is induced by overexpression of thymic stromal lymphopoietin, has provided a reproducible phenotype confirming the histological changes characteristic of this lesion. Therefore, the model has also enabled a variety of hypotheses relating to the part played by a diverse range of mechanisms in the pathogenesis of MPGN to be tested. These include the role of Fcγ receptors, growth factors, and mediators of fibrosis. Furthermore, the model has also been used to interrogate a number of potential therapeutic agents. The second model of cryoglobulinaemic MPGN results from the presence of IgG3 produced either by murine hybridoma cells or overexpressed by the mouse itself. The transgenic murine model has been utilised to define which part of the IgG3 molecule confers its renal pathogenicity, whilst the use of murine hybridoma cells has shown the effects of an acute rise in cryoglobulin levels, and in conjunction with various complement-deficient mice, has illustrated the importance of the alternative pathway and C5a generation in this setting.

Experimental Models of Type II MPGN

Type II MPGN (also termed dense deposit disease) is associated with complement dysregulation. The majority of individuals with MPGN type II have hypocomplementaemia at some stage during their illness [32], and the condition is strongly associated with the presence of C3 nephritic factor (C3NeF) [32]. C3NeF is an IgG autoantibody that binds to the C3 cleaving enzyme complex (termed a C3 convertase) of the complement alternative pathway [33]. To understand the action of C3NeF, it is necessary to outline the key features of the complement alternative pathway (fig. 1).

Unlike the classical and lectin complement activation pathways, which are triggered by immune complexes and bacterial sugars respectively (fig. 1), the alternative pathway of complement activation is in a state of constant activation. This is a consequence of spontaneous hydrolysis in plasma of an internal thioester bond in complement C3. When the thioester bond is hydrolysed (denoted by C3i), C3 is able to interact with another complement plasma protein factor B. The complex C3iB is then a substrate for the serine protease factor D which cleaves factor B to form a complex (denoted by C3iBb). This complex has C3 convertase activity, i.e. it is able to cleave intact C3 to form C3b. C3b, in exactly the same manner as C3i, is able to generate a C3 convertase (C3bBb) through interaction with factor B and D. Therefore, in plasma there is continuous low-level generation of C3b. This pathway has been termed the C3 tick-over pathway. The generation of C3b proceeds at a low level because of the presence of two plasma complement regulatory proteins: factor H and factor I. Factor H inhibits the formation of the C3 convertases whilst factor I inactivates C3b by proteolysis. The importance of both factor H and factor I is seen by studying the complement profile of rare individuals with complete deficiencies of either of these proteins. Deficiency of factor H results in unhindered tick-over C3 activation and consequent severe secondary C3 deficiency. Deficiency of factor I results in accumulation of C3b in plasma.

It is now evident that MPGN type II is strongly associated with dysregulation of the complement alternative pathway. C3NeF binds to the C3 convertase and potentiates its C3 cleaving function [33]. The result is secondary C3 depletion in plasma. Furthermore, other distinct causes of excessive C3 convertase activity, all of which result in secondary C3 activation, have been associated with MPGN type II. These include dysfunctional C3 molecules [34, 35], an anti-factor H autoantibody [36], a dysfunctional factor H molecule [37] and factor H deficiency [38, 39].

Despite these strong associations between C3 dysregulation and MPGN type II, the evidence in humans is predominantly circumstantial. Furthermore, C3NeF and C3 depletion can be found among individuals without renal disease. Whether or not C3 nephritic factors are directly pathogenic in MPGN type II or epiphenomenon has consequently been hotly debated. Insights from two experimental models of MPGN type II have reinforced the intimate association between C3 dysregulation and MPGN type II and, in the case of the mouse, provided evidence that C3 activation is an absolute requirement for this type of renal lesion to develop.

In order to develop a mouse model of MPGN type II, it was thought critical to create a situation of uncontrolled C3 activation in vivo. This was achieved by generating a mouse with complete factor H deficiency. A serendipitous development was the identification of a breed of pigs with spontaneous factor H deficiency. Both the factor H-deficient pig and mouse develop MPGN with features akin to human MPGN type II. The key aspects of both animal models will now be discussed.

Porcine Complement Factor H Deficiency and MPGN

Lethal MPGN has been described in a strain of pigs that were subsequently shown to have complete deficiency of complement factor H [4, 40–44]. Affected piglets appeared healthy at birth, but soon developed a rapidly progressive glomerulonephritis with a median survival of 37 days.

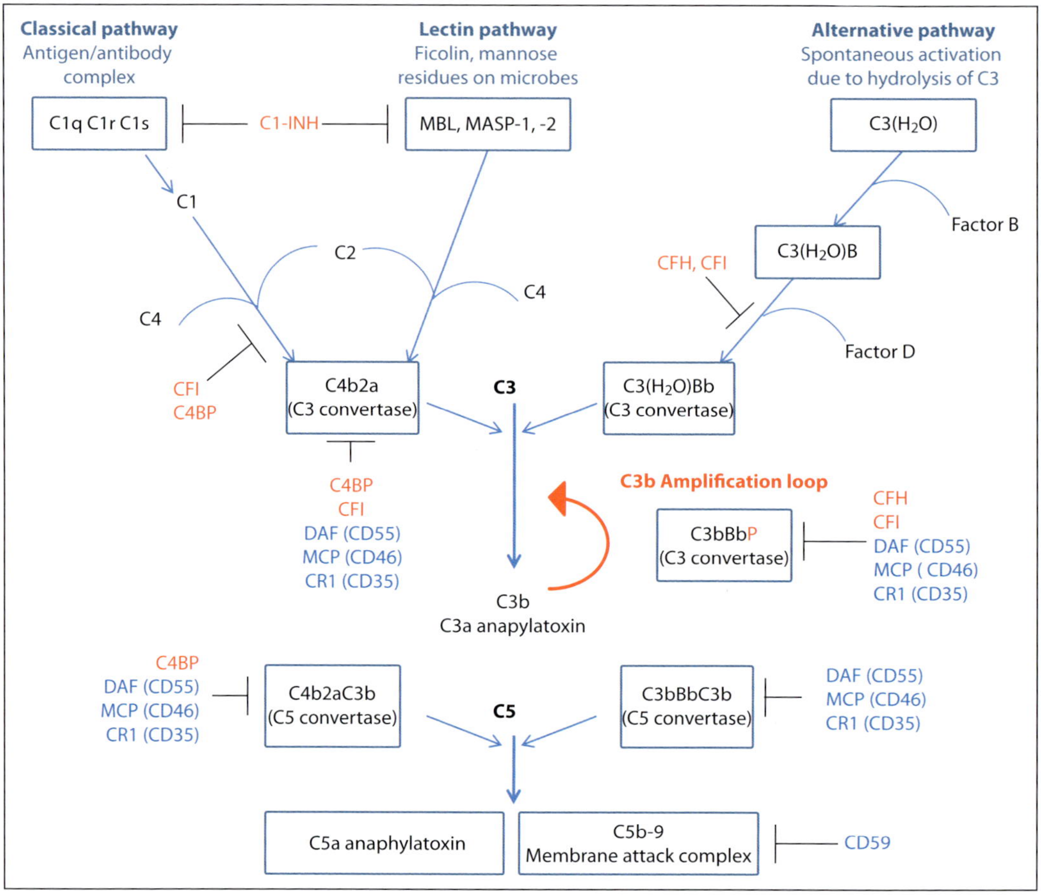

Fig. 1. Schematic depicting the complement system. Complement activation can be triggered through three pathways termed the classical, lectin and alternative pathways. Activation of each of these pathways results in the generation of a complex that is capable of cleaving complement C3. These enzyme complexes are termed C3 convertases and comprise C4b2a (classical and lectin pathway) and C3bBb (alternative pathway). Cleavage of C3 results in the generation of the opsonin C3b and the anaphylatoxin C3a. C3 nephritic factor (C3NeF) potentiates the alternative pathway C3 convertase, C3bBb. This results in uncontrolled conversion of C3 to C3b. Hence, C3NeF is commonly associated with low levels of intact plasma C3. Regardless of which pathway resulted in the initial cleavage of C3, rapid amplification of C3b is achieved through a positive feed-back loop termed the C3b amplification loop. C3 convertases can, through the addition of a single molecule of C3b, generate larger complexes that are capable of cleaving complement C5. These complexes are termed C5 convertases and comprise C4b2aC3b (classical and lectin pathway) and C3bBbC3b (alternative pathway). C5 activation results in the generation of the potent anaphylatoxin C5a and generation of C5b which is the first activation protein of the complement terminal pathway. Activation of the terminal pathway triggers the assembly of the membrane attack complex (MAC). The MAC is formed by the sequential addition to C5b of C6, C7, C8 and multiple C9 molecules. Complement regulatory proteins, in both plasma (depicted in blue) and attached to cell membranes (depicted in red), inhibit complement activation at multiple steps. Important examples include: C4-binding protein for the classical pathway, factor H for the alternative pathway and CD59 for the terminal pathway. Properdin stabilises the alternative pathway C3 convertase and is therefore considered a positive regulator. MBL = mannose-binding lectin; MASP = MBL-associated serine protease; C1-INH = C1 inhibitor; C4BP = C4 binding protein; DAF = decay-accelerating factor (CD55); MCP = membrane cofactor protein (CD46); CFH = complement factor H; CFI = complement factor I; CR1 = complement receptor 1 (CD35); P = properdin.

All affected piglets were deficient in factor H [4, 42] due to a mutation in SCR20 which prevented secretion of the mutant protein [40]. The disease was 100% penetrant in homozygous factor H-deficient pigs, whilst no heterozygous carriers developed disease.

The renal lesion in affected piglets was characterised by MPGN [42–44]. Positive glomerular staining for C3 and the MAC in association with vitronectin was present, whilst immunoglobulin and factor B staining was negative. Ultrastructural changes included subendothelial and intramembranous dense deposits. In older lesions, subepithelial and mesangial dense deposits were also present [43]. Unlike human DDD, electron dense deposits were not seen in the basement membranes of either tubules or Bowman's capsule. Immune electron microscopic studies showed that the GBM deposits stained strongly for both C3 and MAC. Notably, no pathology was evident in any of the other tissues examined (synovium, lungs, spleen, choroid plexus and ciliary body).

The chronology of the glomerulonephritis was extensively documented [43]. Massive glomerular deposits of C3 and the MAC were present in factor H-deficient fetuses before the onset of mesangial hypercellularity or GBM deposits. In fetuses, the C3 and MAC staining was initially linear, along capillary walls, without mesangial staining. However. mesangial staining became evident from 1 week. After 3 weeks, abundant and granular mesangial C3 staining developed with increasingly coarse linear capillary staining. The earliest light microscopic changes were evident in piglets aged 4 days and showed mild mesangial hypercellularity and expansion together with increased glomerular neutrophil numbers. By 3 weeks, these changes were marked. GBM electron-dense deposits developed initially within the subendothelial side of the lamina densa (earliest changes evident at 4 days) and gradually progressed to occupy more of the GBM, including the subepithelial side. After the age of 4 weeks, laminar disintegration of the dense deposits created the effect of intramembranous lucencies with double contours.

In these animals two therapeutic manoeuvres were studied: plasma therapy and renal transplantation. Firstly the effect of plasma therapy (i.e. aiming to reconstitute the missing porcine factor H by infusion of normal porcine plasma) was studied [4]. A single 30ml/kg plasma infusion increased plasma C3 levels and reduced soluble plasma terminal complement complex levels for approximately 3 days. A comparable effect was seen following a single infusion of 5ml/kg purified porcine factor H. Administration of weekly plasma infusions to affected piglets (from the age of 2 weeks) resulted in improved survival (median survival was 37 and 82 days, treated versus non-treated, respectively) [45]. Renal histology in the treated pigs showed the persistence of membranoproliferative changes, although the severity of the lesions was less when compared to the untreated animals. Secondly, following transplantation of a normal pig kidney into a factor H-deficient pig, complement deposits appeared in the transplant kidney within 24 hours (h) [43].

Key Points Learnt from the Factor H-Deficient Pig Model

The factor H-deficient pig strain has now been eradicated [41], but its elegant characterisation has provided important insights into MPGN associated with factor H deficiency, including: (1) the abnormal accumulation of complement along the glomerular basement membrane is the initial renal abnormality, (2) factor H reconstitution restored C3 regulation transiently and was therapeutically beneficial, and (3) the rapid development of the disease in a transplanted normal pig kidney indicated that the disease cannot be prevented by local factor H production within the kidney.

Murine Complement Factor H Deficiency and MPGN

These animals were developed by targeted gene deletion of exon 3 of the murine factor H gene in embryonic stem cells. Similar to homozygous factor H-deficient humans, these animals have extremely low levels of plasma C3 as a result of uncontrolled alternative pathway activation [46].

Spontaneous Renal Disease in the Factor H-Deficient Mouse

At 8 months, all of the homozygous factor H-deficient mice had developed light microscopic features of MPGN [46]. Hence, like homozygous factor H-deficient pigs, MPGN in factor H-deficient mice was fully penetrant. Similarly, no evidence of spontaneous renal disease was evident in heterozygous factor H-deficient mice. However, in contrast to the uniform lethality seen in the factor H-deficient pigs, mortality in murine factor H deficiency was 23% [46].

Although MPGN in the factor H-deficient mice had many similarities with human MPGN type II, particularly the presence of hypocomplementaemia and deposition of C3 and electron-dense material along the GBM, there are (as with factor H-deficient pigs) some important differences. The glomerular basement membrane electron-dense deposits in the factor H-deficient mice are located predominantly in a subendothelial location and linear dense transformation of the basement membrane, as in DDD, is not seen. Deposits did not develop in the mesangium or in the basement membranes of the tubules or Bowman's capsule. Similar to factor H-deficient pigs and humans, the glomerular basement membrane deposits stain directly for C3 and C9, but not immunoglobulin using immune electron microscopy [46].

Sequential analysis of the renal histopathology in these mice showed an identical evolution to that described in the factor H-deficient pigs. Hence, glomerular C3 deposition preceded the appearance of GBM dense deposits, which in turn, preceded the histological light microscopic appearances of MPGN. Glomerular C3 deposition was evident as early as the first week of life, a time when both the light and electron microscopic appearance of the glomerulus was normal [46].

Novel mechanistic insights derived from the factor H-deficient mouse model include the very important observation that MPGN in this animal was totally dependent on C3 activation. Hence, preventing C3 activation (achieved by crossing factor H-deficient mice with factor B-deficient animals) completely prevented the development of MPGN [46]. Furthermore, preventing C5 activation (achieved by crossing factor H-deficient mice with C5-deficient animals) did not prevent the development of the glomerular basement membrane deposits [47]. However, the inability to activate C5 was associated with a reduction in glomerular cellularity, serum creatinine levels and mortality [47]. This suggests that whilst it would not be expected to prevent the development of glomerular basement membrane deposits, chronic inhibition of C5 activation in humans with MPGN associated with factor H dysfunction may be beneficial by reducing the glomerular inflammatory response. Whether this would also be applicable to factor H-sufficient individuals with DDD and C3NeF is not clear.

Similar to the pig transplantation experiments, C3 deposition developed rapidly in wild-type kidneys transplanted into factor H-deficient mice [48, 49]. Conversely, glomerular basement membrane C3 accumulation appeared to fully resolve when factor H-deficient kidneys were transplanted into wild-type mice [48].

A novel observation has recently come from studies of animals with combined deficiency of both factor I and factor H [49]. Factor I-deficient mice develop uncontrolled C3 activation through the alternative pathway together with renal abnormalities characterised by mesangial C3 deposits with nodular mesangial expansion. Importantly however, there was no evidence of GBM abnormalities. Human factor I deficiency is similarly associated with uncontrolled C3 activation through the alternative pathway and renal disease is uncommon, and no association with DDD has been described [50–53].

Remarkably, factor H-deficient mice with deficiency of factor I do not develop MPGN. In these animals the glomerular lesion is identical to that seen in mice deficient in factor I alone.

Administration of a source of factor I to mice with combined deficiency of H and I resulted in the appearance of glomerular C3 staining in a pattern identical to that seen in mice with deficiency of factor H alone. These data show that in the setting of factor H deficiency, factor I is required for the deposition of GBM C3 to develop with morphological changes of MPGN. These animal data suggest that plasma C3b targets the mesangium whilst plasma C3b metabolites target the GBM. The implication then is that in situations where GBM C3 is present, strategies that sequester C3b metabolites in the circulation will be therapeutically beneficial.

Experimentally-Triggered Renal Disease in the Factor H-Deficient Mouse

Factor H-deficient mice have an increased susceptibility to renal injury following the induction of experimental nephritis [46, 47]. This exaggerated response was evident in young animals prior to the development of light microscopic changes of MPGN, suggesting that it is the complement dysregulation rather than pre-existing renal disease that renders these animals susceptible to nephrotoxic agents. Experimental heterologous nephrotoxic nephritis develops following the intravenous injection of heterologous anti-mouse glomerular basement membrane antibody. Binding of the heterologous antibody along the GBM typically results in a transient glomerular neutrophil influx in wild-type animals which resolves within 24 h. However, in factor H-deficient animals, increased glomerular neutrophils remained present at 24 h and were still detectable 6 days after administration of the nephrotoxic antibody [47]. This was also associated with evidence of increased glomerular C3 deposition indicating that, despite profound fluid-phase consumption of complement, factor H-deficient mice retained the ability to deposit further C3 within the kidney following the binding of antibody along the GBM.

The abnormal response to heterologous nephrotoxic nephritis was absent in factor H-deficient mice with deficiency of C5, but persisted in animals with combined deficiency of factor H and C6 [47]. This demonstrated that it was the production of C5a, not the formation of the MAC, that was responsible for enhanced glomerular neutrophil response in this model. Furthermore, administration of an anti-mouse C5 antibody to factor H-deficient mice prior to the induction of heterologous nephrotoxic nephritis prevented both glomerular neutrophil accumulation and proteinuria [47]. These data indicate that, in the setting of factor H dysfunction, (1) the threshold for renal injury is significantly lowered following exposure to a complement-activating nephrotoxic insult and (2) C5a production is a major mediator of renal injury in this setting.

What is the relevance of these observations to human MPGN type II? It is known that patients with MPGN type II may remain stable clinically for long periods. However, relapses associated with acute inflammatory changes within the kidney often result in a significant decline in renal function or threaten transplant kidney survival. Increased glomerular neutrophil numbers, often with crescent formation, can occur in human DDD [54–56] and in DDD lesions developing in transplant kidneys in the absence of transplant-related pathology [57]. Moreover during rapidly progressive disease, all glomerular deposits including paramesangial deposits, react strongly with anti-C5 antibodies [58], suggesting C5 activation is a particularly important component of renal injury during acute inflammatory episodes. These observations suggest that anti-C5 therapy could be effective in preserving renal function during disease flares in human MPGN type II.

Key Points Learnt from the Factor H-Deficient Mouse Model

MPGN associated with factor H deficiency is absolutely dependent on both the ability to activate C3 and on the ability to cleave C3b. There also appears to be an important role for C5 activation in the development of glomerular inflammation in this type of MPGN. The abnormal C3 accumulation within the kidney was reversed when affected kidneys were transplanted into wild-type hosts. Factor H dysfunction is associated with an increased susceptibility to complement-activating nephrotoxic insults. In these scenarios C5 activation appears to play a major role in mediating glomerular injury.

Conclusion

To date, experimental models of MPGN have been used to try and answer a variety of questions regarding the mechanisms involved in the pathogenesis of the condition, and also to investigate possible therapeutic interventions. Establishing reproducible models of MPGN will enable the processes involved in the initiation and progression of this disease to be further defined. Furthermore, they should facilitate not only the identification of potential therapeutic targets, but also provide a means for their efficacy and safety to be tested.

References

1 Cork LC, Morris JM, Olson JL, Krakowka S, Swift AJ, Winkelstein JA: Membranoproliferative glomerulonephritis in dogs with a genetically determined deficiency of the third component of complement. Clin Immunol Immunopathol 1991;60:455–470.

2 Frelier PF, Pritchard J, Armstrong DL, Nagge WT, Lewis RM: Spontaneous mesangiocapillary glomerulonephritis in Finn cross lambs from Alberta. Can J Comp Med 1984;48:215–218.

3 Sabnis SG, Gunson DE, Antonovych TT: Some unusual features of mesangioproliferative glomerulonephritis in horses. Vet Pathol 1984;21:574–581.

4 Hogasen K, Jansen JH, Mollnes TE, Hovdenes J, Harboe M: Hereditary porcine membranoproliferative glomerulonephritis type II is caused by factor H deficiency. J Clin Invest 1995;95:1054–1061.

5 Johnson JP, McLean RH, Cork LC, Winkelstein JA: Genetic analysis of an inherited deficiency of the third component of complement in Brittany spaniel dogs. Am J Med Genet 1986;25:557–562.

6 Winkelstein JA, Cork LC, Griffin DE, Griffin JW, Adams RJ, Price DL: Genetically determined deficiency of the third component of complement in the dog. Science 1981;212:1169–1170.

7 Winkelstein JA, Johnson JP, Swift AJ, Ferry F, Yolken R, Cork LC: Genetically determined deficiency of the third component of complement in the dog: in vitro studies on the complement system and complement-mediated serum activities. J Immunol 1982;129:2598–2602.

8 Nuttall W: New sheep breeds bring new diseases. Surveillance 1991;18:23–24.

9 Angus KW, Sykes AR, Gardiner AC, Morgan KT, Thomson D: Mesangiocapillary glomerulonephritis in lambs. I. Clinical and biochemical findings in a Finnish Landrace flock. J Comp Pathol 1974;84:309–317.

10 Young GB, Sales DI, Waddington D, Angus KW, Gardiner AC: Genetic aspects of mesangiocapillary glomerulonephritis in Finnish sheep. Br Vet J 1981;137:368–373.

11 Botto M, Walport MJ: Hereditary deficiency of C3 in animals and humans. Int Rev Immunol 1993;10:37–50.

12 Steinsson K, Erlendsson K, Valdimarsson H: Successful plasma infusion treatment of a patient with C2 deficiency and systemic lupus erythematosus: clinical experience over forty-five months. Arthritis Rheum 1989;32:906–913.

13 Taneda S, Hudkins KL, Cui Y, Farr AG, Alpers CE, Segerer S: Growth factor expression in a murine model of cryoglobulinemia. Kidney Int 2003;63:576–590.

14 Taneda S, Hudkins KL, Muhlfeld AS, Kowalewska J, Pippin JW, Shankland SJ, Alpers CE: Protease nexin-1, tPA, and PAI-1 are upregulated in cryoglobulinemic membranoproliferative glomerulonephritis. J Am Soc Nephrol 2008;19:243–251.

15 Taneda S, Segerer S, Hudkins KL, Cui Y, Wen M, Segerer M, Wener MH, Khairallah CG, Farr AG, Alpers CE: Cryoglobulinemic glomerulonephritis in thymic stromal lymphopoietin transgenic mice. Am J Pathol 2001;159:2355–2369.

16 Friend SL, Hosier S, Nelson A, Foxworthe D, Williams DE, Farr A: A thymic stromal cell line supports in vitro development of surface IgM+ B cells and produces a novel growth factor affecting B and T lineage cells. Exp Hematol 1994;22:321–328.

17 Sims JE, Williams DE, Morrissey PJ, Garka K, Foxworthe D, Price V, Friend SL, Farr A, Bedell MA, Jenkins NA, Copeland NG, Grabstein K, Paxton RJ: Molecular cloning and biological characterization of a novel murine lymphoid growth factor. J Exp Med 2000;192:671–680.

18 Funabiki K, Horikoshi S, Tomino Y, Nagai Y, Koide H: Immunohistochemical analysis of extracellular components in the glomerular sclerosis of patients with glomerulonephritis. Clin Nephrol 1990;34:239–246.

19 Oomura A, Nakamura T, Arakawa M, Ooshima A, Isemura M: Alterations in the extracellular matrix components in human glomerular diseases. Virchows Arch A Pathol Anat Histopathol 1989;415:151–159.
20 Eddy AA: Plasminogen activator inhibitor-1 and the kidney. Am J Physiol Renal Physiol 2002;283:F209–F220.
21 Fogo AB: Renal fibrosis: not just PAI-1 in the sky. J Clin Invest 2003;112:326–328.
22 Huang Y, Noble N: An unexpected role of plasminogen activator inhibitor-type 1 (PAI-1) in renal fibrosis. Kidney Int 2005;67:2502–2503.
23 Guo S, Muhlfeld AS, Wietecha TA, Peutz-Kootstra CJ, Kowalewska J, Yi K, Spencer M, Pichaiwong W, Nimmerjahn F, Hudkins KL, Alpers CE: Deletion of activating Fcgamma receptors does not confer protection in murine cryoglobulinemia-associated membranoproliferative glomerulonephritis. Am J Pathol 2009;175:107–118.
24 Muhlfeld AS, Segerer S, Hudkins K, Carling MD, Wen M, Farr AG, Ravetch JV, Alpers CE: Deletion of the fcgamma receptor IIb in thymic stromal lymphopoietin transgenic mice aggravates membranoproliferative glomerulonephritis. Am J Pathol 2003;163:1127–1136.
25 Muhlfeld AS, Segerer S, Hudkins K, Farr AG, Bao L, Kraus D, Holers VM, Quigg RJ, Alpers CE: Overexpression of complement inhibitor Crry does not prevent cryoglobulin-associated membranoproliferative glomerulonephritis. Kidney Int 2004;65:1214–1223.
26 Segerer S, Hudkins KL, Taneda S, Wen M, Cui Y, Segerer M, Farr AG, Alpers CE: Oral interferon-alpha treatment of mice with cryoglobulinemic glomerulonephritis. Am J Kidney Dis 2002;39:876–888.
27 Iyoda M, Hudkins KL, Becker-Herman S, Wietecha TA, Banas MC, Guo S, Meyer-Bahlburg A, Kowalewska J, Liu G, Ziegler SF, Rawlings DJ, Alpers CE: Imatinib suppresses cryoglobulinemia and secondary membranoproliferative glomerulonephritis. J Am Soc Nephrol 2009;20:68–77.
28 Iyoda M, Hudkins KL, Wietecha TA, Banas MC, Guo S, Liu G, Wang L, Kowalewska J, Alpers CE: All-trans-retinoic acid aggravates cryoglobulin-associated membranoproliferative glomerulonephritis in mice. Nephrol Dial Transplant 2007;22:3451–3461.
29 Gyotoku Y, Abdelmoula M, Spertini F, Izui S, Lambert PH: Cryoglobulinemia induced by monoclonal immunoglobulin G rheumatoid factors derived from autoimmune MRL/MpJ-lpr/lpr mice. J Immunol 1987;138:3785–3792.
30 Kikuchi S, Pastore Y, Fossati-Jimack L, Kuroki A, Yoshida H, Fulpius T, Araki K, Takahashi S, Lemoine R, Reininger L, Izui S: A transgenic mouse model of autoimmune glomerulonephritis and necrotizing arteritis associated with cryoglobulinemia. J Immunol 2002;169:4644–4650.
31 Trendelenburg M, Fossati-Jimack L, Cortes-Hernandez J, Turnberg D, Lewis M, Izui S, Cook HT, Botto M: The role of complement in cryoglobulin-induced immune complex glomerulonephritis. J Immunol 2005;175:6909–6914.
32 Cameron JS, Turner DR, Heaton J, Williams DG, Ogg CS, Chantler C, Haycock GB, Hicks J: Idiopathic mesangiocapillary glomerulonephritis. Comparison of types I and II in children and adults and long-term prognosis. Am J Med 1983;74:175–192.
33 Daha MR, Fearon DT, Austen KF: C3 Nephritic factor (C3NeF): stabilization of fluid phase and cell-bound alternative pathway convertase. J Immunol 1976;116:1–7.
34 Linshaw MA, Stapleton FB, Cuppage FE, Forristal J, West CD, Schreiber RD, Wilson CB: Hypocomplementemic glomerulonephritis in an infant and mother. Evidence for an abnormal form of C3. Am J Nephrol 1987;7:470–477.
35 Marder HK, Coleman TH, Forristal J, Beischel L, West CD: An inherited defect in the C3 convertase, C3b,Bb, associated with glomerulonephritis. Kidney Int 1983;23:749–758.
36 Meri S, Koistinen V, Miettinen A, Tornroth T, Seppala IJ: Activation of the alternative pathway of complement by monoclonal lambda light chains in membranoproliferative glomerulonephritis. J Exp Med 1992;175:939–950.
37 Licht C, Heinen S, Jozsi M, Loschmann I, Saunders RE, Perkins SJ, Waldherr R, Skerka C, Kirschfink M, Hoppe B, Zipfel PF: Deletion of Lys224 in regulatory domain 4 of Factor H reveals a novel pathomechanism for dense deposit disease (MPGN II). Kidney Int 2006;70:42–50.
38 Levy M, Halbwachs-Mecarelli L, Gubler MC, Kohout G, Bensenouci A, Niaudet P, Hauptmann G, Lesavre P: H deficiency in two brothers with atypical dense intramembranous deposit disease. Kidney Int 1986;30: 949–956.
39 Lopez-Larrea C, Dieguez MA, Enguix A, Dominguez O, Marin B, Gomez E: A familial deficiency of complement factor H. Biochem Soc Trans 1987;15:648–649.
40 Hegasy GA, Manuelian T, Hogasen K, Jansen JH, Zipfel PF: The molecular basis for hereditary porcine membranoproliferative glomerulonephritis type II: point mutations in the factor H coding sequence block protein secretion. Am J Pathol 2002;161:2027–2034.
41 Hogasen K, Jansen JH, Harboe M: Eradication of porcine factor H deficiency in Norway. Vet Rec 1997;140:392–395.
42 Jansen JH, Hogasen K, Grondahl AM: Porcine membranoproliferative glomerulonephritis type II: an autosomal recessive deficiency of factor H. Vet Rec 1995;137:240–244.
43 Jansen JH, Hogasen K, Harboe M, Hovig T: In situ complement activation in porcine membranoproliferative glomerulonephritis type II. Kidney Int 1998;53:331–349.
44 Jansen JH, Hogasen K, Mollnes TE: Extensive complement activation in hereditary porcine membranoproliferative glomerulonephritis type II (porcine dense deposit disease). Am J Pathol 1993;143:1356–1365.
45 Hogasen K, Jansen JH, Mollnes TE, Hovdenes J, Harboe M: Hereditary porcine membranoproliferative glomerulonephritis type II is caused by factor H deficiency. J Clin Invest 1995;95:1054–1061.
46 Pickering MC, Cook HT, Warren J, Bygrave AE, Moss J, Walport MJ, Botto M: Uncontrolled C3 activation causes membranoproliferative glomerulonephritis in mice deficient in complement factor H. Nat Genet 2002;31:424–428.

47 Pickering MC, Warren J, Rose KL, Carlucci F, Wang Y, Walport MJ, Cook HT, Botto M: Prevention of C5 activation ameliorates spontaneous and experimental glomerulonephritis in factor H-deficient mice. Proc Natl Acad Sci USA 2006;103:9649–9654.
48 Alexander JJ, Wang Y, Chang A, Jacob A, Minto AW, Karmegam M, Haas M, Quigg RJ: Mouse podocyte complement factor H: the functional analog to human complement receptor 1. J Am Soc Nephrol 2007;18:1157–1166.
49 Rose KL, Paixao-Cavalcante D, Fish J, Manderson AP, Malik TH, Bygrave AE, Lin T, Sacks SH, Walport MJ, Cook HT, Botto M, Pickering MC: Factor I is required for the development of membranoproliferative glomerulonephritis in factor H-deficient mice. J Clin Invest 2008;118:608–618.
50 Amadei N, Baracho GV, Nudelman V, Bastos W, Florido MP, Isaac L: Inherited complete factor I deficiency associated with systemic lupus erythematosus, higher susceptibility to infection and low levels of factor H. Scand J Immunol 2001;53:615–621.
51 Genel F, Sjoholm AG, Skattum L, Truedsson L: Complement factor I deficiency associated with recurrent infections, vasculitis and immune complex glomerulonephritis. Scand J Infect Dis 2005;37:615–618.
52 Sadallah S, Gudat F, Laissue JA, Spath PJ, Schifferli JA: Glomerulonephritis in a patient with complement factor I deficiency. Am J Kidney Dis 1999;33:1153–1157.
53 Vyse TJ, Spath PJ, Davies KA, Morley BJ, Philippe P, Athanassiou P, Giles CM, Walport MJ: Hereditary complement factor I deficiency. QJM 1994;87:385–401.
54 Habib R, Gubler MC, Loirat C, Maiz HB, Levy M: Dense deposit disease: a variant of membranoproliferative glomerulonephritis. Kidney Int 1975;7:204–215.
55 Vargas R, Thomson KJ, Wilson D, Cameron JS, Turner DR, Gill D, Chantler C, Ogg CS: Mesangiocapillary glomerulonephritis with dense 'deposits' in the basement membranes of the kidney. Clin Nephrol 1976;5:73–82.
56 Walker PD, Ferrario F, Joh K, Bonsib SM: Dense deposit disease is not a membranoproliferative glomerulonephritis. Mod Pathol 2007;20:605–616.
57 Andresdottir MB, Assmann KJ, Hoitsma AJ, Koene RA, Wetzels JF: Renal transplantation in patients with dense deposit disease: morphological characteristics of recurrent disease and clinical outcome. Nephrol Dial Transplant 1999;14:1723–1731.
58 West CD, Witte DP, McAdams AJ: Composition of nephritic factor-generated glomerular deposits in membranoproliferative glomerulonephritis type 2. Am J Kidney Dis 2001;37:1120–1130.

H.T. Cook
Centre for Complement and Inflammation Research
Department of Medicine, Imperial College
London W12 0NN (UK)
Tel. +44 208 383 2009, Fax +44 208 383 2379, E-Mail t.cook@imperial.ac.uk

Herrera GA (ed): Experimental Models for Renal Diseases: Pathogenesis and Diagnosis.
Contrib Nephrol. Basel, Karger, 2011, vol 169, pp 211–220

Experimental Models of Vasculitis and Glomerulonephritis Induced by Antineutrophil Cytoplasmic Autoantibodies

J. Charles Jennette[a] · Hong Xiao[a] · Ronald Falk[b] · Adil Mohamad Hussein Gasim[a]

Departments of [a]Pathology and Laboratory Medicine and [b]Department of Medicine, University of Chapel Hill, Chapel Hill, N.C., USA

Abstract

Antineutrophil cytoplasmic autoantibodies (ANCA) are closely associated with systemic small vessel vasculitis characterized by segmental vessel wall necrotizing inflammation and a paucity of immunoglobulin deposition. Clinically, in vitro and experimental animal model observations indicate a direct pathogenic role for ANCA. This review focuses on the results of experiments utilizing a mouse model of ANCA disease induced by transfer of mouse anti-MPO IgG or anti-MPO lymphocytes into recipient mice, which causes small vessel vasculitis and glomerulonephritis that closely mimics human disease. Evidence for the following conclusion about this model, and by implication about human ANCA disease, will be summarized as follows: (1) anti-MPO IgG is sufficient even in the absence of functional T cells to cause disease and anti-MPO T lymphocytes are not sufficient to cause acute injury; (2) neutrophils are required; (3) ANCA antigens in bone marrow-derived cells are sufficient targets; (4) increased circulating pro-inflammatory cytokines and microbial products exacerbate disease, and concurrent viral infection exacerbates and modulates the phenotype of disease; (5) Fcγ receptor engagement is required for disease induction, and Fcγ receptor repertoire modulates the phenotype of disease, especially pulmonary disease; (6) activation of the alternative pathway of complement is required, complement is activated by factors released by neutrophils stimulated by ANCA IgG and engagement of C5a receptors is a primary event in complement-mediated amplification; and (7) genetic background has a marked influence on the severity and outcome of disease, and modified gene expression in bone marrow-derived cells is the primary basis for genetically determined differences in disease susceptibility. Investigations using this animal model of ANCA disease have provided important insights into the cellular, molecular and genetic factors involved in the pathogenesis of ANCA disease which are likely to lead to the identification of improved markers of disease activity and response to therapy, as well as more effective and less toxic therapies.

Antineutrophil cytoplasmic autoantibodies (ANCA) are closely associated with systemic small vessel vasculitis characterized by segmental vessel wall necrotizing inflammation and a paucity of immunoglobulin deposition [1]. Necrotizing and crescentic glomerulonephritis is frequent in patients with ANCA disease, and is the most common form of aggressive (crescentic) glomerulonephritis, especially in adults. Patients with ANCA disease can be classified on the basis of clinical and pathologic features as microscopic polyangiitis, Wegener's granulomatosis, Churg-Strauss syndrome or renal-limited vasculitis [1]. This chapter will focus on animal models of

ANCA-disease that elucidate the pathogenesis of this aggressive form of disease. Current therapy is toxic and not optimally effective, thus a better understanding of the etiology and pathogenesis of this aggressive disease could lead to better diagnosis, prognostication, treatment and even prevention.

The major ANCA antigen specificities are for myeloperoxidase (MPO-ANCA) [2] and proteinase 3 (PR3-ANCA) [3]. The most direct clinical evidence that ANCA are pathogenic is the report of transplacental transfer of MPO-ANCA to a neonate who developed glomerulonephritis and pulmonary hemorrhage [4]. Many laboratories have used in vitro experiments to demonstrate that ANCA IgG causes neutrophils and monocytes, both of which contain MPO and PR3, to become activated and release inflammatory mediators [5]. MPO and PR3 are predominantly in the cytoplasm of unstimulated neutrophils, although there may be small amounts of antigen on the surface. Neutrophils that have been stimulated by cytokines or other pro-inflammatory molecules (e.g. C5a) express more ANCA antigens at their surfaces and undergo activation when exposed to ANCA IgG [6]. Neutrophils in ANCA patients have a unique increased expression of ANCA antigens (MPO and PR3), which may facilitate disease induction [7]. ANCA IgG causes neutrophil activation by ligation of Fc receptors [8, 9] as well as by direct Fab'2 binding to ANCA antigens on or near cell surfaces [10, 11]. Activation of neutrophils by ANCA results in the release of inflammatory mediators that can injure vessel walls [7–15]. Thus, in vitro experimental data clearly demonstrate that ANCA IgG can activate neutrophils and monocytes. If this happens in vivo, pathogenic inflammatory vascular effects will be a likely result, especially in small vessels where leukocytes come in close contact with endothelial cell surfaces.

The first convincing animal model of ANCA-disease was reported in 2002 [16]. This model is induced by transfer of mouse anti-MPO IgG or anti-MPO lymphocytes from MPO knockout ($Mpo^{-/-}$) mice immunized with mouse MPO [16]. All mice develop necrotizing and crescentic glomerulonephritis and some develop systemic vasculitis, e.g. cutaneous leukocytoclastic angiitis, necrotizing arteritis, pulmonary capillaritis and, occasionally, pulmonary necrotizing granulomatous inflammation (fig. 1).

A subsequent mouse model was developed by immunizing $Mpo^{-/-}$ mice with mouse MPO, followed by irradiation and transplantation with $Mpo^{+/+}$ bone marrow (BM) cells [17]. Engraftment of $Mpo^{+/+}$ cells into $Mpo^{-/-}$ mice with circulating anti-MPO resulted in glomerulonephritis in all mice and pulmonary capillaritis and necrotizing arteritis in some. A third animal model of ANCA disease was reported by Little et al. [18] who produced glomerulonephritis and focal pulmonary capillaritis in rats by immunization with human MPO, which induced anti-MPO antibodies that cross-react with rat MPO. To date, there is no convincing animal model of vasculitic disease induced by anti-PR3 antibodies.

Kain et al. [19] reported the presence of antibodies to lysosomal membrane protein-2 (LAMP-2) in a high proportion of patients with ANCA disease. They reported that anti-LAMP-2 antibodies are present in almost all individuals with ANCA. A human LAMP-2 epitope that is recognized by these antibodies is homologous to peptides of the bacterial adhesin FimH, and human LAMP-2 antibodies cross-react with FimH. Furthermore, anti-LAMP-2 antibodies injected into WKY rats and rats immunized with FimH develop necrotizing and crescentic glomerulonephritis. Kain et al. [19] hypothesize that an immune response to FimH triggers the anti-LAMP-2 immune response by molecular mimicry, and that anti-LAMP-2 antibodies are major players in the pathogenesis of ANCA disease. However, these observations have not been confirmed by additional studies at the time this chapter was written, and this animal model will not be reviewed further.

This chapter will focus primarily on observations made in a mouse model of ANCA disease induced by transfer of mouse anti-MPO IgG or anti-MPO lymphocytes into recipient mice.

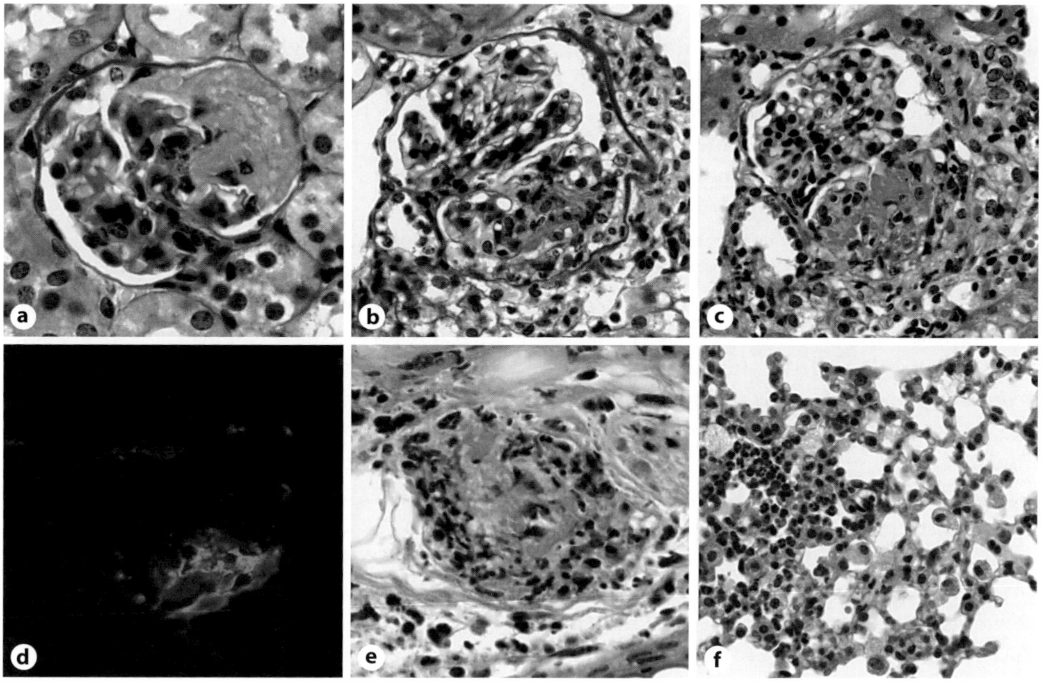

Fig. 1. Vasculitic lesions in WT B6 mice 6 days after they received anti-MPO IgG. **a** Glomerulus with segmental fibrinoid neurosis (periodic acid Schiff stain). **b** Glomerulus with segmental fibrinoid necrosis and crescent formation (periodic acid Schiff stain). **c** Glomerulus with segmental fibrinoid necrosis and crescent formation (H&E stain). **d** Immunofluorescence microscopy for fibrin showing prominent staining corresponding to segmental necrosis and crescent formation. **e** Necrotizing arteritis with leukocytoclasia in the dermis of the ear (H&E stain). **f** Pulmonary alveolar capillaritis on the left and more normal lung on the right. Reproduced with permission from [16].

Induction of Glomerulonephritis and Vasculitis by Transfer of Mouse Anti-MPO IgG or Anti-MPO Lymphocytes

Glomerulonephritis and vasculitis are induced by injecting mice with anti-MPO IgG or anti-MPO splenocytes derived from $MPO^{-/-}$ mice that have been immunized with mouse MPO [16]. Mouse MPO is purified from WEHI-3 mouse myeloid cells. $Mpo^{-/-}$ mice are immunized intraperitoneally with 10 μg of purified murine MPO in complete Freund's adjuvant. The IgG fraction is isolated from serum by 50% ammonium sulfate precipitation and protein G affinity chromatography. Anti-MPO lymphocytes are harvested in splenocytes from MPO-immunized $Mpo^{-/-}$ mice by disrupting the spleens into cold RPMI 1640 medium, rinsed, and red blood cells lysed. For anti-MPO antibody transfer experiments, mice are injected intravenously with 50 μg/g body weight of anti-MPO or control IgG in PBS, and sacrificed on day 6. In splenocyte transfer experiments, $Rag2^{-/-}$ mice are injected intravenously with MPO or control splenocytes in PBS and sacrificed on day 13.

In one set of experiments [16], $Rag2^{-/-}$ mice given 1×10^8 or 5×10^7 anti-MPO splenocytes developed renal insufficiency, but those that received 1×0^7 anti-MPO splenocytes or anti-BSA control splenocytes did not. This indicates a dose-dependent pathogenic effect by an anti-MPO immune response, which was further supported by the relative circulating levels of anti-MPO detected by ELISA. However, the urine analysis data were problematic because they showed the induction of proteinuria, hematuria and pyuria not only by anti-MPO splenocytes, but also by anti-BSA and normal control splenocytes.

The basis for abnormal urine findings in all three groups at the higher doses was immune complex localization in glomeruli that resulted from transfer of immune competent splenocytes into $Rag2^{-/-}$ mice, irrespective of any immunization of the donor mice. However, it is important that only the mice which received the two higher doses of anti-MPO splenocytes developed severe necrotizing and crescentic glomerulonephritis and vasculitis that mimicked human ANCA-disease.

In addition to severe glomerulonephritis, mice that received 1×10^8 or 5×10^7 splenocytes developed hemorrhagic pulmonary capillaritis, and necrotizing arteritis in many organs, including the spleen, lymph nodes and skin. Rare mice developed focal necrotizing granulomatous inflammation in spleens and lungs. Thus, not only does the splenocyte transfer model have crescentic glomerulonephritis as in human ANCA-disease, but also small vessel vasculitis.

The greatest drawback of this splenocyte transfer model is the presence of immune complex deposits in control and anti-MPO lymphocyte treated mice. This is problematic because human ANCA-disease is usually pauci-immune, i.e. the inflammatory vascular lesions have a paucity of staining for immunoglobulins. This problem was overcome by injecting purified anti-MPO IgG rather than splenocytes.

Pauci-immune crescentic glomerulonephritis and vasculitis is induced by intravenous infusion of anti-MPO IgG either into $Rag2^{-/-}$ mice that have no functioning T or B cells of their own, or into wild-type mice [16]. The disease in these mice is pathologically identical to ANCA-associated pauci-immune crescentic glomerulonephritis and vasculitis in humans. All mice (100%) that received a nephritogenic dose of anti-MPO IgG developed necrosis and crescents in glomeruli (fig. 1), whereas no control mice that received anti-BSA IgG developed glomerular lesions. Immunofluorescence microscopy demonstrated a paucity of immunoglobulin and complement in the glomeruli of mice with anti-MPO-induced glomerulonephritis, thus modeling the pauci-immune characteristic of human ANCA glomerulonephritis. There is fibrin formation at sites of necrosis and crescent formation (fig. 1).

In addition to pauci-immune glomerulonephritis, mice that received anti-MPO IgG also developed systemic small vessel vasculitis, such as pulmonary alveolar capillaritis and systemic arteritis affecting small arteries (fig. 1), that is pathologically identical to human ANCA vasculitis. As will be discussed in more detail later, the strain of mouse used as recipients of anti-MPO IgG has a major influence on the severity of disease [20, 21]. For example, the same dose of murine anti-MPO induces on average 10% crescents in C57BL/6 (B6) mice and 70% crescents in 129S6 mice.

Neutrophils Are Required

The importance of neutrophils in the induction of glomerulonephritis in mice by anti-MPO IgG was documented by administering cytotoxic anti-murine neutrophil antibody (NIMP-R14) that caused depletion of neutrophils prior to injection of a nephritogenic dose of anti-MPO IgG [22]. NIMP-R14 caused a severe neutropenia that persisted for 5 days. All mice that received anti-MPO IgG without prior neutrophil depletion developed hematuria, proteinuria, leukocyturia and glomerular necrosis and crescents by 6 days. None of the mice that received anti-MPO following depletion of neutrophil developed renal disease.

For documenting glomerular infiltration by neutrophils during lesion induction, wild-type B6 mice were given anti-MPO IgG or anti-BSA IgG intravenously. Immunoenzyme staining for neutrophils was performed on paraffin sections of kidney using a biotin-labeled anti-mouse neutrophil rat monoclonal antibody (Cedarlane Laboratories Ltd., Ontario, Canada). Neutrophil accumulation was quantified by direct counting of stained cells in glomeruli. All B6 mice that received anti-MPO IgG developed hematuria, proteinuria, leukocyturia and elevated blood urea nitrogen in contrast to the mice that received anti-BSA IgG that had no elevation in hematuria, proteinuria, leukocyturia or blood urea nitrogen. By light microscopy, all mice that received anti-MPO IgG had glomerular necrosis (average 18% of glomeruli) and crescents (average 12% of glomeruli), whereas mice that received anti-BSA IgG had no renal histologic abnormalities. Immunoenzyme staining for neutrophils demonstrated that induction of glomerulonephritis by injection of anti-MPO IgG was accompanied by glomerular influx of neutrophils. Neutrophils were concentrated at sites of segmental necrotizing glomerular injury, and also were identified in Bowman's space in glomeruli with necrosis or crescents, and in a few afferent arterioles. In mice that had received anti-MPO IgG, 30% of glomeruli had positive staining for neutrophils with an average of three neutrophils per glomerular cross-section. In contrast, mice that received anti-BSA IgG had only 6% of glomeruli with neutrophils averaging only one neutrophil per glomerular cross-section [22].

Another set of experiments has shown that BM-derived cells (probably primarily neutrophils) are necessary and sufficient to cause disease, even in the absence of MPO in other cell types (e.g. endothelial cells) [17]. $Mpo^{-/-}$ mice immunized with mouse MPO were exposed to irradiation and transplanted with $Mpo^{+/+}$ or $Mpo^{-/-}$ BM cells. Engraftment in mice with circulating anti-MPO resulted in development of glomerulonephritis in all mice, and pulmonary capillaritis and splenic necrotizing arteritis in some. Anti-MPO IgG also was introduced intravenously into chimeric mice created by transplanting $Mpo^{+/+}$ BM into irradiated $Mpo^{-/-}$ mice or $Mpo^{-/-}$ BM into irradiated $Mpo^{+/+}$ mice. Chimeric $Mpo^{-/-}$ mice with $Mpo^{+/+}$ neutrophils developed glomerulonephritis, whereas chimeric $Mpo^{+/+}$ mice with circulating $Mpo^{-/-}$ neutrophils did not, thus indicating that BM-derived cells are not only sufficient, but also necessary, for induction of anti-MPO disease.

In addition, unpublished observations indicate that anti-MPO T lymphocyte are not capable of causing acute vascular inflammation [23]. Rag $2^{-/-}$ mice injected with unfractionated splenocytes that contained approximately 25% T cells and 65% B cells developed crescents and necrosis in approximately 80% of glomeruli, whereas mice injected with the preparations that contained 80% T cells and 10% B cells developed crescents and necrosis in only 5% of glomeruli, and mice injected with preparations with >99% T cells developed no crescents or necrosis. ELISpot assays documented successful transfer of MPO-reactive T cells into all three experimental groups. These data show that anti-MPO T lymphocytes alone cannot induce glomerulonephritis.

Thus, neutrophils are the primary effector cell of the acute necrotizing and crescentic glomerular injury induced by anti-MPO antibodies. Anti-MPO T cells do not induce acute injury and expression of MPO by BM-derived cells is a sufficient antigen target for the anti-MPO antibodies to be pathogenic.

Influence of Cytokines and Infections

Patients with ANCA disease often have a flu-like illness near the time of onset of vasculitic symptoms, and in vitro studies show a synergistic role for cytokines in the activation of neutrophils by ANCA [5–15]. In accord with these observations, cytokines [24] and viral respiratory tract infection [25] exacerbate murine disease induced by anti-MPO IgG. Huugen et al. [24] demonstrated that systemic administration of bacterial lipopolysaccharide results in increased severity of anti-MPO-induced glomerulonephritis in this model. This is accompanied by increased circulating tumor necrosis factor (TNF)-α, which is known to enhance neutrophil activation by ANCA in vitro. In addition, administration of an antagonist to TNF-α reduced the severity of the lipopolysaccharide-mediated aggravation of anti-MPO IgG-induced glomerulonephritis. These animal model studies confirm the earlier conclusion from in vitro studies that ANCA and pro-inflammatory stimuli act synergistically to induce ANCA disease.

Unpublished studies demonstrate a synergistic effect by concurrent infections in this animal model [25]. B6 mice were inoculated intranasally with influenza A/WSN/33 virus on day 0, and injected with anti-MPO IgG on day 1. Control groups included mice that received virus alone or anti-MPO alone. All mice that received anti-MPO developed glomerular crescents and necrosis. All mice that received both anti-MPO and influenza had significant alveolar capillaritis, but mice that received anti-MPO alone or virus alone did not. Thus, concurrent influenza infection worsened ANCA disease, and concurrent ANCA worsened influenza. These observations provide additional support for the concept that inflammatory stimuli have a synergistic effect on the mediation of vasculitis by ANCA IgG.

Importance of Alternative Pathway Complement Activation

One of the most unexpected and potentially very important discoveries using the anti-MPO mouse model is the major pathogenic role of complement [26–28]. Anti-MPO IgG-induced glomerulonephritis is completely blocked by complement depletion with cobra venom factor, which abrogates all complement pathways [26]. The role of specific complement activation pathways has been investigated using mice with knockout of the final common pathway component C5, classical and lectin binding pathway component C4, and alternative pathway component factor B. After injection of anti-MPO IgG, $C4^{-/-}$ mice developed disease comparable to wild-type disease; however, $C5^{-/-}$ and factor $B^{-/-}$ mice developed no disease, thus implicating the alternative pathway in the pathogenesis of this model [26].

To implicate a role for complement in human ANCA disease and to elucidate the mechanism, IgG was isolated from patients with MPO-ANCA or PR3-ANCA, and from healthy controls. Incubation of MPO-ANCA IgG or PR3-ANCA IgG with normal human neutrophils primed with TNF-α caused the release of factors that activated complement in serum with generation of C3a [26, 28]. IgG from healthy controls did not produce this effect. In vitro experiments using human material also have shown that ANCA-activated neutrophils release factors that activate complement, which in turn primes neutrophils for further activation by ANCA [28].

These effects and other ANCA-mediated pathogenic events depend on generation of C5a by alternative pathway activation and engagement of C5a receptors on neutrophils [27, 28]. Blockade of this critical pathogenic step by blocking C5a or C5aR abrogates disease induction in the mouse model, which suggests a possible novel therapeutic strategy in humans based on currently available complement inhibitors as well as novel therapies soon to be available [29, 30].

The Role of Fcγ Receptors

In mice, engagement by IgG of FcγRI, FcγRIII and FcγRIV activate immunologic and inflammatory processes, whereas engagement of FcγRIIb inhibits immunologic and inflammatory processes. Unpublished data indicate that Fcγ receptors are involved not only in the induction of anti-MPO-induced glomerulonephritis and vasculitis (probably primarily through engagement of the activating receptor FcγRIV), but also in the modulation of disease phenotype (apparently primarily through variable signaling from the inhibitory receptor FcγRIIb) [31]. Preliminary data suggest that the absence of FcγR common γ-chain completely prevents disease induction, the absence of FcγRI or FcγRIII has no effect, the absence of FcγRIV has partial effect, and the absence of FcγRIIb augments induction of pulmonary disease (including granulomatous disease).

An intriguing and potentially very important discovery is that FcγRIIb receptors influence the phenotype of ANCA disease in our anti-MPO model. For example, the absence of inhibitory FcγRIIb receptors results in more severe pulmonary disease with a greater incidence of granulomatous inflammation, and pulmonary disease induced by anti-MPO is more severe in the presence of influenza A infection and even more severe in FcγRIIb$^{-/-}$ mice. Thus, the FcγRIIb gene influences the pulmonary phenotype of ANCA disease in our mouse model. In essence, the absence of FcγRIIb results in a phenotype that is more like Wegener's granulomatosis, as opposed to the usual phenotype in mice that is more like microscopic polyangiitis or renal-limited disease.

Based on these observations in mice, we hypothesized that pulmonary involvement in human ANCA disease might be modulated by the FcγRIIb gene. To test this hypothesis, we genotyped the –386 promotor site of the FcγRIIb gene in 200 patients with ANCA disease. ANCA disease patients with pulmonary involvement were compared to ANCA disease patients without pulmonary involvement. We observed that the –386C allele was associated with the pulmonary phenotype of ANCA disease ($p = 0.02$). Carriers of the –386C allele had a 2.8 times greater risk (95% CI: 1.4–6.84) of lung disease compared to those homozygous for the –386G allele. This suggests that polymorphisms in FcγRIIb affect the incidence of pulmonary involvement in ANCA disease in patients [32].

Demonstration for a role for FcγR in ANCA disease is important because there is mounting evidence that blockade of FcγRs can be an effective treatment in autoimmune disease, not only through blockade of activating FcγRs, but also through blockade of complement activation [33, 34], both of which play major roles in the pathogenesis of disease in the anti-MPO mouse model and possibly in ANCA disease patients.

Genetic Control

Anti-MPO IgG induces disease which is much more severe in 129S6 mice compared to B6 mice [20]. In general, a standard batch of our anti-MPO IgG induces crescents in 5–10% of the glomeruli in 100% of B6 mice 6 days after injection. In marked contrast, injection of the same batch of anti-MPO IgG into 129S6 mice induces disease which is much more severe. In preliminary experiments, 129S6 mice (Taconic) and C57BL/6 (B6) mice (Jackson Laboratory), received the same intravenous dose of anti-MPO IgG. 129S6 mice that received anti-MPO IgG developed severe proteinuria and renal failure, but B6 mice that received the same amount of anti-MPO IgG did not. B6 mice that received anti-MPO IgG developed an average of 9% crescents. In contrast, 129S6 mice that received the same amount of anti-MPO IgG developed an average of 69% crescents.

To assess the role of genetic background, 129S6 mice and B6 mice were used to generate F1 mice by B6 backcross with 129S6 mice. F1 (B6 × 129S6) mice that received anti-MPO IgG developed an average of 13.5% crescents. F2 mice were generated by F1 intercross. 60 F2 mice that received anti-MPO IgG had a wide spectrum of severity of glomerulonephritis. 30 mice had <10%, 24 mice 10–25%, 5 mice 26–50% and 1 mouse >50% glomerular crescents.

To determine the importance of genetic difference in BM-derived cells, chimeric mice were produced with genetic differences between donor BM-derived cells and recipients. Rag2$^{-/-}$ B6 or 129S6 mice were lethally irradiated followed by intravenous injection of 129S6 or B6 BM cells, respectively. Four weeks after BM transfer when engraftment was established, anti-MPO IgG was administrated intravenously. Disease induction was assessed at day 6. In these preliminary experiments, Rag2$^{-/-}$ B6 mice that received BM from 129S6 mice and anti-MPO IgG had an average of 79% crescents similar to wild-type 129S6 mice that received anti-MPO antibodies. However, Rag2$^{-/-}$ 129S6 mice that received the BM from B6 mice and the same dose of anti-MPO IgG had an average of 17% crescents, similar to wild-type B6 mice that receive anti-MPO IgG. Thus, genetically determined characteristics of BM-derived cells (most likely neutrophils and monocytes) are responsible for the marked differences in disease onset and outcome among genetically diverse mice.

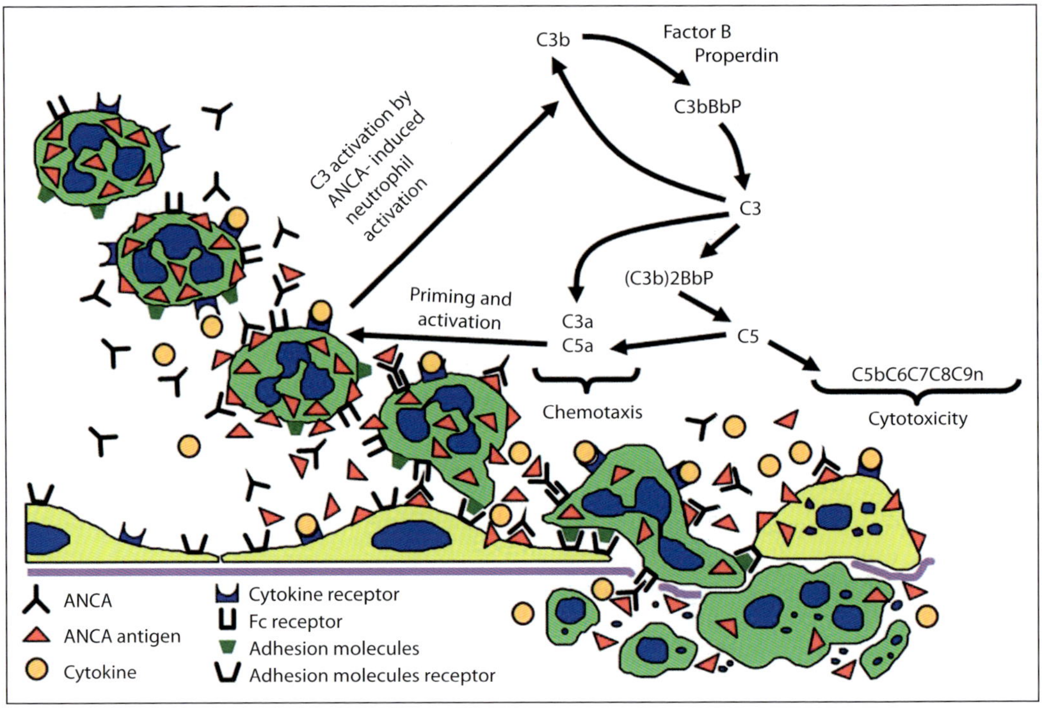

Fig. 2. Depiction of a hypothesis for the pathogenic events in ANCA vasculitis and glomerulonephritis. Beginning in the upper left, cytokines or other priming factors, such as TNF-α and C5a, induce neutrophils to express more ANCA antigens at the cell surface where they are available for binding to ANCA. ANCA engagement of antigen activates neutrophils by both Fc receptor engagement and direct Fab'2 binding to antigen. Activated neutrophils release toxic factors that cause apoptosis and necrosis of endothelial cells and other vessel wall components. Neutrophils that have been activated by ANCA also release factors that trigger the alternative complement pathway, which generates mediators such as C5a and C3a that amplify the intensity of ANCA-induced inflammation, in part by additional priming additional neutrophils for interaction with ANCA. Modified with permission from [26].

From these preliminary studies on the influence of genetic factors, we conclude that (1) there is strong genetic influence on the regulation of pathogenic events in this model of ANCA disease; (2) this genetic influence acts primarily through effects on BM-derived cells, including the responsivity of neutrophils to activation by ANCA; (3) the 129S6 model is more appropriate than the B6 model for experimental studies of pathogenesis and therapeutic strategies because of greater severity; and (4) genomic analysis of the basis for the genetic influence should identify genes and gene products that have important roles in the pathogenesis of ANCA disease, would be candidates for markers of disease activity and outcome, and might also be targets for therapy.

Conclusions

Observations with the anti-MPO-induced mouse model indicate that (1) anti-MPO IgG is sufficient even in the absence of functional T cells to cause disease and anti-MPO T lymphocytes are not sufficient to cause acute injury; (2) neutrophils are required, (3) ANCA antigens in BM-derived cells are sufficient targets; (4) increased circulating pro-inflammatory cytokines and microbial products exacerbate disease, and concurrent viral infection exacerbates and modulates the phenotype of disease; (5) Fcγ receptor engagement is required for disease induction, and Fcγ receptor repertoire modulates the phenotype of disease, especially pulmonary disease; (6) activation of the alternative pathway of complement is required, complement is activated by factors released by neutrophils stimulated by ANCA IgG, and engagement of C5a receptors is a primary event in complement-mediated amplification; and (7) genetic background has a marked influence on the severity and outcome of disease, and modified gene expression in BM-derived cells is the primary basis for genetically determined differences in disease susceptibility. These experimental animal observations, which are substantiated by in vitro experimental observations as well, support the pathogenic pathway diagramed in figure 2. This insight on the pathogenesis of ANCA disease is likely to lead to the identification of improved markers of disease activity and response to therapy, as well as more effective and less toxic therapies.

References

1 Jennette JC, Falk RJ: Small-vessel vasculitis. N Engl J Med 1997;337:1512–1523.
2 Falk RJ, Jennette JC: Anti-neutrophil cytoplasmic autoantibodies with specificity for myeloperoxidase in patients with systemic vasculitis and idiopathic necrotizing and crescentic glomerulonephritis. N Engl J Med 1988;318:1651–1657.
3 Jennette JC, Hoidal JR, Falk RJ: Specificity of anti-neutrophil cytoplasmic autoantibodies for proteinase 3. Blood 1990;75:2263–2264.
4 Schlieben DJ, Korbet SM, Kimura RE, Schwartz MM, Lewis EJ: Pulmonary-renal syndrome in a newborn with placental transmission of ANCAs. Am J Kidney Dis 2005;45:758–761.
5 Jennette JC, Xiao H, Falk RJ: Pathogenesis of vascular inflammation by anti-neutrophil cytoplasmic antibodies. J Am Soc Nephrol 2006;17:1235–1242.
6 Schreiber A, Luft FC, Kettritz R: Membrane proteinase 3 expression and ANCA-induced neutrophil activation. Kidney Int 2004;65: 2172–2183.
7 Yang JJ, Pendergraft WF, Alcorta DA, Nachman PH, Hogan SL, Thomas RP, Sullivan P, Jennette JC, Falk RJ, Preston GA: Circumvention of normal constraints on granule protein gene expression in peripheral blood neutrophils and monocytes of patients with antineutrophil cytoplasmic autoantibody-associated glomerulonephritis. J Am Soc Nephrol 2004;15:2103–2114.
8 Porges AJ, Redecha PB, Kimberly WT, Csernok E, Gross WL, Kimberly RP: Anti-neutrophil cytoplasmic antibodies engage and activate human neutrophils via Fc gamma RIIa. J Immunol 1994;153:1271–1280.
9 Mulder AH, Heeringa P, Brouwer E, Limburg PC, Kallenberg CG: Activation of granulocytes by anti-neutrophil cytoplasmic antibodies (ANCA): a Fc gamma RII-dependent process. Clin Exp Immunol 1994;98:270–278.
10 Kettritz R, Jennette JC, Falk RJ: Crosslinking of ANCA-antigens stimulates superoxide release by human neutrophils. J Am Soc Nephrol 1997;8:386–394.
11 Williams JM, Ben-Smith A, Hewins P, Dove SK, Hughes P, McEwan R, Wakelam MJ, Savage CO: Activation of the G(i) heterotrimeric G protein by ANCA IgG F(ab')2 fragments is necessary but not sufficient to stimulate the recruitment of those downstream mediators used by intact ANCA IgG. J Am Soc Nephrol 2003;14:661–669.
12 Falk RJ, Terrell RS, Charles LA, Jennette JC: Anti-neutrophil cytoplasmic autoantibodies induce neutrophils to degranulate and produce oxygen radicals in vitro. Proc Natl Acad Sci USA 1990;87:4115–4119.
13 Charles LA, Caldas ML, Falk RJ, Terrell RS, Jennette JC: Antibodies against granule proteins activate neutrophils in vitro. J Leukoc Biol 1991;50:539–546.
14 Ewert BH, Jennette JC, Falk RJ: Anti-myeloperoxidase antibodies stimulate neutrophils to damage human endothelial cells. Kidney Int 1992;41:375–383.
15 Ewert BH, Becker ME, Jennette JC, Falk RJ: Antimyeloperoxidase antibodies induce neutrophil adherence to cultured human endothelial cells. Ren Fail 1995;17:125–133.
16 Xiao H, Heeringa P, Hu P, Liu Z, Zhao M, Aratani Y, Maeda N, Falk RJ, Jennette JC: Antineutrophil cytoplasmic autoantibodies specific for myeloperoxidase cause glomerulonephritis and vasculitis in mice. J Clin Invest 2002;110:955–963.
17 Schreiber A, Xiao H, Falk RJ, Jennette JC: Bone marrow-derived cells are sufficient and necessary targets to mediate glomerulonephritis and vasculitis induced by anti-myeloperoxidase antibodies. J Am Soc Nephrol 2006;17:3355–3364.
18 Little MA, Smyth L, Salama AD, Mukherjee S, Smith J, Haskard D, Nourshargh S, Cook HT, Pusey CD: Experimental autoimmune vasculitis: an animal model of anti-neutrophil cytoplasmic autoantibody-associated systemic vasculitis. Am J Pathol 2009;174:1212–1220.
19 Kain R, Exner M, Brandes R, Ziebermayr R, Cunningham D, Alderson CA, Davidovits A, Raab I, Jahn R, Ashour O, Spitzauer S, Sunder-Plassmann G, Fukuda M, Klemm P, Rees AJ, Kerjaschki D: Molecular mimicry in pauci-immune focal necrotizing glomerulonephritis. Nat Med 2008;14:1088–1096.

20 Xiao H, Ciavatta D, Zeng Y, Johnson LA, Pardo-Manuel de Villena F, Falk RJ, Jennette JC: Genetic control of the severity of experimental anti-MPO necotizing and crescentic glomerulonephritis (Abstract). APMIS 2009; 117:90.
21 Xiao H, Zeng YW, Pardo-Manuel De Villena F, Ciavatta D, Falk R, Jennette JC: Genetic modulation of anti-myeloperoxidase induced murine crescentic glomerulonephritis. Lab Invest 2010;90:348A.
22 Xiao H, Heeringa P, Liu Z, Huugen D, Hu P, Maeda N, Falk RJ, Jennette JC: The role of neutrophils in the induction of glomerulonephritis by anti-myeloperoxidase antibodies. Am J Pathol 2005;167:39–45.
23 Xiao H, Peter P, Hu P, Liu Z, Falk R, Jennette JC: Neutrophils but not T lymphocytes are important in the induction of necrotizing and crescentic glomerulonephritis by antineutrophil cytoplasmic autoantibodies specific for myeloperoxidase (MPO-ANCA) in mice. J Am Soc Nephrol 2003;14:634A.
24 Huugen D, Xiao H, van EA, Falk RJ, Peutz-Kootstra CJ, Buurman WA, Tervaert JW, Jennette JC, Heeringa P: Aggravation of anti-myeloperoxidase antibody-induced glomerulonephritis by bacterial lipopolysaccharide: role of tumor necrosis factor-alpha. Am J Pathol 2005;167:47–58.
25 Xiao H, Morrison TE, Heise MT, Falk RJ, Jennette JC: MPO-ANCA IgG and influenza: a viral infection acts in concert to induce severe hemorrhagic pulmonary capillaritis (Abstract). Clin Exp Rheumatol 2007;25: S-91.
26 Xiao H, Schreiber A, Heeringa P, Falk RJ, Jennette JC: Alternative complement pathway in the pathogenesis of disease mediated by anti-neutrophil cytoplasmic autoantibodies. Am J Pathol 2007;170:52–64.
27 Huugen D, van EA, Xiao H, Peutz-Kootstra CJ, Buurman WA, Tervaert JW, Jennette JC, Heeringa P: Inhibition of complement factor C5 protects against anti-myeloperoxidase antibody-mediated glomerulonephritis in mice. Kidney Int 2007;71:646–654.
28 Schreiber A, Xiao H, Jennette JC, Schneider W, Luft FC, Kettritz R: C5a receptor mediates neutrophil activation and ANCA-induced glomerulonephritis. J Am Soc Nephrol 2009; 20:289–298.
29 Monk PN, Scola AM, Madala P, Fairlie DP: Function, structure and therapeutic potential of complement C5a receptors. Br J Pharmacol 2007;152:429–448.
30 Pellas TC, Wennogle LP: C5a receptor antagonists. Curr Pharm Des 1999;5:737–755.
31 Xiao H, Heeringa P, Schreiber A, Falk RJ, Jennette JC: The role of Fcγ receptors in the induction of glomerulonephritis and pulmonary granulomatous inflammation in mice by anti-myeloperoxidase antibodies (Anti-MPO). J Am Soc Nephrol 2004;15:37A.
32 Nester CM, Lionaki S, Chin H, Edberg J, Kimberly R, Wilhelmsen K, Jennette JC, Falk RJ: Polymorphisms associated with the pulmonary phenotype of antineutrophil cytoplasmic autoantibody vasculitis (Abstract). J Am Soc Nephrol 2007;18:784A.
33 Nimmerjahn F, Ravetch JV: Anti-inflammatory actions of intravenous immunoglobulin. Annu Rev Immunol 2008;26:513–533.
34 Jordan SC, Toyoda M, Vo AA: Intravenous immunoglobulin a natural regulator of immunity and inflammation. Transplantation 2009;88:1–6.

J. Charles Jennette, MD
Brinkhous Distinguished Professor and Chair
Department of Pathology and Laboratory Medicine
308 Brinkhous-Bullitt Building, University of North Carolina
Chapel Hill, NC 27599–7525 (USA)
Tel. +1 919 966 4676, Fax +1 919 966 4542, E-Mail jcj@med.unc.edu

Herrera GA (ed): Experimental Models for Renal Diseases: Pathogenesis and Diagnosis.
Contrib Nephrol. Basel, Karger, 2011, vol 169, pp 221–231

Monoclonal Immunoglobulin Light and Heavy Chain Deposition Diseases: Molecular Models of Common Renal Diseases

Pierre Ronco[a–c] · Emmanuelle Plaisier[a–c] · Pierre Aucouturier[a,d]

[a]UPMC Univ Paris 06, [b]INSERM, UMR_S 702, [c]AP-HP, Hôpital Tenon, Service de Néphrologie et Dialyses, [d]INSERM, UMR_S 938, Hôpital St-Antoine, Paris, France

Abstract

Light, light and heavy, and heavy chain deposition diseases (LCDD, LHCDD and HCDD, respectively) belong to a family of diseases featuring deposition in the kidney of a monoclonal immunoglobulin (Ig) or its isolated subunits, which also includes light chain amyloidosis, non-amyloid fibrillary and immunotactoid glomerulonephritis, and cryoglobulinemic glomerulonephritis. In clinical and pathologic terms, LCDD, LHCDD and HCDD display essentially similar characteristics, such as involvement of multiple organs, prominent renal involvement with severe renal failure, diabetes-like nodular glomerulosclerosis, marked thickening of tubular basement membranes, and monotypic deposits of Ig light chains (mostly κ) and/or heavy chains (mostly γ) that feature a non-organized granular, electron-dense appearance by electron microscopy. The most common cause is myeloma, although in a fair proportion of cases there is no (clinically) patent hematological disease. Recent progress has been made in understanding the molecular mechanisms of Ig-chain deposition and extracellular matrix accumulation, which opens up new therapeutic avenues in addition to eradication of the Ig-secreting plasma cell clone. Because these diseases represent a model of glomerular and interstitial fibrosis that is induced by a single molecule species, a better understanding of their pathophysiological mechanism may help unraveling the pathomechanisms of kidney fibrosis and renal disease progression.

Because the renal plasma flow represents 20% of the total plasma flow, the kidney is a prominent target for deposition of abnormal proteins. Among these proteins, monoclonal Ig or their isolated heavy chain (HC) or light chain (LC) subunits are responsible for a wide spectrum of kidney diseases that can be classified into two categories by electron microscopy. The first category is those with organized deposits and includes diseases with fibril formation, mainly amyloidosis, and diseases with microtubule formation, including cryoglobulinemic kidney and immunotactoid glomerulonephritis. Only amyloid deposits stain for Congo red, a tinctorial property related to the cross-β sheet organization of amyloid fibrils that is shared by all forms of amyloid.

The second category of diseases is characterized by non-organized electron-dense granular deposits that are mainly localized along basement membranes in most tissues, especially the kidney, and define a disease now referred to as monoclonal Ig deposition disease (MIDD) [1, 2], a term that by convention excludes (immunoglobulinic) amyloidosis. MIDD differs from amyloidosis in that the deposits do not have a fibrillar organization and, as a consequence, lack affinity for

Congo red. The distinction also relates to different molecular processes of deposition: while amyloid implicates a one-dimensional elongation of a pseudocrystalline structure, MIDD more likely results from a one-step amorphous precipitation of Ig chains.

It was known from the late 1950s that non-amyloidotic forms of glomerular disease resembling diabetic glomerulosclerosis could occur in multiple myeloma [3, 4]. The presence of monoclonal LC in these lesions was recognized in 1973 by Antonovych et al. [5] and confirmed by Randall et al. [6], who in 1976 published the first description of light chain deposition disease (LCDD). Although the deposits in AL-amyloidosis contain monoclonal LC or LC fragments, the term light chain deposition disease is used only for the disease that features non-amyloid electron-dense granular deposits as described by Randall et al. [6]. Monoclonal HCs were found together with LC in the tissue deposits from certain patients [6, 7], and the term light and heavy chain deposition disease (LHCDD) [1] was proposed. More recently, deposits containing monoclonal HC in the absence of detectable LC were observed in patients who were affected by otherwise typical Randall's disease [heavy chain deposition disease (HCDD)] [8], and three series of similar patients were published later [9–11]. In clinical and pathologic terms, LCDD, LHCDD and HCDD essentially are similar and are now gathered under the generic term of MIDD because their quite similar pathological features suggest common pathophysiological mechanisms.

In most kidney diseases that lead to the development of tissue fibrosis and end-stage renal failure, the initial pathogenic events are mostly unknown, and the mechanisms of fibrosis are often elusive. For instance, in diabetic nephropathy, a wealth of intermingled factors including alterations of hemodynamics, the diabetic milieu, and genetic and epigenetic factors induce fibrosis of kidney structures, particularly glomerulosclerosis. In contrast, in patients with MIDD that typically mimics diabetic glomerulosclerosis, a single molecular species, i.e. a monoclonal Ig subunit, triggers the whole spectrum of kidney lesions that also occur in diabetes and other nephropathies. Therefore, unraveling the mechanisms responsible for the deposition of the monoclonal Ig subunit and the ensuing extracellular matrix (ECM) accumulation may substantially increase our knowledge of the pathogenesis of kidney diseases.

Renal Pathology and Clinical Features

Lesions observed in MIDD are the key to understanding the pathogenesis of the disease.

Light Microscopy

MIDD is not only a glomerular disease, and tubular lesions may be more conspicuous than the glomerular damage (fig. 1). Tubular lesions are characterized by the deposition of a refractile, eosinophilic, periodic acid-Schiff (PAS)-positive, ribbon-like material along the outer part of the tubular basement membrane (fig. 1a). The deposits predominate around the distal tubules, the loops of Henle, and, in some instances, around the collecting ducts whose epithelium is flattened and atrophied. In advanced stages, a marked interstitial fibrosis including refractile deposits is frequently associated with tubular lesions.

Glomerular lesions are heterogeneous, with nodular glomerulosclerosis being the most characteristic (fig. 1b). It is found in 30–100% of patients with LCDD and in virtually all HCDD patients reported so far. Mesangial nodules are composed of PAS-positive, membrane-like material and are often accompanied by mild mesangial hypercellularity. Lesions resemble diabetic nodular glomerulosclerosis, the distribution of the nodules is fairly regular in a given glomerulus, and the nodules are often poorly argyrophilic. Exudative lesions as fibrin caps and extensive hyalinosis of the efferent arterioles are not observed. Milder forms of LCDD show increased mesangial matrix and/or cells, and a modest basement membrane

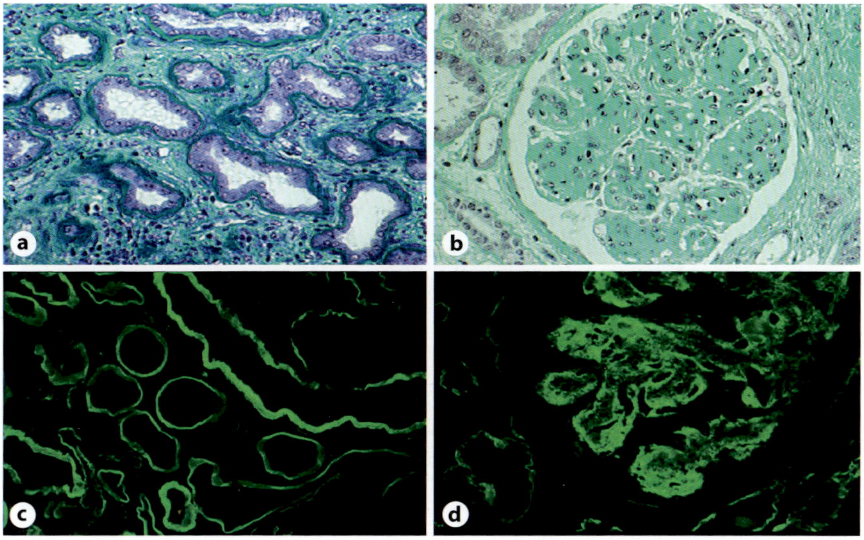

Fig. 1. Patient with the nephrotic syndrome and κLCDD, light and immunofluorescence microscopy. **a** Marked thickening of tubular basement membranes associated with significant interstitial fibrosis. **b** Nodular glomerulosclerosis. Bright staining of tubular basement membranes (**c**) and of the periphery of mesangial nodules (**d**) with anti-κ LC antibody. Magnification: ×312 (Masson's trichrome stain in **a** and **b**; immunofluorescence in **c** and **d**). From Ronco et al. [14].

thickening, with abnormal brightness and rigidity. Glomerular lesions may only be detectable by immunostaining or ultrastructural examination in early stages or if induced by light chains with a weak pathogenic potential.

Arteries, arterioles and peritubular capillaries all may contain PAS-positive deposits in close contact with their basement membranes. Deposits do not show the staining characteristics of amyloid, but Congo red-positive amyloid deposits co-occur in approximately 10% of patients [10].

Immunohistology

Immunohistology is central in the diagnosis of the various forms of MIDD. A criterion required for the diagnosis of MIDD is a monotypic LC [mostly κ (λ in ~20%); fig. 1c, d] and/or HC (mostly γ) fixation along tubular basement membranes. The tubular deposits stain strongly (fig. 1c) and predominate along the loops of Henle and the distal tubules, but they are also often detected along the proximal tubules. In contrast, glomerular immunohistology patterns display marked heterogeneity.

In patients with nodular glomerulosclerosis, deposits of monotypic Ig chains are usually found along the peripheral glomerular basement membranes and, to a lesser extent, in the nodules themselves (fig. 1d). Glomerular staining is typically weaker than that observed along the tubular basement membranes. This may not relate to the actual amount of deposited material because glomerular immunohistology may be negative despite prominent granular glomerular deposits by electron microscopy, which could relate to impaired accessibility of light chains antigenic determinants due to local modifications. In patients without nodular lesions, glomerular staining occurs along the basement membrane and less frequently in the mesangium. A linear staining usually decorates Bowman's capsule.

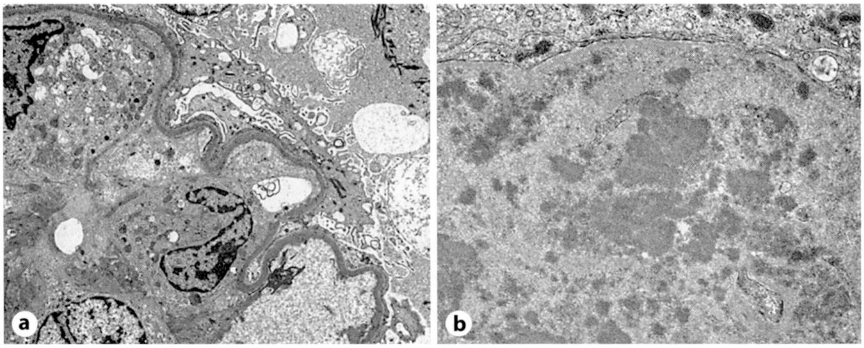

Fig. 2. Electron microscopy. Patient with the nephrotic syndrome and κLCDD. **a** A continuous dense layer of dark material is seen under the glomerular endothelium along the glomerular basement membrane. **b** On higher magnification in mesangial nodules, deposits have a non-fibrillar, granular appearance. Magnification: ×2,500 in **a**; ×10,000 in **b**. From Ronco et al. [14].

Deposits are frequent in vascular walls and the interstitium.

In patients with HCDD, immunohistology with anti-LC antibodies is negative despite typical nodular glomerulosclerosis. Monotypic deposits of γ, α or μ HCs may be identified. Any main Ig HC class and subclass may be observed, but there is a striking overrepresentation of γ3 and, to a lesser extent γ4, as compared with the isotype frequencies in normal plasma cells as well as in immunoproliferative disorders. Analysis of the kidney biopsy with monoclonal antibodies specific for the constant domains of the γ HC allowed identification of a deletion of the C_H1 domain in all tested cases. The hinge and C_H2 may also be deleted [8, 12]. In most cases of HCDD, especially when a γ1 or γ3 chain is involved, complement components including C1 could be demonstrated in a granular or pseudolinear pattern. Complement deposits were often associated with serum hypocomplementemia [11, 13].

Electron Microscopy

The most characteristic ultrastructural features are finely-to-coarsely granular electron-dense deposits along the outer (interstitial) aspect of the tubular basement membranes. In the glomerulus, they predominate in a subendothelial position along the glomerular basement membrane and are located mainly along and in the lamina rara interna (fig. 2). They can also be found in mesangial nodules, Bowman's capsule and the wall of small arteries between the myocytes. Non-amyloid fibrils have been reported in a few patients with LCDD or HCDD.

Clinical Features

MIDD is a systemic disease with Ig chain deposition in a variety of organs leading to various clinical manifestations, but visceral Ig chain deposits may be totally asymptomatic and found only at autopsy [14].

Renal involvement is a constant feature of MIDD, and renal symptoms, mostly proteinuria and chronic kidney disease, often dominate the clinical presentation [10, 14]. In most patients with LCDD, albuminuria is associated with the nephrotic syndrome. However, in about one fourth, it is <1 g/day, and these patients exhibit mainly a tubulointerstitial syndrome. The high prevalence, early appearance and severity of chonic kidney disease are other salient features of LCDD. In most cases, GFR declines rapidly, which is a main reason for referral. The prevalence of hypertension is variable, but it must be interpreted according to associated medical history. Renal features of patients with HCDD are basically similar to those seen in LCDD and LHCDD, with a higher prevalence of hypertension and hematuria, while extrarenal manifestations are apparently less frequent. Importantly, a patent hematological tumor is much less frequent in HCDD.

Liver and cardiac involvement occurs in about 25% of patients with LCDD and LHCDD. Liver deposits are constant. They are either discrete and confined to the sinusoids and basement membranes of biliary ductules without associated parenchymal lesions, or massive with marked dilatation and multiple ruptures of sinusoids, resembling peliosis. Hepatomegaly with mild alterations of liver function tests is the most common symptom, but patients may also develop life-threatening hepatic insufficiency and portal hypertension. Cardiac involvement is also frequent and may be responsible for cardiomegaly and severe heart failure. Arrhythmias, conduction disturbances and congestive heart failure are also seen. Echocardiography and catheterization may reveal diastolic dysfunction and reduction in myocardial compliance similar to that found in cardiac amyloidosis. Deposits may also occur along the nerve fibers and in the choroid plexus, as well as in the lymph nodes, bone marrow, spleen, pancreas, thyroid gland, submandibular glands, adrenal glands, gastrointestinal tract, abdominal vessels, lungs and skin. They may be responsible for peripheral neuropathy (20% of the reported cases), gastrointestinal disturbances, pulmonary nodules, amyloid-like arthropathy and sicca syndrome. Extrarenal deposits are less common in patients with HCDD.

Myeloma is diagnosed in about 50% of the patients with LCDD or LHCDD and in about 25% of those with HCDD. MIDD, like AL-amyloidosis, is often the presenting disease that leads to the discovery of myeloma at an early stage. MIDD may occasionally complicate Waldenström's macroglobulinemia, chronic lymphocytic leukemia and nodal marginal zone lymphoma. It often occurs in the absence of a detectable malignant process, even after prolonged (>10 years) follow-up. A monoclonal bone marrow plasma cell population is then easily detectable by immunohistologic examination.

Structure Peculiarities of Ig LC and/or HC Associated with MIDD

It is now well established that intrinsic properties of Ig components are responsible for the renal lesions observed in patients with MIDD. The hypothesis of unusual properties of LC is supported by the recurrence of the disease in the transplanted kidney, and biochemical anomalies such as abnormal size of LC observed on biosynthesis experiments by bone marrow plasma cells [1, 15]. However, LC deposition does not mean pathogenicity because after injection into mice, one third of LCs from patients with myeloma or AL-amyloidosis became deposited in basement membranes [16]. In vivo expression of a LCDD LC in a murine plasmacytoma also clearly demonstrated the intrinsic property of the monoclonal LC to deposit in the kidney [12]. However, this experimental model was too transient to permit studies of the pathogenic events consecutive to LC deposition. Singular properties of LC most likely are required for completion of the pathogenetic process that leads to kidney fibrosis.

Ig Chain Structure

Ig is made up of two pairs of subunits, two HC and two LC, each of which includes a variable domain implicated in antigen recognition, and a constant region that itself includes one or several domains and that activates a variety of effector systems. Binding to antigen and activating effector cells or complement (i.e. antibody function) is made possible by the association of two HC and two LC. LC may be of κ or λ type, with one constant and one variable domain (termed C_L and V_L). HC may be of nine isotypes (μ, γ1, γ2, γ3, γ4, α1, α2, δ and ε) that define Ig classes and

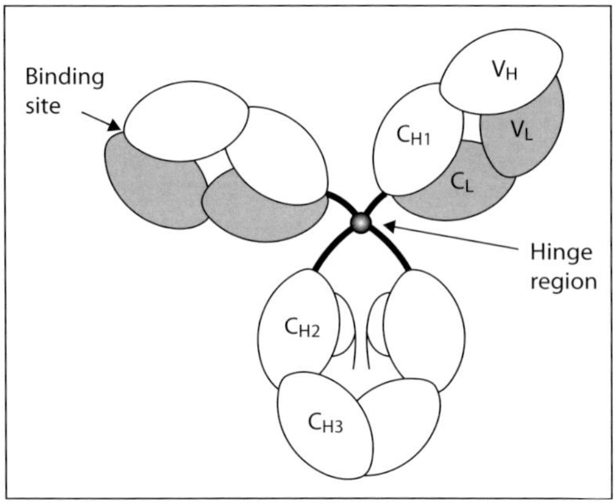

Fig. 3. Domain structure of an IgG. The LC and HC domains are shown in gray and white, respectively. V_L = LC variable domain; C_L = LC constant domain; V_H = HC variable domain; C_H = HC constant domain. From Ronco et al. [27].

subclasses: IgM, IgG (including IgG1 to IgG4), IgA (including IgA1 and IgA2), IgD and IgE. HC include from the $-NH_2$ to the $-COOH$ terminus one variable domain (V_H) from a common repertoire for all isotypes and three (IgG, IgA and IgD) or four (IgM and IgE) constant domains, termed C_H1 to C_H4 (fig. 3).

Normal Ig display a striking structural heterogeneity that is borne essentially by the variable domains and is related to the diversity of the antigenic determinants they are supposed to recognize. These regions of the Ig chains are encoded by numerous gene segments, the rearrangement of which occurs during B cell differentiation and is mandatory for Ig production. Alignments of V_H and V_L sequences allowed definition of variability subgroups on the basis of their homologies: HC express six V_H subgroups, κ chains express four Vκ subgroups and λ chains express six Vλ subgroups. The variability of V_H and V_L is mostly borne by three restricted peptide portions called complementarity determining regions (CDR1, CDR2 and CDR3) that form three loops at one end of the domain that constitute the antibody binding site.

At variance with normal polyclonal Ig, monoclonal Ig are secreted by a single clone of differentiated B cells that expands excessively, either in a tumoral context (myeloma, Waldenström's disease, etc.) or without patent hematologic malignancy (monoclonal gammapathy of undetermined significance). A monoclonal Ig may be secreted as free LC (or rarely, HC) that, in certain cases, displays structural anomalies causing tissue deposition.

Structure and Glycosylation of LC in LCDD

In LCDD, there is a modest but significant over-representation of κ chains occurring in approximately 80% of cases (versus ~2/3 among polyclonal Ig), which contrasts with the increased λ to κ ratio observed in AL-amyloidosis. The rare $V_{\kappa IV}$ variability subgroup is frequent [17], which is worth noting because this subgroup features a strikingly long complementarity-determining region 1 (CDR1) loop that contains several hydrophobic residues. The determining role of the LC variable region V_L was supported by the fact that tissue deposition in mice grafted with a plasmacytoma expressing a human LCDD κ chain was abolished by amino acid changes in the V_L [12].

The first complete primary structure of a LC in LCDD was determined by Cogné et al. [18]. An abnormally heavy (30-kDa) κ chain was extracted from the kidney and was presumably identical to that secreted by the malignant plasma cells since they shared the same apparent molecular mass and 13-amino acid N-terminal sequence. It was encoded by a normal-sized κ mRNA and was N-glycosylated, which is quite rare in Ig LCs. The C region was entirely normal, and the V region belonged to the $V\kappa_{IV}$ subgroup. Eight mutations were observed, including replacement of Pro95 (considered as essential for canonical conformations of the CDR3 loop). Replacement of Asp70 by Asn determined an N-glycosylation site. The primary structures of a few additional LCDD precursors were analyzed at the complementary DNA [19, 20] and protein levels [21]. The C_L domains from all studied LCDD LC had entirely normal sequences. Most peculiarities were clustered in the CDRs, which could suggest that a first step of the pathogenesis might be a LC tropism for extracellular components that behave as antigen-like structures. Unusual hydrophobic residues were found at positions where they could either modify the conformation of the V_L domain, potentially leading to LC aggregation or be involved in interactions with other hydrophobic molecules in the extracellular space [22]. Molecular modeling simulations suggested that the overall three-dimensional structures were normal.

When pathogenic LC could not be detected in the serum and urine, they were N-glycosylated in all tested cases [17, 23]. In vitro biosynthetic labeling experiments on short-term plasma cell cultures showed that LC which were absent in the urine actually were secreted by the bone marrow plasma cells [1, 15, 18]. Together with the presence of exposed hydrophobic residues, N-glycosylation might increase the LC propensity to precipitate in tissues and displace the equilibrium from soluble toward deposited amorphous forms.

As in AL-amyloidosis, extrinsic conditions also may contribute to aggregation of the LC. The same LC can form granular aggregates or amyloid fibrils, depending on the environment, and different partially folded intermediates of this protein may be responsible for amorphous or fibrillar aggregation pathways [24].

HCDD: A Disease Featuring Deletions of HC Domains

A deletion of the first constant domain C_H1 was found in the deposited and circulating HC in the 11 patients with γ-HC deposition disease, where it was searched for [8–11, 13, 25]. This was also suggested in a patient with αHCDD [26]. A larger deletion that included the C_H1, hinge and C_H2 domains was found in one case [8]. In the blood, the deleted HC could either circulate in small amounts as a free unassembled subunit or be associated with a LC, mostly of the λ isotype [9]. The C_H1 deletion is required for secretion of free HC, which are cleared rapidly from the circulation by organ deposition [27]. Deletion of the C_H1 also is found in classical HC disease, a lymphoproliferative disorder with free HC secretion without renal tissue deposition, and in the exceptional cases of heavy-chain (AH)-amyloidosis in which deposits have a fibrillar organization. In HC disease, however, the V_H domain is also partially or completely deleted or replaced with non-Ig peptide portions, which suggests that the V_H domain of HCDD HCs participates in the tissue precipitation process. Sequence analyses of two HCDD proteins indeed revealed quite unusual amino acid substitutions in the V_H, which might change their physicochemical properties, including surface charge and hydrophobicity [28]. For better understanding how structures influence the pathogenicity, additional sequences of HCDD HCs are required.

Pathophysiology of ECM Accumulation

A striking feature of MIDD is the dramatic accumulation of ECM, whose pathogenesis is beginning to be understood thanks to in vitro models developed by Herrera and colleagues [29–32]. Nodules are made of normal ECM constituents (collagen type IV, laminin, fibronectin and tenascin) [31, 33] and stain weakly for the small proteoglycans decorin and biglycan [34]. Tenascin-C is an ECM protein that is a component of the mesangium in normal and, most importantly, pathologic conditions [35]. Tenascin-C was shown to accumulate in the center of mesangial nodules in patients with LCDD and other pathologies featuring nodular glomerulosclerosis [31]. Normally, tenascin-C is destroyed predominantly by metalloproteinase-7 (MMP-7) in combination with minor contributions from other MMPs, most notably MMP-1 and MMP-3. Deposition of tenascin-C might be involved in the perpetuation and irreversibility of the glomerular lesions in LCDD.

A role for transforming growth factor-β (TGF-β) is supported by its strong expression in glomeruli of patients with LCDD and by in vitro experiments using cultured mesangial cells [36]. Incubation of mesangial cells with LC from patients with LCDD induced cell changes, increased expression of platelet-derived growth factor-β and its receptor, increased production of monocyte chemoattractant protein-1, and increased expression of the proliferation marker protein Ki-67, whereas tubulopathic LC had no effect [37]. The resulting phenotype is close to the myofibroblastic phenotype, including activation of ECM synthesis and decreased MMP production and activity [29, 31]. In contrast, mesangial cells that are incubated with LC from patients with AL-amyloidosis undergo a macrophage-like transformation, with increased MMP activity and decreased ECM production [29]. Those divergent phenotypic transformations may relate to distinct cellular trafficking of LCDD and AL-amyloid LCs after binding to a yet unidentified, common caveolae-associated receptor [30]. AL-amyloid LCs are delivered to the mature lysosomal compartment, where amyloid formation is thought to occur, whereas LCDD LCs do not show significant internalization [30]. Whether these events occur in vivo remains to be established. Understanding the signaling pathways that are triggered by LC binding to mesangial cells may contribute substantially to our knowledge on the mechanisms of glomerulosclerosis.

Recently, Keeling and Herrera [32] developed an elegant model of nodular glomerulosclerosis by growing human mesangial cells in three-dimensional matrices and incubating the cells with LCs purified from the urine of patients with LCDD, or control LCs. Compared with control LCs from patients with cast nephropathy, LCDD LCs induced a sequence of crucial events that included the initial activation of platelet-derived growth factor-β. This resulted in mesangial cell proliferation, followed by TGF-β activation, leading to an increase in ECM production and a negative feedback loop inhibiting cellular proliferation (fig. 4). Addition of active MMP-7 degraded the excess tenascin-C in vitro, a process that could be prevented by an exogenous MMP inhibitor. MMP-7 appears to be adequately produced by mesangial cells in LCDD, but its release from the cells is impaired, preventing it from reaching the matrix where it normally acts on the deposited tenascin. Tenascin-C and LCDD LC co-localize in the center of mesangial nodules, leading to nodular glomerulosclerosis, which represents the classical lesion found in advanced LCDD.

The recent advances made in the pathophysiology of LCDD-related glomerulosclerosis provide new insights into the pathobiology of mesangial cells and, thus, into the molecular mechanisms of frequent kidney diseases involving mesangial cells such as diabetic and IgA nephropathies. There are also lessons to be learned about the progression of kidney injury leading to end-stage renal failure, since in LCDD, fibrosis affects all renal compartments, including the tubules, vessels and interstitium (in addition to glomerulosclerosis). MIDD indeed is the only sclerotic kidney disease in which the offending molecule, although different from one patient to another, can be known and defined at its sequence level.

Furthermore, Keeling and Herrera [32] showed that exogenous active MMP-7 added to the culture media in the three-dimensional matrix model, markedly decreased tenascin-C expression. This observation suggests that nodular glomerulosclerosis can partially resolve, as also observed in some patients after long-term chemotherapy [38].

Therapeutic Implications and Perspectives

As in AL-amyloidosis, treatment should be aimed at reducing Ig production. Clearance of the LC deposits has been demonstrated in a few patients after intensive chemotherapy supported by syngeneic bone marrow transplantation or blood stem cell autografting. In a retrospective study of 11 LCDD and LHCDD patients (younger than 65 years) treated by high-dose therapy with the support of autologous blood stem cell transplantation, no treatment-related death occurred [39]. A decrease in the monoclonal Ig level was observed in eight patients, with complete disappearance from serum and urine in six cases. Clinical

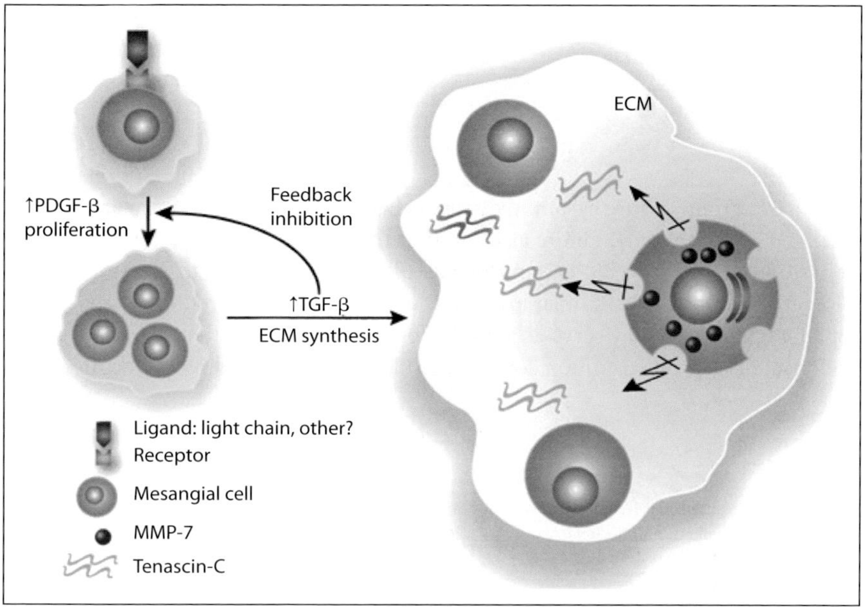

Fig. 4. Proposed pathway leading to glomerular sclerosis in LCDD. PDGF-β = platelet-derived growth factor-β. From Picken [45].

improvement was observed in six patients, and histologic regression was documented in cardiac, hepatic and skin biopsies. These results were confirmed in two small North American series [40, 41]. Reversal of dialysis dependency and sustained improvement in renal function were also occasionally noted. Whether high-dose chemotherapy with blood stem cell transplantation provides benefits compared to conventional chemotherapy, including high-dose dexamethasone (with or without bortezomib), remains to be established. As with AL-amyloidosis, monitoring of LC production should rely on serum-free LC assay, particularly when a blood or urine monoclonal Ig component cannot be detected by conventional methods. Kidney transplantation in a few patients with MIDD usually led to recurrence of the disease and should therefore not be performed in LCDD patients unless measures have been taken to efficiently reduce LC production [42].

Of note, the pattern of Ig chain isotypes in HCDD is strikingly different from that of classical immunoproliferative diseases: in particular, the γ3 subclass is remarkably frequent in HCDD, where the monoclonal plasma cells also predominantly express a λ LC. In myeloma and lymphoplasmacytic lymphomas, a much higher proportion of IgG2, IgA, IgM and κ LC are encountered. This raises questions about the precise nature of the underlying hematological disease. In any case, treatment should aim at eradicating the secreting cell clone, either benign or malignant, in order to reduce the production of the nephritogenic HC as much as possible.

New therapies aimed at targeting key steps in the cascade of events leading to kidney fibrosis are to be designed for use in combination with chemotherapy. They include anti-TGF-β [43] and anti-platelet-derived growth factor therapy [44], as well as inhibitors of LC-mesangial cell interactions. As discussed above, the interaction with LCDD LC is governed by purported receptors located on the mesangial cell membrane [30]. These receptors, which remain to be identified, may be shared by other mediators of glomerular injury. We need to learn how to control the responsible receptors triggering the pathways to sclerosis.

Conclusion

MIDD is a group of rare systemic diseases characterized by severe renal failure as a result of the deposition of a monotypic LC and/or HC of Ig. Glomerular lesions are so similar to diabetic nephropathy that MIDD may serve as a model for the understanding of this common condition. MIDD can be considered the only model of sclerotic kidney disease where the initial injury is defined at the level of a single molecule, most often a monoclonal LC of Ig. Although significant advances have been made in understanding the pathophysiology both from clinical studies and experimental models, the fine molecular characteristics of the LC or HC responsible for pathogenicity are still elusive. In addition to chemotherapy aimed at eradicating the secreting plasma cell clone, new strategies based on a better understanding of the cascade of events leading to fibrosis are promising.

References

1 Buxbaum JN, Chuba JV, Hellman GC, Solomon A, Gallo GR: Monoclonal immunoglobulin deposition disease: light chain and light and heavy chain deposition diseases and their relation to light chain amyloidosis. Clinical features, immunopathology, and molecular analysis. Ann Intern Med 1990;112:455–464.

2 Preud'homme JL, Aucouturier P, Touchard G, Striker L, Khamlichi AA, Rocca A, Denoroy L, Cogne M: Monoclonal immunoglobulin deposition disease (Randall type). Relationship with structural abnormalities of immunoglobulin chains. Kidney Int 1994;46:965–972.

3 Kobernick SD, Whiteside JH: Renal glomeruli in multiple myeloma. Lab Invest 1957;6:478–485.

4 Sanchez LM, Domz CA: Renal patterns in myeloma. Ann Intern Med 1960;52:44–54.

5 Antonovych TT, Lin RC, Parrish E, Mostofi K: Light chain deposits in multiple myeloma (abstract). Lab Invest 1974;30:370A.

6 Randall RE, Williamson WC Jr, Mullinax F, Tung MY, Still WJ: Manifestations of systemic light chain deposition. Am J Med 1976;60:293–299.

7 Preud'homme JL, Morel-Maroger L, Brouet JC, Mihaesco E, Mery JP, Seligmann M: Synthesis of abnormal heavy and light chains in multiple myeloma with visceral deposition of monoclonal immunoglobulin. Clin Exp Immunol 1980;42:545–553.

8 Aucouturier P, Khamlichi AA, Touchard G, Justrabo E, Cogne M, Chauffert B, Martin F, Preud'homme JL: Brief report: heavy chain deposition disease. N Engl J Med 1993;329:1389–1393.

9 Moulin B, Deret S, Mariette X, Kourilsky O, Imai H, Dupouet L, Marcellin L, Kolb I, Aucouturier P, Brouet JC, Ronco PM, Mougenot B: Nodular glomerulosclerosis with deposition of monoclonal immunoglobulin heavy chains lacking C_H1. J Am Soc Nephrol 1999;10:519–528.

10 Lin J, Markowitz GS, Valeri AM, Kambham N, Sherman WH, Appel GB, D'Agati VD:Renal monoclonal immunoglobulin deposition disease: the disease spectrum. J Am Soc Nephrol 2001;12:1482–1492.

11 Kambham N, Markowitz GS, Appel GB, Kleiner MJ, Aucouturier P, D'agati VD: Heavy chain deposition disease: the disease spectrum. Am J Kidney Dis 1999;33:954–962.

12 Khamlichi AA, Rocca A, Touchard G, Aucouturier P, Preud'homme JL, Cogne M: Role of light chain variable region in myeloma with light chain deposition disease: evidence from an experimental model. Blood 1995;86:3655–3659.

13 Soma J, Sato K, Sakuma T, Saito H, Sato H, Sato T, Abbas A, Aucouturier P: Immunoglobulin gamma3-heavy-chain deposition disease: report of a case and relationship with hypocomplementemia. Am J Kidney Dis 2004;43:E10–E16.

14 Ronco P, Plaisier E, Mougenot B, Aucouturier P: Immunoglobulin light (heavy)-chain deposition disease: from molecular medicine to pathophysiology-driven therapy. Clin J Am Soc Nephrol 2006;1:1342–1350.

15 Preud'homme JL, Morel-Maroger L, Brouet JC, Cerf M, Mignon F, Guglielmi P, Seligmann M: Synthesis of abnormal immunoglobulins in lymphoplasmacytic disorders with visceral light chain deposition. Am J Med 1980;69:703–710.

16 Solomon A, Weiss DT, Kattine AA: Nephrotoxic potential of Bence Jones proteins. N Engl J Med 1991;324:1845–1851.

17 Denoroy L, Deret S, Aucouturier P: Overrepresentation of the VkappaIV subgroup in light chain deposition disease. Immunol Lett 1994;42:63–66.

18 Cogné M, Preud'homme JL, Bauwens M, Touchard G, Aucouturier P: Structure of a monoclonal kappa chain of the VkappaIV subgroup in the kidney and plasma cells in light chain deposition disease. J Clin Invest 1991;87:2186–2190.

19 Khamlichi AA, Aucouturier P, Silvain C, Bauwens M, Touchard G, Preud'homme JL, Nau F, Cogne M: Primary structure of a monoclonal kappa chain in myeloma with light chain deposition disease. Clin Exp Immunol 1992;87:122–126.

20 Rocca A, Khamlichi AA, Aucouturier P, Noel LH, Denoroy L, Preud'homme JL, Cogne M: Primary structure of a variable region of the VkappaI subgroup (ISE) in light chain deposition disease. Clin Exp Immunol 1993;91:506–509.

21 Bellotti V, Stoppini M, Merlini G, Zapponi MC, Meloni ML, Banfi G, Ferri G: Amino acid sequence of k Sci, the Bence Jones protein isolated from a patient with light chain deposition disease. Biochim Biophys Acta 1991;1097:177–182.
22 Deret S, Chomilier J, Huang DB, Preud'homme JL, Stevens FJ, Aucouturier P: Molecular modeling of immunoglobulin light chains implicates hydrophobic residues in non-amyloid light chain deposition disease. Protein Eng 1997;10:1191–1197.
23 Ganeval D, Noel LH, Preud'homme JL, Droz D, Grunfeld JP: Light-chain deposition disease: its relation with AL-type amyloidosis. Kidney Int 1984;26:1–9.
24 Khurana R, Gillespie JR, Talapatra A, Minert LJ, Ionescu-Zanetti C, Millett I, Fink AL: Partially folded intermediates as critical precursors of light chain amyloid fibrils and amorphous aggregates. Biochemistry 2001;40: 3525–3535.
25 Husby G, Blichfeldt P, Brinch L, Brandtzaeg P, Mellbye OJ, Sletten K, Stenstad T: Chronic arthritis and gamma heavy chain disease: coincidence or pathogenic link? Scand J Rheumatol 1998;27:257–264.
26 Cheng IK, Ho SK, Chan DT, Ng WK, Chan KW: Crescentic nodular glomerulosclerosis secondary to truncated immunoglobulin alpha heavy chain deposition. Am J Kidney Dis 1996;28:283–288.
27 Ronco PM, Alyanakian MA, Mougenot B, Aucouturier P: Light chain deposition disease: a model of glomerulosclerosis defined at the molecular level. J Am Soc Nephrol 2001;12:1558–1565.
28 Khamlichi AA, Aucouturier P, Preud'homme JL, Cogne M: Structure of abnormal heavy chains in human heavy chain deposition disease. Eur J Biochem 1995;229:54–60.
29 Keeling J, Teng J, Herrera GA: AL-amyloidosis and light-chain deposition disease light chains induce divergent phenotypic transformations of human mesangial cells. Lab Invest 2004;84:1322–1338.
30 Teng J, Russell WJ, Gu X, Cardelli J, Jones ML, Herrera GA: Different types of glomerulopathic light chains interact with mesangial cells using a common receptor but exhibit different intracellular trafficking patterns. Lab Invest 2004;84:440–451.
31 Keeling J, Herrera GA: Matrix metalloproteinases and mesangial remodeling in light chain-related glomerular damage. Kidney Int 2005;68:1590–1603.
32 Keeling J, Herrera GA: An in vitro model of light chain deposition disease. Kidney Int 2009;75:574–576.
33 Bruneval P, Foidart JM, Nochy D, Camilleri JP, Bariety J: Glomerular matrix proteins in nodular glomerulosclerosis in association with light chain deposition disease and diabetes mellitus. Hum Pathol 1985;16:477–484.
34 Stokes MB, Holler S, Cui Y, Hudkins KL, Eitner F, Fogo A, Alpers CE: Expression of decorin, biglycan, and collagen type I in human renal fibrosing disease. Kidney Int 2000;57:487–498.
35 Truong LD, Pindur J, Barrios R, D'Agati V, Lechago J, Suki W, Majesky M: Tenascin is an important component of the glomerular extracellular matrix in normal and pathologic conditions. Kidney Int 1994;45:201–210.
36 Zhu L, Herrera GA, Murphy-Ullrich JE, Huang ZQ, Sanders PW: Pathogenesis of glomerulosclerosis in light chain deposition disease: role of transforming growth factor-beta. Am J Pathol 1995;147:375–385.
37 Russell WJ, Cardelli J, Harris E, Baier RJ, Herrera GA: Monoclonal light chain–mesangial cell interactions: Early signaling events and subsequent pathologic effects. Lab Invest 2001;81 :689 –703.
38 Komatsuda A, Wakui H, Ohtani H, Kodama T, Miki K, Imai H, Miura AB: Disappearance of nodular mesangial lesions in a patient with light chain nephropathy after long-term chemotherapy. Am J Kidney Dis 2000;35:E9.
39 Royer B, Arnulf B, Martinez F, Roy L, Flageul B, Etienne I, Ronco P, Brouet JC, Fermand JP: High dose chemotherapy in light chain or light and heavy chain deposition disease. Kidney Int 2004;65:642–648.
40 Hassoun H, Flombaum C, D'Agati VD, Rafferty BT, Cohen A, Klimek VM, Boruchov A, Kewalramani T, Reich L, Nimer SD, Comenzo RL: High-dose melphalan and auto-SCT in patients with monoclonal Ig deposition disease. Bone Marrow Transplant. 2008;42:405–412.
41 Lorenz EC, Gertz MA, Fervenza FC, Dispenzieri A, Lacy MQ, Hayman SR, Gastineau DA, Leung N: Long-term outcome of autologous stem cell transplantation in light chain deposition disease. Nephrol Dial Transplant 2008;23:2052–2057.
42 Leung N, Lager DJ, Gertz MA, Wilson K, Kanakiriya S, Fervenza FC: Long-term outcome of renal transplantation in light-chain deposition disease. Am J Kidney Dis 2004;43:147–153.
43 Dussaule JC, Chatziantoniou C: Reversal of renal disease: is it enough to inhibit the action of angiotensin II? Cell Death Differ 2007;14:1343–1349.
44 Ostendorf T, Rong S, Boor P, Wiedemann S, Kunter U, Haubold U, van Roeyen CR, Eitner F, Kawachi H, Starling G, Alvarez E, Smithson G, Floege J: Antagonism of PDGF-D by human antibody CR002 prevents renal scarring in experimental glomerulonephritis. J Am Soc Nephrol 2006;17:1054–1062.
45 Picken MM: The role of mesangial homeostasis in glomerular injury progression: hope for mesangial sclerosis reversal. Kidney Int 2009;75:574–576.

Pierre Ronco
Service de Néphrologie et Dialyses, Hôpital Tenon
4 rue de la Chine
FR–75020 Paris (France)
Tel. +33 1 56 01 66 39, Fax +33 1 56 01 69 99, E-Mail pierre.ronco@tnn.aphp.fr

Herrera GA (ed): Experimental Models for Renal Diseases: Pathogenesis and Diagnosis.
Contrib Nephrol. Basel, Karger, 2011, vol 169, pp 232–246

Renal Amyloidosis: Current Views on Pathogenesis and Impact on Diagnosis

Guillermo A. Herrera[a] · Jiamin Teng[a] · Elba A. Turbat-Herrera[b]

[a]Nephrocor and [b]Bostwick Laboratories, Tempe, Ariz., USA

Abstract

The amyloidoses constitute a group of diseases in which misfolding of extracellular proteins plays a fundamental role. The aggregation of normally soluble proteins into insoluble unbranching fibrils is the basic underlying pathology in amyloidosis. The process of amyloid formation generates toxic insoluble (in saline) protein aggregates that are deposited in tissues in the form of β-pleated sheets of fibrillary material. The amyloidoses are considered to be part of the so-called protein storage diseases (protein thesauroses). In addition, due to the unusual protein folding associated with amyloid, this group of diseases has been referred to as conformational and protein folding disorders. For many years amyloidosis was considered an extremely rare, somewhat mysterious disease. However, in recent years its pathogenesis, particularly that of renal amyloidosis, has been carefully dissected in the research laboratory using in vitro and, to a lesser extent, in vivo models. These have provided a molecular understanding of sequential events that take place in the renal mesangium leading to the formation of amyloid fibrils and eventual extrusion into the mesangial matrix, which itself becomes seriously damaged and, in due time, replaced by the fibrillary material. Amyloid, once considered to be an 'inert' substance, has been proven to be involved in crucial biological processes that result in the destruction and eventual replacement of normal renal constituents. Although there are more than two dozen recognized amyloid precursor proteins (and new ones being added to the list) that can be involved in the genesis of amyloid fibrils, the pathophysiologic mechanisms that occur in the renal mesangium are likely to be very similar, if not the same, regardless of the type of amyloidosis. Likewise, the same is true of amyloid formation in the renal vasculature. Mesangial cells are essentially smooth muscle cells and the events that take place in the mesangium and vasculature (where smooth muscle cells and/or pericytes are present) in the entire body responsible for the formation of amyloid are the same. In the renal interstitium, fibroblasts likely participate in the formation of amyloid, following a similar sequence of events as smooth muscle cells. Although much of the information gathered has been from in vitro systems, an in vivo model of renal amyloidosis has recently been designed to study renal amyloidogenesis. Crucial steps in the cascade of events that result in the formation of amyloid fibrils have been elucidated in the laboratory. The information that has been gathered regarding the pathogenesis of amyloidosis has been translated to the clinical arena where implementation of new therapeutic approaches is beginning to occur. Additional molecular-based therapies will be implemented in the near future.

Amyloid: Conceptual Understanding and Early Research Developments

The term amyloid was coined by Schleiden in 1838 to describe a normal component of plants [1]. Rudolf Virchow adopted the term amyloid a few years later to refer to a waxy substance that reacted with iodine and sulfuric acid, which at the time was considered a reaction characteristic of starch [2, 3]. Therefore, amyloid, which means starch-like, was deemed an appropriate term for the material that he had found. This material when deposited in the spleen imparted to it a consistency that led to the descriptive term lardaceous (lard-like). In 1842, Rokitansky described a lardaceous appearance in organs of individuals suffering of malaria and tuberculosis [4], which was the beginning of an association between amyloid and chronic inflammatory diseases. Virchow criticized the term lardaceous to refer to organs with amyloid and indicated that the consistency was rather similar to wax and referred to the change as 'waxy-like' [5], a term that has remained popular when describing organs with extensive amyloid infiltration.

In 1859, Freidrich and Kelule recognized that the waxy spleen did not have any cellulose constituents and suggested that it was proteinaceous in nature [6]. In 1902, Jochmann and Schumm [7] reported a case of myeloma and amyloidosis, and 15 years later Glaus [8] reported a second case. Bernhold introduced the Congo red stain in 1922, first as a diagnostic test, and then used it as a stain to detect amyloid in tissue sections [9–11]. In 1933, Bell [12] compiled 90 cases of previously reported renal disease in patients with myeloma and found 3 cases with amyloidosis. In 1949, Sikl [13] was the first to indicate a rather frequent association of myeloma with what he described as 'atypical amyloidosis'. He collected 40 cases from the literature. Nine years later, Grabar and Williams [14] demonstrated abnormal serum proteins in patients with amyloidosis.

In 1928, apple green birefringence associated with Congo red-stained amyloid material was first recognized by Divry and Florkin in the brains of patients with Alzheimer's disease [15, 16].

In 1927, Smetana [17] proposed the concept that reticuloendothelial cells were the main cells engaged in the production of amyloid. Encompassed within the reticuloendothelial system are macrophages/histiocytes and Kupffer cells in the liver. Organs with a prominent reticuloendothelial system include the spleen, liver and lungs, all of which are commonly recognized sites of amyloid deposition.

Interest in amyloidosis research in general reached a significant level of interest from the 1960s to the early 1980s. Some seminal work was carried out during that time which solidified our understanding of amyloidosis. Some of this work will be detailed here in order to better understand how this field developed and how renal amyloidosis research has advanced.

In 1962, Edelman and Gally [18] characterized what was recognized at the time as Bence Jones proteins as immunoglobulin light chains. This work was the beginning of the recognition of the association of renal amyloidosis occurring in association with an underlying plasma cell dyscrasia.

In 1963, Gueft (at the time a post-doctoral research fellow) and Guidoni in 1963 stated, 'We suggest that the amyloid fibrils are formed at the moment that a histiocyte, etc., delivers the precursor (protein) to the outside', supporting the role of the reticuloendothelial system in the process of amyloid formation [19]. In the late 1960s and early 1970s, Shirahama and Cohen [20, 21] emphasized the role of lysosomes in amyloidogenesis. They conducted seminal research in the pathogenesis of AA-amyloidosis using macrophages and concluded that lysosomes played an important role in amyloidogenesis. They also demonstrated a role for lysosomes in the breakdown of amyloid fibrils in macrophages [22]. Furthermore, studies by Shirahama and Cohen in the late 1960s and 1970s [23] reaffirmed the fibrillary nature of amyloid, elegantly documenting it using high-resolution electron microscopic analysis. The cross β-pleated sheet configuration pattern of amyloid fibrils by X-ray diffraction was discovered in 1969 [24].

In 1970, Glenner et al. [25, 26] demonstrated the derivation of a type of amyloid from certain light chains; thus, they are responsible for developing the concept of AL-amyloidosis. Also in the early 1970s, the elucidation of the type of amyloid later recognized as AA-amyloid took place [27]. In the early 1970s, Linke et al. [28] reported the formation of amyloid fibrils by digesting light chains with proteolytic enzymes. The

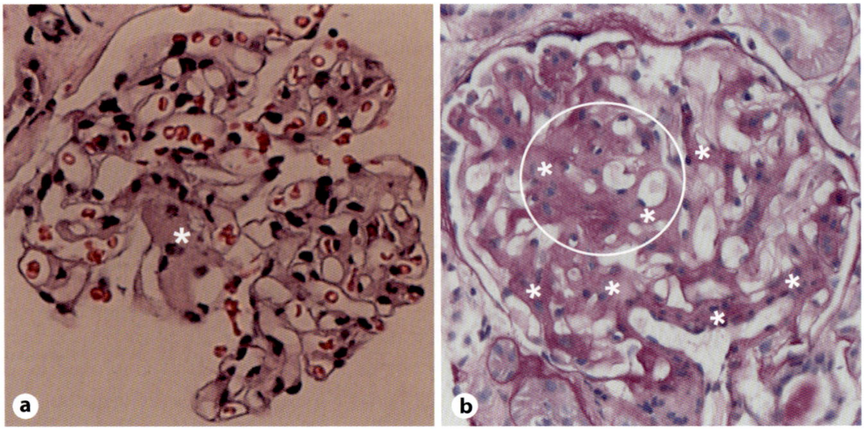

Fig. 1. Amyloidosis. Hematoxylin and eosin stain. Magnification: ×500 (**a**) and ×500 (**b**). Deposition of eosinophilic, amorphous material in a focal mesangial area (asterisk) representing the earliest manifestation of glomerular amyloidosis (**a**). Note more advanced glomerular amyloidosis with variable deposition in different mesangial areas (asterisks) (**b**).

following year, Epstein et al. [29] reported digestion of human light chains with kidney lysosomal enzymes resulting in the formation of amyloid fibrils, further advancing the idea that lysosomes are integral elements in the sequence of events that lead to the genesis of amyloid fibrils. It has remained a matter of controversy whether amyloid is ever located intracytoplasmically before it is extruded from the cells [30]. For years the definition of amyloid indicated that it was an extracellular material, although this definition has been recently changed.

In 1981, Gise et al. [31] observed that glomerular amyloidosis begins in the mesangium (fig. 1 and 2a, b). The following year, Solomon et al. [32] showed the preferential association of AL-amyloid with lambda light chains, and in doing so, began to hint at the concept that only certain light chains engaged in amyloidogenesis. Although there are more than 26 precursor proteins capable of engaging in amyloid formation, the main types of amyloid found in the kidney, which are recognized today, are AA and AL types. The discussion that follows addresses the pathogenesis of these two types of amyloidosis. Most of the basic research that has been conducted in renal amyloidosis has been of the AL type. Hereditary forms of amyloidosis can also involve glomeruli, but are commonly characterized by preferential amyloid deposition in the medulla.

Unfortunately, efforts at understanding amyloidosis remained essentially dormant for approximately 10 years (from the early 1980s to the early 1990s). The concept that amyloid was an 'inert' substance which deposited in the affected organs leading to irreversible damage and organ failure by pressure atrophy of normal tissues prevailed during that time [33]. This perception created an aura of fatalism that virtually abolished all interest in researching the subject. Research in amyloidosis was incorrectly perceived as essentially a 'waste of time and effort'. Scientific contributions essentially ceased for at least one decade.

Understanding Renal Amyloidosis: The 1990s

One of the first, but seminal, works on renal amyloidosis culminated with a publication by Solomon et al. in 1991 [34]. They reported findings in the kidneys of mice that were injected intraperitoneally with Bence Jones proteins isolated from 40 patients with myeloma and amyloidosis (the mice were killed 48 h after injection).

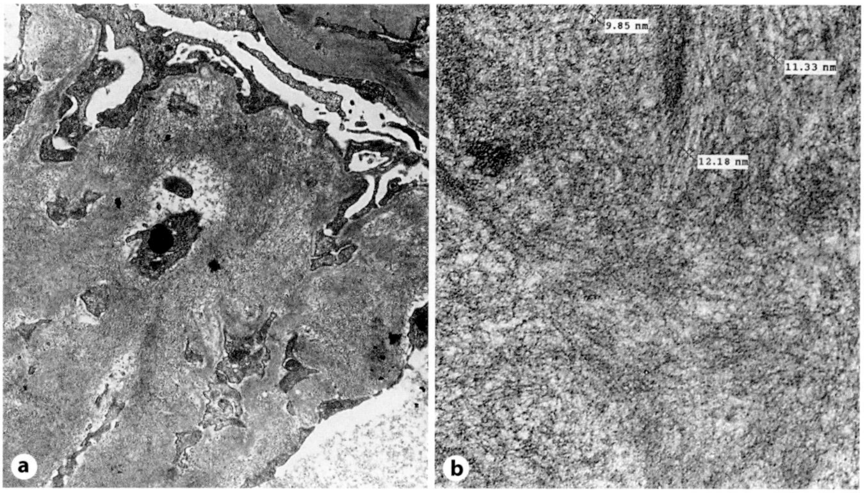

Fig. 2. Mesangial amyloidosis. Transmission electron microscopy, uranyl acetate and lead citrate stain. Magnification: ×4,500 (**a**) and ×15,500 (**b**). Mesangial matrix is completely replaced (**a**) by randomly disposed, non-branching, 8- to 12-nm in diameter fibrils, ultrastructural characteristics typical of amyloid (**b**). Note that there is no normal mesangial matrix left in **a**.

In 26 of these mice, monoclonal proteins similar to those found in the sera of the affected patients were found deposited in association with amyloid deposits in the kidneys of the experimental mice. This contribution cemented the concept that abnormal circulating light chains are primarily responsible for producing renal amyloidosis. In this study, the light chain deposits were only documented to be present in the renal vasculature, not in glomeruli or in the interstitium.

Further studies to elucidate the pathogenesis of amyloidosis were carried out by Kluve-Beckerman et al. [35] who incubated cultures of peritoneal macrophages with recombinant serum amyloid-A protein. The precursor protein was shown to be internalized into the macrophages and intracellular amyloid formation was documented to occur 5 days after incubation. Clustering of the macrophages was shown to occur prior to amyloid formation. This work clearly showed the importance of internalization of the precursor protein prior to formation of amyloid and revealed intracellular amyloid in the macrophages, corroborating the work done some 20 years earlier [22].

Previous work suggested that the mesangium was the primary glomerular compartment involved in early amyloid formation [31]. One of the fundamental questions that had to be answered was how amyloidosis actually occurred in the renal parenchyma. It seemed logical that this research would be centered on the mesangium and renal vasculature, where amyloid is first detected. It, therefore, became logical to approach the subject of amyloidogenesis using mesangial cell cultures. Beginning with rat mesangial cells and eventually moving to human mesangial cells was the plan.

The first experiments performed in the late 1990s simply explored how cultured mesangial cells reacted to incubation with isolated and purified light chains obtained from the urine of patients with biopsy-proven glomerular AL-amyloidosis. The initial results were spectacular as it was demonstrated in 1996 by Tagouri et al. [36] that mesangial cells grown in monolayers incubated with monoclonal light chains obtained from the urine of patients with amyloidosis produced amyloid in vitro (fig. 3a–e). Additional studies with mesangial cells cultured

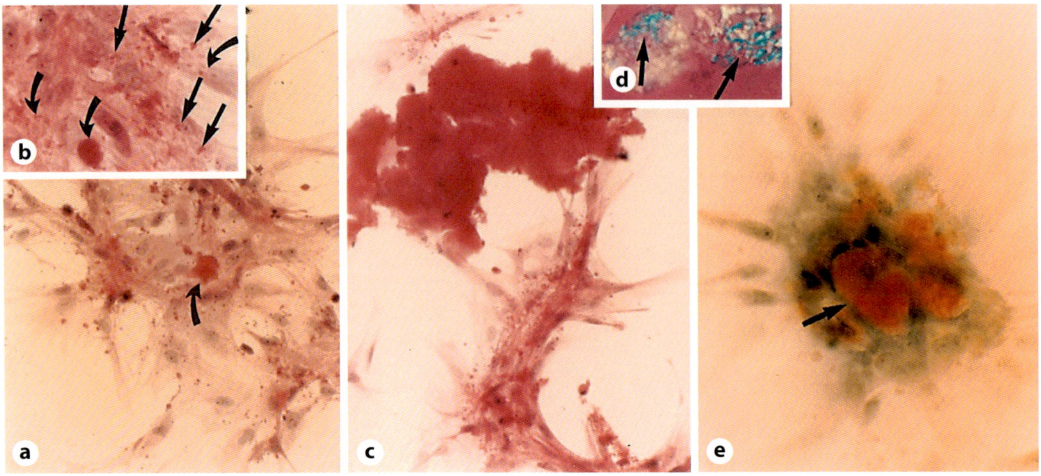

Fig. 3. Amyloid formation in culture. Monolayers of mesangial cells incubated with amyloidogenic light chains purified from urine of patients with biopsy-proven AL-amyloidosis. **a**, **b**, and **c** Hematoxylin and eosin stains. Magnification: ×150 (**a**), ×500 (**b**) and ×350 (**c**). Polarized Congo red stain (**d**), and Congo red stain (**e**). **a**, **b** and **c** Mesangial cells incubated with amyloidogenic light chains form amyloid with morphologic and tinctorial characteristics identical to that seen in tissue sections. Material produced is Congo red positive (salmon pink color) shown in **e** and when polarized revealed characteristic apple green birefringence confirming its amyloid nature.

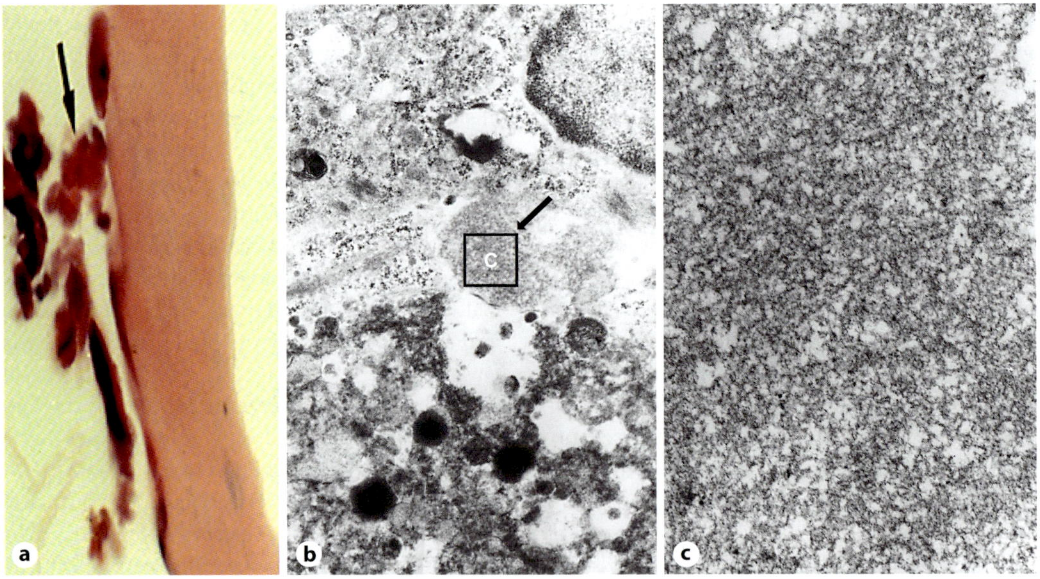

Fig. 4. Mesangial cells growing on a matrix (Amgel) incubated with amyloidogenic light chains purified from the urine of patients with biopsy proven AL-amyloidosis. **a** Hematoxylin and eosin stain (magnification: ×350). **b**, **c** Transmission electron microscopy, uranyl acetate and lead citrate stain (magnification: ×12,500 (**b**) and ×25,500 (**c**). In **a,** the arrow points toward amyloid formed by mesangial cells, and **b** and **c** reveal the site where amyloid is formed. The mesangial cells engaged in the formation of amyloid have acquired a macrophage phenotype exhibiting numerous lysosomes and show gaps in their cell membranes. **c** Ultrastructural details (typical of amyloid) of the fibrillary material produced by mesangial cells.

on a matrix recapitulated these findings and highlighted morphologic alterations in mesangial cells engaged in amyloidogenesis, indicating a change from a smooth muscle to a macrophage phenotype (fig. 4a–c) in these cells.

In 1999, Solomon et al. [37] designed a transgenic mouse model of AA-amyloidosis using mice carrying the IL-6 gene under the control of the metallothionein-I promoter which led to the formation of amyloid in the spleen, liver and kidneys. The mice died at 6 months with extensive amyloid deposits. Although the kidneys were described as being involved (glomeruli and medulla showed congophilia), no detailed studies of the process of renal amyloidogenesis were carried out, in spite of the fact that renal failure was what was determined to be the cause of death in the experimental animals.

The Recent Era, Entering the New Millenium: Unraveling the Mechanisms Involved in Renal Amyloidogenesis

Studies performed on the in vitro model of AL-amyloidogenesis provided insights into the step-by-step process involved in renal amyloidogenesis [38–41]. The in vitro work revealed a number of fascinating findings. A receptor on mesangial cells for glomerulopathic light chains was identified, although it has not yet been entirely characterized [38, 39]. This finding was crucial to explain why some light chains, as a result of their peculiar amino acid composition and overall physicochemical characteristics, interact with mesangial cells, as light chains due to their low molecular weight are typically filtered through the capillary walls into the urinary space to be processed in proximal tubules [42].

These light chains with affinity for the mesangium are recognized as glomerulopathic light chains. Different types of glomerulopathic light chains compete for the purported receptor on the mesangial cells. One very important event that occurs right after the light chains interact with the mesangial cells is that the latter transform from their normal smooth muscle to a macrophage phenotype, endowing them with the machinery to actively participate in endocytosis and processing of the endocytosed amyloid precursor proteins [39, 43]. This is not only seen in vitro, but has also been confirmed in renal biopsies from patients with amyloidosis (fig. 5a–c). Internalization and trafficking of amyloidogenic light chains in mesangial cells with eventual delivery to the mature lysosomal system where processing led to the formation of amyloid fibrils was demonstrated. The process of internalization was found to be clathrin-mediated.

Lysosomal processing of these light chains with partial proteolysis was shown to be a crucial step in the genesis of amyloid fibrils [39–41]. For example, increasing the pH of the lysosomes decreased amyloid formation. Other maneuvers, including some which changed the mesangial milieu, also altered amyloid formation [44]. The one step(s) that has not been completely elucidated is what happens after the lysosomes process the light chains forming fibrils and the eventual delivery of these to the extracellular domain. Transmission and scanning electron microscopic studies suggest that the amyloid is extruded via small gaps in the membranes of mesangial cells from an intracytoplasmic location or that lysosomes abutting the cell membranes release to the extracellular matrix the fibrils using the same membrane gaps (fig. 4b, c). In experimental material (fig. 4b, c), as well as in human renal biopsies, the amyloid fibrils are noted to be closely associated with the cell membranes of mesangial cells (fig. 6), in many instances intimately related to lysosomes. Only light chains from patients with AL-amyloidosis (not those with light chain deposition disease or light chain cast nephropathy) were associated with amyloid formation.

The speed of amyloid generation and quantity of amyloid produced depended on the monoclonal light chain used, lending support to the fact that the physicochemical structure of the light chains is in great part responsible for determining the kinetics of the process. However, how the addition of glycosaminoglycan moieties to the amyloid fibrils occurs remains unclear. One possibility is that this occurs after the amyloid fibrils

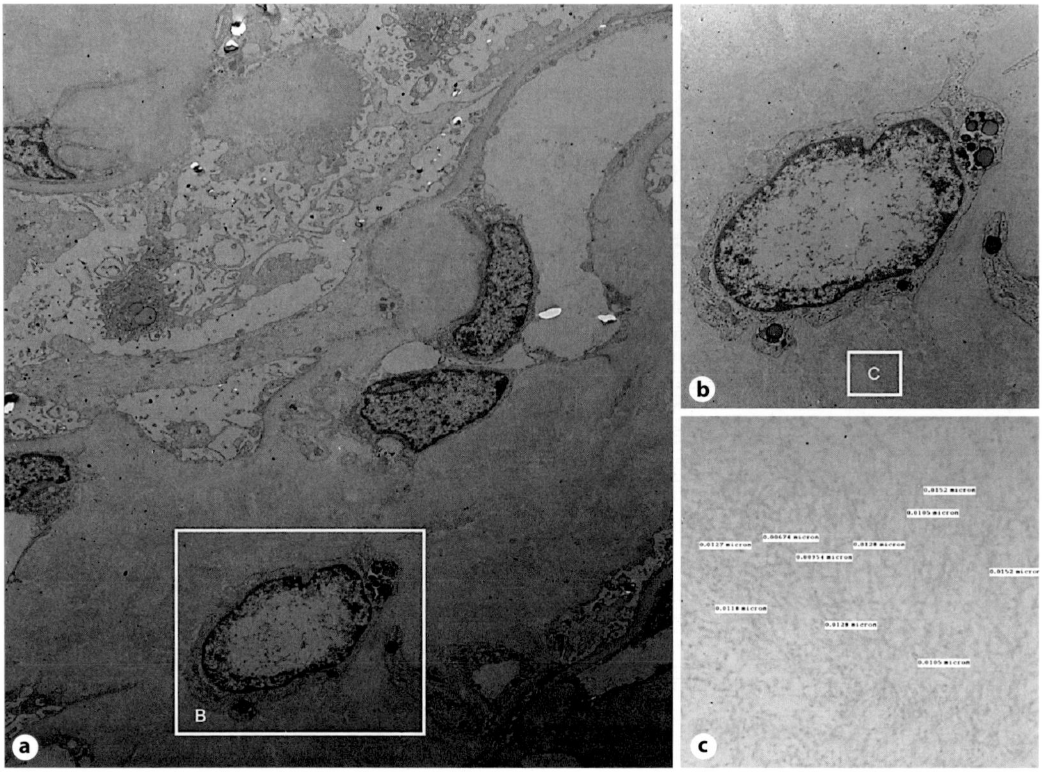

Fig. 5. AL-amyloidosis, renal biopsy. **a**, **b** and **c** Transmission electron microscopy, uranyl acetate and lead citrate stain. Magnification: ×5,500 (**a**) ×18,500 (**b**) and ×25,500 (**c**). Note in **a** abundant mesangial amyloid surrounding a transformed mesangial cell (better seen in **b**) with a macrophage phenotype exhibiting numerous lysosomes in their cytoplasm and loss of intracytoplasmic filaments and attachment plaques, features seen in normal mesangial cells (smooth muscle differentiation). Amyloid fibrils with typical characteristics are illustrated in **c**. The findings in renal biopsy specimens are similar to what is seen in the in vitro model (fig. 4).

have been released to the extracellular matrix. Once the amyloid is extruded to the extracellular matrix, activation of metalloproteinases and destruction with eventual replacement of the matrix by amyloid fibrils becomes a crucial pathologic event which leads to eventual glomerular destruction and complete replacement by amyloid deposition. Not only the actual deposition of fully formed amyloid fibrils results in pathologic damage, but some of the non-fibrillar intermediary forms are also toxic to tissues.

Animal models for AA-amyloidosis have been used in the past, including some transgenic mice models [20–22, 45–47], but no detailed studies pertaining to the kidneys were reported. Furthermore, AA-amyloidosis is found to occur spontaneously in a number of animals such as dogs, cheetahs, tigers, hares, stone martens, mink, cattle, horses, sheep, mountain gazelles, rabbits, hamsters, guinea pigs, bats, chickens and ducks. While some researchers took advantage of these naturally occurring instances where amyloid formation occurs to study various issues in amyloidosis, there were no serious attempts to study renal the mechanisms involved in amyloid fibril formation specifically.

Until recently, no good animal model of renal AL-amyloidogenesis had been described.

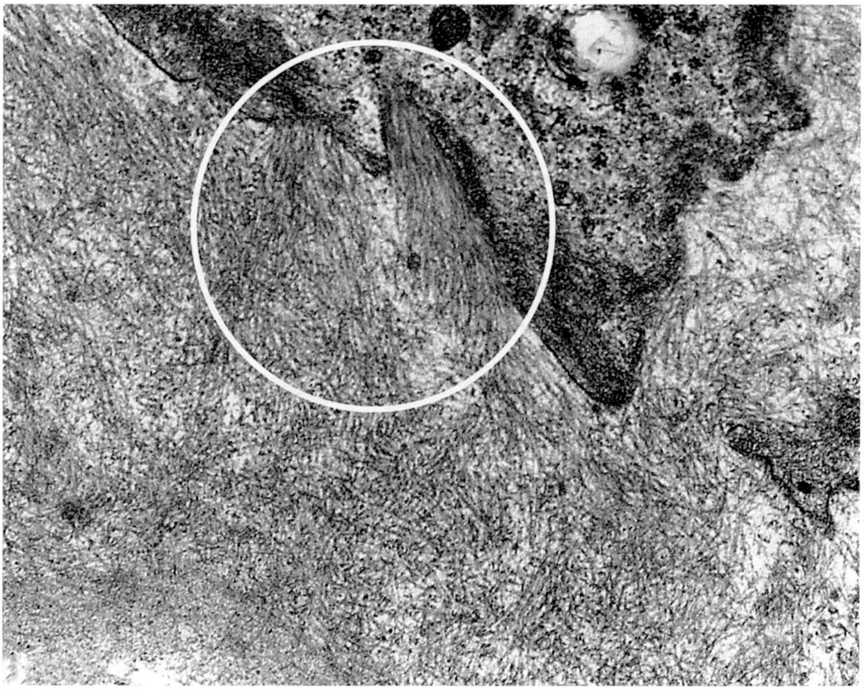

Fig. 6. AL-amyloidosis, renal biopsy. Transmission electron microscopy, uranyl acetate and lead citrate stain (magnification: ×13,500). Note intimate association of amyloid fibrils with the surface of a mesangial cell with small gaps in cell membrane in renal biopsy specimen.

Herrera et al. [48] in 2001 described an isolated kidney perfusion model whereby purified light chains from the urine of patients with renal biopsy-proven AL-amyloidosis were perfused ex vivo into the mice kidney to study the early steps in mesangial amyloid formation [41]. This model had the great drawback of only permitting studies for approximately 3–4 h, as the perfused kidneys could only be maintained physiologically intact for that period of time.

In the last few years an animal model of AL-amyloidogenesis was created permitting extended studies on the process of glomerular AL-amyloidogenesis. Injection of amyloidogenic light chains isolated from the urine of patients with renal biopsy-proven AL-amyloidosis into the penile vein of mice resulted in the formation of amyloid in the glomerular mesangium (fig. 7a, b) and renal vasculature. Scanning electron microscopic studies have shown details of the process of extrusion of the amyloid fibrils into the extracellular matrix. In general, the sequence of events seen in the in vitro model have been reaffirmed in the in vivo model. This represents the main contribution of this experimental model [49].

Mechanisms of interstitial amyloid formation has also been a focus of interest. It is postulated that interstitial fibroblasts transform into macrophages and intake the precursor protein engaging in the formation of amyloid fibrils in a manner similar to that of mesangial cells and cardiac fibroblasts [50]. However, this has not yet been proven. Furthermore, there have been some studies demonstrating that amyloid formation can occur intratubularly, leading to the formation of tubular casts composed of amyloid fibrils [51, 52].

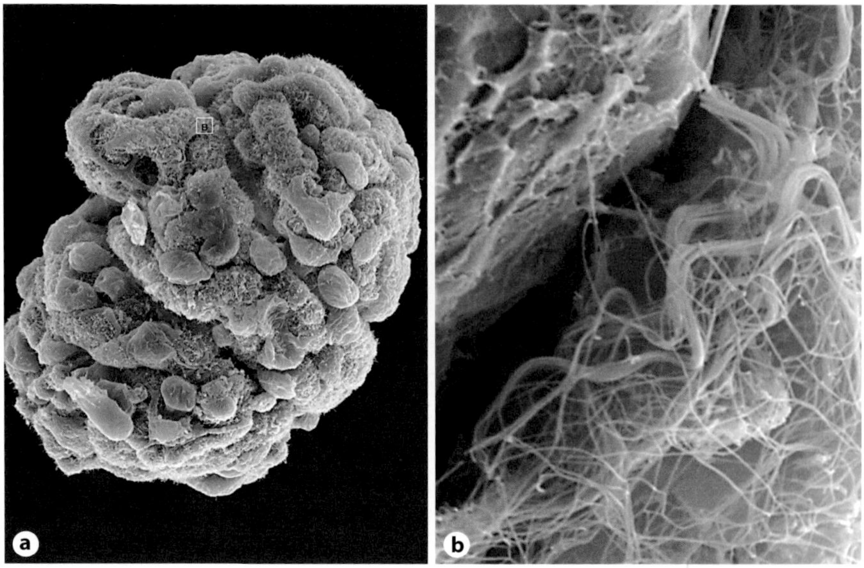

Fig. 7. In vivo model of amyloidogenesis (**a**) and scanning electron microscopy (**b**). **a** Whole glomerulus preparation (magnification: ×1,000). Details of mesangial amyloid fibrils (magnification: ×50,000). Note in **a** amyloid fibrils in intact glomerular preparation with details of these fibrils highlighted on **b**.

From the Research Laboratory to the Diagnostic Arena: What Have We Learned that Has Improved the Diagnosis of Renal Amyloidosis and Clinical Impact

The initial observations in renal specimens led to a definition of specific findings that permitted a diagnosis of amyloidosis essentially based on evaluation of hematoxylin and eosin stained sections. Because this material reacts with iodine and sulfuric acid (in a manner similar to starch), this was used as a diagnostic test many years ago to support a suspected diagnosis of amyloidosis [53, 54]. Amyloid was characterized morphologically as an amorphous eosinophilic material when seen in hematoxylin and eosin sections and was referred to generically as hyaline material.

About 70 years later, Divry and associates discovered that the amyloid deposits demonstrated apple green birefringence when specimens were stained with Congo red and viewed under polarized light. This remains today as the main criterion for a diagnosis of amyloidosis [15, 16, 55]. However, there are problems with using Congo red staining as the sole criterion to confirm a diagnosis of amyloidosis [56–58]. Thioflavin T and S staining of amyloid deposits is considered by some to be more sensitive than Congo red, although there are also pitfalls with these fluorescence stains [59]. It was not until 1959 that the fibrillary nature of amyloid was elucidated by Cohen and Calkins [60] using transmission electron microscopy. Ultrastructural evaluation represents the most sensitive and specific method for diagnosing amyloidosis. Amyloid was found to be composed of randomly disposed non-branching fibrils measuring 8–12 nm in diameter and non-branching. The ultrastructural appearance of all types of amyloid is the same.

Amyloid fibrils are composed of a central core containing the precursor protein (accounting for 95% of the amyloid composition) which is covered by glycosaminoglycans and serum amyloid-P component (remaining 5%). The function of the latter remains elusive. Typing of amyloid is the standard of care and determining the precursor protein associated with the amyloid fibrils is used in the typification of amyloid present [61–63]. This is very important as the clinical implications depend directly on the type of amyloid detected. Therefore, once amyloidosis is diagnosed in a renal specimen, amyloid typing is imperative.

While there are good fluorescence antibodies to immunoglobulin kappa and lambda light chains, not all cases of AL-amyloidosis can be detected with the commercially available antibodies used in the renal biopsies. This is due to the fact that some of the light chain proteins present in the amyloid fibrils are fragmented or so abnormal that they cannot be detected with the existing antibodies as they lack the antigenic epitopes detected by the commercially available antibodies [64]. There are also commercially available fluorescent antibodies to immunoglobulin heavy chains [65, 66] and fibrinogen. Fluorescent stains for light and heavy chains and fibrinogen are performed as part of the routine work-up of a renal biopsy. Immunohistochemical antibodies to serum AA-protein, B2-microglobulin, calcitonin, leukocyte chemotactic factor 2 (LECT2) and transthyretin (TTR) are also available.

Unfortunately, antibodies to the precursor proteins associated with most hereditary amyloidosis are only in the hands of researchers. It is important to recognize that more than one type of amyloid may be seen in the same case and that a diagnosis of a specific type of amyloidosis should not be made by exclusion [62]. Mass spectroscopy is possibly the most sensitive and specific method to characterize amyloid, but is only available in a few laboratories [61]. A small amount of paraffin embedded tissue is all that is required for typing amyloid using mass spectroscopy. The limitations and pitfalls of this technique are not yet well recognized.

Organs involved in amyloidosis acquire a rubbery consistency and a waxy, pink-to-grayish appearance. The most commonly involved organs are the liver, spleen, heart and kidney. The vasculature may be involved in any organ or site. In the case of the liver and spleen, these are organs rich in macrophages. In the case of heart and kidney, phenotypic transformation of existing cell types (predominantly fibroblasts in the heart and mesangial cells in the kidney) to a macrophage phenotype represents an early essential step for amyloidogenesis to occur [39, 40].

Phenotypic transformation of smooth muscle cells to macrophages in the vasculature all over the body is also crucial for amyloidogenesis to take place. Once extruded into the extracellular space, amyloid activates metalloproteinases destroying normal matrix constituents and eventually replacing them completely [40, 41, 67, 68]. In clinical material, this is evident in the silver stain where the normal argyrophilia of the mesangium is progressively lost as the matrix is destroyed [69, 70]. Normal mesangial matrix is silver-positive. This has been demonstrated clearly in the in vitro model of AL-amyloidosis. With time, and likely when the mesangial compartment is full, amyloid fibrils deposit along peripheral capillary walls resulting in loss of the permeability barrier to the filtration of proteins of certain size and leaking of large amounts of protein occur. These patients may exhibit massive proteinuria. Amyloid can also be associated with spikes along peripheral capillary walls when it extends into them [70].

Once the normal constituents of the various renal components are destroyed, other consequences occur. For example, the vasculature loses its turgor and integrity. This is responsible for the hypotension and the propensity to form aneurysms that may be seen in some of the patients with systemic amyloidosis. There are two other important effects of amyloid deposition in the extracellular compartment: amyloid enhances apoptosis and mesangial cells eventually disappear. In addition, amyloid inhibits transforming growth factor-β, the growth factor involved in rebuilding mesangial matrix [71].

As a result of these three crucial actions of amyloid once it is deposited extracellularly, the normal mesangial matrix is destroyed by active metalloproteinases, mesangial cells disappear by apoptosis [72] and the mesangial matrix is unable to repair since transforming growth factor-β is inhibited. Renal biopsies of patients with advanced glomerular amyloidosis show complete replacement of the mesangial matrix by the amyloid fibrils, as a consequence of the above-mentioned pathologic mechanisms. Apoptosis of MCs results from the combined effect of the action of glomerulopathic light chains on mesangial cells and a secondary, but just as important, direct apoptotic effect associated with the amyloid deposition in extracellular tissues.

Final Considerations

The advances that have taken place in the field of renal amyloidosis have been exciting. Elucidation of the sequence of events that is involved in the processing of precursor proteins by mesangial cells and formation of amyloid fibrils has provided a solid and comprehensive conceptual framework to understand the process of glomerular amyloidogenesis. These studies, although restricted to AL-amyloidosis, have resulted in a much better understanding of findings that can be seen in renal biopsies and have helped in defining specific guidelines for the earlier diagnosis of amyloidosis in renal biopsies. There are still challenges dealing with the mechanism involved in extrusion of the amyloid fibrils from cells that produce it and steps that occur in the assembly of the amyloid fibrils, as these may influence the design of therapeutic approaches [73–76]. Typification of the type of renal amyloidosis is now the norm which provides a rather useful guide for management and treatment of patients with this disease. Proteomics appears to be the way to address amyloid speciation in the most accurate manner at this time [77].

The treatment and management of patients with amyloidosis has improved considerably. Information regarding mechanisms involved in the degradation of amyloid fibrils once deposited in tissues is beginning to emerge. Advances in therapy have taken place as a result of our improved understanding of structural and physicochemical characteristics of the amyloid producing proteins, mechanisms of protein deposition in tissues and fibril formation, and the various pathological processes involved that have been dissected for the various sites.

Typing of the amyloid is essential, as therapies vary according to the type. In some instances, simply maintaining a certain conformation in the precursor protein, as is the case with the therapeutic use of the non-steroidal anti-inflammatory drug diflunisal (which maintains TTR as a homotetramer that is thermodynamically more stable and less amyloidogenic than its monomeric counterpart), or employing small molecules to stabilize precursor proteins to prevent generation of misfolded moieties suffices. Unfortunately in TTR amyloidosis, amyloid deposits sometimes persists because wild-type TTR can deposit as amyloid at pre-existing body sites where amyloid was present. TTR amyloidosis is rather infrequent in the kidney. The ability to make earlier diagnoses (often in renal biopsies) has led to the finding of small plasma cell clones responsible for the production of renal amyloidosis and the ability to treat these patients with AL-amyloidosis earlier and with the intent to achieve a 'cure' or, alternatively, optimal patient management [78–80]. Aggressive treatment of the plasma cell proliferative process followed by bone marrow transplantation or stem cell transplants is now a recognized avenue to treat patients with AL-amyloidosis.

Eprodisate has been shown to improve prognosis in patients with AA-amyloidosis by slowing the rate of progression of renal function deterioration with an acceptable safety profile [81], but not that of other types of amyloidosis involving the renal parenchyma. In TTR and α-fibrinogen-amyloidosis, liver transplantation is considered the definitive and curative treatment [79, 80]. However, even if the primary disease can be controlled or abolished, the end organ damage must be corrected. While transplantation may be a logical approach in at least some of these cases, repairing the damaged organs is the preferred route to take if at all possible and practical.

Amyloidogenicity is dependent on mutations in precursor proteins that destabilize them, favoring misfolding and aggregation. A series of key steps in the amyloidogenesis process have been defined in vitro and confirmed to occur in vivo models. These are amenable to therapeutic targeting to stop progression of amyloid fibril formation, deposition and eventual destruction of the native mesangium. In the case of AL-amyloidosis, the physicochemical characteristic of some light chains (a minority) makes them capable of engaging in amyloid formation [82–86], suggesting that particular structural and conformational aspects in these light chains facilitate interactions with receptors on mesangial cells that lead to internalization and processing in the mature lysosomal compartment. Comenzo et al. [82] clearly demonstrated lambda VI light chain tropism for the kidney and its crucial role in renal amyloidosis, solidified with in vitro data. Post-translational modifications of some glomerulopathic light chains, such as glycosylation, increases their amyloidogenicity [87]. Host factors are also important and very little work has been done in this area [88]. Amyloid enhancing factor, a somewhat mysterious (because of its unknown nature and derivation) and purported facilitator (accelerator) of amyloid formation in early investigational studies, has proven not to be necessary factor for mesangial cells to actively engage in renal amyloidogenesis in vitro and in vivo studies [47, 89, 90], but it certainly accelerates the process of amyloid formation. The specific nature of this factor remains unclear.

The repair of the damaged mesangium must first involve the cleaning up of the deposited amyloid. Metalloproteinases and catalytic enzymes may play a role in the degradation of amyloid fibrils [91, 92], but, obviously, they are unable to do so effectively in many instances as the amyloid eventually progressively takes over the normal mesangium. One of the main reasons for this is that there are generally few or no mesangial cells left (due to apoptotic removal) to engage in the active production of metalloproteinases and enzymes capable of degrading amyloid fibrils. If the disorder responsible for the production of the amyloid precursor protein can be abolished or controlled early in the process, it is likely that the mesangium would be able to heal itself. One of the recent areas of research interest has focused on repairing the altered mesangium by providing stem cells capable not only to clear amyloid, but to engage in the production of a new healthy mesangial matrix [93, 94].

References

1 Virchow R: Zur Cellulose-Frage. Virchows Arch 1854;6:416–426.
2 Haas LF: Rudolph Ludwig Carl Virchow (1821–1902). J Neurosurg Psychiatry 1996;61:578.
3 Tildsley GJ, Lakhani S: Early clinical pathologists 4: Rudolf Virchow (1821–1902). J Clin Pathol 1992;45:6–7.
4 Mathis H: Rokitansky, Virchow, and Heschl on the problem of amyloidosis (in German). Zentrablatt fur Allgeneine Pathologie and Pathologische Anatomie (Jena) 1968;111:103–107.
5 Virchow RLK (Chance F, trans): Cellular Pathology. New York, RM DeWitt, 1860, p 236.
6 Ruske W: August Kekule and the development of the theory of chemical structure (in German). Naturwissenschaften 1965;52:485–489.
7 Jochmann G, Schumm O: Zur kenntniss des myeloms und der sogenannten Kahler'schen Krankheit (multiple myeloma, einhergehend mit Bence Jones'scher albuminosurie. Ztschr Klin Med 1902;14:445–502.
8 Glaus A: Ueber multiples myelozytom mit eigenartigen zum teil kristallahnlichen zelleinlagerungers kombinieri mit elastolyse und ausgedehnter amyloidose und verkalkung (Basel, in German). Virchows Arch 1916–1917;223:301.
9 Bock HE: 80th anniversary of Prof. Dr. Hans Hermann Bennhold on Sept 11, 1973 (in German). Medizinishe Welt 1973;24:1341–1342.
10 Elghetany MT, Saleen A, Barr K: The Congo red stain revisited. Ann Clin Lab Sci 1989;19:190–195.
11 Kallee E: The 110th birthday of Hans Hermann Bennhold (in German). Dtsch Med Wochenschr 1993;118:1336–1338.
12 Bell ET: Amyloid disease of the kidneys. Am J Pathol 1933;9:185–203.
13 Sikl H: A case of diffuse plasmacytosis with deposition of protein crystals in the kidneys. J Pathol Bacteriol 1949;61:149–163.
14 Grabar P, Williams CA: Methode permettant l'etude conjugee des proprietes electrophoretiques et immunochimiques d'un mélange de proteins: Application au serum sanguine. Biochim Biophys Acta 1953;10:193–194.
15 Oifa, AI: Paul Divry- Founder of the concept of cerebral amyloidosis [Russian]. Zhurral Neuropathologii I Psikhiatrii Imeni S S Krsakova 1973;73:1078–1082.

16 Reznik M: Paul Divry: The discovery of cerebral amyloidosis (in French). Acta Neurol Belg 1989;89:168–178.
17 Smetana H: The relation of the reticulo-endothelial system to the formation of amyloid. J Exp Med 1927;45:619–632.
18 Edelman GM, Gally JA: The nature of Bence Jones protein: chemical similarities to polypeptide chains of myeloma globulins and normal gamma globulins. J Exp Med 1962; 116:202–227.
19 Gueft G, Ghidoni JJ: The site of formation and ultrastructure of amyloid. Am J Pathol 1963;43:837–854.
20 Shirahama T, Cohen AS: An analysis of the close relationship of lysosomes to early deposits of amyloid. Ultrastructural evidence in experimental mouse amyloidosis. Am J Pathol 1973;73:97–114.
21 Shirahama T, Cohen AS: Intralysosomal formation of amyloid fibrils. Am J Pathol 1975;81:101–116.
22 Shirahama, T, Cohen, AS: Lysosomal breakdown of amyloid fibrils by macrophages. Am J Pathol 1971;63:463–486.
23 Shirahama T, Cohen AS: Fine structure of the glomerulus in human and experimental amyloidosis. Am J Pathol 1973;73:97–114.
24 Bonar L, Cohen AS, Skinner MM: Characterization of the amyloid fibril as a cross-beta protein. Proc Soc Exp Biol Med 1969;131:1373–1375.
25 Glenner GG, Ein D, Eanes ED, Bladen HA, Tery W, Page DL: Creation of 'amyloid' fibrils from Bence Jones proteins in vitro. Science 1971;174:712–714.
26 Glenner GG, Terry W, Harada M, Isersky C: Amyloid fibril proteins: proof of homology with immunoglobulin light chains by sequence analysis. Science 1971;172:1150–1151.
27 Rocken C, Shakespeare A: Pathology, diagnosis, and pathogenesis of AA-amyloidosis. Virchows Arch 2002;440:111–122.
28 Linke RP, Zucker-Franklin D, Franklin EC: Morphologic, chemical, and immunologic studies of amyloid-like fibrils formed from Bence Jones proteins proteolysis. J Immunol 1973;111:10–23.
29 Epstein WC, Tran M, Wood IS: Formation of 'amyloid' fibrils in vitro by action of human kidney lysosomal enzymes on Bence Jones proteins. J Lab Clin Med 1974;84:107–110.
30 Zucker-Franklin D, Franklin EC: Intracellular localization of human amyloid by fluorescence and electron microscopy. Am J Pathol 1970;59:23–42.
31 Gise H, Christ H, Bohle A: Early glomerular lesions in amyloidosis. Virchows Arch A Pathol Anat Histol 1981;390:259–272.
32 Solomon A, Frangione B, Franklin EC: Bence Jones proteins and light chains of immunoglobulins. Preferred association of the V lambda VI subgroup of human light chains with amyloidosis. J Clin Invest 1982;70:453–460.
33 Kisilevsky R: Amyloids: Tombstones or triggers? Nature Med 2000;6:633–634.
34 Solomon A, Weiss DT, Kattine AA: Nephrotoxic potential of Bence Jones proteins. N Engl J Med 1991;324:1845–1851.
35 Kluve-Beckerman B, Manaloor JJ, Liepnieks JJ: A pulse-chase study tracking the conversion of macrophage-endocytosed serum amyloid A into extracellular amyloid. Arthritis Rheum 2003;46:1905–1913.
36 Tagouri Y, Sanders PW, Picken MM, Siegal GP, Kerby JD, Herrera GA: In vitro AL-amyloid formation by rat and human mesangial cells. Lab Invest 1996;74:290–302.
37 Solomon A, Weiss DT, Schell M, Hrncic R, Murphy C, Wall J, McGavin MD, Pan HJ, Kabalka GW, Paulus MJ: Transgenic mouse model of AA amyloidosis. Am J Pathol 1999;154:1267–1272.
38 Russell WJ, Cardelli J, Harris E, Baier RJ, Herrera GA: Monoclonal light chain-mesangial cells interactions: early signaling events and subsequent pathologic effects. Lab Invest 2001;81:689–703.
39 Teng J, Russell WJ, Gu X, Cardelli J, Jones ML, Herrera GA: Different types of glomerulopathic light chains interact with mesangial cells using a common receptor but exhibit different intracellular trafficking patterns. Lab Invest 2004;84:440–451.
40 Keeling J, Herrera GA: Matrix metalloproteinases and mesangial remodeling in light chain-related glomerular damage. Kidney Int 2005;68:1590–1603.
41 Teng J, Turbat-Herrera EA, Herrera GA: Role of translational research in advancing the understanding of the pathogenesis of light chain-mediated glomerulopathies. Pathol Int 2007;57:398–412.
42 Herrera GA: Low molecular weight proteins and the kidney: physiologic and pathologic considerations. Ultrastruct Pathol 1994;18:89–98.
43 Herrera GA: Plasticity of mesangial cells: a basis for understanding pathological alterations. Ultrastruct Pathol 2006;30:471–479.
44 Isaac J, Kerby JD, Russell WJ, Dempsey SC, Sanders PW, Herrera GA: In vitro modulation of AL-amyloid formation by human mesangial cells exposed to amyloidogenic light chains. Amyloid 1998;5:238–246.
45 Tagoe C, Gallo G, Buxbaum J: Amyloid and nonfibrillar deposits in mice transgenic for wild-type human transthyretin: a possible model for senile systemic amyloidosis. Lab Invest 2001;81:385–396.
46 Sawashita J, Mori M, Higuchi K: Amyloidosis in transgenic mice expressing murine amyloidogenic apolipoprotein A-II (Apo 2c). Lab Invest 2007;87:633–643.
47 Varga J, Flinn MSM, Shirahama T, Rodgers OG, Cohen AS: The induction of accelerated murine amyloid with human splenic extract. Probable role of amyloid enhancing factor. Virchows Arch B Cell Pathol Incl Mol Pathol 1986;51:177–185.
48 Herrera GA, Welbourne TC, Russell WJ: Isolated perfused rat kidney: a new model to evaluate light chain nephrotoxicity (abstract). Lab Invest 2001;81:188A.
49 Herrera GA, Turbat-Herrera EA, Teng J: AL-amyloidogenesis by mesangial cells: insights from an in-vivo animal model (abstract). J Am Soc Nephrol 2009;web http://www.abstracts2view.com/asn/view.php?nu = ASN09L160a&terms =.
50 Trinkhaus-Randall V, Walsh MT, Steeves S, Monis G, Connors LH, Skinner M: Cellular response of cardiac fibroblasts to amyloidogenic light chains. Am J Pathol 2005;166: 197–208.

51 Miura K, Shirasawa H: Amyloid A (AA) fibril formation in renal tubules occurs intracytoplasmically, possibly at the site of membrane assembling structures. Amyloid 1994;1:107–113.
52 El-Zoghby Z, Lager D, Gregorie J, Lewin M, Sethi S: Intra-tubular amyloidosis. Kidney Int 2007;72:1282–1288.
53 Puchtler H, Sweat F: A review of early concepts of amyloid in context with contemporary chemical literature from 1839 to 1859. J Histochem Cytochem 1966;14:123–143.
54 Aterman K: A historical note on the iodine-sulphuric acid reaction of amyloid. Histochemistry 1976;49:131–143.
55 Linke RP: Congo red staining of amyloid: improvements and practical guide for a more precise diagnosis of amyloid and the different amyloidoses; in Uversky VN, Fink AL (eds): Protein Misfolding, Aggregation, and Conformational Diseases. Protein Reviews. New York, Springer Science and Business Media, 2006, vol 4, pp 239–276.
56 Puchtler H, Waldrop FS, McPolan SN: A review of light, polarization and fluorescence microscopic methods for amyloid. Appl Pathol 1985;3:5–17.
57 Klastskin G: Non-specific green birefringence in Congo-red stained tissues. Am J Pathol 1969;56:1–13.
58 Carson FL, Kingsley WB: Non-amyloid green birefringence following Congo red staining. Arch Pathol Lab Med 1980;104:333–335.
59 Hobbs JR, Morgan AD: Fluorescence microscopy with thioflavin-T in the diagnosis of amyloid. J Pathol Bacteriol 1963;86:437–442.
60 Cohen AS, Calkins F: Electron microscopic observations on a fibrous component in amyloid of diverse origins. Nature 1959;183:1202–1203.
61 Murphy CL, Eulitz M, Hrncic R, Sletten K, Westermark P, Williams T, Macy SD, Wooliver C, Wall J, Weiss DT, Solomon S: Chemical typing of amyloid protein contained in formalin-fixed paraffin-embedded biopsy specimens. Am J Clin Pathol 2001;116:135–142.
62 Picken MM, Herrera GA: The burden of 'sticky' amyloid. Arch Pathol Lab Med 2007;131:850–851.
63 Satoskar AA, Burdge K, Cowden DJ, Nadasdy GM, Hebert LA, Nadasdy T: Typing of amyloidosis in renal biopsies. Diagnostic pitfalls. Arch Pathol Lab Med 2007;131:917–923.
64 Novak L, Cook WJ, Herrera GA, Sanders PW: AL-amyloidosis is underdiagnosed in renal biopsies. Nephrol Dial Transplant 2004;19:3050–3053.
65 Eulitz M, Weiss DT, Solomon A: Immunoglobulin heavy chain-associated amyloidosis. Proc Nat Acad Sci USA 1990;87:6542–6546.
66 Mai HL, Sheikh-Hamad D, Herrera GA, Gu X, Truong LD: Immunoglobulin heavy chain can be amyloidogenic. Am J Surg Pathol 2003;27:541–545.
67 Keeling J, Herrera GA: Human metalloproteinases: characteristics and pathologic role in altering mesangial homeostasis. Microsc Res Tech 2008;71:371–379.
68 Keeling J, Herrera GA: Matrix metalloproteinases and mesangial remodeling in light chain-related glomerular damage. Kidney Int 2005;68:1590–1603.
69 Herrera GA, Lott RL: Silver stains in diagnostic renal pathology. J Histotechnol 1996;19:219–223.
70 Ge F, Yao J, Fu X, Guo Z, Yan J, Zhang B, Zhang H, Tomozawa H, Miyazaki J, Dikman SH, Churg J, Kahn T: Morphologic and clinical correlates in renal amyloidosis. Human Pathol 1981;12:160–169.
71 Stetler-Stevenson WG: Dynamics of matrix turnover during pathologic remodeling of extracellular matrix. Am J Pathol 1996;148:1345–1350.
72 Tagouri Y, Sanders PW, Zhu I, Picken MM, Winokur TS, Soong S-J, Herrera GA: Glomerulopathic monoclonal light chains potentiate mesangial cell apoptosis in vitro (abstract). Lab Invest 1995;92:161A.
73 Teng M, Yin J, Vidal R, Ghiso J, Kuman A, Rabenou R, Shah A, Jacobson DR, Dealwis C, Wall J: Towards understanding the structure-function relationship of human amyloid disease. Curr Drug Targets 2004;5:159–171.
74 Ionescu-Zanetti C, Khurana R, Gillespie JR, Petrick JS, Trabachino LC, Minert LJ, Carter SA, Fink AL: Monitoring the assembly of Ig light-chain amyloid fibrils by atomic force microscopy. Proc Natl Acad Sci USA 1999;96:13175–13179.
75 Taubes G: Misfolding the way to disease. Science 1996;271:1493–1495.
76 Stevens FJ, Myatt EA, Chang CH, Westholm FA, Eulitz M, Weiss DT, Murphy C, Solomon A, Schiffer M: A molecular model for self assembly of amyloid fibrils: immunoglobulin light chains. Biochemistry 1995;34:10697–10702.
77 Lavatelli F, Perlman DH, Spencer B, Prokaeva T, McComb ME, Theberge R, Connors LH, Bellotti V, Seldin DC, Merlini G, Skinner M, Costello CE: Amyloidogenic and associated proteins in systemic amyloidosis proteome of adipose tissue. Moll Cell Proteom 2008;7:1570–1583.
78 Merlini G, Palladini G: Amyloidosis: is a cure possible? Ann Oncol 2008;19(Suppl 4):iv63-iv66.
79 Seldin DC: AL-amyloidosis, a curable disease? Amyloid 2002;9:203–206.
80 Dember LM: Amyloidosis-associated kidney disease. J Am Soc Nephrol 2006;17:3458–3471.
81 Dember LM: Eprodisate for the treatment of renal disease in AA-amyloidosis. N Eng J Med 2007;356:2349–2360.
82 Comenzo RL, Zhang Y, Martinez C, Osman K, Herrera GA: The tropism of organ involvement in primary systemic amyloidosis. Contributions of the Ig VL germline gene use and clonal plasma cell burden. Blood 2001;98:714–720.
83 Merlini G, Belloti V: Molecular mechanisms of amyloidosis. New Engl J Med 2003;349:583–596.
84 Belloti V, Mangione P, Merlini G: Review: immunoglobulin light chain amyloidosis – the archetype of structural and pathogenic variability. J Struct Biol 2000;130:280–289.
85 Stevens FJ: Four structural risk factors identify most fibril-forming kappa light chains. Amyloid 2000;7:200–211.
86 Wall J, Schell M, Murphy C, Hrncic R, Stevens F, Solomon A: Thermodynamic instability of human lambda 6 light chains: correlation with fibrillogenicity. Biochemistry 1999;38:14101–14108.
87 Omtvedt LA, Bailey D, Renouf DV, Davies MJ, Paramonov NA, Haavik S, Husby G, Sletten K, Hounsell EF: Glycosylation of immunoglobulin light chains associated with amyloidosis. Amyloid 2000;7:227–244.

88 Solomon A, Weiss DT: Protein and host factors implicated in the pathogenesis of light chain amyloidosis (AL amyloidosis). Amyloid 1995;2:269–279.
89 Kisilevsky R, Bourdeau L: Kinetics of amyloid deposition. I. The effects of amyloid-enhancing factor and splenectomy. Lab Invest 1983;48:53–59.
90 Cohen AS, Connors LH: The pathogenesis and biochemistry of amyloidosis. J Pathol 1987;151:1–10.
91 Stix B, Kahne T, Sletten K, Raynes J, Roessner A, Rocken C: Proteolysis of AA amyloid fibril proteins by matrix metalloproteinases-1, -2, and -3. Am J Pathol 2001;159:561–570.
92 Rocken C, Stix B, Bromme D, Ansorge B, Roessner A, Buhling F: A putative role for cathepsin K in degradation of AA and AL amyloidosis. Am J Pathol 2001;158:1029–1038.
93 Teng J, Herrera GA: Replacement of injured rat mesenchymal mesangial cells by glomerulopathic light chains (GLCs) by rat mesenchymal cells (RMSCs): a 6D living cell model of mesangial repair (abstract). J Am Soc Nephrol 2008;19:435A.
94 Herrera GA, Turbat-Herrera EA, Teng J: Repair of damaged mesangium by rat mesenchymal stem cells (RMSCs) in a 6D living cell model (abstract). Amyloid. J Prot Folding Disord 2010;17(Suppl 1):44.

Guillermo A. Herrera, MD
Nephrocor
1700 North Desert Drive
Tempe, AZ 85 281 (USA)
Tel. +1 602 443 6119, Fax +1 602 443 6102 ext. 3106, E-Mail nephronpath@gmail.com

Herrera GA (ed): Experimental Models for Renal Diseases: Pathogenesis and Diagnosis.
Contrib Nephrol. Basel, Karger, 2011, vol 169, pp 247–261

Toward Understanding Renal Fanconi Syndrome: Step by Step Advances through Experimental Models

Christophe Sirac[a] · Frank Bridoux[a,b] · Marie Essig[c] · Olivier Devuyst[d] · Guy Touchard[b] · Michel Cogné[a]

[a]Laboratoire d'immunologie, CNRS UMR6101, Université de Limoges, Limoges, [b]Service de Néphrologie, CHU Jean Bernard, Poitiers, [c]Service de Néphrologie, CHU Dupuytren, Limoges, France; [d]Division of Nephrology, Université catholique de Louvain Medical School, Brussels, Belgium

Abstract

Renal Fanconi syndrome (FS) is a generalized dysfunction of proximal tubular epithelial cells leading to the urinary leak of essential metabolites like phosphate, uric acid, glucose, amino acids and low molecular weight proteins. From inherited forms involving mutations on apparently unrelated genes to acquired forms induced by drugs, heavy metals or monoclonal immunoglobulin (Ig) light chains (LC), heterogeneous causalities of FS have complicated the understanding of this pathology for a long time. Experimental models of FS have allowed researchers to face the challenge and have helped unravel the main mechanisms disturbing proximal tubule reabsorption. Administration of cadmium to animals first demonstrated an inhibition of Na/K/ATPase activity, highlighting how a single toxic component could induce the general sodium-linked transport defect observed in FS. Today, genetically modified mice allow the development of reliable and reproducible experimental models for inherited or acquired forms of FS. One of the most exciting advances offered by these models is the unexpected major role of endocytosis in the function of the proximal tubule revealed by megalin and ClC-5 knockout mice. Using gene-targeted insertion, a transgenic mouse for LC-associated FS, the most frequent adult form of FS, has also been recently developed and represents a major step in the development of models of this pathology. Beyond deciphering molecular and cellular events at the origin of FS, these models also represent essential tools for the development of therapeutic strategies.

Ions and low molecular weight solutes filtered through the glomerulus are actively reabsorbed in the proximal tubule of the kidney. This reuptake process, essential to avoid the wasting of important solutes in the urine, plays a pivotal role in body fluid homeostasis [1]. In proximal tubular epithelial cells (PTEC), which are mainly involved in this process, the electrochemical gradient created by the basolateral Na^+/K^+ ATPase facilitates the activity of several sodium dependent co-transporters expressed at the apical side, allowing the reuptake of glucose, amino acids and ions [2, 3]. In addition, an extensive apical endocytotic apparatus, including brush border-shaped membranes, early and late endosomes, lysosomes, and specialized endocytotic receptors like megalin and cubilin, favors the reabsorption and degradation of freely filtered small plasma proteins (<70 kDa) [4–7]. These include plasma carrier proteins (retinol-binding protein,

Table 1. Models for renal FS and associated human diseases

Models	Human disease	Effect on PTEC functions	Phenotype	Reference
ClC-5 KO (mouse)	Dent's disease	impaired lysosome acidification	full-blown FS	[22, 23, 52, 53, 54]
OCRL1 KO (mouse)	Lowe's syndrome	none	none	[61]
Cubilin mutation (dog)	Imerslund-Grasbeck disease	inhibition of Vit-B12 reabsorption, inhibition of protein reabsorption	low molecular weight proteinuria, vitamin D deficiency	[42, 43]
CTNS KO (mouse)	cystinosis	intralysosomal cystine accumulation	ocular damages, bone defects and behavioral anomalies	[62, 63]
Transgenic Ig LC CHEB (mouse)	LC-associated FS	LC crystal accumulation within lysosomes	partial FS (no amino aciduria, no glucosuria)	[95]
Cadmium/lead administration (rat/mouse)	heavy metals-induced FS	inhibition of Na/K/ATPase activity and decreased ATP production	full-blown FS	[16, 18]
Megalin KO (kidney specific, mouse)	Donnai-Barrow and facio-oculo-acoustico-renal syndrome (DB/FOAR)	inhibition of protein reabsorption and general defect of endocytosis	low molecular weight proteinuria, vitamin D and A deficiency, hypocalcemia, osteomalacia, defective forebrain	[21, 40]
HNF1-α KO (mouse)	MODY3	inhibition of sodium-glucose transporter type 2 and type 1 sodium-phosphate co-transporter gene expression	full-blown FS, wasting syndrome, liver dysfunction, phenylketonuria	[58, 59, 60]

transferrin, vitamin D-binding protein), peptide hormones (insulin, parathyroid hormone), lysozyme and immunoglobulin (Ig) light chains (LC).

Renal Fanconi syndrome (FS) corresponds to generalized dysfunction of sodium-linked active transport in PTEC, leading to the urinary leak of phosphate, uric acid, glucose, amino acids and low molecular weight proteins (β_2-microglobulin, retinol-binding protein and α_1-macroglobulin), which are often associated with proximal (type II) renal tubular acidosis [2]. FS may result from various conditions (table 1), including hereditary diseases such as mitochondrial cytopathies [8], Dent's disease [9, 10], Lowe's syndrome [11, 12] and cystinosis [13, 14], or following the administration of certain drugs or heavy metal poisoning [15–18]. In adults, FS is frequently related to the urinary secretion of a monoclonal Ig LC, produced by either malignant or non-malignant plasma cells [19, 20].

In the last decades, in vitro and in vivo experimental models have highlighted some of the mechanisms leading to the generalized proximal tubule dysfunction (fig. 1 and table 1). Among them, direct or indirect inhibition of Na/K/

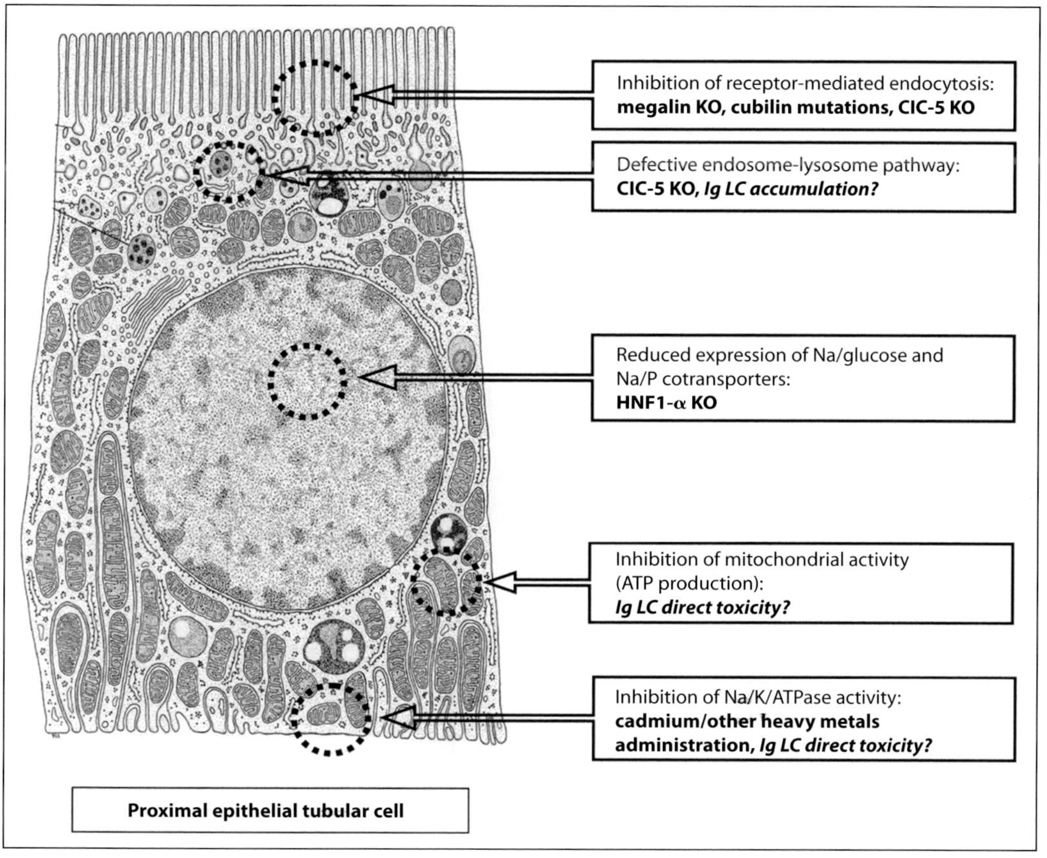

Fig. 1. Experimental models for renal FS. Conditions for the reabsorption of solutes: integrity of cell structure (polarity, brush border, endocytic apparatus), apical multiligand receptors (megalin and cubilin), transport systems (apical – endosomes/lysosomes – basolateral), energy production and transport (mitochondria), Na^+/K^+/ATPase basolateral (driving force for Na^+-coupled transport) and paracellular pathway. They all represent targets for disease. In experimental models (bold), disturbance of any of these conditions is able to induce a general or partial dysfunction of PTEC characterized by renal FS. In italics, putative targets in the LC-associated FS model. Modified from Devuyst and Pirson [54].

ATPase activity, a hallmark of proximal tubular reabsorption defect, was first described in animals exposed to repetitive administration of toxic components [2]. The more recent advent of genetically modified mice allowed the development of reliable and reproducible experimental models of FS, particularly for the inherited forms. One of the most exciting advances offered by these transgenic models is the unexpected major role of endocytosis in the function of the proximal tubule [5]. Accordingly, models where the endocytic pathway is disturbed, either by genetic invalidation of essential endolysosomal proteins (megalin and cubilin receptors, ClC-5 exchanger, etc.) [21–23] or by the abnormal accumulation of a toxic agent (cadmium, Ig LC) [20, 24, 25], display the major phenotypic characteristics of FS. This chapter will describe some of the major advances provided by animal models in deciphering molecular events at the origin of the FS, focusing on genetically modified mice. In particular the recent development of LC-associated

FS transgenic murine models from pioneer molecular studies and in vitro experiments will be reviewed.

Defective Endocytosis: A Common Feature of FS Revealed by Genetically Modified Mouse Models

Numerous experimental models for FS have been developed over the last 40 years, and have been based on the administration of maleic acid or cadmium to mice, rats or rabbits [15, 16, 26], or in vivo perfusion of isolated rat tubules [18, 27, 28]. One major step in the understanding of the pathophysiological mechanisms of FS was made in the early 1980s by Gonick et al. [16], using a model of cadmium-induced FS in rats. They showed a direct inhibition of Na/K/ATPase activity in renal tubules after repetitive intraperitoneal administration of cadmium chloride and, thus, demonstrated how a single toxic component could induce the general defect in tubular reabsorption characteristic of FS. This discovery was later confirmed by in vitro studies based on the exposure of PTEC to heavy metals, maleic acid or Ig LC [29–31]. Whether or not these toxic components exert a direct or indirect inhibition on Na/K/ATPase activity remains largely debated [32].

It was not until the late 1990s and the advent of transgenic mouse models that defective proximal tubular endocytosis was shown as another major mechanism of FS. In PTEC, clathrin-mediated endocytosis represents the predominant pathway for low molecular weight protein uptake across the apical membrane, with an endocytic pathway consisting of five main interrelated compartments: (i) microvilli and clathrin-coated pits, (ii) early endosomes, (iii) dense apical tubules responsible for apical recycling, (iv) late endosomes and (v) lysosomes [33]. The process essentially requires two multiligand receptors, megalin and cubilin, which are expressed at the brush border of PTEC [34]. Ligand binding and interactions between both receptors induce their internalization into clathrin-coated vesicles and their subsequent delivery to endosomes and lysosomes for ligand processing and receptor degradation or recycling (fig. 2).

Receptor-mediated endocytosis depends on the integrity of the actin cytoskeleton and the microtubules [35], whereas progression along the endocytic apparatus requires a sustained vesicular acidification from early to late endosomes and finally to lysosomes [36, 37]. In PTEC, the endosomal acidification is driven by the electrogenic vacuolar H^+-ATPase (V-ATPase). The translocation of H^+ from the cytoplasm into the endosomes generates a transmembrane electrical potential ($\Delta\Psi$) resulting in a rapid inhibition of V-ATPase activity. Thus, in order to ensure progressive vesicular acidification, either anions have to enter vesicles or cations have to leave. In most cases, vesicular acidification seems dependent on a parallel Cl^- conductance [38]. Furthermore, the intravesicular Cl^- concentration itself could directly affect the V-ATPase activity [39] as well as the vesicle recycling independently of its effect on pH [36].

Using transgenic models, it is now clearly demonstrated that any dysfunction of this endosomal pathway can induce renal FS (fig. 1) and the first outstanding animal model was the megalin knockout mouse [21, 40]. Megalin (lrp2) is expressed in abundance at the brush border membrane and in all parts of the endocytic apparatus of the PTEC [41]. Mice lacking megalin exhibit lung and kidney abnormalities with severe malformations of the forebrain, most dying within minutes after birth. However, about 2% of the animals survive to adulthood and display low molecular weight proteinuria, including vitamin D-binding protein, retinol-binding protein, α(1)-microglobulin, transthyretin, hemoglobin and transferrin [21, 42–44]. Calcidiol and vitamin A complexed with their carrier proteins are significantly lost in the urine of megalin-invalidated mice, thereby disturbing calcium and vitamin A homeostasis [45–47]. Moreover, megalin modifies the adequate steady-state expression of the sodium-phosphate co-transporter and its capacity to be internalized by PTEC in response to PTH [48].

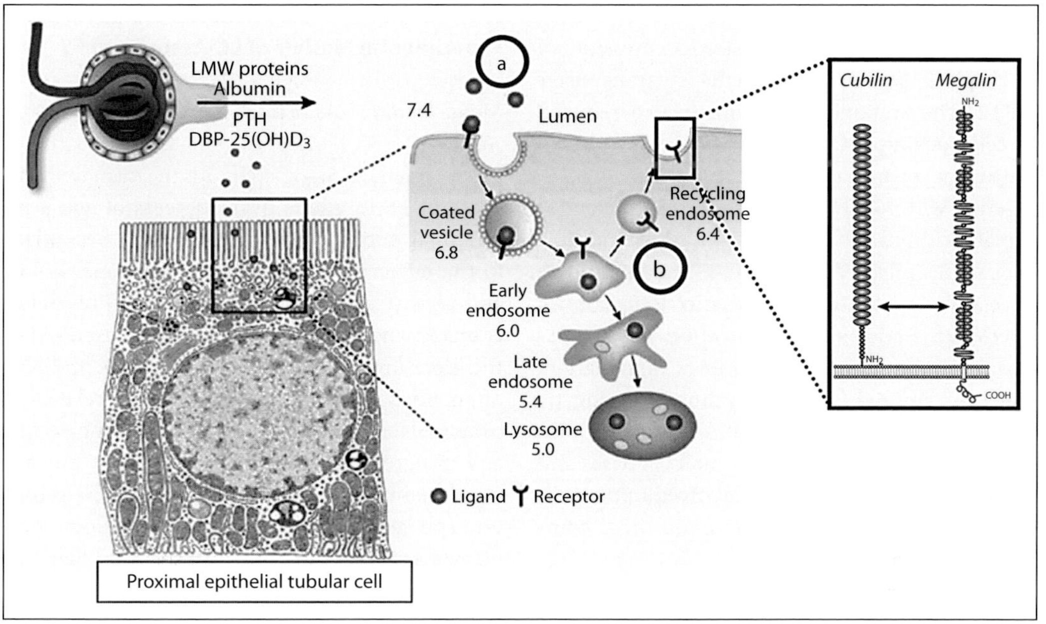

Fig. 2. Reabsorption of low molecular weight proteins by proximal tubule cells. Low molecular weight (LMW) proteins that are naturally filtered by the glomerulus into the primary urine are endocytosed by PTEC via the megalin/cubilin receptor pathway. Following internalization in coated vesicles, the receptor-ligand complexes progress along the endocytic pathway. The endosomes undergo a progressive acidification that results in the dissociation of the receptor-ligand complexes, with megalin and cubilin (inset) being recycled to the apical membrane, whereas the ligand is directed to lysosomes for degradation. Genetic mutations affecting the receptors megalin or cubilin (Imerslund-Grasbeck disease or Donnai-Barrow and facio-oculo-acoustico-renal syndrome) (**a**) or any proteins involved in the endosomal acidification (and consequently in receptor recycling to the apical membrane) (**b**), like the ClC-5 exchanger (Dent's disease), result in a general decrease in LMW protein reabsorption. Modified from Devuyst and Pirson [54] and Devuyst and Guggino [100].

A similar mechanism of large endocytic dysfunction of PTEC is also observed in cubilin-deficient dogs, another multiligand-binding receptor which interacts with megalin [49, 50]. Cubilin was first identified as the cobalamin-intrinsic factor receptor and its mutation causes hereditary megaloblastic anemia 1 (Imerslund-Grasbeck disease) [51]. Of note, proteinuria has been detected in patients with selective vitamin B12 malabsorption due to cubilin deficiency [52]. Certainly, due to the essential role of megalin and cubilin during development, no human disease associated with the complete absence of these proteins has been described in humans.

A second major model of renal FS was provided by mice invalidated for the renal endosome-associated chloride-proton exchanger, ClC-5 [22, 23]. These mice reproduce Dent's disease, an X-linked disorder caused by inactivating mutations in the *CLCN5* gene that encodes ClC-5 [53, 54]. Patients with Dent's disease show a complex phenotype combining the urinary loss of low molecular weight proteins, phosphate and calcium with kidney stones, nephrocalcinosis, and progressive renal failure in some cases. ClC-5-deficient mice display decreased endocytosis of numerous proteins physiologically reabsorbed by the proximal tubule, such as β2-microglobulin, vitamin D-binding protein and retinol-binding protein, together with decreased expression of the apical sodium phosphate co-transporter NPT2. The amount of megalin is also reduced in cells lacking ClC-5, which also show altered trafficking of megalin and cubilin [55]. ClC-5, together with V-ATPase, is involved in the endosomal acidification in PTEC. Cortical kidney endosomes purified from ClC-5-knockout mice have a lower acidification capacity than control mice [22]. Endosomal acidification seems necessary for receptor-ligand complex dissociation, which is required for the recycling of endocytic receptor to the apical membrane [56]. It also allows the recruitment of Arf small GTPases and other cytosolic coat proteins involved in the trafficking between early and late endosomal compartments [57].

Some of the mouse models of FS differ by their inability to reproduce all the features of proximal tubular dysfunction. For example, mice invalidated for HNF1-α, a transcription factor involved in the regulation of a large set of hepatic genes, have a dramatic reduction in glucose, amino acid and phosphate reabsorption, but without apparent low molecular weight proteinuria [58]. In this model, the proximal tubule dysfunction results mostly from a reduced expression of the Na^+/glucose co-transporter SGLT2 (SLC5A2) and the Na^+/phosphate co-transporter NPT1 (SLC17A1), respectively [59, 60]. This raises the possibility that full-blown FS is only present when the endocytic pathway is affected.

Strikingly, some attempts to develop experimental models of inherited human FS through gene disruption failed to induce proximal tubular dysfunction. This was due either to an unsuspected functional overlap between two murine gene products in the mouse model for Lowe's syndrome (OCRL1 knockout mice) [61] or to the influence of the genetic background in the mouse model of cystinosis (CTNS knockout mice) [62, 63]. In these models, the mechanisms by which mice overcome the defects observed in humans should be explored to develop alternative therapeutic approaches [64].

From Proteins to Transgenes: Development of Experimental Models of LC-Associated FS

Monoclonal LC-Associated FS: An Overview

Monoclonal LC-associated FS is the most frequent cause of FS in adults over 50 years of age, with proximal tubule dysfunction being secondary to the urinary secretion of a monoclonal Ig LC, and almost always of the κ-isotype. Prevalence is unknown, with around 100 cases described in the literature thus far, but the disease is probably often misdiagnosed [65]. Bone pain related to osteomalacia secondary to hypophosphatemia, and progressive chronic renal failure are the usual manifestations. In most cases, these symptoms precede the diagnosis of a slow progressive plasma cell disorder by several years. Signs of proximal tubular dysfunction include hypophosphatemia, hypouricemia and hypokalemia due to renal leakage (associated with normoglycemic glucosuria), generalized aminoaciduria, low molecular weight proteinuria and proximal renal tubular acidosis. Some of these signs may be absent in case of severe renal failure. FS is most commonly associated with low-grade lymphoplasmacytic neoplastic disorder [19, 20, 65].

The first descriptions of FS associated with myeloma were reported 50 years ago [66, 67]. Based on careful immunofluorescence and electron microscopic studies of kidney biopsies, Maldonado et al. [19], later defined the characteristic pathological features of monoclonal LC-associated FS. Examination of toluidine blue-stained kidney biopsy samples by light microscopy typically reveals intracytoplasmic inclusions in the proximal tubule epithelium, which, in most cases, stain with the anti-κ-conjugate on immunofluorescence analysis. Ultrastructural studies reveal rhomboid or needle-shaped crystals, located in endolysosomal vesicles and displaying an 80 Å periodic striation. Similar crystals are commonly detected in the cytoplasm of macrophages in the renal interstitium and within bone marrow plasma cells, where they probably slow the progression rate of the hematological disease [68–70]. Even if this typical clinical presentation of monoclonal LC-associated FS has been shown as inconstant by some recent observations [20, 71–75], it remains largely predominant.

Since the first complete sequence published by Aucouturier et al. [76], molecular studies carried out in a dozen case [20, 25, 72, 77–79] have greatly contributed to depict the pathophysiological mechanisms of monoclonal LC-associated FS by revealing a striking genetic and biochemical homogeneity. Except for few exceptions [25, 72, 75], typical monoclonal LC-associated FS, characterized by crystalline inclusions within PTEC and low-grade lymphoplasmacytic neoplastic disorder, belongs to the Vκ1 subgroup of variability and is derived from only two germline genes: O2/O12 and O8/O18. The crystallization properties are associated with the resistance of the 12-kDa Vκ domain to proteolysis by various enzymes, trypsin, pepsin and cathepsin B, the main lysosomal enzyme involved in degradation of proteins in PTEC [76, 79]. Peculiarities in the primary structure of the variable domain induced by the somatic hypermutation process are frequently found in the CDR1 loop, especially at position 30 where the polar serine is replaced by non-polar or hydrophobic residues in all the O2/O12 V domains studied to date. Other mutations of interest have been localized within all the LC studied as a result of sequences analysis and 3D models of the proteins [25, 79]. They all induce a change in charge, polarity or structure of the LC and are physiologically very rarely observed in the V domains of κ-LC from the same subgroups.

LC-associated FS has also been described in crystal-storing histiocytosis (CSH), a rare condition characterized by accumulation of immunoglobulinic crystals within lysosomes of histiocytes in the bone marrow or other organs. CSH-associated FS is associated with characteristic proximal tubular lesions including intracellular crystals. In a recent study, it was shown that the ultrastructural pattern of proximal tubule crystalline inclusions in CSH-associated FS differs from that observed in myeloma-associated FS, with rectangular/needle-shaped or hexagonal/diamond-shaped patterns, respectively. These morphological differences were associated with distinct molecular structure peculiarities of the κ-LC variable domains, but led to the same typical resistance of the V domain to trypsin digestion. Comparison of the 3D-structures of the CSH/FS Vκ1 and Vκ3 domains with the germline-encoded structures and those from patients with myeloma-associated FS revealed distinct hydrophobic residues exposed to the solvent, likely favoring the formation of a variant form of crystals that may further resist degradation after phagocytosis [80].

Finally, based on histological, biochemical and molecular findings, a tempting scenario has been proposedl. Physiologically, circulating free LCs, after being freely filtered by the glomerulus, are reabsorbed in the proximal tubule by a mechanism of endocytosis involving LC binding to the multiligand receptors cubilin and megalin [41]. LCs are degraded by lysosomal proteases into amino acids that are recycled through the basolateral membrane [81]. Because of their structural peculiarities in the V domain, FS-associated monoclonal κ-LCs are resistant to complete proteolysis and self-aggregate to form crystals within the endolysosomal compartment, leading to congestion and functional impairment of PTEC. However, the link between overload in the endosome-lysosome apparatus and functional impairment in PTEC is not obvious as seen in the ClC-7 KO mice, which do not show any renal FS despite major lysosomal accumulation [82]. In LC-associated FS, this attractive scenario was also challenged by some observations of full-blown FS devoid of crystalline inclusions in the tubular cells [20, 72]. The absence of crystals is often associated with normal sensitivity to proteolysis, suggesting that reabsorption of monoclonal LCs might also interfere with proximal tubule function through direct toxic effects.

From in vitro Experiments to Transgenic Animal Models: A Necessary Evolution

In the past 20 years, several experimental models have been developed in order to explore the renal physiologic disturbances induced by monoclonal LC. Pioneer experimental approaches consisted of rapid in vitro or in vivo exposition of renal tubule to human monoclonal LC. In 1986, Batuman et al. [83] were the first to demonstrate the direct nephrotoxicity of monoclonal LC. In an in vitro model of rat brush border membrane cultures exposed to different physiologic concentrations of purified Bence Jones proteins, they showed an inhibition of alanine and glucose uptake by the PTEC. Thereafter, similar experiments, using primary cultures or immortalized tubular cell lines, showed that LC inhibit Na^+/K^+/ATPase expression [31] and are processed through the cubilin and megalin-mediated endocytosis pathway [81, 84–86].

Another model, proposed by Sanders et al. [87], consisted of the in vivo perfusion of rat nephrons with monoclonal LC purified from the urine of myeloma patients with renal failure. In an initial study, nephron morphology and function was analyzed after perfusion with an artificial tubule fluid that contained either no protein, or a human monoclonal κ-LC, or bovine serum albumin. The authors showed that, by contrast to albumin, the κ-Bence Jones protein induced morphologic changes of the proximal tubule epithelium together with reduced reabsorption of water, chloride and glucose. As for the in vitro models, this important breakthrough was achieved using a monoclonal LC isolated from a patient with no evidence of FS. Later, using the same experimental approach, the same group demonstrated that LC may induce different physiologic and morphologic alterations of the tubule on the sole basis of their distinct physicochemical characteristics [88]. As these differences could likely arise from V domain composition, it appeared that the next step was to develop more specific experimental models using myeloma LC from patients with a defined renal disease. Moreover, in spite of their past and current invaluable interest [89–91], these models only permit a transitory reproduction of the initial stages of the renal disturbances and they remain worthless for testing new therapeutic approaches.

To circumvent these limits, we established a model closer to human monoclonal LC-associated FS by using tumor grafts [77, 92]. In this model, DNA corresponding to the pathogenic LC was cloned into a vector allowing a strong expression in cells of the B lineage [93, 94]. These constructions were transfected into SP2/O cells (a hybridoma producing no endogenous Ig derived from Balb/C mice). Transfected cells secreting the largest amount of LC were then injected intraperitoneally to Balb/C mice, which developed a tumor producing high levels of the monoclonal LC, mimicking the human plasma cell dyscrasia. Histological studies performed on organs taken after sacrifice (4–8 weeks after injection) revealed renal lesions similar to those observed in patient CHEB from whom the LC sequence was cloned [76], with typical crystalline inclusions at the apical pole of PTEC [77]. Moreover, experiments on tubule suspensions from mice injected with SP2/0 cells transfected with CHEB κ-LC showed impaired proximal tubular reabsorption of phosphate. Using site-directed mutagenesis, we showed that sequential replacement of the residues at position 30 and 94 by their germline counterpart resulted in the disappearance of crystalline inclusions in mice. Therefore, this model demonstrated that limited sequence changes of the V region determined the development of proximal tubular lesions and that the concomitant presence of the hydrophobic amino acid at positions 30 and 94 were needed in order for the CHEB LC to display its propensity to assemble into crystals.

Unfortunately, this model was limited by the short survival of mice due to rapid tumor growth, which prevented complete in vivo metabolic explorations and therapeutic approach. To overcome these problems, we developed a transgenic mouse model with overexpression of the human CHEB κ-LC, but without associated plasma cell dyscrasia. To ensure a high level of expression of the transgene and to more specifically assess the role of the V domain in the tissue deposition process, we targeted insertion of the V domain from CHEB LC in the Ig-κ locus of the mouse [95]. This resulted in the expression of a hybrid κ-LC made up of the human V domain and the mouse constant region.

Despite the replacement of the human C region, animals exhibited proximal tubular lesions typical of FS with characteristic intracellular crystals made from the LC, which are associated with enlargement of mitochondria and alterations of the apical brush border in PTEC (fig. 3). This confirmed that LC aggregation in FS is promoted by the V domain structure. The extent of tubular inclusions was proportional to the production rate and serum levels of CHEB LC. Using an inducible CRE-mediated deletion of the CHEB V domain, we demonstrated that tubular lesions recovered after a few weeks when LC production was stopped (fig. 3).

In contrast with our previous attempts to set up a model of myeloma-associated FS using tumor grafts, transgenic mice showed no obvious alteration of their general condition and were easily amenable to physiologic explorations. Transgenic homozygous animals had significant impairment of uric acid, phosphate and the low molecular weight protein CC16 tubular reabsorption, three cardinal features of FS [23, 96] (table 2). However, there was no evidence of glucosuria and aminoaciduria, which was likely related to the low production rate of free pathogenic LC in mice, compared to patient CHEB [76].

In this model, it appears that human LC are mostly associated with endogenous murine Ig heavy chains and various strategies to increase the rate of free LC are under investigation. We can anticipate that an enhanced free LC production in animals would lead to full-blown FS. Nevertheless, these mice display some major phenotypic characteristics of FS and thus provide the first complete animal model for a monoclonal LC-associated renal disease. As cellular mechanisms by which monoclonal κ-LC induce proximal tubule dysfunction remain poorly understood, this model should be useful for identifying the cellular targets of reabsorbed monoclonal nephrotoxic LC and for testing drugs which can potentially counteract these effects to set up new therapeutic approaches for patients.

Studies of the effect of LC reabsorption on sodium co-transporters expression and trafficking [31, 83, 87, 97], endosomal acidification [55, 98], and mitochondrial energy supply [20] are under investigation in CHEB transgenic mice. In this model, which mimics the most frequent form of monoclonal LC-associated FS, congestion of the endolysosomal compartment of PTEC and the resulting decreased receptor recycling alone might explain proximal tubule dysfunction. However, whether a definitive comprehension of LC toxicity arises from this unique model remains hypothetical, as mechanisms of toxicity could vary with the molecular characteristics of the involved LC. Nevertheless, this transgenic model for LC-associated FS has already questioned an important issue. As monoclonal LC-associated FS generally occurs in the setting of smoldering myeloma, chemotherapy is considered to be of little benefit, despite progression to end-stage renal disease in many patients [20, 99]. Reversibility of proximal tubule morphologic alterations after deletion of the nephrotoxic Vκ gene into mice suggests that reducing the tumor mass and decreasing production of the nephrotoxic monoclonal LCs in patients with FS would not only stop the accumulation of deposits, but also allow their regression.

Finally, the feasibility of this transgenic approach for LC-associated FS highlights the necessity to create similar models for other LC-associated diseases like AL amyloidosis or Randall-type light chain deposition disease.

Conclusion

Over the years, reliable and reproducible experimental models have always been very important in studying human pathologies. This is even more true for rare and multicausal diseases like renal FS since biological materials are scarce. Administration of toxic environmental agents to animals constituted the first real model of FS. Observation of the inhibitory effect of toxic components on Na/K/ATPase activity was an important breakthrough in the understanding of the tubular dysfunction observed in FS, but the other forms of FS long remained difficult to model for a long time. Cell culture and nephron perfusion

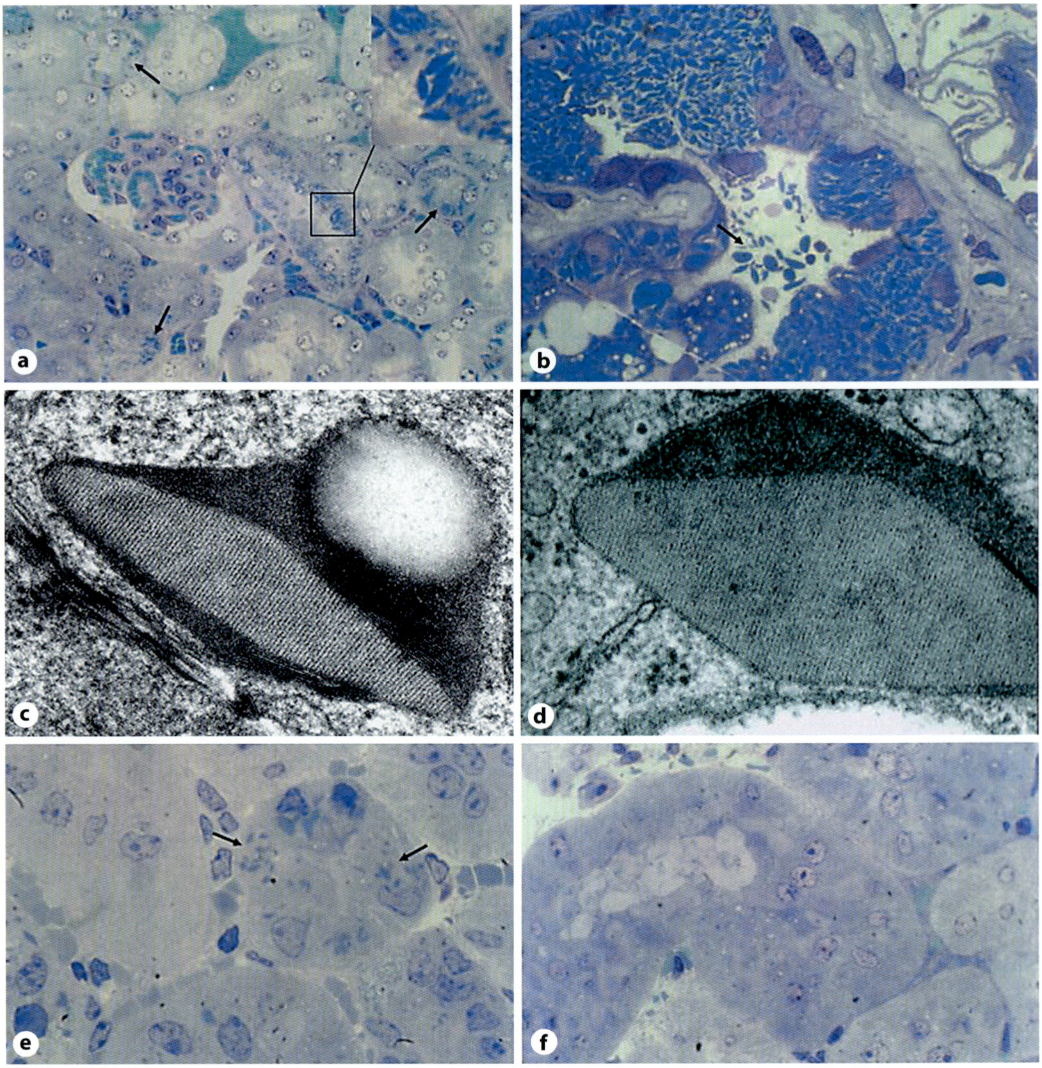

Fig. 3. CHEB transgenic mice kidneys. **a**, **b** Semithin sections of araldite-included samples, toluidine blue staining. **a** Homozygous CHEB mice. Numerous crystalline inclusions in the cytoplasm of proximal tubular cells (arrows and inset), original magnification ×400. **b** Patient CHEB. Massive cytoplasmic crystalline accumulation in proximal tubular cells, with some free crystals in tubular lumen (arrow), original magnification ×1,000. **c**, **d** Electron microscopy examination of homozygous CHEB mice (original magnification ×50,000) (**c**)and patient CHEB (original magnification ×80,000) (**d**). Similar ultrastructural appearance of osmiophilic crystalline inclusions in patient CHEB and CHEB mice, with an 80-Å periodic striation. **e**, **f** Effect of CHEB LC conditional deletion on renal morphology. Light microscopy, araldite inclusion, toluidine blue staining, oil immersion (original magnification ×1,000). **e** Heterozygous CHEB mouse, left nephrectomy, performed before CHEB LC gene deletion. Numerous crystalline inclusions within the cytoplasm of proximal tubular cells (arrows). **f** Same heterozygous CHEB mouse, right nephrectomy, performed 2 months after deletion. No more crystalline inclusions were observed in proximal tubular cells.

Table 2. CHEB transgenic mice serum and urine parameters

Group	A (κ-CHEB/κ-CHEB) (n)	B (κ-CHEB/κ-DEL) (n)	C (controls) (n)	ANOVA	p Unpaired t test or Bonferroni multiple comparisons test A vs. B	 A vs. C	 B vs. C
Diuresis (μl/mn)	1.01±0.6 (13)	ND	0.81±0.47 (7)			0.49	
Serum creatinine (μmol/l)	16.6±5.0 (13)	19.5±4.3 (11)	17.9±4.2 (10)	0.31	n.s.	n.s.	n.s.
Serum uric acid (μmol/l)	43.7±20.1 (13)	57.0±20.7 (11)	107.6±52.6 (10)	0.0002	n.s.	<0.001	<0.01
FE uric acid (%)	14.3±11.0 (10)	10.6±7.3 (11)	4.5±3.5 (10)	0.03	n.s.	<0.05	n.s.
Serum phosphate (mmol/l)	2.1±0.4 (13)	2.4±0.6 (11)	3.0±0.4 (10)	0.003	n.s.	<0.01	n.s.
FE phosphate (%)	23.6±19.1 (9)	17.9±10.3 (8)	7.7±3.5 (9)	0.04	n.s.	<0.05	n.s.
Serum total calcium (mmol/l)	2.38±0.06 (12)	2.47±0.09 (10)	2.43±0.10 (10)	0.11	n.s.	n.s.	n.s.
Urine calcium/creatinine (mmol/mmol)	0.74±0.32 (13)	0.62±0.39 (11)	0.81±0.30 (10)	0.42	n.s.	n.s.	n.s.
FE calcium (%)	0.49±0.21 (13)	0.41±0.20 (10)	0.58±0.19 (10)	0.2	n.s.	n.s.	n.s.
Urine glucose/creatinine (mmol/mmol)	0.15±0.1 (12)	0.15±0.1 (11)	0.09±0.08 (9)	0.64	n.s.	n.s.	n.s.
Urine CC16/creatinine (μg/g)	756±1496 (13)	ND	75±30 (7)			0.008	

FE = Fractional excretion; ND = not done; n.s. = not significant.

with LC purified from patients allowed critical advances in the understanding of the cellular and molecular toxicity of LC-associated FS. But it was only later, with the advent of molecular studies and genetically modified mice, that another crucial step was reached. Genetic invalidation of the megalin receptor or the endosomal ClC-5 exchanger, which yielded the first mouse model of Dent's disease, emphasized the unexpected involvement of endocytosis in FS. More recently, a transgenic mouse model of LC-associated FS was created using a human LC gene from a patient with FS. The development and study of this model took advantage of the advances provided by molecular studies and previous in vitro experiments. Unraveling the molecular and cellular events induced by a toxic LC associated with FS is now under way in living animals mimicking the human pathology. Whether common or specific therapeutic approaches for FS will arise from the study of the different models discussed in this chapter remains hypothetical, but the tools are available and the challenge is on.

References

1 Jacobson HR, Seldin DW: Proximal tubular reabsorption and its regulation. Annu Rev Pharmacol Toxicol 1977;17:623–646.

2 Gonick HC: Pathophysiology of human proximal tubular transport defects. Klin Wochenschr 1982;60:1201–1211.

3 Laski ME, Kurtzman NA: The renal adenosine triphosphatases: functional integration and clinical significance. Miner Electrolyte Metab 1996;22:410–422.

4 Maack T, Johnson V, Kau ST, Figueiredo J, Sigulem D: Renal filtration, transport, and metabolism of low-molecular-weight proteins: a review. Kidney Int 1979;16:251–270.

5 Marshansky V, Bourgoin S, Londono I, Bendayan M, Maranda B, Vinay P: Receptor-mediated endocytosis in kidney proximal tubules: recent advances and hypothesis. Electrophoresis 1997;18:2661–2676.

6 Christensen EI, Birn H, Verroust P, Moestrup SK: Membrane receptors for endocytosis in the renal proximal tubule. Int Rev Cytol 1998;180:237–284.

7 Christensen EI, Verroust PJ, Nielsen R: Receptor-mediated endocytosis in renal proximal tubule. Pflugers Arch 2009;458:1039–1048.

8 Grunfeld JP, Niaudet P, Rotig A: Renal involvement in mitochondrial cytopathies. Nephrol Dial Transplant 1996;11:760–761.

9 Lloyd SE, Pearce SH, Fisher SE, Steinmeyer K, Schwappach B, Scheinman SJ, Harding B, Bolino A, Devoto M, Goodyer P, Rigden SP, Wrong O, Jentsch TJ, Craig IW, Thakker RV: A common molecular basis for three inherited kidney stone diseases. Nature 1996;379:445–449.

10 Thakker RV: Pathogenesis of Dent's disease and related syndromes of X-linked nephrolithiasis. Kidney Int 2000;57:787–793.

11 Lowe CU, Terrey M, Mac LE: Organic-aciduria, decreased renal ammonia production, hydrophthalmos, and mental retardation; a clinical entity. AMA Am J Dis Child 1952;83:164–184.

12 Lowe M: Structure and function of the Lowe syndrome protein OCRL1. Traffic 2005;6:711–719.

13 Gahl WA, Bashan N, Tietze F, Bernardini I, Schulman JD: Cystine transport is defective in isolated leukocyte lysosomes from patients with cystinosis. Science 1982;217:1263–1265.

14 Baum M: The Fanconi syndrome of cystinosis: insights into the pathophysiology. Pediatr Nephrol 1998;12:492–497.

15 Kramer HJ, Gonick HC: Experimental Fanconi syndrome. I. Effect of maleic acid on renal cortical Na-K-ATPase activity and ATP levels. J Lab Clin Med 1970;76:799–808.

16 Gonick H, Indraprasit S, Neustein H, Rosen V: Cadmium-induced experimental Fanconi syndrome. Curr Probl Clin Biochem 1975;4:111–118.

17 Kintzel PE: Anticancer drug-induced kidney disorders. Drug Saf 2001;24:19–38.

18 Gonick HC: Nephrotoxicity of cadmium & lead. Indian J Med Res 2008;128:335–352.

19 Maldonado JE, Velosa JA, Kyle RA, Wagoner RD, Holley KE, Salassa RM: Fanconi syndrome in adults. A manifestation of a latent form of myeloma. Am J Med 1975;58:354–364.

20 Messiaen T, Deret S, Mougenot B, Bridoux F, Dequiedt P, Dion JJ, Makdassi R, Meeus F, Pourrat J, Touchard G, Vanhille P, Zaoui P, Aucouturier P, Ronco PM: Adult Fanconi syndrome secondary to light chain gammopathy. Clinicopathologic heterogeneity and unusual features in 11 patients. Medicine (Baltimore) 2000;79:135–154.

21 Leheste JR, Rolinski B, Vorum H, Hilpert J, Nykjaer A, Jacobsen C, Aucouturier P, Moskaug JO, Otto A, Christensen EI, Willnow TE: Megalin knockout mice as an animal model of low molecular weight proteinuria. Am J Pathol 1999;155:1361–1370.

22 Piwon N, Gunther W, Schwake M, Bosl MR, Jentsch TJ: ClC-5 Cl--channel disruption impairs endocytosis in a mouse model for Dent's disease. Nature 2000;408:369–373.

23 Wang SS, Devuyst O, Courtoy PJ, Wang XT, Wang H, Wang Y, Thakker RV, Guggino S, Guggino WB: Mice lacking renal chloride channel, CLC-5, are a model for Dent's disease, a nephrolithiasis disorder associated with defective receptor-mediated endocytosis. Hum Mol Genet 2000;9:2937–2945.

24 Herak-Kramberger CM, Brown D, Sabolic I: Cadmium inhibits vacuolar H(+)-ATPase and endocytosis in rat kidney cortex. Kidney Int 1998;53:1713–1726.

25 Rocca A, Khamlichi AA, Touchard G, Mougenot B, Ronco P, Denoroy L, Deret S, Preud'homme JL, Aucouturier P, Cogne M: Sequences of V kappa L subgroup light chains in Fanconi's syndrome. Light chain V region gene usage restriction and peculiarities in myeloma-associated Fanconi's syndrome. J Immunol 1995;155:3245–3252.

26 Stowe HD, Wilson M, Goyer RA: Clinical and morphologic effects of oral cadmium toxicity in rabbits. Arch Pathol 1972;94:389–405.

27 Yanase M, Orita Y, Okada N, Nakanishi T, Horio M, Ando A, Abe H: Decreased Na+-gradient-dependent D-glucose transport in brush-border membrane vesicles from rabbits with experimental Fanconi syndrome. Biochim Biophys Acta 1983;733:95–101.

28 Khalil-Manesh F, Gonick HC, Cohen AH, Alinovi R, Bergamaschi E, Mutti A, Rosen VJ: Experimental model of lead nephropathy. I. Continuous high-dose lead administration. Kidney Int 1992;41:1192–1203.

29 Kramer HJ, Gonick HC, Lu E: In vitro inhibition of Na-K-ATPase by trace metals: relation to renal and cardiovascular damage. Nephron 1986;44:329–336.

30 Foreman JW, Bowring MA, Lee J, States B, Segal S: Effect of cystine dimethylester on renal solute handling and isolated renal tubule transport in the rat: a new model of the Fanconi syndrome. Metabolism 1987;36:1185–1191.

31 Guan S, el-Dahr S, Dipp S, Batuman V: Inhibition of Na-K-ATPase activity and gene expression by a myeloma light chain in proximal tubule cells. J Investig Med 1999;47:496–501.

32 Thevenod F, Friedmann JM: Cadmium-mediated oxidative stress in kidney proximal tubule cells induces degradation of Na+/K(+)-ATPase through proteasomal and endo-/lysosomal proteolytic pathways. FASEB J 1999;13:1751–1761.

33 Conner SD, Schmid SL: Regulated portals of entry into the cell. Nature 2003;422:37–44.

34 Christensen EI, Verroust PJ: Megalin and cubilin, role in proximal tubule function and during development. Pediatr Nephrol 2002;17:993–999.
35 Gekle M: Renal tubule albumin transport. Annu Rev Physiol 2005;67:573–594.
36 Faundez V, Hartzell HC: Intracellular chloride channels: determinants of function in the endosomal pathway. Sci STKE 2004;2004: re8.
37 Shi LB, Fushimi K, Bae HR, Verkman AS: Heterogeneity in ATP-dependent acidification in endocytic vesicles from kidney proximal tubule. Measurement of pH in individual endocytic vesicles in a cell-free system. Biophys J 1991;59:1208–1217.
38 Jentsch TJ: CLC chloride channels and transporters: from genes to protein structure, pathology and physiology. Crit Rev Biochem Mol Biol 2008;43:3–36.
39 Moriyama Y, Nelson N: The purified ATPase from chromaffin granule membranes is an anion-dependent proton pump. J Biol Chem 1987;262:9175–9180.
40 Willnow TE, Hilpert J, Armstrong SA, Rohlmann A, Hammer RE, Burns DK, Herz J: Defective forebrain development in mice lacking gp330/megalin. Proc Natl Acad Sci USA 1996;93:8460–8464.
41 Christensen EI, Birn H: Megalin and cubilin: synergistic endocytic receptors in renal proximal tubule. Am J Physiol Renal Physiol 2001;280:F562–F573.
42 Gburek J, Verroust PJ, Willnow TE, Fyfe JC, Nowacki W, Jacobsen C, Moestrup SK, Christensen EI: Megalin and cubilin are endocytic receptors involved in renal clearance of hemoglobin. J Am Soc Nephrol 2002;13:423–430.
43 Kozyraki R, Fyfe J, Verroust PJ, Jacobsen C, Dautry-Varsat A, Gburek J, Willnow TE, Christensen EI, Moestrup SK: Megalin-dependent cubilin-mediated endocytosis is a major pathway for the apical uptake of transferrin in polarized epithelia. Proc Natl Acad Sci USA 2001;98:12491–12496.
44 Sousa MM, Norden AG, Jacobsen C, Willnow TE, Christensen EI, Thakker RV, Verroust PJ, Moestrup SK, Saraiva MJ: Evidence for the role of megalin in renal uptake of transthyretin. J Biol Chem 2000;275:38176–38181.
45 Christensen EI, Moskaug JO, Vorum H, Jacobsen C, Gundersen TE, Nykjaer A, Blomhoff R, Willnow TE, Moestrup SK: Evidence for an essential role of megalin in transepithelial transport of retinol. J Am Soc Nephrol 1999;10:685–695.
46 Nykjaer A, Dragun D, Walther D, Vorum H, Jacobsen C, Herz J, Melsen F, Christensen EI, Willnow TE: An endocytic pathway essential for renal uptake and activation of the steroid 25-(OH) vitamin D3. Cell 1999;96:507–515.
47 Raila J, Willnow TE, Schweigert FJ: Megalin-mediated reuptake of retinol in the kidneys of mice is essential for vitamin A homeostasis. J Nutr 2005;135:2512–2516.
48 Bachmann S, Schlichting U, Geist B, Mutig K, Petsch T, Bacic D, Wagner CA, Kaissling B, Biber J, Murer H, Willnow TE: Kidney-specific inactivation of the megalin gene impairs trafficking of renal inorganic sodium phosphate cotransporter (NaPi-IIa). J Am Soc Nephrol 2004;15:892–900.
49 Nykjaer A, Fyfe JC, Kozyraki R, Leheste JR, Jacobsen C, Nielsen MS, Verroust PJ, Aminoff M, de la Chapelle A, Moestrup SK, Ray R, Gliemann J, Willnow TE, Christensen EI: Cubilin dysfunction causes abnormal metabolism of the steroid hormone 25(OH) vitamin D(3). Proc Natl Acad Sci USA 2001;98:13895–13900.
50 Birn H, Fyfe JC, Jacobsen C, Mounier F, Verroust PJ, Orskov H, Willnow TE, Moestrup SK, Christensen EI: Cubilin is an albumin binding protein important for renal tubular albumin reabsorption. J Clin Invest 2000;105:1353–1361.
51 Aminoff M, Carter JE, Chadwick RB, Johnson C, Grasbeck R, Abdelaal MA, Broch H, Jenner LB, Verroust PJ, Moestrup SK, de la Chapelle A, Krahe R: Mutations in CUBN, encoding the intrinsic factor-vitamin B12 receptor, cubilin, cause hereditary megaloblastic anaemia 1. Nat Genet 1999;21:309–313.
52 Wahlstedt-Froberg V, Pettersson T, Aminoff M, Dugue B, Grasbeck R: Proteinuria in cubilin-deficient patients with selective vitamin B12 malabsorption. Pediatr Nephrol 2003;18:417–421.
53 Devuyst O, Jouret F, Auzanneau C, Courtoy PJ: Chloride channels and endocytosis: new insights from Dent's disease and ClC-5 knockout mice. Nephron Physiol 2005;99:p69–p73.
54 Devuyst O, Pirson Y: Genetics of hypercalciuric stone forming diseases. Kidney Int 2007;72:1065–1072.
55 Christensen EI, Devuyst O, Dom G, Nielsen R, Van der Smissen P, Verroust P, Leruth M, Guggino WB, Courtoy PJ: Loss of chloride channel ClC-5 impairs endocytosis by defective trafficking of megalin and cubilin in kidney proximal tubules. Proc Natl Acad Sci USA 2003;100:8472–8477.
56 Mellman I: The importance of being acid: the role of acidification in intracellular membrane traffic. J Exp Biol 1992;172:39–45.
57 Hurtado-Lorenzo A, Skinner M, El Annan J, Futai M, Sun-Wada GH, Bourgoin S, Casanova J, Wildeman A, Bechoua S, Ausiello DA, Brown D, Marshansky V: V-ATPase interacts with ARNO and Arf6 in early endosomes and regulates the protein degradative pathway. Nat Cell Biol 2006;8:124–136.
58 Pontoglio M, Barra J, Hadchouel M, Doyen A, Kress C, Bach JP, Babinet C, Yaniv M: Hepatocyte nuclear factor 1 inactivation results in hepatic dysfunction, phenylketonuria, and renal Fanconi syndrome. Cell 1996; 84:575–585.
59 Cheret C, Doyen A, Yaniv M, Pontoglio M: Hepatocyte nuclear factor 1 alpha controls renal expression of the Npt1-Npt4 anionic transporter locus. J Mol Biol 2002;322:929–941.
60 Pontoglio M, Prie D, Cheret C, Doyen A, Leroy C, Froguel P, Velho G, Yaniv M, Friedlander G: HNF1alpha controls renal glucose reabsorption in mouse and man. EMBO Rep 2000;1:359–365.
61 Janne PA, Suchy SF, Bernard D, MacDonald M, Crawley J, Grinberg A, Wynshaw-Boris A, Westphal H, Nussbaum RL: Functional overlap between murine Inpp5b and Ocrl1 may explain why deficiency of the murine ortholog for OCRL1 does not cause Lowe syndrome in mice. J Clin Invest 1998;101:2042–2053.

62 Cherqui S, Sevin C, Hamard G, Kalatzis V, Sich M, Pequignot MO, Gogat K, Abitbol M, Broyer M, Gubler MC, Antignac C: Intralysosomal cystine accumulation in mice lacking cystinosin, the protein defective in cystinosis. Mol Cell Biol 2002;22:7622–7632.
63 Nevo N, Chol M, Bailleux A, Kalatzis V, Morisset L, Devuyst O, Gubler MC, Antignac C: Renal phenotype of the cystinosis mouse model is dependent upon genetic background. Nephrol Dial Transplant 2010;25:1059–1066.
64 Elsea SH, Lucas RE: The mousetrap: what we can learn when the mouse model does not mimic the human disease. ILAR J 2002;43:66–79.
65 Ronco PM, Mougenot B, Moulin B, Bouiller M, Messiaen T, Aucouturier P: Adult Fanconi's syndrome due to monoclonal immunoglobulin light chains: an underdiagnosed disease with unexpected clinical and pathologic heterogeneity in G; in Touchard PA, Hermine O, Ronco P (eds): Monoclonal Gammapathies and the Kidney. Dordrecht, Kluwer Academic Publishers, 2003, pp 75–86.
66 Engle RL Jr, Wallis LA: Multiple myeloma and the adult Fanconi syndrome. I. Report of a case with crystal-like deposits in the tumor cells and in the epithelial cells of the kidney. Am J Med 1957;22:5–23.
67 Sirota JH, Hamerman D: Renal function studies in a adult subject with the Fanconi syndrome. Am J Med 1954;16:138–152.
68 Thorner PS, Bedard YC, Fernandes BJ: Lambda-light-chain nephropathy with Fanconi's syndrome. Arch Pathol Lab Med 1983;107:654–657.
69 Orfila C, Lepert JC, Modesto A, Bernadet P, Suc JM: Fanconi's syndrome, kappa light-chain myeloma, non-amyloid fibrils and cytoplasmic crystals in renal tubular epithelium. Am J Nephrol 1991;11:345–349.
70 Truong LD, Mawad J, Cagle P, Mattioli C: Cytoplasmic crystals in multiple myeloma-associated Fanconi's syndrome. A morphological study including immunoelectron microscopy. Arch Pathol Lab Med 1989;113: 781–785.
71 Lacy MQ, Gertz MA: Acquired Fanconi's syndrome associated with monoclonal gammopathies. Hematol Oncol Clin North Am 1999;13:1273–1280.
72 Bridoux F, Sirac C, Hugue V, Decourt C, Thierry A, Quellard N, Abou-Ayache R, Goujon JM, Cogne M, Touchard G: Fanconi's syndrome induced by a monoclonal Vkappa3 light chain in Waldenstrom's macroglobulinemia. Am J Kidney Dis 2005;45:749–757.
73 Decourt C, Bridoux F, Touchard G, Cogne M: A monoclonal V kappa l light chain responsible for incomplete proximal tubulopathy. Am J Kidney Dis 2003;41:497–504.
74 Bate KL, Clouston D, Packham D, Ratnaike S, Ebeling PR: Lambda light chain induced nephropathy: a rare cause of the Fanconi syndrome and severe osteomalacia. Am J Kidney Dis 1998;32:E3.
75 Isobe T, Kametani F, Shinoda T: V-domain deposition of lambda Bence Jones protein in the renal tubular epithelial cells in a patient with the adult Fanconi syndrome with myeloma. Amyloid 1998;5:117–120.
76 Aucouturier P, Bauwens M, Khamlichi AA, Denoroy L, Spinelli S, Touchard G, Preud'homme JL, Cogne M: Monoclonal Ig L chain and L chain V domain fragment crystallization in myeloma-associated Fanconi's syndrome. J Immunol 1993;150:3561–3568.
77 Decourt C, Rocca A, Bridoux F, Vrtovsnik F, Preud'homme JL, Cogne M, Touchard G: Mutational analysis in murine models for myeloma-associated Fanconi's syndrome or cast myeloma nephropathy. Blood 1999;94:3559–3566.
78 Deret S, Denoroy L, Lamarine M, Vidal R, Mougenot B, Frangione B, Stevens FJ, Ronco PM, Aucouturier P: Kappa light chain-associated Fanconi's syndrome: molecular analysis of monoclonal immunoglobulin light chains from patients with and without intracellular crystals. Protein Eng 1999;12:363–369.
79 Leboulleux M, Lelongt B, Mougenot B, Touchard G, Makdassi R, Rocca A, Noel LH, Ronco PM, Aucouturier P: Protease resistance and binding of Ig light chains in myeloma-associated tubulopathies. Kidney Int 1995;48:72–79.
80 El Hamel C TA, Trouillas P, Bridoux F, Carrion C, Quellard N, Goujon JM, Aldigier C, Gombert JM, Cogne M, Touchard G: Crystal-storing histiocytosis with renal Fanconi syndrome: pathological and molecular characteristics compared with classical myeloma-associated Fanconi syndrome. Nephrol Dial Transplant 2010, E-pub ahead of print.
81 Klassen RB, Allen PL, Batuman V, Crenshaw K, Hammond TG: Light chains are a ligand for megalin. J Appl Physiol 2005;98:257–263.
82 Plans V, Rickheit G, Jentsch TJ: Physiological roles of CLC Cl(-)/H (+) exchangers in renal proximal tubules. Pflugers Arch 2009;458:23–37.
83 Batuman V, Sastrasinh M, Sastrasinh S: Light chain effects on alanine and glucose uptake by renal brush border membranes. Kidney Int 1986;30:662–665.
84 Batuman V, Dreisbach AW, Cyran J: Light-chain binding sites on renal brush-border membranes. Am J Physiol 1990;258:F1259– F1265.
85 Batuman V, Guan S: Receptor-mediated endocytosis of immunoglobulin light chains by renal proximal tubule cells. Am J Physiol 1997;272:F521–F530.
86 Batuman V, Verroust PJ, Navar GL, Kaysen JH, Goda FO, Campbell WC, Simon E, Pontillon F, Lyles M, Bruno J, Hammond TG: Myeloma light chains are ligands for cubilin (gp280). Am J Physiol 1998;275:F246–F254.
87 Sanders PW, Herrera GA, Galla JH: Human Bence Jones protein toxicity in rat proximal tubule epithelium in vivo. Kidney Int 1987;32:851–861.
88 Sanders PW, Herrera GA, Chen A, Booker BB, Galla JH: Differential nephrotoxicity of low molecular weight proteins including Bence Jones proteins in the perfused rat nephron in vivo. J Clin Invest 1988;82:2086–2096.
89 Sengul S, Zwizinski C, Batuman V: Role of MAPK pathways in light chain-induced cytokine production in human proximal tubule cells. Am J Physiol Renal Physiol 2003;284:F1245–F1254.
90 Sengul S, Zwizinski C, Simon EE, Kapasi A, Singhal PC, Batuman V: Endocytosis of light chains induces cytokines through activation of NF-kappaB in human proximal tubule cells. Kidney Int 2002;62:1977–1988.

91 Wang PX, Sanders PW: Immunoglobulin light chains generate hydrogen peroxide. J Am Soc Nephrol 2007;18:1239–1245.
92 Khamlichi AA, Rocca A, Touchard G, Aucouturier P, Preud'homme JL, Cogne M: Role of light chain variable region in myeloma with light chain deposition disease: evidence from an experimental model. Blood 1995;86:3655–3659.
93 Chauveau C, Decourt C, Cogne M: Insertion of the IgH locus 3′ regulatory palindrome in expression vectors warrants sure and efficient expression in stable B cell transfectants. Gene 1998;222:279–285.
94 Khamlichi AA, Rocca A, Cogne M: The effect of intron sequences on expression levels of Ig cDNAs. Gene 1994;150:387–390.
95 Sirac C, Bridoux F, Carrion C, Devuyst O, Fernandez B, Goujon JM, El Hamel C, Aldigier JC, Touchard G, Cogne M: Role of the monoclonal kappa chain V domain and reversibility of renal damage in a transgenic model of acquired Fanconi syndrome. Blood 2006;108:536–543.
96 Bernard AM, Thielemans NO, Lauwerys RR: Urinary protein 1 or Clara cell protein: a new sensitive marker of proximal tubular dysfunction. Kidney Int Suppl 1994;47:S34–S37.
97 Batuman V, Guan S, O'Donovan R, Puschett JB: Effect of myeloma light chains on phosphate and glucose transport in renal proximal tubule cells. Ren Physiol Biochem 1994;17:294–300.
98 Nicastri AL, Brandao de Almeida Prado MJ, Sesso A, Brandao de Almeida Prado EB: Defective proximal tubule lysosomal acidification by Bence Jones proteins. An immunoelectron microscopy study. Exp Nephrol 1998;6:514–521.
99 Ma CX, Lacy MQ, Rompala JF, Dispenzieri A, Rajkumar SV, Greipp PR, Fonseca R, Kyle RA, Gertz MA: Acquired Fanconi syndrome is an indolent disorder in the absence of overt multiple myeloma. Blood 2004;104:40–42.
100 Devuyst O, Guggino WB: Chloride channels in the kidney: lessons learned from knockout animals. Am J Physiol Renal Physiol 2002; 283:F1176–F1191.

Christophe Sirac
Laboratoire d'Immunologie, CNRS UMR 6101
2 rue du Dr. Marcland
FR–87025 Limoges (France)
Tel. +33 555 435 924, Fax +33 555 435 897, E-Mail christophe.sirac@unilim.fr

Herrera GA (ed): Experimental Models for Renal Diseases: Pathogenesis and Diagnosis.
Contrib Nephrol. Basel, Karger, 2011, vol 169, pp 262–269

Light Chain-Mediated Tubulopathies

Paul W. Sanders

Division of Nephrology, Department of Medicine, Department of Physiology and Biophysics, and the Nephrology Research and Training Center, University of Alabama at Birmingham, and Department of Veterans Affairs Medical Center, Birmingham, Ala., USA

Abstract

Immunoglobulin light chains are low molecular weight proteins that are filtered through the glomerulus and reabsorbed into the proximal tubular epithelium by binding initially to a heteromeric receptor complex composed of megalin and cubilin. Saturation of this receptor-mediated endocytotic process results in the presence of free light chains in the distal nephron and urine. In the course of metabolism of monoclonal light chains, nephrotoxicity can occur, resulting in clinical manifestations that can include acute kidney injury and progressive chronic kidney disease. Patterns of tubulopathic renal injury include proximal tubular epithelial cell cytotoxicity, tubulointerstitial nephritis and cast nephropathy (also known as 'myeloma kidney'). Research efforts over the past two decades have refined understanding of the molecular mechanisms involved in light chain-mediated tubular injury and are the subject of this review.

Immunoglobulin light chains are of significant interest in nephrology because this family of low molecular weight proteins can promote a wide spectrum of progressive renal disorders. Patterns of tubulopathic renal injury occurring in the setting of plasma cell dyscrasias include isolated proximal tubule epithelial cell cytotoxicity, tubulointerstitial nephritis and cast nephropathy (also known as 'myeloma kidney') [1]. Monoclonal immunoglobulin light chain deposition is directly responsible for the pathology of these renal diseases. Light chains are unlike most endogenous proteins because they have a propensity to produce tubular damage. Research efforts over the past two decades have refined understanding of the molecular mechanisms of light chain-mediated tubular injury.

Experiments that defined the renal handling of light chains were essential in understanding the mechanisms of tubular injury. Light chains, termed Bence Jones proteins when they appear in the urine [2], circulate as monomers (M_r ~22 kDa) and dimers (M_r ~44 kDa). Metabolism of these proteins is directly related to glomerular filtration [3–5]. As the glomerular filtration rate falls, serum concentrations of monoclonal (and polyclonal) κ and λ light chains increase [5, 6]; thus, the tubular fluid concentration will increase in the remaining viable nephrons. Filtered light chains are reabsorbed into the proximal tubule by binding initially to a heteromeric receptor composed of megalin and cubilin [7–10]. After endocytosis, vesicles containing the ligand/receptor complex acidify to release the light chain and other bound proteins and permit recycling of the receptor back to the apical membrane to load more cargo. Addition of lysosomal enzymes permits hydrolysis of the light chains and return of the amino acid components to the circulation [11–13].

About 500 mg of free polyclonal light chains are produced daily by the normal lymphoid system and are catabolized by the kidney, specifically in the proximal tubule. This system is highly efficient and only about 1–10 mg of polyclonal light chains normally appear in the urine each day [2]. In the setting of a plasma cell dyscrasia, light chain production increases significantly, producing circulating levels of monoclonal light chain that can approach 100,000 mg/l [5, 14]. Although saturation of the multi-ligand endocytotic receptor complex of the proximal tubule permits light chains to appear in the tubular fluid of distal nephron segments and ultimately in the urine, production of monoclonal light chains can result in significant accumulation of these proteins within the proximal tubular epithelium.

Epithelial Cell Cytotoxicity

Although the highly efficient protein reclamation system of the proximal tubule provides an important mechanism of conservation of amino acids and other essential molecules, this process of protein reabsorption can produce detrimental results in the setting of overproduction of monoclonal light chain proteins. Along with Dr. Guillermo Herrera's group, this laboratory was among the first to demonstrate the cytotoxic potential of light chains. Functional and morphologic evidence of direct toxicity of human monoclonal light chains was demonstrated by in vivo perfusion of proximal convoluted tubules of Sprague-Dawley rats. Toxicity occurred in a time-dependent fashion and at a time when the lysosomal system was markedly activated [15, 16].

This interesting alteration in proximal tubule function and in lysosomal morphology prompted a study of human specimens. Kidney specimens obtained from patients with monoclonal light chain deposition (7 λ, 4 κ) in the absence of intraluminal cast formation demonstrated varying degrees of proximal tubule damage, manifested by cell vacuolation, desquamation, loss of the luminal brush border and coagulation necrosis, supporting the concept that some light chains promote proximal tubulopathy [17]. Associated with these cytotoxic alterations was engorgement of the lysosomes from accumulated monoclonal proteins. Using proximal tubular cells in culture, Batuman's laboratory demonstrated that endocytosis of human light chains promotes cytotoxicity [18, 19]. While the biochemical property(s) of the monoclonal light chains responsible for epithelial cell injury remain unknown, these in vivo and in vitro experiments have been essential in defining the mechanisms of cytotoxicity and provide the basis for exploring these important structure-function studies in the future. In particular, the description of this unique pattern of proximal tubulopathy with distention of the endolysosomal system in human specimens completes a bench-to-bedside cycle in the understanding of light chain-related renal disorders.

Proximal Tubular Epithelial Cell Activation

While endocytosis of some monoclonal light chains can produce direct toxicity to the proximal tubule epithelium with associated clinical manifestations of renal failure [1, 15–17, 20–22], attention in recent years has turned to the possibility that the process of excessive protein reabsorption by the proximal tubule in pathological states results in chemokine and cytokine release. For light chains, this process has been characterized in some detail. Proximal tubular reabsorption of light chains sparks the activation of NF-κB [23] and mitogen-activated protein kinases [24]; these signaling pathways cooperate to promote chemokine and cytokine production by these cells [23–25]. An important point of these experiments is that the capability of light chains to activate the proximal tubule epithelium greatly exceeds the activating effect of other proteins, such as albumin [23, 24].

Light chains are peculiar in that they can generate the intracellular production of H_2O_2 in proximal tubule cells sufficient to increase production of monocyte chemotactic protein-1 [26]. H_2O_2

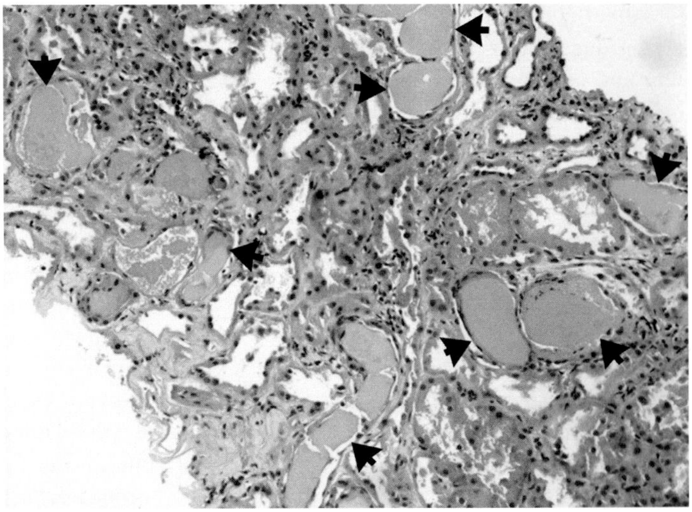

Fig. 1. Example of cast nephropathy in a patient with multiple myeloma. The casts (arrows) occlude the lumen of the nephron. Interstitial inflammation and tubular atrophy commonly accompany the casts.

can overcome cytoplasmic antioxidant defense mechanisms to induce activation of important signaling molecules. For example, H_2O_2 targets reduced sulfhydryl groups of cysteine residues present in the catalytic site of tyrosine and serine/threonine phosphatases, inactivating these enzymes [27]. H_2O_2 is a potent activator of the redox-sensitive transcription factor, NF-κB [28]. Monoclonal light chains therefore produce a relatively unique form of intracellular oxidative stress sufficient to generate important chemokines, such as monocyte chemotactic protein-1, involved in interstitial inflammation and fibrosis.

Activation of proximal tubular epithelium through this mechanism likely plays an important role in acceleration of interstitial inflammation and fibrosis and associated progressive renal disease related to light chain deposition. Some patients may present with isolated tubulointerstitial nephritis that resembles drug-induced interstitial nephritis [1]. Although the pathogenesis of this light chain-mediated lesion is undefined, elaboration of chemokines by the proximal tubule may also contribute to this inflammatory process. Interestingly, Batuman's laboratory recently showed that addition to the medium of pituitary adenylate cyclase-activating polypeptide with 38 residues (PACAP38) with monoclonal light chains dramatically prevented light chain-induced injury of cultured renal proximal tubular cells and suppressed pro-inflammatory cytokine production by inhibiting p38 MAPK and translocation of NF-κB. These findings suggest a novel treatment of light chain-mediated tubulointerstitial kidney disease [25].

Cast Nephropathy

The major mechanism of light chain-mediated tubule damage relates to intratubular obstruction from precipitation of light chains in the distal nephron; interstitial inflammation and fibrosis commonly accompany the cast formation (fig. 1) [1]. By use of in vivo microperfusion studies, two separate mechanisms of tubular damage emerged: proximal tubular cell injury and cast formation. Some light chains used in those experiments produced a dose-dependent alteration in cell structure and function, but most of the proteins formed casts in the distal nephron [16]. The hemodynamic relevance of cast formation was further suggested by intravenous infusion in rats of a nephrotoxic human light chain which elevated proximal tubule pressure and simultaneously decreased the single nephron glomerular filtration rate; intraluminal protein casts were identified in these kidneys [29]. If the casts persist, then permanent nephron loss follows, since isolated mechanical obstruction of the lumen of single nephrons in vivo reduces single-nephron glomerular blood flow to the obstructed nephron and produces atrophy of the nephron proximal to the obstruction [30, 31].

This laboratory developed a single-nephron perfusion technique that permitted perfusion of the nephron in vivo using a pressure-dependent perfusion apparatus while monitoring proximal tubular pressures [32]. Initially, proximal tubule intraluminal pressures were measured at random in four to six nephrons using a servo-null micropressure system, and were averaged. A pipette attached to this micropressure-monitoring system was then inserted into a late surface proximal convolution to continuously monitor intraluminal proximal tubule pressure. A bone wax cast was then inserted into an early proximal segment to obstruct flow through the nephron completely. Just distal to this block, a pipette was inserted into a proximal convolution to perfuse the remainder of the nephron. The perfusion pipette contained green-tinted artificial tubular fluid and one of the test proteins. This perfusion pipette was connected to a pressure-driven perfusion apparatus that consisted of an airtight Erlenmeyer flask attached to a syringe and a pressure-monitoring gauge. Perfusion pressure was adjusted in the initial few seconds of the experiment until the previously determined average proximal tubule pressure was attained; perfusion pressure was then kept constant.

In all experiments, perfusion was stopped arbitrarily if proximal tubule pressure rose to 20 mm Hg, which was the perfusion pressure used in most of the experiments, or if proximal tubule pressure did not increase after 60 min of perfusion. Subsequently, the perfused nephrons were dissected and examined visually with a dissecting microscope. While difficult, this approach permitted detailed understanding of the process of cast formation. When infused directly into the rat nephron in vivo, monoclonal light chains from patients with cast nephropathy produced intraluminal obstruction by precipitating in the distal nephron; protein casts were never identified before the tip of the loop of Henle (fig. 2).

Obstruction occurred in a dose-dependent fashion and was accelerated by the presence of furosemide. Pretreatment of rats with colchicine, which decreased urinary Tamm-Horsfall protein (THP), prevented obstruction and cast formation of perfused nephrons. β-Lumicolchicine, which is a structural isomer of colchicine and does not interfere with microtubule assembly, did not alter THP excretion or prevent obstruction. THP purified from β-lumicolchicine-treated and untreated rats co-aggregated with light chains in vitro, while THP from colchicine-treated rats did not contain sialic acid and did not aggregate with light chains in vitro [32].

Combined with previous data demonstrating binding of cast-forming light chains with THP in vitro [20], these studies established an integral relationship between cast-forming light chains and THP in cast nephropathy. In humans, casts generally are seen primarily in the distal portions of the nephron in humans, although on occasion casts may also be found in proximal tubular segments and even in glomeruli in renal biopsy specimens. When examined, however, these casts generally contain THP, suggesting reflux of co-precipitated THP and light chain into the proximal nephron [33, 34]. THP appears to form the matrix upon which the cast containing the monoclonal light chain is formed in rats and humans (fig. 3).

The critical part of the process of cast formation centers upon the interaction between light chains and THP, which is also known as uromodulin. Only cells of the thick ascending limb of the loop of Henle synthesize THP, the major protein constituent of normal urine [35]. THP belongs to the class of glycophospholipid-anchored proteins, permitting extracellular localization on the apical surface of the cell [36]. A soluble form of THP also appears in tubular fluid of the distal nephron and urine. THP undergoes homotypic aggregation through the zona pellucida domain, producing filamentous molecules that can be observed with electron microscopy [37].

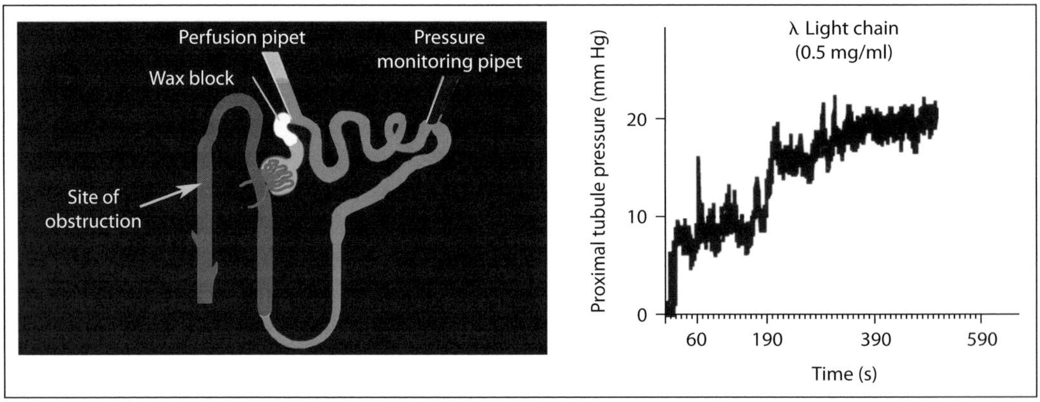

Fig. 2. Left, illustration depicting the in vivo microperfusion technique used to demonstrate the pathogenesis and functional significance of cast formation. Casts uniformly developed in the distal nephron. Right, original tracing of proximal tubule pressures during perfusion with a nephrotoxic human λ light chain. In this experiment, the perfusion solution contained 0.5 mg/ml light chain. Proximal tubule pressure had been determined to be 9 mm Hg prior to initiating the study. Following administration of a wax plug, proximal tubule pressure fell to zero. With re-establishment of flow through the nephron using a pressure-sensitive perfusion apparatus, proximal tubule pressure returned to approximately 9 mm Hg. However, within 3 min of perfusion, proximal pressures acutely rose to stop-flow pressure, which was 20 mm Hg. The experiment was terminated and subsequent dissection of the perfused nephron demonstrated an obstructing lesion in the tubule lumen.

Although constitutively expressed, increased dietary salt intake increases expression of THP in rats [38]. Intraluminal cast formation in vivo is dictated by multiple variables that include the ionic composition of the tubule fluid, tubule fluid flow rates, concentration of THP and light chains, the strength of binding interaction between THP and light chains, and presence of furosemide [16, 20, 32, 38–42]. These observations have direct clinical relevance since many of these factors, except the intrinsic binding interaction between THP and light chains, can be modified with current treatment modalities.

Clinical and laboratory evidence suggests that each light chain has unique toxicity that must therefore be ascribed to the structure of the molecule. Light chains are modular proteins composed of two domains: a variable (V_L) and a constant (C_L) region [43]. Within the globular V_L domain are four framework regions that consist of β-sheets that develop a hydrophobic core [44–47]. Interspersed among the framework regions are three separate hypervariable segments known as complementarity determining regions (CDR1, CDR2 and CDR3). The CDR domains represent those regions of sequence variability in the light chain and constitute part of the antigen-binding site of the immunoglobulin [44–46]. Diversity among the CDR regions occurs because the V_L domain is synthesized by combining V and J gene segments. Because the number of human Vκ and Vλ gene segments are estimated to be 30–50 and 20–30, respectively, and there are probably 5 Jκ and 20–30 Jλ gene segments [48], a large number of combinatorial possibilities exist. Thus, while possessing similar structures, no two light chains are identical, although there are enough similarities among light chains to permit categorizing them according to subgroups. There are 4 κ [49] and 10 λ subgroups [50, 51], although of the λ subgroups, most patients (94%) with multiple myeloma express λI, λII, λIII or λV [52].

Biochemical and molecular analyses of the interaction between THP and light chains has shown that the carbohydrate moiety of THP did

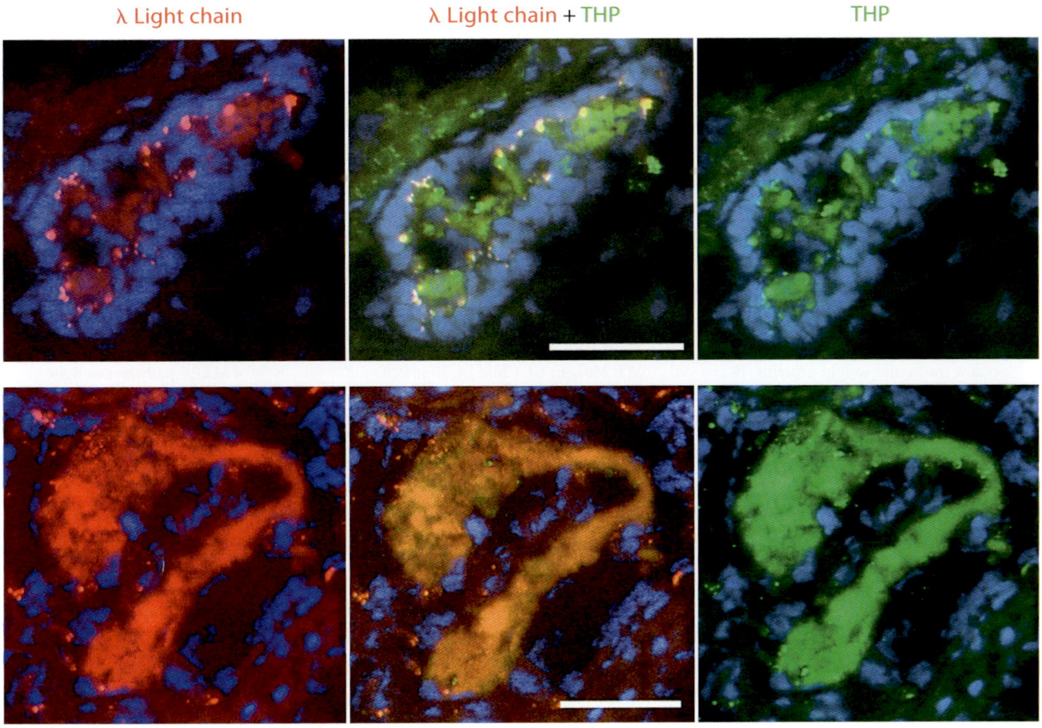

Fig. 3. Immunofluorescence micrographs of two casts in kidney tissue from a patient with cast nephropathy (serum λ light chain concentration was 32 mg/ml), showing co-localization of λ light chain (red) with THP (green). The merged image (middle panel) shows co-localization (yellow) in the casts. The section has been counterstained with Hoescht to identify cells (blue nuclei). White bars = 50 μm.

not alter binding to light chains [40]. A single 9-amino acid (residues 225–233) binding domain for light chains was mapped on THP; all light chains tested bound to this domain [42]. Light chains interact with THP specifically through the CDR3 domain in the variable region of both κ and λ light chains [53]. The findings provide another clue to the molecular interactions between light chains and THP, and now present an opportunity to inhibit the functional consequences in vivo. Differences in binding affinity for THP among light chains are related to the amino acid composition of the CDR3 domain [53].

In summary, remarkable advances have been made in the dissection of the structure-function relationships involved in light chain-mediated tubular damage. Such advances required development of animal models of human disease. With these model systems, the molecular pathogenesis has been detailed, increasing the potential to create novel treatments designed to interrupt the pathologic process and prevent progressive loss of renal function related to deposition of monoclonal light chains.

Acknowledgments

Dr. Sanders is supported by the Medical Research Service of the Department of Veterans Affairs and by grants from the National Institutes of Health (R01 DK46199) and George M. O'Brien Kidney and Urological Research Centers Program (P30 DK079337).

References

1 Sanders PW, Herrera GA, Kirk KA, Old CW, Galla JH: Spectrum of glomerular and tubulointerstitial renal lesions associated with monotypical immunoglobulin light chain deposition. Lab Invest 1991;64:527–537.

2 Berggård I, Peterson PA: Polymeric forms of free normal kappa and lambda chains of human immunoglobulin. J Biol Chem 1969; 244:4299–4307.

3 Waldmann TA, Strober W, Mogielnicki RP: The renal handling of low molecular weight proteins. II. Disorders of serum protein catabolism in patients with tubular proteinuria, the nephrotic syndrome, or uremia. J Clin Invest 1972;51:2162–2174.

4 Wochner RD, Strober W, Waldmann TA: The role of the kidney in the catabolism of Bence Jones proteins and immunoglobulin fragments. J Exp Med 1967;126:207–220.

5 Bradwell AR, Mead GP, Carr-Smith HD: Serum Free Light Chain Analysis, ed 3. Birmingham, The Binding Site Ltd., 2005.

6 Katzmann JA, Clark RJ, Abraham RS, et al: Serum reference intervals and diagnostic ranges for free kappa and free lambda immunoglobulin light chains: relative sensitivity for detection of monoclonal light chains. Clin Chem 2002;48:1437–1444.

7 Batuman V, Dreisbach AW, Cyran J: Light-chain binding sites on renal brush-border membranes. Am J Physiol 1990;258:F1259– F1265.

8 Batuman V, Guan S: Receptor mediated endocytosis of immunoglobulin light chains by renal proximal tubule cells. Am J Physiol 1997;272:F521–F530.

9 Batuman V, Verroust PJ, Navar GL, et al: Myeloma light chains are ligands for cubilin (gp280). Am J Physiol 1998;275:F246–F254.

10 Klassen RB, Allen PL, Batuman V, Crenshaw K, Hammond TG: Light chains are a ligand for megalin. J Appl Physiol 2005;98:257–263.

11 Maack T, Johnson V, Kau ST, Figueiredo J, Sigulem D: Renal filtration, transport, and metabolism of low-molecular-weight proteins: a review. Kidney Int 1979;16:251–270.

12 Christensen EI, Devuyst O, Dom G, et al: Loss of chloride channel ClC-5 impairs endocytosis by defective trafficking of megalin and cubilin in kidney proximal tubules. Proc Natl Acad Sci USA 2003;100:8472–8477.

13 Kozyraki R, Fyfe J, Verroust PJ, et al: Megalin-dependent cubilin-mediated endocytosis is a major pathway for the apical uptake of transferrin in polarized epithelia. Proc Natl Acad Sci USA 2001;98:12491–12496.

14 Mead GP, Carr-Smith HD, Drayson MT, Morgan GJ, Child JA, Bradwell AR: Serum free light chains for monitoring multiple myeloma. Br J Haematol 2004;126:348–354.

15 Sanders PW, Herrera GA, Galla JH: Human Bence Jones protein toxicity in rat proximal tubule epithelium in vivo. Kidney Int 1987;32:851–861.

16 Sanders PW, Herrera GA, Chen A, Booker BB, Galla JH: Differential nephrotoxicity of low molecular weight proteins including Bence Jones proteins in the perfused rat nephron in vivo. J Clin Invest 1988;82:2086–2096.

17 Sanders PW, Herrera GA, Lott RL, Galla JH: Morphologic alterations of the proximal tubules in light chain-related renal disease. Kidney Int 1988;33:881–889.

18 Pote A, Zwizinski C, Simon EE, Meleg-Smith S, Batuman V: Cytotoxicity of myeloma light chains in cultured human kidney proximal tubule cells. Am J Kidney Dis 2000;36:735–744.

19 Li M, Balamuthusamy S, Simon EE, Batuman V: Silencing megalin and cubilin genes inhibits myeloma light chain endocytosis and ameliorates toxicity in human renal proximal tubule epithelial cells. Am J Physiol Renal Physiol 2008;295:F82–F90.

20 Sanders PW, Booker BB, Bishop JB, Cheung HC: Mechanisms of intranephronal proteinaceous cast formation by low molecular weight proteins. J Clin Invest 1990;85:570–576.

21 Kapur U, Barton K, Fresco R, Leehey DJ, Picken MM: Expanding the pathologic spectrum of immunoglobulin light chain proximal tubulopathy. Arch Pathol Lab Med 2007;131:1368–1372.

22 Herrera GA, Sanders PW, Reddy BV, Hasbargen JA, Hammond WS, Brooke JD: Ultrastructural immunolabeling: a unique diagnostic tool in monoclonal light chain-related renal diseases. Ultrastruct Pathol 1994;18:401–416.

23 Sengul S, Zwizinski C, Simon EE, Kapasi A, Singhal PC, Batuman V: Endocytosis of light chains induces cytokines through activation of NF-kappaB in human proximal tubule cells. Kidney Int 2002;62:1977–1988.

24 Sengul S, Zwizinski C, Batuman V: Role of MAPK pathways in light chain-induced cytokine production in human proximal tubule cells. Am J Physiol Renal Physiol 2003;284:F1245–F1254.

25 Arimura A, Li M, Batuman V: Potential protective action of pituitary adenylate cyclase-activating polypeptide (PACAP38) on in vitro and in vivo models of myeloma kidney injury. Blood 2006;107:661–668.

26 Wang PX, Sanders PW: Immunoglobulin light chains generate hydrogen peroxide. J Am Soc Nephrol 2007;18:1239–1245.

27 Kamata H, Hirata H: Redox regulation of cellular signalling. Cell Signal 1999;11:1–14.

28 Li N, Karin M: Is NF-kappaB the sensor of oxidative stress? FASEB J 1999;13:1137–1143.

29 Weiss JH, Williams RH, Galla JH, et al: Pathophysiology of acute Bence-Jones protein nephrotoxicity in the rat. Kidney Int 1981;20:198–210.

30 Tanner GA, Knopp LC: Glomerular blood flow after single nephron obstruction in the rat kidney. Am J Physiol 1986;250: F77–F85.

31 Tanner GA, Evan AP: Glomerular and proximal tubular morphology after single nephron obstruction. Kidney Int 1989;36:1050–1060.

32 Sanders PW, Booker BB: Pathobiology of cast nephropathy from human Bence Jones proteins. J Clin Invest 1992;89:630–639.

33 Start DA, Silva FG, Davis LD, D'Agati V, Pirani CL: Myeloma cast nephropathy: immunohistochemical and lectin studies. Modern Pathol 1988;1:336–347.

34 Winearls CG: Acute myeloma kidney. Kidney Int 1995;48:1347–1361.

35 Serafini-Cessi F, Malagolini N, Cavallone D: Tamm-Horsfall glycoprotein: biology and clinical relevance. Am J Kidney Dis 2003;42: 658–676.
36 Rindler MJ, Naik SS, Li N, Hoops TC, Peraldi M-N: Uromodulin (Tamm-Horsfall glycoprotein/uromucoid) is a phosphatidylinositol-linked membrane protein. J Biol Chem 1990;265:20784–20789.
37 Jovine L, Qi H, Williams Z, Litscher E, Wassarman PM: The ZP domain is a conserved module for polymerization of extracellular proteins. Nat Cell Biol 2002;4:457–461.
38 Ying W-Z, Sanders PW: Expression of Tamm-Horsfall glycoprotein is regulated by dietary salt in rats. Kidney Int 1998;54:1150–1156.
39 Holland MD, Galla JH, Sanders PW, Luke RG: Effect of urinary pH and diatrizoate on Bence Jones protein nephrotoxicity in the rat. Kidney Int 1985;27:46–50.
40 Huang Z-Q, Kirk KA, Connelly KG, Sanders PW: Bence Jones proteins bind to a common peptide segment of Tamm-Horsfall glycoprotein to promote heterotypic aggregation. J Clin Invest 1993;92:2975–2983.
41 Huang Z-Q, Sanders PW: Biochemical interaction of Tamm-Horsfall glycoprotein with Ig light chains. Lab Invest 1995;73:810–817.
42 Huang Z-Q, Sanders PW: Localization of a single binding site for immunoglobulin light chains on human Tamm-Horsfall glycoprotein. J Clin Invest 1997;99:732–736.
43 Day ED: The light chains of immunoglobulins; in: Day ED (ed): Advanced Immunochemistry. New York, Wiley, 1990, pp 3–51.
44 Bruccoleri RE, Haber E, Novotny J: Structure of antibody hypervariable loops reproduced by a conformational search algorithm. Nature 1988;335:564–568.
45 Chothia C, Lesk AM: Canonical structures for the hypervariable regions of immunoglobulins. J Mol Biol 1987;196:901–917.
46 Chothia C, Lesk AM, Tramontano A, et al: Conformations of immunoglobulin hypervariable regions. Nature 1989;342:877–883.
47 Glockshuber R, Steipe B, Huber R, Plückthun A: Crystallization and preliminary X-ray studies of the V_L domain of the antibody McPC603 produced in *Escherichia coli*. J Mol Biol 1990;213:613–615.
48 Solomon A: Light chains of immunoglobulins: structural-genetic correlates. Blood 1986;68:603–610.
49 Hieter PA, Maizel JV Jr, Leder P: Evolution of human immunoglobulin kappa J region genes. J Biol Chem 1982;257:1516–1522.
50 Williams SC, Frippiat JP, Tomlinson IM, Ignatovich O, Lefranc MP, Winter G: Sequence and evolution of the human germline V lambda repertoire. J Mol Biol 1996;264:220–232.
51 Kawasaki K, Minoshima S, Nakato E, et al: One-megabase sequence analysis of the human immunoglobulin lambda gene locus. Genome Res 1997;7:250–261.
52 Ozaki S, Abe M, Wolfenbarger D, Weiss DT, Solomon A: Preferential expression of human λ-light-chain variable-region subgroups in multiple myeloma, AL amyloidosis, and Waldenström's macroglobulinemia. Clin Immunol Immunopathol 1994;71:183–189.
53 Ying W-Z, Sanders PW: Mapping the binding domain of immunoglobulin light chains for Tamm-Horsfall protein. Am J Pathol 2001;158:1859–1866.

Paul W. Sanders, MD
Division of Nephrology/Department of Medicine, University of Alabama at Birmingham
642 Lyons-Harrison Research Building, 1530 Third Avenue
South Birmingham, AL 35294–0007 (USA)
Tel. +1 205 934 3589, Fax +1 205 975 6288, E-Mail psanders@uab.edu

Herrera GA (ed): Experimental Models for Renal Diseases: Pathogenesis and Diagnosis.
Contrib Nephrol. Basel, Karger, 2011, vol 169, pp 270–285

HIV-Associated Nephropathy: Experimental Models

M. Carmen Avila-Casado[a] · Teresa I. Fortoul[b] · Sumant S. Chugh[c]

[a]Department of Pathology, Instituto Nacional de Cardiologia Ignacio Chavez, Department of Cellular and Tissue Biology, and [b]Basic Sciences Coordinator, Universidad Nacional Autonoma de Mexico, School of Medicine, Mexico City, Mexico; [c]Glomerular Disease Therapeutics Laboratory, University of Alabama, Birmingham, Ala., USA

Abstract

Since 1984 reports of renal involvement in AIDS patients have been presented in the literature. Different forms of renal disease were noted in the AIDS population including those related to systemic and local renal infections, tubulointerstitial disease, renal involvement by neoplasm and glomerular disease including collapsing glomerulopathy (CG). HIV-associated nephropathy (HIVAN) has been demonstrated to be more severe in the black population. HIVAN is the most common cause of renal failure in HIV-1–seropositive patients. The term HIVAN is reserved for the typical histopathological form of focal and segmental glomerulosclerosis (FSGS) characterized by the findings of coexistent glomerular and severe tubulointerstitial disease. In both humans and the murine model, glomerular lesions include FSGS, glomerular collapse and podocyte hyperplasia. The tubulointerstitial damage as well as the glomerular collapse can also be seen in non-HIV primary collapsing GN, raising the question of common mechanisms to HIV and other non-identified viral agents related to the development of the disease. Although controversial, increasing evidence supports a direct effect of the virus on renal cells either as a result of exposure to viral proteins or direct renal parenchyma infection. The use of a HIV-1 transgenic mouse model has demonstrated a direct etiologic link between HIV-1 expression in kidney and the development of HIVAN with unique viral-host interactions, which depend at the same time on stimulating features of the virus and the individual nature of the host response. The infection of renal cells by HIV-1 could be detected by reverse transcription-polymerase chain reaction (RT-PCR) of gag RNA at a low level. Some studies using an HIV-1 transgenic mouse model have demonstrated that expression of HIV-1 in the kidney is required for the development of HIVAN. The final common pathway in the development of HIV-associated nephropathy is likely to involve alterations in the patterns of gene expression of renal parenchyma cells by cytokines and growth factors, leading to interstitial fibrosis and enhanced glomerular matrix synthesis. The nature of the host response to viral infection is critical to the development of nephropathy. HLA-linked responses particular to a subset of blacks may explain some of the epidemiologic features of HIVAN. There may also be biological heterogeneity in the strains of HIV-1 that could account for a particular renotropic strain. HIV strains from different parts of the world may vary by as much as 15% at the level of nucleotide sequence. The infectivity of human immunodeficiency virus (HIV-1) in human glomerular cells has been evaluated by exposing homogeneous cultures of human glomerular capillary endothelial, mesangial and epithelial cells to HIV in vitro. The mechanism of access of HIV into glomerular endothelial and mesangial cells is unknown up to now; HIV is generally infectious for cells expressing the CD4 antigen in their cell membrane. Other modes of HIV entry into cells independent of the CD4 receptor are possible through mechanisms involving Fc-receptors or co-infection with other enveloped viruses such as HTLV-I. Our understanding of the pathogenesis of HIVAN has been aided by the development of a transgenic model. The curious fact that only 3 of 8 founded transgenic lines developed nephropathy emphasizes that the expression of viral gene products per se is not sufficient to produce nephropathy. Human renal epithelium does not express CD4 receptors and in vitro attempts to infect glomerular epithelial cells using laboratory strains of HIV-1 have proven fruitless. The striking morphologic and phenotypic similarities between HIVAN and collapsing idiopathic FSGS raise the question whether the altered podocyte gene expression in collapsing idiopathic FSGS may also be due to a viral infection. This hypothesis is further supported by de novo occurrence of collapsing idiopathic FSGS in immunosuppressed renal transplantation patients and by epidemiologic data. In conclusion, there are likely to be common mechanisms in the pathogenesis for collapsing idiopathic glomerulosclerosis and HIVAN. A primary injury of the podocyte leading to dysregulation of the cellular phenotype appears to mediate the glomerular tuft collapse in both conditions. Primary collapsing glomerulopathy recurs post-transplantation, raising the possibility of circulating factors implicated in the pathogenesis of visceral epithelial cell damage in steroid-resistant minimal change disease or recurrent FSGS. Recurrence of CG can occur hours after transplantation, suggesting that the plasma of CG patients contains one or more factors capable of inducing proteinuria due to the damage of the podocyte that results in the increase in glomerular permeability. In a rat model of CG developed by our group, the injection of serum from CG patients resulted in proteinuria, glomerular tuft retraction and podocyte damage at the ultrastructural level (visceral epithelial cell foot-process effacement). No ultrastructural or light microscopy abnormalities were seen in rats injected with serum from non-collapsing FSGS or healthy subjects. Based on the experience of our group, circulating factors play a dominant role in the pathogenesis of idiopathic CG.

HIV-Associated Nephropathy

Since the advent of the AIDS pandemic in the late 1970s and early 1980s, an estimated 31.3 million adults and 2.1 million children worldwide live with human immunodeficiency virus (HIV) infection, and 2.7 million adults and 430,000 children under 15 years of age have become infected with HIV (table 1) [1]. At the beginning of the HIV pandemic in the early l980s, it appeared that the kidney might be spared from major complications of the disease. However, in 1984, reports of renal involvement in AIDS patients first appeared. An early report documented the spectrum of renal involvement in HIV-infected patients [2, 3]. Different forms of renal disease were noted in the AIDS population, including those related to systemic and local renal infections, tubulointerstitial disease, renal involvement by neoplasm and glomerular disease including collapsing glomerulopathy (CG) [4]. In a small group of AIDS patients with proteinuria or renal insufficiency, a mixture of renal lesions was found, including acute tubular necrosis, focal interstitial nephritis, nephrocalcinosis, renal cell carcinoma, cryptococcal abscess and a variety of glomerular lesions [4]. HIV-associated nephropathy (HIVAN) is also more severe in the black population; the cause of this strong racial predilection among blacks is unknown.

Pathology of HIVAN

HIVAN is the most common cause of renal failure in HIV-1-seropositive patients. The term HIVAN should be reserved for the typical histopathological form of focal and segmental glomerulosclerosis (FSGS) and recently proposed mesangiopathies [e.g. DMH and minimal change disease (MCD)] which are potentially related. It should not be applied indiscriminately to the diverse immune deposit glomerular diseases and non-immune deposit glomerular diseases that have been reported in this population.

Typically, HIVAN occurs late in HIV-1 infection, with a prognosis similar to that of other AIDS-defining conditions [5, 6]. What is remarkable of this last entity is the severe tubulointerstitial disease.

Upon histopathological examination, HIVAN is characterized by the findings of coexistent glomerular and tubulointerstitial disease. The tubulointerstitial changes include microcystic dilation, interstitial fibrosis, tubular cell proliferation and apoptosis [3, 4]. In both humans and the murine model, glomerular lesions include FSGS, glomerular collapse and podocyte hyperplasia [7]. The term 'collapse' refers to the remarkable retraction of the glomerular basement membrane with resulting occlusion of glomerular capillary lumina, either segmental or more

Table 1. Global summary of the AIDS epidemic

December 2008	
Number of people living with HIV in 2008	
Total: 33.4 million	(31.1–35.8 million)
Adults: 31.3 million	(29.2–33.7 million)
Women: 15.7 million	(14.2–17.2 million)
Children under 15 years: 2.1 million	(1.2–2.9 million)
People newly infected with HIV in 2008	
Total: 2.7 million	(2.4–3.0 million)
Adults: 2.3 million	(2.0–2.5 million)
Children under 15 years: 430 000	(240,000–610,000)
AIDS-related deaths in 2008	
Total: 2.0 million	(1.7–2.4 million)
Adults: 1.7 million	(1.4–2.1 million)
Children under 15 years: 280,000	(150,000–410,000)

The ranges around the estimates in this table define the boundaries in which the actual numbers lie, based on the best available information.

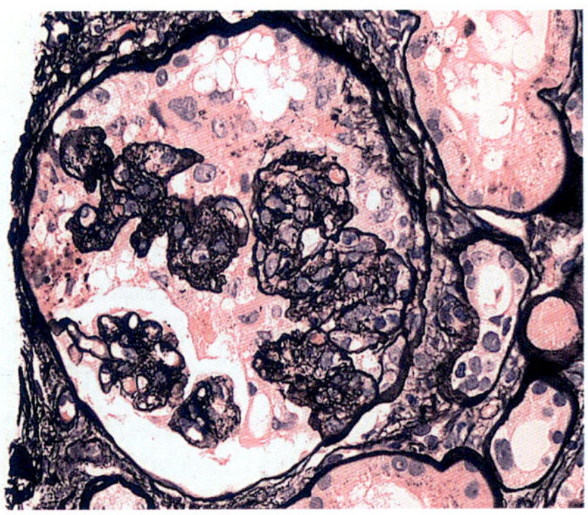

Fig. 1. Glomerulus showing global collapse of the capillary loops with glomerular tuft retraction. Visceral epithelial cells appear hypertrophied and fill the urinary space, showing numerous vacuoles and reabsorption droplets.

commonly global in distribution. An important component of this lesion is striking hypertrophy and hyperplasia of the overlying visceral epithelial cells, which composes a cellular corona over the collapsed lobules [2, 6]. The visceral cells may display enlarged vesicular nuclei, mitotic figures and numerous intracytoplasmic protein resorption droplets (fig. 1). Studies using antibody to PCNA (proliferation cell nuclear antigen) in experimental models developed by our group have revealed a high proliferative rate for podocytes, which may even display mitotic figures (fig. 2) Tubulointerstitial disease includes tubular necrosis, interstitial nephritis and nephrocalcinosis [4]. The renal findings of CG and severe tubulointerstitial injury can be seen in patients with asymptomatic HIV infection as well as in those with AIDS or AIDS-related complex.

Pathogenesis

Although some researchers have proposed that infections per se do not appear to play a role in the development of the nephropathy – since HIVAN commonly occurs before opportunistic infections are manifest – it is interesting to point out that the tubulointerstitial damage as well as the glomerular collapse can also be seen in non-HIV primary collapsing GN, raising the question

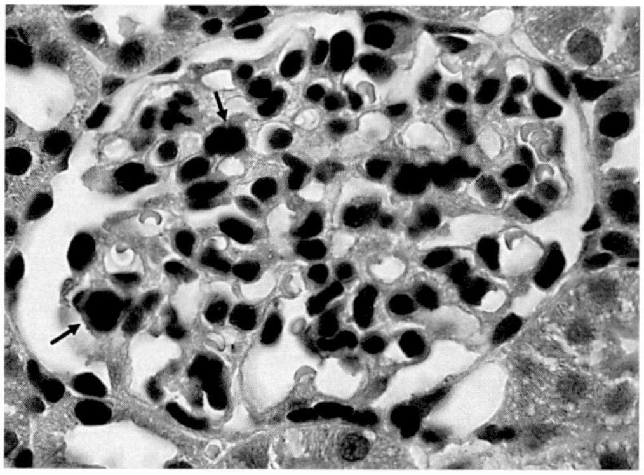

Fig. 2. Glomerulus from a mouse with visceral epithelial cell damage in a PASN model showing podocytes proliferation marked with PCNA immunostaining. Positive cells have a brown nuclear stain. Arrows show mitotic figures and binucleation. PCNA, 400×.

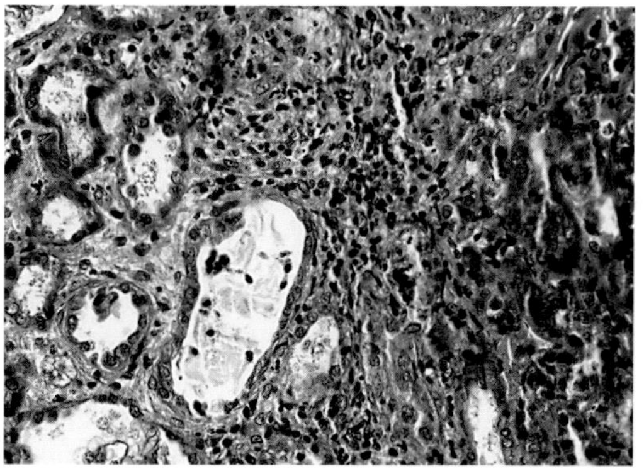

Fig. 3. Tubulointerstitial damage is evident with the presence of interstitial infiltration by an inflammatory infiltrate rich in CD8 T cells. The presence of cystic dilation of tubules with numerous regenerative changes can also be appreciated. Masson trichrome stain, 250×.

of common mechanisms to HIV and other non-identified viral agents related with the development of the disease. The role of HIV-1 infection in glomerular and tubular pathogenesis is not completely understood, but increasing evidence supports a direct effect of the virus on renal cells either as a result of exposure to viral proteins or direct renal parenchyma infection.

When compared with biopsies of patients with heroin nephropathy and idiopathic FSGS, there are more prevalent and severe tubular degenerative and regenerative changes (fig. 3) [8]. The interstitial infiltrate consists largely of lymphocytes, with a predominance of CD8+ T cells, and fewer plasma cells, monocytes and B cells. A distinctive form of acute interstitial nephritis known as DILS (diffuse infiltrative lymphocytosis syndrome) has been described in HIV-infected patients. This is a systemic illness affecting a minority of HIV-infected patients characterized by interstitial infiltration of salivary glands, lung, gastrointestinal tract and kidney by an inflammatory infiltrate rich in CD8 T cells [9, 10]. Renal manifestations are often accompanied by type IV renal tubular acidosis. A genetic predisposition to the syndrome has been provided by the finding of HLA DR5 allele DRB1*1102 or the HLA DR6 allele DRB1*1301 in 67% of 38 individuals, compared with 29% of control subjects (odds ratio = 8.2) [11]. It has been demonstrated that a component of the disease process is the direct infection of renal parenchymal cells, including all tubular, parietal and visceral epithelial cell types [2].

Experimental Models of HIVAN

The use of a HIV-1 transgenic mouse model has demonstrated a direct etiologic link between HIV-1 expression in kidney and the development of HIVAN [12]. Due to the fact that only a minority of HIV-infected patients develop nephropathy, a particular subset appears to be at risk. These observations suggest unique viral-host interactions, which depend at the same time on stimulating features of the virus and the individual nature of the host response. Conaldi et al. [13] have recently demonstrated that HIV-1 infection of human renal cells is possible in vitro. They demonstrated that infection of renal cells by HIV-1 could be detected by reverse transcription-polymerase chain reaction (RT-PCR) of gag RNA at a low level [13]. Other recent studies using an HIV-1 transgenic mouse model have demonstrated that expression of HIV-1 in the kidney is required for the development of HIVAN [5]. What has remained unclear, however, is the renal cell type responsible for the pathogenesis and the essential pathological process. Up to now, the following mechanisms of HIV nephropathy have been proposed: (1) direct injury to renal epithelial (visceral and tubular cells) by cytopathic effects of viral infection of renal parenchyma cells, (2) indirect injury to the kidney by renal cellular uptake of circulating virally encoded molecules, and (3) indirect injury to the kidney through release of cytokines by infected lymphocytes or monocytes in the circulation or infiltrating the kidney [2].

The final common pathway in the development of HIVAN is likely to involve alterations in the patterns of gene expression of renal parenchyma cells by cytokines and growth factors, leading to interstitial fibrosis and enhanced glomerular matrix synthesis. The nature of the host response to viral infection is critical to the development of nephropathy. HLA-linked responses particular to a subset of blacks may explain some of the epidemiologic features of HIVAN. For example, considerable genetic heterogeneity has been described among blacks in North America and the Caribbean and may be traced to different geographic origins in Africa [2]. There may also be biological heterogeneity in the strains of HIV-1 that could account for a particular renotropic or 'nephritogenic' strain. For example, isolates of HIV-1 may vary in their tropism, replication rates and syncytium-producing capacity [14, 15].

Strains isolated from asymptomatic individuals are more often monocyte-tropic and slowly replicating, whereas isolates from patients with fully developed AIDS are more often T-tropic or monocyte/T-tropic and generally replicate rapidly [16, 17]. It may be that monocyte-tropic strains characteristic of early disease states have greater capacity to produce nephropathy. It is also possible that genetic differences in HIV strains may be critical. HIV strains from different parts of the world may vary by as much as 15% at the level of nucleotide sequence [16].

Differences in the genetic sequence of HIV-1 could account for the existence of strains with greater renal tropism. There is also evidence that HIV infection does not progress uniformly in different organs within an infected individual [2].

Although controversial, it has been proposed that HIV may directly infect renal parenchymal cells, specifically the glomerular mesangial cell which shares many features with monocytes and has been shown to bear possible HIV membrane receptors: the CD4 antigen and FcR [17]. HIV has been revealed to infect and replicate in glomerular endothelial cells and, to a lesser degree, in mesangial cells, but not in visceral epithelial cells [18].

In another study, Alpers et al. [19] were unable to demonstrate productive infection of human mesangial cells by either T-tropic or monocyte-tropic strains of HIV-1 or HIV-2. Although one study demonstrated the presence of viral genome in tubular and visceral epithelial cells of renal biopsies with HIVAN by in situ hybridization using a c-DNA probe and immunostaining for p24 antigen, these data have not been confirmed by others [20, 21]. To test the hypothesis that HIV can infect and replicate in glomerular cells, in an elegant experiment, cell culture systems were developed from normal human kidney to determine the infectivity of HIV in specific glomerular cell types [18]. The infectivity of human immunodeficiency virus (HIV-1) in human glomerular cells was evaluated by exposing homogeneous cultures of human glomerular capillary endothelial, mesangial and epithelial cells to HIV in vitro. Infectivity and HIV expression was assessed by: (1) the measurement of p24 antigen production from culture supernatants, (2) the presence of intracellular p24 antigen by immunofluorescence, (3) levels of p24 antigen production or syncytial formation following the co-cultivation of glomerular cells exposed to HIV with normal human peripheral blood mononuclear cells or MT-2 lymphocytes, and (4) the presence of intracellular HIV DNA by PCR. Homogeneous cultures of endothelial, mesangial and epithelial cells were prepared from glomeruli isolated from normal human kidneys deemed unsuitable for transplantation. Glomeruli were isolated by differential sieving of renal cortical tissue. After 40 days of exposure to HIV, the culture medium from the mesangial cells was used to infect normal human peripheral blood mononuclear cells and MT-2 cells. In rodents, two subtypes of glomerular mesangial cells have been described: the majority smooth muscle cells while a minority, of about 5%, are bone marrow-derived macrophages [22]. The latter have not been convincingly described in human glomeruli.

In this study it was not demonstrated whether a subtype of mesangial cells became infected or whether infection involved a small number of cells from a single cell type. In contrast, the same studies with cultures of human glomerular epithelial cells exposed to HIV were negative for evidence of HIV infection. The results indicate that HIV can infect and replicate in glomerular capillary endothelial cells and in a small percentage of mesangial cells, but not in human glomerular epithelial cells in vitro [18]. The mechanism of access of HIV into glomerular endothelial and mesangial cells is unknown. The CD4 molecule is the primary receptor for the HIV glycoprotein envelope (GP 120), thus HIV is generally infectious for cells expressing the CD4 antigen in their cell membrane. Preliminary immunohistochemical studies using murine monoclonal antibodies against the CD4 molecule showed a cytoplasmic staining of CD4 antigen in normal human glomerular cells [23]. This data showing partial inhibition of endothelial cell infection with soluble CD4 suggest that the CD4 receptor may play a role in this process [24]. However, complete inhibition could not be achieved suggesting other mechanisms may also be important. Other modes of HIV entry into cells independent of the CD4 receptor are possible through mechanisms involving Fc-receptors [25] or co-infection with other enveloped viruses such as HTLV-l [26].

Specific Animal Models

Transgenic Mice

Our understanding of the pathogenesis of HIVAN has been aided by the development of a transgenic model that lacks a 3.1-kilobase fragment of the HIV genome, overlapping gag that encodes for the structural (nucleocapsid core) proteins and pol, which encodes for the enzymes (reverse transcriptase, integrase and ribonuclease), but preserves the genomic structure of the HIV envelope and regulatory genes [27]. Heterozygous transgenic mice from 3 of 8 founders developed a form of nephropathy characterized by focal segmental sclerosis, with mesangial hypercellularity, visceral cell hypertrophy, interstitial inflammatory infiltrates and tubular microcysts. The presence of proteinuria is first detectable at 24 days of age and progresses to nephrotic syndrome and end-stage renal disease. Modest expression of viral RNA was demonstrated in renal tissue at early stages before the onset of nephropathy, and became almost undetectable at 60 days. Rev protein was the only virally encoded protein that could be identified in glomeruli, although it is not certain whether this resulted from endogenous synthesis or trapping in the course of glomerular filtration.

These findings suggest that productive virus in the kidney may not be necessary to the development of nephropathy. Rather, renal uptake of viral gene products may induce nephropathy through transactivation of host genes, either through direct effects of viral gene products on target cells or indirectly through the release of cytokines [27]. Subsequent work implicates upregulation of basic fibroblast growth factor and transforming growth factor-beta (TGF-β) in this model [28]. It is remarkable that the transgenic renal tubular epithelial cells grown in vitro manifest features of transformed cells, with greatly enhanced proliferative capacity that could mediate tubular microcyst formation [28].

The viral gene products that are operative in this model have not been identified yet. Candidate viral proteins include tat, which can be released by virally infected cells and endocytosed by various cell types [29]. Tat induces bone marrow macrophages to upregulate production of TGF-β [30] and glial cells to increase synthesis of collagen and fibronectin in vitro [29]. In other systems, HIV infection can dysregulate a variety of host cytokines, including IL-1, IL-6, TNF-α and granulocyte macrophage colony-stimulating factor, any of which could be important in the development of nephropathy [31]. The curious fact that only 3 of 8 founded transgenic lines developed nephropathy emphasizes that expression of viral gene products per se is not sufficient to produce nephropathy, and susceptibility to nephropathy may depend on individual host responses.

Other studies on HIV-1 transgenic mice support the hypothesis that HIV-1 can injure renal epithelium even in the absence of immunosuppressant or productive viral infection. Human renal epithelium does not express CD4 receptors and in vitro attempts to infect glomerular epithelial cells using laboratory strains of HIV-1 have proven fruitless. Other strains of HIV-1 have been shown capable of infecting CD4-negative human cells, indicating that other molecules can facilitate the infection like the G protein-coupled receptor named CXCR4 and the chemokine receptors CCR5, CCR2b and CCR3, a group of recently described co-factors necessary for viral entry, even in the presence of CD4, that may play a role in HIV-1 infection in CD4-negative cells [32]. Therefore, it is possible that HIV-1 may adapt to host tissues by acquiring adhesion molecules from host cells [33] and gaining a selected tropism to infect these cells.

Some authors have hypothesized that the inability of HIV-1 to infect renal epithelial cells may be due to a limited tropism of cloned laboratory strains, as opposed to wild-type virus, and HIV-1 virus derived from children with HIVAN are capable of infecting renal tubular epithelial cells derived from the same patient, supporting the hypothesis that HIV-1 may cause direct cytotoxicity of renal epithelial cells from children with HIVAN [33]. The HIV-isolates were obtained from peripheral blood-mononuclear cells from 3 patients and from the peritoneal fluid from another patient. Primary renal tubular epithelial cells were isolated from renal biopsies, autopsies and urine of 5 children with HIV-associated renal disease. Renal epithelial cells were cultured by cloning homogeneous microcolonies of epithelial cells on collagen I-coated dishes and grown in a selective renal culture medium. To determine that the epithelial cells were not previously infected at the outset of the experiments, several studies were performed, such as PCR for HIV-1 proviral DNA, the measurement of p24 antigen in culture supernatants and the detection of viral particles by electron microscopy. The cultured renal epithelial cells were divided in two groups: one group was exposed to HIV-1 and the other was used as a control. Cultures were followed for a period of 3 weeks and, finally, they were washed and harvested for HIV-1 studies.

The results of this study provide strong evidence to support the hypothesis that HIV-1 is capable of infecting and injuring renal tubular epithelial cells. It has been proposed that HIV-1 can enter epithelial cells by three possible different ways: binding of cell-free virus to specific receptors, fusion with infected mononuclear cells and direct transfer during intimate contact between productively infected mononuclear cells and epithelial cells [33]. Up to now, the mechanism(s) responsible for HIV-1 infection of renal tubular epithelial cells is not completely understood. It is important to point out that under experimental conditions, very high viral titers are needed to infect cultured visceral epithelial cells and it has been proposed that in vivo renal tubular epithelial cells are more likely to be infected from intimate contact with mononuclear cells [33].

Infection of renal tubular epithelial cells was not significantly inhibited by anti-CD4 antibodies suggesting that this process is mediated by CD4-independent pathways. Moreover, in the previous study CD4 receptors within the renal tubular epithelial cells were not found and more studies will be required to identify the right receptors implicated in this process. A major issue in the pathogenesis of HIVAN is whether renal tubular epithelial cells can actually replicate HIV-1. The presence of mature virions within cytoplasmic vacuoles in renal tubular epithelial cells has been demonstrated by transmission electron microscopy [33]. Nevertheless, caution is necessary before extrapolating these results to in vivo situations, and many other factors may play a relevant role in the pathogenesis of HIVAN. The high number of regenerative renal tubular epithelial cells present in the urine and renal sections of patients with HIVAN may suggest that infected cells may be rapidly replaced [33]. Mice made transgenic with a replication defective subgenomic proviral HIV-1 construct, manifest renal tubular epithelial cell injury in correlation with the induction of high levels of HIV gene expressions [7].

A major unresolved issue in the pathogenesis of HIVAN is whether the kidney disease is due to renal cell infection or a 'bystander' phenomenon mediated by systemically dysregulated cytokines [12]. To address this issue, Bruggeman et al. [12] used two different experimental approaches and an HIV-1 transgenic mouse line that develops a progressive renal disease histologically similar to HIVAN in humans. They used two approaches to the HIV-1 transgenic mouse model to demonstrate that expression of the viral transgene in renal cells and not exposure to the circulating milieu is sufficient to induce the features of HIVAN. Using ex vivo fetal transgenic kidneys, it was demonstrated that transgene expression induces apoptosis in renal tubular structures. Using reciprocal transplantation of transgenic and normal kidneys into normal and transgenic mice, they demonstrated that the histological features of HIVAN develop only when the kidneys are transgenic. These studies support the hypothesis that a direct effect of HIV-1 gene products or even possibly HIV-1 infection of the kidney causes HIVAN.

The development of the transgenic mouse line TgN(pNL43d14)26Lom ('Tg26') has been reported previously [7, 33]. These mice contained 10 copies of the HIV-1 proviral DNA pNL4-3d1443 under the transcriptional control of the native long terminal repeat. For histopathological analysis, kidneys from adult normal and heterozygous transgenic mice with and without disease were prepared for light microscopy, fixed in 4% paraformaldehyde and stained with hematoxylin and eosin. Apoptotic nuclei were counted in 3 sections per 5 high-power fields per each section from normal mice, transgenic mice with proteinuria and transgenic mice with edema. For embryo cultures, Day 12 fetuses were dissected from pregnant Tg26 heterozygous mice and both kidneys were removed. HIV-1 transgenic mice develop a progressive renal disease similar to HIVAN with microcystic tubular interstitial disease. On light microscopy, dilated tubules in adult transgenic mice had evidence of epithelial cells undergoing apoptosis [33].

At the time of death, the blood urea nitrogen was 302 mg/dl and the transplanted kidney manifested marked histological changes that were typical for HIVAN, including severe microcystic tubular dilation, proteinacious casts and sclerotic glomeruli. The other animal in the group survived to study termination at 24 weeks and was killed with proteinuria. Similarly, the transplanted kidney also showed progressive changes as compared with the residual kidney from the transgenic donor, although not to the extent as seen in the animal that was in renal failure. Kidneys from normal nontransgenic donors were then transplanted into HIV-1 transgenic mice. The HIV-1 transgenic mice that received transplants from nontransgenic littermates remained healthy for 24 weeks after transplant. These animals did not manifest abnormal proteinuria. At the time of death, the histology of the nontransgenic kidney transplant was normal without histological features of HIVAN. These results suggest renal transgene expression is a necessary component in the development of kidney disease and the milieu associated with systemic transgene expression in itself does not appear to cause HIVAN. This does not exclude the possibility, however, that circulating factors may contribute to the rate of progression of this disease in the transgenic animal model or in humans [27].

Podocytes in Primary and HIV-Associated CG

Podocytes are highly differentiated, postmitotic cells, whose function is largely based on their complex cytoarchitecture. Based on their particular cytomorphology, they may be divided into three structurally and functionally different segments:

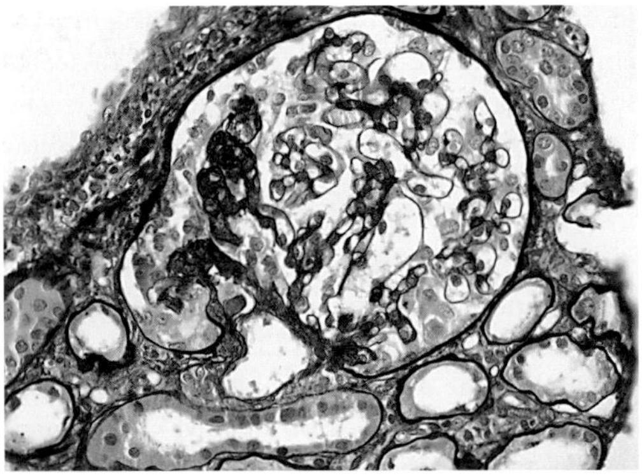

Fig. 4. Segmental area of glomerular collapse surrounded by hypertrophied visceral epithelial cells with reabsorption droplets. Prominence of podocytes is also seen in areas of open capillary loops as well. Silver methenamine stain, 400×.

cell body, major processes and foot processes. Segmentation of podocytes is also observed at the level of the cytoskeleton. Microtubules and intermediate filaments predominate in cell bodies and major processes. Foot processes contain an actin-based contractile apparatus, which is linked to the glomerular basement membrane at focal contact with the 3-β1-integrin complex [34].

The expression of these specialized cellular features is developmentally regulated. Maturation is associated with a loss of proliferative activity and progressive expression of distinctive podocyte markers, including the Wilms' tumor protein WT-1, CALLA (common acute lymphoblastic leukemia antigen), C3b receptor, GLEPP-1 (glomerular epithelial protein-1), podocalyxin and synaptopodin. WT-1 is a critical protein in nephrogenesis, as indicated by total failure of metanephric development in WT-1-deficient mice [35]. On their apical surface, podocytes are equipped with a negatively charged glycocalyx which plays a dual role in permselectivity and maintenance of foot process cytoarchitecture. GLEPP-1, a podocyte-specific protein tyrosine phosphatase, is another integral apical membrane protein that first appears in the S-shaped body stage. It has been proposed that signal transduction through GLEPP-1 is involved in regulation of foot process dynamics, cell differentiation and contact inhibition [36].

Because of its association with highly developed cytoskeletal architecture, synaptopodin is an important marker of the mature podocyte phenotype. During rat kidney development, synaptopodin first appears at the capillary loop stage. Thus, the expression of synaptopodin coincides with the formation of foot processes [37]. In glomerular diseases with massive proteinuria, podocytes undergo dramatic structural alterations. Among the many forms of FSGS, the glomerular degeneration in collapsing idiopathic glomerulosclerosis and HIVAN differs from that of other forms by its more severe clinical presentation, prognosis and morphologic features [38]. In these two diseases, the glomerular alterations are characterized by segmental or global tuft collapse, and most striking, the crowding of epithelial cells on the outer aspect of the tuft (fig. 4). It also differs in its lack of synechiae, at least in the early stages. It has been suggested that a primary podocyte injury plays a central role in the pathogenesis of certain forms of FSGS [39], although the nature of this injurious factor remains to be discovered.

In an interesting study based on the loss of all specific structural features as well as the loss of specific markers of differentiated podocytes, Barisoni et al. [6] proposed a pathogenetic mechanism of a dysregulated podocyte phenotype for both collapsing idiopathic glomerulosclerosis and HIVAN. In this study, in order to characterize the expression profiles of podocyte marker proteins in glomerular development and in mature kidney, sections of normal fetal and adult human kidney were studied by immunohistochemistry using the full panel of podocyte-specific antibodies just mentioned. In addition, the proliferative activity of podocytes was analyzed using an antibody directed to the cell cycle protein Ki-67. The percentage of glomeruli with 1–5 cells expressing Ki-67 was calculated in each category of disease (minimal change disease, primary CG, HIV-associated collapsing nephropathy and membranous nephropathy).

The percentage of nonsclerotic/noncollapsed glomeruli with decreased expression of three maturity markers (GLEPP-1, podocalyxin, synaptopodin) was also calculated in each category. Expression of Ki-67 was found in up to 90% of cells comprising the glomerular vesicles, the earliest stage of glomerular development, with progressive reduction in the percentage of Ki-67-expressing cells in the course of glomerular maturation. In the normal adult kidney, no Ki-67-expressing podocytes were observed. WT-1 was weakly expressed in all cells of renal vesicles, showed increased staining during the S-shaped body stage, and became restricted to podocytes in the capillary loop stage of development and in the adult kidney, where strong nuclear labeling of podocytes was found. The persistence of the mature podocyte phenotype Ki-67 was observed despite severe foot process effacement and high levels of proteinuria in patients with minimal change disease or membranous nephropathy. The results in collapsing idiopathic FSGS and HIVAN were virtually identical [6, 39]. In contrast to MCD and membranous glomerulonephritis (MGN), in both collapsing idiopathic glomerulosclerosis and HIVAN, Ki-67-positive cells were seen in regions where the number of epithelial cells on the outer aspect of the tuft was obviously increased. Eleven percent of glomeruli in collapsing idiopathic FSGS and 10% of glomeruli in HIVAN showed 1–5 podocytes with nuclear staining for Ki-67.

Extensive loss of all podocyte markers (CALLA, C3b receptor, GLEPP-1, podocalyxin, synaptopodin and WT-1) in all glomeruli with collapsing sclerosis was also observed [6]. A reduction in synaptopodin expression (in contrast to the other markers) was also frequently observed in unaffected portions of the tuft of segmentally injured glomeruli with FSGS, or even in glomerular profiles without any visible collapse or other histologic abnormalities. In collapsing idiopathic FSGS and HIVAN, 16% of the histologically unaffected (i.e. nonsclerotic and noncollapsed) glomeruli showed marked reduction of synaptopodin expression versus only 3% in collapsing idiopathic FSGS, 4% in HIVAN for GLEPP-1 and 0% in both categories of podocalyxin.

The loss of specific podocyte markers in these two diseases was accompanied by characteristic structural changes, among which the most prominent and characteristic were tuft collapse and the accumulation of epithelial cells on the outer aspect of the glomerular basement membrane in the collapsed areas. In the study [6], it was proposed that in early stages of glomerular collapse, these cells can be identified as podocytes or podocyte-like cells by their attachment to the outer aspect of the glomerular basement membrane: the visceral and parietal cells are still separated by a urinary space. However, in later stages, such a space disappears rendering a distinction between parietal and visceral epithelial cells impossible. Intermediates between a normal podocyte and the fully dysregulated phenotype point to a gradual loss of characteristic cellular features, which is different from what is usually seen when podocytes (e.g. in secondary forms of glomerular sclerosis) undergo simplification and degeneration [40, 41]. In addition to the loss of foot processes, they found that these cells lose their ultrastructurally visible actin cytoskeleton. The results of the study provide direct support for the concept that collapsing forms of FSGS (both idiopathic and related to HIV infection) have a common pathomechanism leading to a dysregulated podocyte phenotype. They arrived at this concept based on three lines of evidence. First, both diseases showed identical changes in podocyte phenotype as reflected by the loss of all specific podocyte markers examined. Second, in both diseases, a comparable number of podocytes were recruited into the cell cycle. Finally, the morphologic alterations of podocyte cytoarchitecture were indistinguishable between collapsing idiopathic glomerulosclerosis and HIVAN.

These findings were in sharp contrast to those obtained from the other two diseases studied, MCD and MGN, where severe podocyte damage also occurs. Data were consistent with a previous report that mentioned that in MCD and MGN, expression of GLEPP-1 is preserved despite intracellular redistribution [42]. Unlike in secondary forms of FSGS [42], in collapsing forms of FSGS the key event is not the loss of podocytes

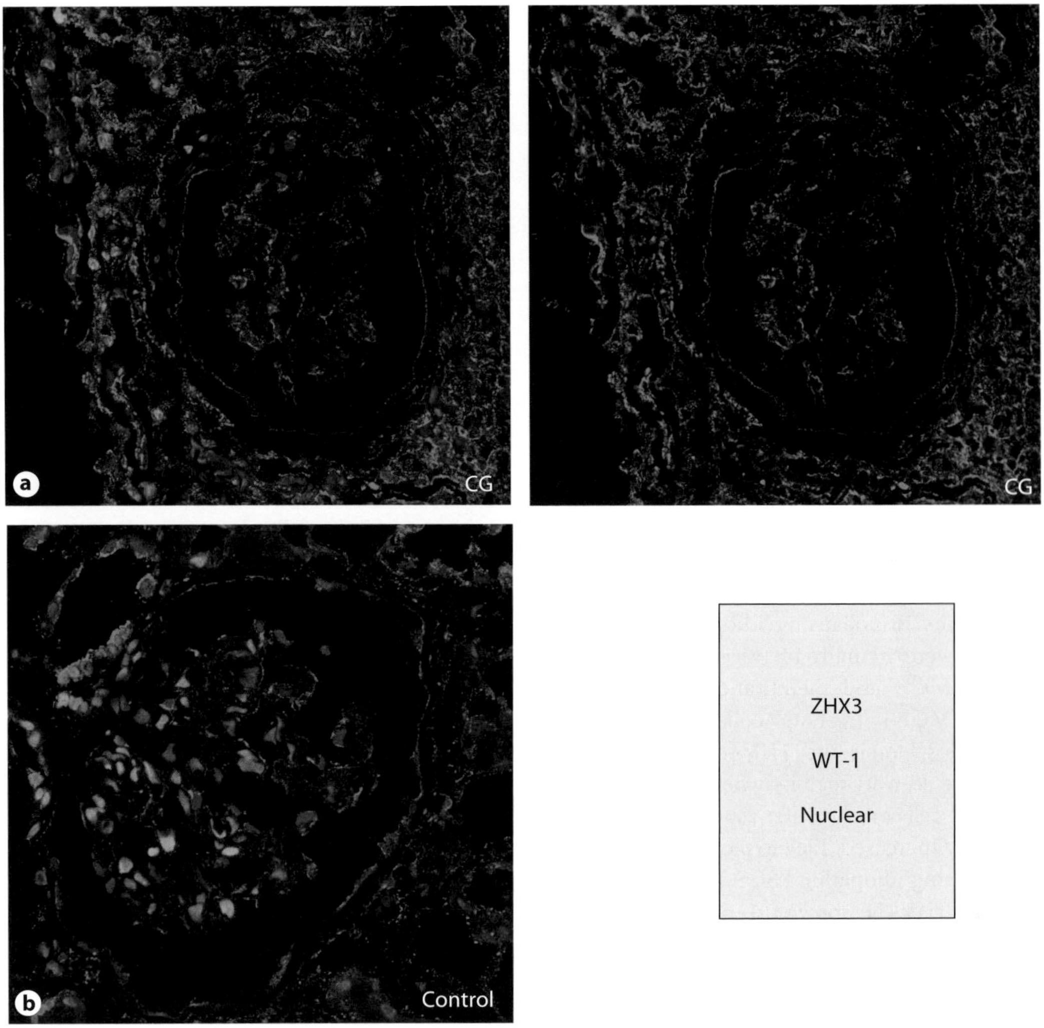

Fig. 5. **a** Immunomarkers in frozen tissue performed by our group studied by confocal microscopy for WT-1 in normal kidney used as a control. Podocyte nuclear staining is shown in blue. **b** In contrast, the glomerulus of a patient with collapsing glomerulopathy where lacking of WT-1 normal staining is clearly appreciated. WT-1 staining, 400×. G. = Groups.

themselves, but particular alterations of the podocyte phenotype that include loss of molecular markers and mature structural features. The loss in collapsing sclerosis of WT-1, which is present during nephrogenesis from the very beginning of podocyte ontogeny, makes it very difficult to assign these cells the status of 'dedifferentiated' podocytes, but lends strong support to the concept of a dysregulated phenotype. The presence of normal WT-1 in podocytes in CG patients is drastically reduced or absent as we have been shown by confocal immunostaining (fig. 5).

There is an ongoing debate in the literature as to whether podocytes in the mature glomerulus are capable of undergoing cell division [42–44]. In the collapsing forms of FSGS, the 'crowding' of cells in Bowman's space, together with the increased number of cycling cells clearly demonstrated in the study just presented, provides strong evidence that podocytes do indeed proliferate in association with an altered cellular phenotype (fig. 1 and 2). One central question in collapsing forms of FSGS is whether the podocyte damage is primary to the collapse of the tuft or a consequence of the collapse.

Three lines of evidence support the hypothesis that a severe and direct injury of the podocyte represents the primary injury, which then leads to the collapse of the tuft. First, as shown in this study, synaptopodin is markedly reduced in noncollapsed glomeruli; podocalyxin, which is expressed by both podocytes and endothelial cells, is selectively lost from podocytes but retained by the endothelial cells of the affected glomeruli. Second, in the glomerulus, podocytes are the only cells that produce vascular endothelial growth factor (VEGF). Kitamoto et al. [45] have recently shown that application of a neutralizing antibody to VEGF in newborn mice may effectively block the development of the glomerular capillary tuft. It would be interesting to investigate whether the dysregulation of the podocyte phenotype leads to a loss of VEGF production, which in turn causes tuft collapse. Third, strong support for the concept that the collapse is secondary to podocyte damage comes from a transgenic mouse model of HIVAN carrying a construct lacking the gag and pol genes, but retaining the other HIV regulatory elements [46]. In this model, which shows HIV transgene expression in podocytes, a similar dysregulation of podocyte phenotype as in the human disease has been observed [6]. Based on the experience of our group, circulating factors play a dominant role in the pathogenesis of idiopathic CG [47]. At present, the pathogenesis of collapsing idiopathic glomerulosclerosis remains unknown, although possible viral mediators have been proposed [47]. The striking morphologic and phenotypic similarities between HIVAN and collapsing idiopathic FSGS raise the question whether the altered podocyte gene expression in collapsing idiopathic FSGS may also be due to a viral infection. This hypothesis is further supported by de novo occurrence of collapsing idiopathic FSGS in immunosuppressed renal transplantation patients [47] and by epidemiologic data.

In summary, there are likely to be common mechanisms in the pathogenesis of collapsing idiopathic glomerulosclerosis and HIVAN. A primary injury of the podocyte leading to dysregulation of the cellular phenotype appears to mediate the glomerular tuft collapse in both conditions. The availability of an HIV-transgenic mouse model mimicking the human disease [36] should allow further avenues to identify factors responsible for altered podocyte gene expression in collapsing forms of glomerulosclerosis. It will be of particular interest to unravel the cellular signaling pathways that mediate podocyte dysregulation with the hope of designing therapeutic interventions to block their activation.

Experimental Models of Glomerular Collapse

Primary CG recurs post-transplantation, raising the possibility of circulating factors implicated in the pathogenesis of visceral epithelial cell damage in steroid-resistant minimal change disease or recurrent FSGS [47] since it has been reported that 15–55% of the patients with FSGS that undergo renal transplantation have recurrent disease [48]. Such factors are able to induce functional alterations in the renal issue in vitro [48] and/or proteinuria in vivo when injected to the rat [49]. Patients with CG present severe damage to the podocyte that cause heavy proteinuria, usually >10 g in 24 h. Both recurrent and de novo CG have been described in the renal allograft with histopathological features indistinguishable from CG in the native kidney [47]. Recurrence of CG can occur hours after transplantation, raising the possibility that the plasma of CG patients contains one or more factors capable of inducing proteinuria due to the damage of the visceral epithelial cell that results in the increase in glomerular permeability [50]. In order to demonstrate

Healthy subjects	Collapsing glomerulopathy	FSGS
Total Serum *G 3*	Total Serum *G 4**	Total Serum *G 5*
Serum without IgG *G 6*	Serum without IgG *G 7**	Serum without IgG *G 8*
Isolated IgG *G 9*	Isolated IgG *G 10**	Isolated IgG *G 11*

Fig. 6. Eleven groups (n = 6) of rats received serum from healthy subjects, CG or non-collapsing FSGS patients in 3 different modalities: total serum, serum without IgG or isolated IgG. As the control, Group 1 received no injections and Group 2 received injections of saline.

that humoral factors are involved in the pathogenesis of the visceral epithelial cell damage in CG and other forms of FSGS, we developed an experimental model where we injected serum and isolated IgG from patients with CG to Sprague-Dawley rats in order to demonstrate that serum from CG is capable of producing proteinuria and ultrastructural changes in the visceral epithelial cells, similar to those abnormalities seen in patients with CG [49]. To this effect we tested serum from CG patients and non-collapsing FSGS patients in an in vivo assay (fig. 6). Eleven groups (n = 9) of female rats (Sprague-Dawley) received serum from CG patients (idiopathic non-collapsing FSGS or healthy subjects in its native form) isolated IgG or serum without IgG as follows: Group 1 (control) received no injections, Group 2 received saline solution (NaCl 0.9%), Group 3 was given serum from healthy subjects in its native form, Group 4 received serum from CG in its native form, Group 5 was given serum from non-collapsing FSGS, Group 6 received serum from healthy subjects without IgG, Group 7 received serum from CG patients without IgG, Group 8 was given serum from non-collapsing FSGS without IgG, Group 9 received purified IgG from healthy subjects, Group 10 received purified IgG from CG patients and Group 11 was injected with purified IgG from non-collapsing FSGS patients. All rats were randomly distributed and placed in metabolic cages with water and food ad libitum. Serum was administered once intravenously, daily, in the vein of the tail for 5 days in a concentration of 25 mg/ml in saline to a final volume of 1 ml, with previous sedation in a narcosis chamber. Urine was collected every 24 h for 6 days prior to the first injection (baseline, Day 0), and 24, 48, 72, 96 and 120 h after the first injection (Days 1–5, respectively). The presence of proteinuria was measured using the method of Biuret [51]. The urine of Day 5 was also tested for creatinine determination [49]. Rats were sacrificed 24 h after the last injection. Blood samples were taken for sCr determination. Renal tissue was processed for light microscopy, immunofluorescence and electron microscopy. CG rats developed proteinuria while rats injected with serum from non-collapsing FSGS patients and healthy subjects did not. Rats injected with serum of CG in its native form developed marked proteinuria (99.2 ± 42 mg/24 h at Day 5, $p = 0.0001$, compared to the baseline), and decreased in creatinine clearance. Rats receiving isolated IgG or serum without IgG from CG developed mild proteinuria (46.5 ± 8.4 and 30.9 ± 11 mg/24 h, respectively) at the end of the observation period (Day 5 $p = 0.0001$). Glomerular tuft retraction and podocyte damage at the ultrastructural level (visceral epithelial cell foot-process effacement) were seen only in CG rats (fig. 7). No histological abnormalities by light/electron microscopy were found in rats injected with serum from non-collapsing FSGS or healthy subjects. In conclusion, circulating factors present in the serum of CG patients produce podocyte damage, whereas such factors are not present in non-collapsing forms of FSGS. IgG eluates from CG produce proteinuria when injected into the rat. Such factors remain in the circulation when serum of patients is adsorbed into protein A, supporting the possibility that there are more than one circulating factors present in patients with CG [49]. Moreover, studies in the utility of plasmapheresis in the treatment of recurrent FSGS have shown that this method reduces proteinuria and glomerular injury [52]. The demonstration of circulating factors in the pathogenesis of the glomerular damage in a

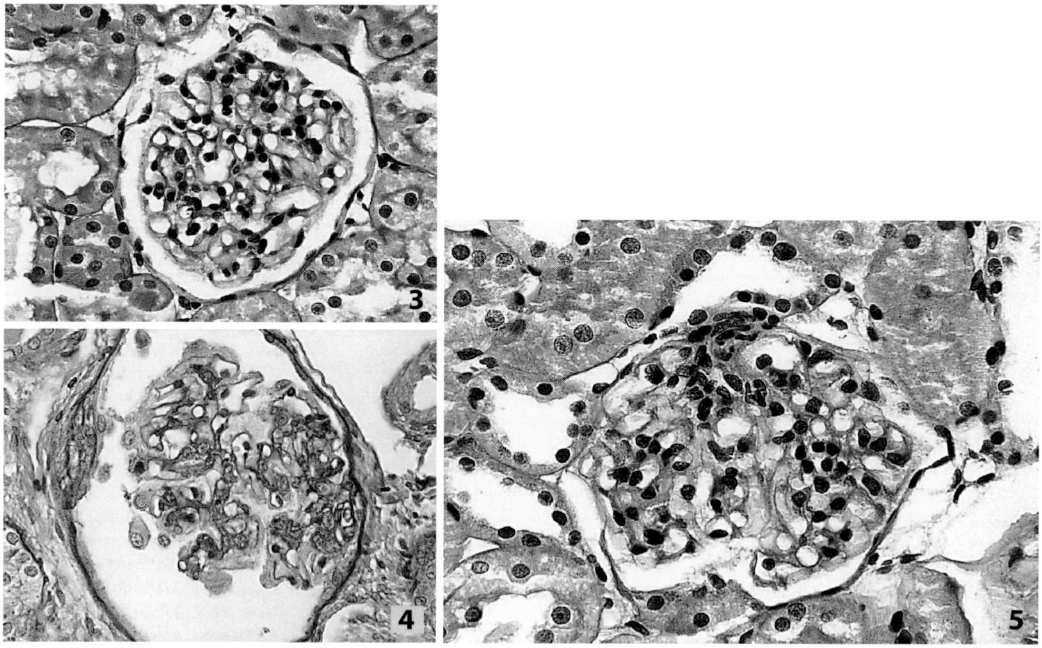

Fig. 7. Groups of rats injected with total native serum of collapsing glomerulopathy patients showed glomerular tuft retraction (4) while rats injected with serum from non-collapsing FSGS or healthy subjects showed almost no abnormalities by light microscopy (3 and 5; 400×, PAS staining).

particular variant of FSGS (CG) and its absence in non-collapsing variants of FSGS raises the possibility of different pathogenesis among the different variants of FSGS, recently supported by other authors [52]. The identification of different pathogenesis of podocyte damage would support the idea of different forms of treatment directly to the specific mechanisms of disease. Based on the experience of our group, circulating factors play a dominant role in the pathogenesis of idiopathic CG [47, 49]. Treatments like plasmapheresis could help not only to reduce proteinuria, but if such treatment is initiated early after diagnosis, patients can have lasting remission as we presented in the retrospective evaluation of 22 Mexican transplanted patients with positive circulating factors [53]. Recurrent CG was seen in 8 patients and the diagnosis was confirmed by a renal biopsy in 5. The other 3 patients had massive proteinuria. Time of recurrence varied from 24 h to 20 days post-transplantation. Three cases were considered de novo as they did not have proven CG in the native kidneys. Three cases without recurrent disease had received plasmapheresis: two of them pre- and post-transplantion and prophylactic pretransplant plasmapheresis in another patient. All the resting 5 non-plasmapheresed patients had recurrent CG and finished up with end-stage renal disease with the requirement of replacement therapy.

References

1 World Health Organization: Global Summary of the HIV/AIDS Statistics. UNAIDS Progress Report 2009 (Epiupdate). Geneva, WHO, 2009.

2 D'Agati V, Appel GB: HIV infection and the kidney. J Am Soc Nephrol 2005;16:2403–2411.

3 Gardenswartz MH, Lerner CW, Seligson OR, Zabetakis PM, Rotterdam H, Tapper ML, Michelis MF, Bruno MS: Renal disease in patients with AIDS: a clinicopathologic study. Clin Nephrol 1984;21:197–204.

4 Soriano Rosas J, Avila- Casado MC, Carrera- González E, Chavez-Mercado L, Cruz Ortiz H, Rojo J: AIDS–associated nephropathy: 5-year retrospective morphologic analysis of 87 cases. Pathol Res Pract 1998;194:567–570.

5 Barisoni L, Bruggeman LA, Mundel P, D'Agati VD, Klotman PE: HIV-1 induces renal epithelial dedifferentiation in a transgenic model of HIV-associated nephropathy. Kidney Int 2000;58:173–181.

6 Barisoni L, Kriz W, Mundel P, D'Agati V: The dysregulated podocyte phenotype: a novel concept in the pathogenesis of collapsing idiopathic focal segmental glomerulosclerosis and HIV-associated nephropathy. J Am Soc Nephrol 1999;10:51–61.

7 Dickie P, Felser J, Eckhaus M, Bryant J, Silver J, Marinos N, Notkins AL: HIV-associated nephropathy in transgenic mice expressing HIV-1 genes. Virology 1991;185:109–119.

8 D'Agati V, Suh JI, Carbone L, Cheng IT, Appel G: The pathology of HIV nephropathy: a detailed morphologic and comparative study. Kidney Int 1989;35:1358–1370.

9 Itescu S, Brancato LI, Buxbaum I, Oregersen PK, Rizk CC, Croxson S, Solomon O, Winchester R: A diffuse infiltrative CD8 lymphocytosis syndrome in human (HIV) infection: a host immune response associated with HLADR5 Ann Intern Med 1990;12:3–10.

10 Itescu S, Brancato LI, Winchester R: A sicca syndrome in HIV infection: association with HLA-DR5 and CD8 lymphocytosis. Lancet 1989;2:466–468.

11 Itescu S, Rose 5, Dwyer E, Winchester R: Certain HLA-DR5 and -DR6 major histocompatibility complex class II alleles are associated with a CD8 lymphocytic host response to human immunodeficiency virus type I characterized by low lymphocyte viral strain heterogeneity and slow disease progression. Proc Natl Acad Sci USA 1994;91:11472–11476.

12 Bruggeman LA, Dikman S, Meng C, Quaggin SE, Coffman TM, Klotman PE: Nephropathy in human immunodeficiency virus-1 transgenic mice is due to renal transgene expression. J Clin Invest 1997;100:84–92.

13 Conaldi PG, Biancone L, Bottelli A, Wade-Evans A, Racusen LC, Boccellino M, Orlandi V, Serra C, Camussi G, Toniolo A: HIV-1 kills renal tubular epithelial cells in vitro by triggering an apoptotic pathway involving caspase activation and Fas upregulation. J Clin Invest 1998;102:2041–2049.

14 Albert I, Olausson E, AsJo B: Distinct replicative and cytopathic characteristics of human immunodeficiency virus isolates. J Virol 1988;62:4414–4419.

15 Cboyd NW, Moore BE: Spectrum of biological properties of human immunodeficiency virus (HIV-l) isolates. J Virol 1990;174:103–106.

16 Zhu T, Mo H, Wang N, Nam DS, Cao Y, Koup RA, Ho DD: Oenotypic and phenotypic characterization of HIV-1 patients with primary infection. Science 1993;261:1179–1181.

17 Karlsson-Parra A, Dimeney E, Fellstrom B, Klarkeskog L: HIV receptors (CD4 antigen) in normal human glomerular cells. N Engl J Med 1989;320:741–747.

18 Green DF, Resnick L, Bourgoignie JJ: HIV infects glomerular endothelial and mesangial but not epithelial cells in vitro. Kidney Int 1992;4l:956–960.

19 Alpers CE, McClure I, Bursten SL: Human mesangial cells are resistant to productive infection by multiple strains of human immunodeficiency virus types 1 and 2. Am J Kidney Dis 1992;19:126–130.

20 Nadasdy T, Hanson-Painton O, Davis LD, Miller KW, DeBault L, Burns D, Silva FG: Conditions affecting the immunohistochemical detection of HIV in fixed and embedded renal and nonrenal tissues. Mod Pathol 1992; 5:283–291.

21 Pardo V, Shapshak P, Yoshioka M, et al: HIV-associated nephropathy (HIVN): Direct viral invasion or indirect glomerular involvement (abstract). FASEB J 1991;5:907A.

22 Striker GE, Mannik M, Tung MY: Role of marrow-derived monocytes and mesangial cells in removal of immune complexes from renal glomeruli. J Exp Med 1979;149:127–136

23 Nagata M, Kriz W: Glomerular damage after uninephrectomy in young rats. II. Mechanical stress of podocytes as a pathway to sclerosis. Kidney Int 1992;42:148–160.

24 Clapham PR, Weber in, Whitby D, Mcintosh K, Daloleish AG, Maddon PT, Deen KC, SwetRW, Weiss RA: Soluble CD4 blocks the infectivity of diverse strains of HIV and SlY for T cells and monocytes but not for brain and muscle cells. Nature 1989;337:368–370.

25 Mckeating JA, Griffiths PD, Weiss RA: HIV susceptibility conferred to human fibroblasts by cytomegalovirus-induced Fe receptor. Nature 1990;343:659–661.

26 Raffanti S, Svennigsson A, Resnick L: HIV-l infection of CD4 negative cells via HTLV pseudoviruses. AIDS 1991;5:769–772.

27 Kajiyama W, Kopp JB, Marinos NJ, Klotman PE, Dickie P: Glomerulosclerosis and viral gene expression in HIV-transgenic mice. Kidney Int 2000;58:1148–1159.

28 Bruggeman LA, Dikman S, Meng Ch, Quaggin SE, Coffman TM, Klotman PE: Nephropathy in human immunodeficiency virus-1 transgenic mice is due to renal transgene expression. J Clinical Invest 1997;100:84–92.

29 Dantal J, Bigot E, Bogers W, et al: Effect of plasma protein adsorption on protein excretion in kidney-transplant recipient with recurrent nephrotic syndrome. N Engl J Med 1994;30:7–14.

30 Cheung PK, Klok PA, Bakker WW: Minimal change-like alterations induced by a human plasma factor. Nephron 1996;74:586–593.

31 Marras D, Bruggeman LA, Gao F, et al: Replication and compartmentalization of HIV-1 in kidney epithelium of patients with HIV-associated nephropathy. Nat Med 2002;8:522–524.
32 Ray PE, Liu XH, Henry D, Dye L, Xu L, Orenstein JM, Schutzbank TE: Infection of human primary renal epithelial cells with HIV-1 from children with HIV-associated nephropathy. Kidney Int 1998:53;1217–1229
33 Kopp JB, Klotman ME, Adler SH, Bruggeman LA, Dickie P, Marinos NJ, Eckhaus M, Bryant JL, Notkins AL, Klotman PE: Progressive glomerulosclerosis and enhanced renal accumulation of basement membrane components in mice transgenic for HIV-1 genes. Proc Natl Acad Sci USA 1992;89:1577–1582.
34 Kemeny E, Durmuller U, Nickeleit V, Gudat F, Mihatsch MJ: Distribution of podocyte protein (44 kD) in different types of glomerular diseases. Virchows Arch 1997;431:425–430.
35 Kreidberg JA, Sariola H, Loring JM, Maeda M, Pelletier J, Housman D, Jaenisch R: WT-1 is required for early kidney development. Cell 1993;74:679–691.
36 Thomas PE, Wharram BL, Goyal M, Wiggins JE, Holzman LB, Wiggins RC: GLEPP-1, a renal glomerular epithelial (podocyte) membrane protein tyrosine phosphatase: identification, molecular cloning, and characterization in rabbit. J Biol Chem 1994;269: 19953–19962.
37 Mundel P, Gilbert P, Kriz W: Podocytes in glomerulus of rat kidney express a characteristic 44 kD protein. J Histochem Cytochem 1991:39:1047–1056.
38 D'Agati V: The many masks of focal segmental glomerulosclerosis. Kidney Int 1994;46:1223–1241.
39 Rennke HG: How does glomerular epithelial cell injury contribute to progressive glomerular damage? Kidney Int 1994;45:S58–S63.
40 Kriz W: Progressive renal failure-inability of podocytes to replicate and the consequences for development of glomerulosclerosis. Nephrol Dial Transplant 1996;11:1738–1742.
41 Kriz W, Gretz N, Lemley KV: Progression of glomerular diseases: is the podocyte the culprit? Kidney Int 1998;54:687–697.
42 Sharif K, Goyal M, Kershaw D, Kunkel R, Wiggins R: Podocyte phenotypes as defined by expression and distribution of GLEPP-1 in the developing glomerulus and in nephrotic glomeruli from MCD, CNF, and FSGS: a dedifferentiation hypothesis for the nephrotic syndrome. Exp Nephrol 1998;6:234–244.
43 Kriz W, Hähnel B, Rösener S, Elger M: Long-term treatment of rats with FGF-2 results in focal segmental glomerulosclerosis. Kidney Int 1995;48:1435–1450.
44 Shankland SJ: Cell-cycle control and renal disease. Kidney Int 1997;52:294–308.
45 Kitamoto Y, Tokunaga H, Tomita K: Vascular endothelial growth factor is an essential molecule for mouse kidney development: glomerulogenesis and nephrogenesis. J Clin Invest 1997;99:2351–2357.
46 Weiss MA, Daquioag E, Margolin EG, Pollak VE: Nephrotic syndrome, progressive irreversible renal failure, and glomerular 'collapse': a new clinicopathologic entity? Am J Kidney Dis 1986;7:20–28.
47 Avila-Casado C, Galvanek EG, Rennke HG: Collapsing Glomerulopathy (CG) developing in the renal allograft (abstract). J Am Soc Nephrol 1994;5:994.
48 Stokes MB, Davis CL, Alpers CE: Collapsing glomerulopathy in renal allografts: a morphological pattern with diverse clinicopathological associations. Am J Kidney Dis 1999;33:658–666.
49 Avila-Casado MC, Perez-Torres I, Auron A, Soto V, Fortoul TI, Herrera-Acosta J: Proteinuria in rats induced by serum from patients with collapsing glomerulopathy. Kidney Int 2004;66:133–143.
50 Artero ML, Sharma R, Savin VI, et al: plasmapheresis reduces proteinuria and serum capacity to injure glomeruli in patients with recurrent focal glomerulosclerosis. Am J Kidney Dis 1994;23:574–581.
51 Gornall AG, Bardawill CJ, Davis M: Determination of serum proteins by means of Biuret reaction. J Biol Chem 1949;177:751–766.
52 Smith SR, Suetkey LP, Dennis VA: Racial differences in the incidence and progression of renal diseases. Kidney Int 1991;40:815–822.
53 Avila-Casado C, Perez-Torres I, Bochicchio T, Torres-Rosales A, Lopez E, Niño JA, Fortoul TI: Clinical characteristics and outcomes in FSGS patients with circulating factors (abstract). J Am Soc Nephrol 2007;18:565A.

M.C. Avila-Casado, MD, PhD
Chief Department of Pathology,
Instituto Nacional de Cardiologia Ignacio Chavez
Juan Badiano 1, Col. Seccion XVI
Mexico City, 14050 (Mexico)
Tel. +52 55 55 73 2911, Ext. 1217, Fax +52 55 5606 1422,
E-Mail mcavilacasado@gmail.com

Herrera GA (ed): Experimental Models for Renal Diseases: Pathogenesis and Diagnosis.
Contrib Nephrol. Basel, Karger, 2011, vol 169, pp 286–296

Acute Kidney Injury: Lessons from Experimental Models

Samuel N. Heyman[a] · Christian Rosenberger[b] · Seymour Rosen[c]

[a]Department of Medicine, Hadassah Hospital, Mt. Scopus and the Hebrew University Medical School, Jerusalem, Israel; [b]Nephrology and Renal Transplantation, Charité Universitaetsmedizin, Berlin, Germany; [c]Department of Pathology, Beth Israel Deaconess Medical Center and Harvard Medical School, Boston, Mass., USA

Abstract

For decades severe tubular necrosis has been the hallmark of experimental models of acute renal failure (ARF), such as prolonged ischemia and reflow. This fits well with the still widely used traditional clinical term 'acute tubular necrosis'. Nevertheless, the rareness of tubular necrosis in human kidney biopsies in the background of hypoxic, toxic and septic AKI led to the adoption of a new term, 'acute kidney injury' (AKI), which refers to such clinical scenarios irrespective of the renal morphology. Indeed, experimental AKI models, which have more limited acute renal parenchymal compromise, underscore the focal and regional tissue injury patterns that range from adaptive stress response, through cellular dysfunction, apoptotic cell death and frank acute tubular necrosis. Such stress- and injury patterns, short of necrosis, may go unnoticed morphologically and even functionally, and may even confer resistance to subsequent insults. Herein we describe the spectrum of what we call 'sublethal injury', referring to a condition, which by itself is insufficient to produce cellular death, but may or may not produce organ failure. Such sublethal injury can be detected by overt morphological changes, by the upregulation of cell survival factors, and by recently developed biomarkers and imaging techniques of organ physiology and dysfunction. The use of combined sublethal insults in animal models is an attempt to replicate the clinical situation of comorbidities, a circumstance which underlies most situations of AKI. This review will discuss the definition of such sublethal injury and the modes of its detection.

The hallmark and essential end point of the most widely used experimental models of acute kidney injury (AKI, see definition below) is the development of clear-cut, usually widespread renal tubular cell death. This fits well with the traditional clinical term 'acute tubular necrosis' (see definition below), which is still widely used. Indeed, as reviewed elsewhere in depth [1, 2], for the sake of convenience, simplicity, reproducibility and consistency, many experimental models of hypoxic or toxic AKI apply drastic renal insults, such as protracted warm ischemia and reflow, or the administration of mega-doses of nephrotoxins. Such models often culminate with extensive tubular necrosis. Nevertheless, the little we know about the human morphology of AKI is quite confusing and does not fit well with this type of change. Indeed, tubular necrosis is rarely seen in kidney biopsies [3]. One might speculate whether this is the true phenotype or a mistake due to timing and sampling.

If tubular necrosis is rather the exception than the rule in human AKI, are we using the wrong experimental models? For decades, severe tubular necrosis has been the hallmark of experimental AKI, with numerous interventions being able to ameliorate it. Yet, none of these interventions has reached clinical application [3–5].

The three terms ARF, acute tubular necrosis and AKI are often used as synonyms. They may be regarded as a mere evolution of semantics across different nephrology eras, but initially each definition has been assigned to a distinct clinical-pathologic syndrome, as defined below. For simplicity, only the term AKI will be used in this manuscript.

Though often overlooked, sublethal renal cell injury may play an important role in human AKI. Some animal models of AKI reveal a spectrum of conditions that might be considered as sublethal injury patterns (table 1). In this review we discuss possible definitions, detection methods and functional impact of sublethal renal cell injury, and its relevance to the experimental and human AKI syndrome.

Definitions

Hypoxia

Within the kidney there are regions of poor oxygen supply [6]. This is a tribute to the physiologic urine concentration process, requiring intense oxygen consumption for tubular transport, and sparse regional blood supply [6]. Oxygen homeostasis is preserved due to fine tuning of vascular tone and oxygen consumption for tubular transport. It is a matter of debate whether (in the absence of demonstrable cell injury) such renal zones with low oxygen tension (like the papilla, the outer medulla and medullary rays) can be regarded as 'physiologically hypoxic' or not. Hypoxia might be defined as a mismatch between oxygen supply and demand, leading to a regional sub-physiologic concentration of oxygen. Some researchers suggest that rising tissue levels of NADH/NADPH would depict cellular oxygen insufficiency [7].

Historically, AKI is subdivided into hypoxic and non-hypoxic (toxic) forms. The recent demonstrations of renal hypoxia in various forms of so-called toxic AKI [8–10] prompt us to revise this view. Indeed, cell toxicity can lead to increased renal oxygen consumption or to reduced regional blood flow, with subsequent regional intrarenal hypoxia. Thus, hypoxia seems to be a common phenomenon in AKI, probably irrespective of its genesis.

Acute Renal Failure

Classically, acute renal failure was considered a rapid fall in glomerular filtration rate, although the required extent of renal functional impairment has never been defined, and more than 30 interpretations have been used [11]. The term has been replaced by AKI (see below).

Acute Kidney Injury

AKI is a clinical condition clearly defined as either a rise in serum creatinine by 50% or 0.3 mg/dl within 48 h, or by urinary output less than 0.5 ml/kg × h during a 6-hour period (provided proper hydration). AKI has been coined by critical care and nephrology disciplines, generating the RIFLE and AKIN criteria [12, 13]. Stages termed 'risk', 'injury' and 'failure' are based on the extent of serum creatinine and urine output changes. It is noteworthy that neither markers of acute tubular dysfunction nor morphologic changes are included in the AKI definition. Probably, the renal pathology of AKI ranges from normal appearance over sublethal injury (see below) and apoptosis to tubular necrosis.

AKI encompasses a spectrum of etiologies, including pre-and post-renal causes and diverse renal hemodynamic, glomerular and tubulointerstitial pathophysiologies. Our discussion is restricted to the hypoxic, toxic and septic forms which are usually considered 'intrinsic' AKI and lack specific therapy.

Acute Tubular Necrosis

Acute tubular necrosis may simply be defined as the appearance of necrotic tubules. It has been thought that tubular necrosis and AKI go in

Table 1. Commonly used experimental models of acute renal failure

Model Class	Model	Principal location	Comments
Whole kidney ischemia	warm ischemia-reflow	OS	– usually prolonged ischemic periods – consistent outcome – extensive acute tubular necrosis
	cold ischemia reflow	IS, IM	– replicates renal transplantation – difficult to perform in rodents
	isolated perfused kidneys	IS	– ex vivo whole organ model – good example for sublethal injury – allows variation of oxygen supply and consumption – allows online functional data
Regional renal ischemia/ hypoxia	distal nephron models	IS, IM	– combined sublethal perturbations – induce regional renal hypoxia – selective medullary damage
Toxic-(hypoxic) models	cisplatin nephropathy	OS	– end points usually assessed by days 3–5 (extensive injury) – earlier time points may reveal sublethal changes
	glycerol-induced rhabdomyolysis	C	– combined hypoxic/toxic and pre-renal failure
Sepsis models	endotoxin (LPS) administration	IS	– low-dose LPS + perturbations causing hypoxia – mostly sublethal damage
	cecal ligation-perforation	C	– complex model – standardized cecal pole ligation and perforation – additional fluid resuscitation and antibiotics needed to prolong survival and allow renal injury to develop

Injury patterns, location and extent largely depend on the severity and duration of the applied insult, where the more intense and protracted stress produces more acute tubular necrosis and less sublethal injury. Sublethal injury of different types may occur, depending on the experimental set-up and cell type involved. Proximal tubular injury includes cell swelling and diminution of brush border (typically in ischemia), vacuolar changes (sepsis), autophagy (cisplatin) and basolateral membranal invagination (radiocontrast-induced). Mitochondrial swelling and nuclear pyknosis characterizes medullary thick limbs sublethal injury. Collecting ducts disclose cell swelling, vacuolar formation and loss of cellular attachments. C = Cortex; OS = outer stripe of the outer medulla; IS = inner stripe of the outer medulla; IM = inner medulla.

parallel. Nevertheless, AKI might exist without overt morphologic changes [4, 5]. In fact, literal/extensive tubular necrosis in human biopsies is extremely rare [4, 5]. It is usually evident after a profound catastrophe, in which blood flow literally stops to multiple organs, followed by resuscitation and reflow [5]. In such a situation, it is probably associated with inflammatory 'cross-talk' between injured organs as a part of multi-organ failure [14]. In contrast, 'minimal/limited parenchymal injury, compatible with acute tubular necrosis' is a common pathological interpretation and diagnosis in many cases of human AKI [5]. Unfortunately, biopsies from human AKI are rare, mostly from later disease stages, and for safety reasons usually miss deep medullary regions.

Apoptosis

Apoptosis is a distinct form of a controlled, energy-consuming programmed cell death, often noticed in compromised human kidneys [15]. The absence of an accompanying inflammatory response makes it a favorable pattern of cell involution, which is part of normal tissue renewal. In contrast with necrosis, apoptosis requires residual cellular function and energy. Apoptosis is enhanced in various forms of experimental AKI [16–20]. Theoretically, sepsis would be a case for apoptosis triggered by membrane death receptors. However, tubular cell apoptosis was not demonstrated in a mice sepsis model [21].

Sublethal Cell Injury

To the best of our knowledge, there is no acknowledged definition of sublethal injury in the pathophysiology of AKI. Usually the term means any kind of damage short of cell death, where both morphologic and functional changes may occur. Other disciplines, e.g. cardiology and neurology, have already included sublethal tissue injury into everyday clinical practice. Hibernating or stunned myocardium [22, 23] and the penumbra at the border of ischemic brain lesions [24] are prime examples.

Morphology of Sublethal Injury in the Kidney

Typical sublethal tubular morphologic changes are loss of brush border, cell and mitochondrial swelling (fig. 1a), cytoplasmic vacuolization, and nuclear pyknosis [25]. Some of these changes were shown to be reversible in hypoxic isolated perfused kidneys, following the restoration of oxygen supply [25]. As morphologic changes evolve over time, progression to apoptosis and frank necrosis or resolution of damage can occur. Cell damage of different categories can coexist within the kidney. For instance, in a model of hypoxic medullary injury induced by iodinated radiologic contrast media (CM) and other perturbations, focal-to-extensive tubular necrosis [26, 27] may co-exist with apoptosis and sublethal injury [8, 16, 17, 28]. Depending on the degree of cell attachment to the basal membrane, damaged tubular cells can be shed into the urine. At least some of these cells are viable, as has been shown by cell culture [29]. Elimination of injured cells from the tubular lining epithelium (single cell dropout) and their replacement by regenerating cells can lead to the illusion of an intact parenchyma. Finally, autophagy represents another form of sublethal cell injury, reflecting intracellular degredation process of damaged cytoplasmic components [30].

Methods for Detecting Sublethal Renal Tubular Injury

Sublethal tubular injury might not be detectable on routine paraffin sections, but may require semi-thin plastic sections or electron microscopy. Since even such sophisticated methods can fail, there is a need for additional sublethal injury markers. Several candidates are under investigation, among which are the following:

- Pimonidazole (PIM), initially designed as a radiosensitizer, binds to tissues at a pO_2 below 10 mm Hg. Following delivery in vivo, PIM can be visualized at the single cell level with help of commercially available antibodies [31, 32]. Hypoxic or toxic/hypoxic rodent AKI models have clearly shown that PIM accumulates in the cell types at risk for injury. Moreover, PIM accumulation precedes overt cellular damage. This has been demonstrated in warm ischemia and reperfusion (mostly in the outer stripe of outer medulla) [3], in glycerol-induced rhabdomyolysis (both in the cortex and medulla) [9] and in so-called distal tubular models [3, 8]. The latter can be induced by various combinations of the following: CM, inhibition of prostaglandin or nitric oxide (NO)-synthesis, heart failure, endotoxemia, diabetes, pre-existing kidney disease and the acute enhancement of glomerular filtration rate [26–28, 33].
- Regional renal hypoxia is a common denominator of various AKI types, and hypoxia rapidly leads to activation of hypoxia-inducible

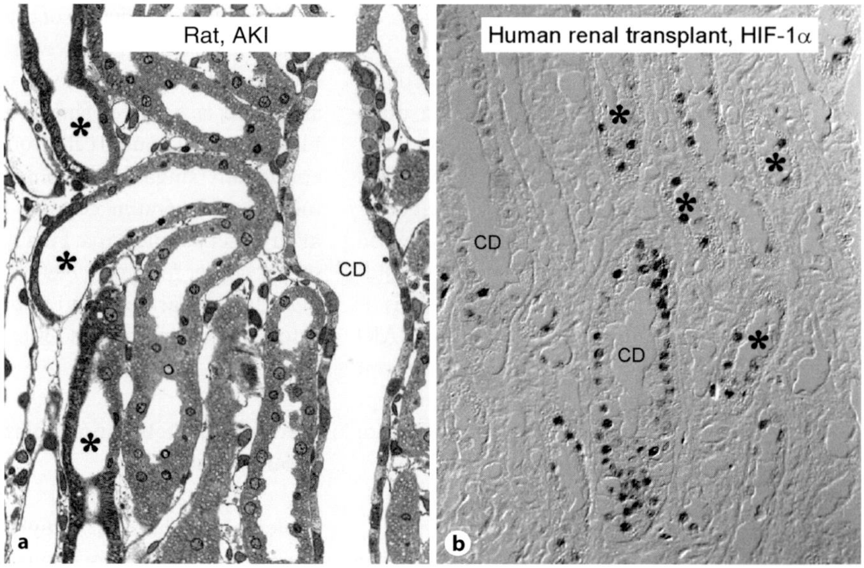

Fig. 1. Two examples of sublethal injury. **a** Isolated rat kidney perfused with red blood cell-free Krebs Henseleit solution, supplemented with albumin and amino acids. Under these conditions, in the renal medulla, hypoxic injury becomes manifest as a gradient. mTALs adjacent to collecting ducts (CD) show sublethal injury (mitochondrial swelling) in contrast to the preserved cells of the mTAL (mitochondria barely perceptible as dark rods) immediately adjacent to the vasa recta (*left). If an oxygen carrier is added to the perfusate, this sublethal injury pattern is reversed. Magnification: 1,000×. **b** Renal allograft protocol biopsy (human) done 2 weeks after transplantation (normal creatinine). Prominent HIF-1α nuclear staining is seen in collecting ducts and adjacent tubules (*mTALS). Three months after transplantation, such staining is not seen, suggesting that hypoxic injury has diminished and there is normal hemodynamic regulation. Magnification: 1,000×.

transcription factors (HIFs) [34–37]. HIFs seem to be a useful marker of early hypoxic cell injury, but some details deserve attention: (i) different cell types vary significantly in their ability to activate HIF, (ii) HIF activation is transient and (iii) a window of opportunity seems to exist for HIF activation, i.e. HIFs are activated in moderate hypoxia when cell survival can likely be achieved, but not in severe or protracted hypoxia. For these reasons, HIFs likely have a high positive predictive value and a low negative predictive value for detecting cellular hypoxia in vivo. HIF-1α activation has been demonstrated in rodent AKI models [8, 9, 38, 39], as well as in human renal transplants shortly after engraftment [40], as illustrated in figure 1b.

- Non-invasive blood and/or urinary biomarkers of kidney injury are a rapidly evolving field. Cytokines (IL-6, IL-8, IL-18), neutrophil gelatinase-associated lipocalin, N-acetyl-β-(D)-glucosaminidase activity and kidney injury molecule 1 [41, 42] are currently under intensive clinical investigation. Some might serve as cell-specific indicators of AKI, such as kidney injury molecule 1 for proximal tubules or neutrophil gelatinase-associated lipocalin for medullary thick ascending limbs and collecting ducts. Correlations with morphology, as well as the proper thresholds for test validity need to be determined in experimental AKI models.
- Blood oxygen level-dependent magnetic resonance imaging (BOLD-MRI) detects

Table 2. End points in experimental AKI

Method	Invasive	Discrimination between lethal/ sublethal injury	Serial evaluation	Resolution
Renal glomerular function tests ($Crea_{Pl}$, $Urea_{Pl}$, $Clear_{Crea}$)	–	–	+	whole organ
Tubular function tests (TR_{Na}, FE_K)	–	–	+	whole organ
Morphology (conventional, semi-thin sections, electron microscopy)	+	+	–	single cell
HIF-immunohistochemistry	+	+	–	single cell
Apoptosis immunostaining (e.g.TUNEL)	+	+	–	single cell
BOLD-MRI	–	–	+	renal zone
Sodium MRI	–	–	+	renal zone
Confocal video-microscopy	+	+?	+?	single cell?
Plasma and urine biomarkers (NGAL, KIM-1, IL-6, cystatin-C)	–	+?	+	whole organ/ nephron segment?

$Clear_{Crea}$ = Creatinine clearance; $Crea_{Pl}$ = plasma creatinine; FE_K= fractional excretion of potassium; GFR = glomerular filtration rate; KIM-1 = kidney injury molecule 1; NGAL = neutrophil gelatinase-associated lipocalin; TR_{Na} = tubular sodium reabsorption; TUNEL = TdT-mediated dUTP-biotin nick end labeling; $Urea_{Pl}$ = plasma urea.

increased deoxyhemoglobin concentration within hypoxic tissues. This noninvasive test has proven sensitive for rodent and human AKI [43–47].

- Na MRI, recently used for the detection of evolving AKI in rodents, disclosed an early tubular functional loss, leading to the disappearance of the physiologic corticomedullary sodium gradient [48, 49]. This is an early phenomenon in AKI, when medullary tubular damage is still focal and limited [48]. Conceivably, Na MRI can detect sublethal and potentielly reversible tubular injury. Indeed, in short-term warm ischemia and reflow, characterized by subethal tubular injury, a transient change of corticomedullary sodium gradient was noted [49]. This technology has already been extended to the human native- and transplanted kidney [50, 51], and represents a promising noninvasive method for early AKI detection.
- Metabolic changes likely take place in the hypoxic kidney. Such changes have been well documented in the hypoxic myocardium (increased glucose utilization and reduced fatty acid oxidation [52, 53]), but are speculative with regard to the kidney. Nevertheless, since so-called metabolic imaging is already available for the myocardium, it could become a complementary detection technique for AKI (table 2).

Combination of Different Sublethal Injuries

In humans the likelihood of developing AKI during an acute insult (such as CM) increases with the number of pre-existing risk factors like diabetes, age, chronic renal failure, anemia, effective blood volume depletion and the co-administration of other nephrotoxic agents, like nonsteroidal anti-inflammatory drugs (NSAID) [54]. Most of these risk factors disturb renal oxygen homeostasis [55, 56]. This was documented by oxygen microelectrodes or with PIM and HIF immunostaining in animal studies, and with BOLD-MRI in both humans [46] and experimental settings [54]. Subclinical sublethal tubular injury probably occurs in these conditions, and it is tempting to assume that the clinical syndrome of AKI can derive from a combination of such sublethal insults.

In the rat, single insults of either CM, the NO synthase inhibitor L-NAME or the NSAID indomethacin probably lead to sublethal cell damage (as suggested by normal renal morphology and function, but presence of HIF-1α and PIM identifies the injury [8]). By contrast, all three interventions together lead to tubular necrosis and reduced glomerular filtration rate [8].

Other experimental examples for combined renal injury are low-dose gentamicin nephrotoxicity in the setting of acute renal hypoperfusion [57], endotoxin injections potentiating experimental ischemic or hypoxic injury [58, 59], subclinical gentamicin- [60] or cisplatin toxicity augmenting LPS responsiveness [61], or renal subclinical compromise, induced by heart failure [62] or myoglobinuria [63], and augmented by prostaglandin- or NO-synthesis inhibition. Other reports have shown that mice with AKI resulting from ischemia-reperfusion, myoglobinuria or urinary tract obstruction have marked intrarenal cytokine (TNF-α; IL-10), chemokine (MCP-1) and NO production elevations when challenged with specific Toll-like receptor ligands (endotoxin, lipoteichoic acid) [64].

The isolated perfused kidney (fig. 1a) has provided insights into the stepwise evolution from sublethal to lethal tubular damage. With an oxygenated but red blood cell-free perfusate, medullary thick ascending limbs develop sublethal injury (mitochondrial swelling) [25]. This injury pattern is reversible upon addition of red blood cells, or by inhibition of oxygen consumption for tubular transport. On the other hand, sublethal injury progresses to necrosis, with additional conditions that either enhance tubular oxygen consumption (CM, amphothericin, renal hypertrophy) or cause renal vasoconstriction (NSAID, NO synthase inhibition) [6, 65].

Possible Advantage of Sublethal Injury

In a thorough review, Bonventre [66] examined the resistance to further injury induced by a number of insults, such as uranium nephrotoxicity diminished by previous uranium administration, glycerol-induced acute injury protecting against subsequent glycerol reinjection or mercuric chloride administration, and gentamicin toxicity ameliorated by previous gentamicin exposure. Multiple studies of short-term ischemic preconditioning in the mouse and rat demonstrate both short- and long-term protection (up to 15 days) against repeated insults. Zager et al. [67, 68] actually were the first to examine the effects of prior exposure to ischemia on the effects of subsequent ischemic injury.

It is important to note that this sort of ischemic preconditioning occurs in very short-term warm ischemia reflow, a situation which has very limited morphological consequences [69], but it does produce downregulation of certain transport systems [70]. If the ischemic time is extended, there is still protection from acute hypoxic or toxic insults, but at a cost of extensive injury and repair from the previous challenge [71].

Interestingly, studies of short-term ischemia in large animals, such as the dog and pig [72–74], have failed to replicate the protection shown in rodents. The reason for such species difference has yet to be defined.

Mechanisms of Adaptive Responses to Sublethal Injury

A number of mechanisms have been proposed to explain the adaptive response to sublethal injury, including altering the pattern of stress-related kinase activation (reducing cytokine/chemokine elaboration and/or the expression of adhesion molecules, thereby blunting inflammatory processes) and increasing the levels of heat shock proteins (stabilizing actin cytoskeleton) [66]. Recently, such cytoprotection has been shown to be transferable by serum administration [75].

Sublethal injury has also been shown to induce adaptational changes in the cellular plasma membrane [76] which confer resistance to various nephrotoxic, ischemic and obstructive injuries. While several mechanisms may contribute to this state, an upregulation of membrane cholesterol content seems to be pivotal in this protection [76]. These changes lead to a decrease in membrane fluidity, which by itself is cytoprotective [76]. In addition, there is an injury-evoked perturbation of cholesterol cycling within the cell, culminating in the excess production of cholesterol esters [76]. The latter may then be directly cytoprotective by positively influencing mitochondrial energy production and decreasing the generation of oxygen free radicals [77, 78].

As described above, sublethal injury likely is a case for HIF activation. HIFs transactivate over 100 target genes that regulate a multitude of biologic processes like cell metabolism, cell cycle, angiogenesis, vascular tone, erythropoiesis, scavenging of free radicals, inflammation, immune responses and apoptosis [79]. Many HIF target genes have cell protective properties. HIF activation by means of a hypoxic chamber, pharmacologic blockade of HIF proteolysis or transgenic models improves AKI due to warm ischemia reflow [39, 80], isolated kidney perfusion [81], cisplatin [82] or cold storage and transplantation [83]. However, the HIF target genes involved in AKI protection are insufficiently clear. Possible candidates are erythropoietin (EPO) and heme oxygenase-1 (HO-1). EPO ameliorates AKI induced by warm ischemia and reflow and CM [84–86]. HO-1 degrades heme to iron, biliverdin and carbon monoxide (CO) [87]. The latter two molecules have vasorelaxant, anti-apoptotic and anti-inflammatory effects. However, the process of HO-1 induction may be species-dependent and is complex, e.g. factors other than HIF may be involved as well, such as Nrf2, NF-κB and inactivation of the repressor protein (Bach1) [87].

Conclusions

AKI is a clinically defined syndrome with largely different etiologies and renal pathologic changes that have not been well characterized. However, in contrast with widespread belief, existing data support the view that acute tubular necrosis rarely occurs in AKI. Instead, it appears that sublethal renal tubular injury may lead to functional impairment. Furthermore, multiple sublethal injuries will produce the clinical syndrome (comorbidities). Paradoxically, sublethal injury may elicit adaptive responses which confer resistance to subsequent insults, a phenomenon known as hypoxic or ischemic preconditioning. Deciphering the mechanism involved in tissue preconditioning likely would help to develop strategies for the prevention/treatment of AKI. Animal models of AKI mostly rely on extensive single insults which reproducibly cause marked tubular necrosis. More recent experimental models offer the possibility to study sublethal injury in a variety of ways and are more analogous to the human situation. AKI injury markers are a rapidly evolving field whose usefulness will probably depend on their ability to detect sublethal injury in the early stages of AKI.

Acknowledgements

This work was supported in part by the Harvard Medical Faculty Physicians at Beth Israel Deaconess Medical Center, Boston, Mass., USA, and by the Israeli Science Foundation (1473/08).

References

1 Heyman SN, Rosen S, Rosenberger C: Critical assessment of experimental models of acute renal failure; in Ronco C, Bellomo R, Kellum J (eds): Critical Care Nephrology, ed 2. Philadelphia, Saunders/Elsevier, 2009, pp 237–250.

2 Heyman SN, Rosen S, Rosenberger C. Animal models of renal dysfunction: acute kidney injury. Expert Opin Drug Discov 2009;4:629–641.

3 Heyman SN, Rosenberger C, Rosen S. Experimental ischemia-reperfusion – biases and myths: the proximal vs. distal hypoxic tubular injury debate revisited. Kidney Int 2010;77:9–16.

4 Rosen S, Heyman SN: Difficulties in understanding human 'acute tubular necrosis': limited data and flawed animal models. Kidney Int 2001;60:1220–1224.

5 Rosen S, Stillman IE: Acute tubular necrosis is a syndrome of physiologic and pathologic dissociation. J Am Soc Nephrol 2008;19:871–875.

6 Brezis M, Rosen S: Hypoxia of the renal medulla – its implications for disease. N Engl J Med 1995;332:647–655.

7 Clanton TL: Hypoxia-induced reactive oxygen species formation in skeletal muscle. J Appl Physiol 2007;102:2379–2388.

8 Rosenberger C, Heyman SN, Rosen S, et al: Up-regulation of HIF in experimental acute renal failure: evidence for a protective transcriptional response to hypoxia. Kidney Int 2005;67:531–542.

9 Rosenberger C, Goldfarb M, Shina A, Bachmann S, Frei U, Eckardt KU, Schrader T, Rosen S, Heyman SN: Evidence for sustained renal hypoxia and transient hypoxia adaptation in experimental rhabdomyolysis-induced acute kidney injury. Nephrol Dial Transpl 2008;23:1135–1143.

10 Yasuda H, Yuen PS, Hu X, Zhou H, Star RA: Simvastatin improves sepsis-induced mortality and acute kidney injury via renal vascular effects. Kidney Int 2006;69:1535–1542.

11 Kellum JA, Levin N, Bouman C, Lameire N: Developing a consensus classification system for acute renal failure. Curr Opin Crit Care 2002;8:509–514.

12 Bellomo R, Ronco C, Kellum JA, Mehta RL, Palevsky P, ADQI Workgroup: Acute renal failure: definition, outcome measures, animal models, fluid therapy and information technology needs. The Second International Consensus Conference of the Acute Dialysis Quality Initiative(ADQI) Group. Crit Care 2004;8:R204-R212.

13 Bagshaw SM, George C, Bellomo R: A comparison of the RIFLE and AKIN criteria for acute kidney injury in critically ill patients. Nephrol Dial Transpl 2008;23:1569–1574.

14 Li X, Hassoun HT, Santora R, Rabb H: Organ crosstalk: the role of the kidney. Curr Opin Crit Care 2009;15:481–487.

15 Jaffe R, Ariel I, Beeri R, Paltiel O, Hiss Y, Rosen S, Brezis M: Frequent apoptosis in human kidneys after acute renal hypoperfusion. Exp Nephrol 1997;5:399–403.

16 Beeri R, Symon Z, Brezis M, Ben-Sasson SA, Baehr PH, Rosen S, Zager RA: Rapid DNA fragmentation from hypoxia along the thick ascending limb of rat kidneys. Kidney Int 1995;47:1806–1810.

17 Yokomaku Y, Sugimoto T, Kume S, et al: Asialoerythropoietin prevents contrast-induced nephropathy. J Am Soc Nephrol 2008;19:321–328.

18 Kindt N, Menzebach A, Van de Wouwer M, Betz I, De Vriese A, Conway EM: Protective role of the inhibitor of apoptosis protein, survivin, in toxin-induced acute renal failure. FASEB J 2008;22:510–521.

19 Weidemann A, Bernhardt WM, Klanke B, Daniel C, Buchholz B, Campean V, Amann K, Warnecke C, Wiesener MS, Eckardt KU, Willam C: HIF activation protects from acute kidney injury. J Am Soc Nephrol 2008;19:486–494.

20 Langworthy M, Zhou B, de Caestecker M, Moeckel G, Baldwin HS: NFATc1 identifies a population of proximal tubule cell progenitors. J Am Soc Nephrol 2009;20:311–321.

21 Yasuda H, Leelahavanichkul A, Tsunoda S, Dear JW, Takahashi Y, Ito S, Hu X, Zhou H, Doi K, Childs R, Klinman DM, Yuen PS, Star RA: Chloroquine and inhibition of Toll-like receptor 9 protect from sepsis-induced acute kidney injury. Am J Physiol Renal Physiol 2008;294:F1050–F1058.

22 Rahimtoola S: The hibernating myocardium. Am Heart J 1989;117:211–221.

23 Elsaesser A, Vogt AM, Nef H, Kostin S, Moellmann H, Skwara W, Bode C, Hamm, C, Schaper J: Human hibernating myocardium is jeopardized by apoptotic and autophagic cell death. J Am Coll Cardiol 2004;43:2191–2199.

24 Kato H, Kogure K: Biochemical and molecular characteristics of the brain with developing cerebral infarction. Cell Mol Neurobiol 1999;19:93–108.

25 Brezis M, Kopolovic J, Rosen S: Hypoxic injury to medullary thick ascending limbs in perfused rat kidneys: reversible and irreversible phases. J Electron Microsc Tech 1988;9:293–298.

26 Heyman SN, Brezis M, Reubinoff CA, et al: Acute renal failure with selective medullary injury in the rat. J Clin Invest 1988;82:401–412.

27 Agmon Y, Peleg H, Greenfeld Z, et al: Nitric oxide and prostanoids protect the renal outer medulla from radiocontrast toxicity in the rat. J Clin Invest 1994;94:1069–1075.

28 Heyman SN, Brezis M, Epstein FH, Spokes K, Silva P, Rosen S: Early renal medullary hypoxic injury from radiocontrast and indomethacin. Kidney Int 1991;40:632–642.

29 Racusen LC: Epithelial cell shedding in acute renal injury. Clin Exp Pharmacol Physiol 1998;25:273–275.

30 Bolisetty S, Traylor AM, Kim J, et al: Heme Oxygenase-1 Inhibits Renal Tubular Macroautophagy in Acute Kidney Injury. J AM Soc Nephrol 2010;21:1702–1712.

31 Terada N, Ohno N, Saitoh S, Ohno S: Immunohistochemical detection of hypoxia in mouse liver tissues treated with pimonidazole using 'in vivo cryotechnique'. Histochem Cell Biol 2007;128:253–261.

32 Rosenberger C, Rosen S, Paliege A, Heyman SN: Pimonidazole adduct immunohistochemistry in the rat kidney: detection of tissue hypoxia. Methods Mol Biol 2009;466:161–174.
33 Rosenberger C, Rosen S, Heyman SN: Renal parenchymal oxygenation and hypoxia adaptation in acute kidney injury. Clin Exp Pharmacol Physiol 2006;33:980–988.
34 Webb JD, Coleman ML, Pugh CW: Hypoxia, hypoxia-inducible factors (HIF), HIF hydroxylases and oxygen sensing. Cell Mol Life Sci 2009;66:3539–3554.
35 Maxwell PH: HIF-1's relationship to oxygen: simple yet sophisticated. Cell Cycle 2004;3:156–159.
36 Metzen E, Ratcliffe PJ: HIF hydroxylation and cellular oxygen sensing. Biol Chem 2004;385:223–230.
37 Semenza GL: Hydroxylation of HIF-1: oxygen sensing at the molecular level. Physiology 2004;19:176–182.
38 Rosenberger C, Shina A, Rosen S, Goldfarb M, Eckardt K, Heyman SN: Hypoxia inducible factors and tubular cell survival in isolated perfused kidneys. Kidney Int 2006;70:60–70.
39 Hill P, Shukla D, Tran MG, et al: Inhibition of hypoxia inducible factor hydroxylases protects against renal ischemia-reperfusion injury. J Am Soc Nephrol 2008;19:39–46.
40 Rosenberger C, Pratschke J, Rudolph B, et al: Immunohistochemical detection of hypoxia-inducible factor-1alpha in human renal allograft biopsies. J Am Soc Nephrol 2007;18: 343–351.
41 Molitoris BA, Melnikov VY, Okusa MD, Himmelfarb J: Technology insight: biomarker development in acute kidney injury – what can we anticipate? Nat Clin Pract Nephrol 2008;4:154–165.
42 Nguyen MT, Devarajan P: Biomarkers for the early detection of acute kidney injury. Pediatr Nephrol 2008;23:2151–2157.
43 Prasad PV, Edelman RR, Epstein FH: Noninvasive evaluation of intrarenal oxygenation with BOLD MRI. Circulation 1996;94:3271–3275.
44 Ries M, Basseau F, Tyndal B, et al: Renal diffusion and BOLD MRI in experimental diabetic nephropathy. Blood oxygen level-dependent. J Magn Reson Imaging 2003;17: 104–113.
45 Hofman L: BOLD-MRI for the assessment of renal oxygenation in humans: acute effect of nephrotoxic xenobiotics. Kidney Int 2006;70:144–150.
46 Prasad P, Epstein FH: Changes in renal medullary pO_2 during water diuresis as evaluated by blood oxygenation level-dependent magnetic resonance imaging: effects of aging and cyclooxygenase inhibition. Kidney Int 1999;55:294–298.
47 Prasad PV, Priatna A, Spokes K, Epstein FH: Changes in intrarenal oxygenation as evaluated by BOLD MRI in a rat kidney model for radiocontrast nephropathy. J Magn Reson Imaging 2001;13:744–747.
48 Maril N, Margalit R, Rosen S, et al: Detection of evolving acute tubular necrosis with renal ^{23}Na MRI: studies in rats. Kidney Int 2006;69:765–768.
49 Atthe BK, Babsky AM, Hopewell PN, Phillips CL, Molitoris BA, Bansal N: Early monitoring of acute tubular necrosis in the rat kidney by ^{23}Na-MRI. Am J Physiol Renal Physiol 2009;297:F1288–F1298.
50 Maril N, Rosen Y, Reynolds GH, Ivanishev A, Ngo L, Lenkinski RE: Sodium MRI of the human kidney at 3 Tesla. Magn Reson Med 2006;56:1229–1234.
51 Rosen Y, Lenkinski RE: Sodium MRI of a human transplanted kidney. Acad Radiol 2009;16:886–889.
52 Miyabe H, Ohte N, Iida A, Narita H, Yoshida T, Kimura G: Evaluation of fatty acid beta-oxidation in patients with prior myocardial infarction in relation to myocardial blood flow, total oxidative metabolism, and left ventricular wall motion. Circ J 2005;69:1459–1465.
53 Sorrentino AR, Acampa W, Petretta M, Mainolfi C, Salvatore M, Cuocolo A: Comparison of the prognostic value of SPECT after nitrate administration and metabolic imaging by PET in patients with ischaemic left ventricular dysfunction. Eur J Nucl Med Mol Imaging 2007;34:558–562.
54 Heyman SN, Rosen S, Rosenberger C: Renal parenchymal hypoxia, hypoxia adaptation and the pathogenesis of radiocontrast nephropathy. Clinical J Am Soc Nephrol 2008;3:288–296.
55 Palm F, Cederberg J, Hansell P, Liss P, Carlsson PO: Reactive oxygen species cause diabetes-induced decrease in renal oxygen tension. Diabetologia 2003;46:1153–1160.
56 De Vriese AS, Stoenoiu MS, Elger M, Devuyst O, Vanholder R, Kriz W, Lameire NH: Diabetes-induced microvascular dysfunction in the hydronephrotic kidney: role of nitric oxide. Kidney Int 2001;60:202–210.
57 Zager RA: Gentamicin nephrotoxicity in the setting of acute renal hypoperfusion. Am J Physiol 1988;254:F574–F581.
58 Zager RA: *Escherichia coli* endotoxin injections potentiate experimental ischemic renal injury. Am J Physiol 1996;251:F988–F994.
59 Heyman SN, Rosen S, Darmon D, Goldfarb M, Bitz H, Shina A, Brezis M: Endotoxin-induced renal failure: II. A role for tubular hypoxic damage. Exp Nephrol 2000;8:275–282.
60 Zager RA: 'Subclinical' gentamicin nephrotoxicity: a potential risk factor for exaggerated endotoxin-driven TNF-alpha production. Am J Physiol Renal Physiol 2007;293:F325–F342.
61 Ramesh G, Zhang B, Uematsu S, et al: Endotoxin and cisplatin synergistically induce renal dysfunction and cytokine production in mice. Am J Physiol Renal Physiol 2007;293:F325–F332.
62 Goldfarb M, Abassi Z, Rosen S, Shina A, Brezis M, Heyman SN: Chronic heart failure predisposes to hypoxic outer medullary tubular injury: studies in rats. Kidney Int 2001;60:607–613.
63 Heyman SN, Rosen S, Fuchs S, Epstein FH, Brezis M: Myoglobinuric acute renal failure in the rat: a role for medullary hypoperfusion, hypoxia and tubular obstruction. J Am Soc Nephrol 1996;7:1066–1074.
64 Zager RA, Johnson AC, Hanson SY, Lund S: Acute nephrotoxic and obstructive injury primes the kidney to endotoxin-driven cytokine/chemokine production. Kidney Int 2006;69:1181–1188.
65 Heyman SN, Brezis M, Rosen S: The isolated perfused rat kidney model in experimental renal injury; in Bennett WM, DeBroe ME, Porter GA, Verpooten GA (eds): Clinical Nephrotoxins: Renal Injury from Drugs and Chemicals. Dordrecht, Kluwer Academic Publishers, 1998, pp 77–82.
66 Bonventre JV: Kidney ischemic preconditioning. Curr Opin Nephrol Hypertens 2002;11:43–48.

67 Zager RA, Baltes LA, Sharma HM, Jurkowitz MS: Responses of the ischemic acute renal failure kidney to additional ischemic events. Kidney Int 1984;26:689–700.
68 Zager RA, Iwata M, Burkhart KM, Schimpf BA: Post-ischemic acute renal failure protects proximal tubules from O_2 deprivation injury, possibly by inducing uremia. Kidney Int 1994;45:1760–1768.
69 Shanley PF, Rosen MD, Brezis M, et al: Topography of focal proximal tubular necrosis after ischemia with reflow in the rat kidney. Am J Pathol 1986;122:462–468.
70 Kwon TH, Frokiaer J, Han JS, et al: Decreased abundance of major Na(+) transporters in kidneys of rats with ischemia-induced acute renal failure. Am J Physiol Renal Physiol 2000;278:F925–F939.
71 Goldfarb M, Rosenberger C, Abassi Z, et al: Acute-on-chronic renal failure in the rat: functional compensation and hypoxia tolerance. Am J Nephrol 2006;26:22–33.
72 Orvieto MA, Zorn KC, Mendiola FP, et al: Ischemia preconditioning does not confer resilience to warm ischemia in a solitary porcine kidney model. Urology 2007;69:984–987.
73 Behrends M, Walz MK, Kribben A, et al: No protection of the porcine kidney by ischaemic preconditioning. Exp Physiol 2000;85:819–827.
74 Kosieradzki M, Ametani M, Southard JH, Mangino MJ: Is ischemic preconditioning of the kidney clinically relevant? Surgery 2003;133:81–90.
74 Reutzel-Selke A, Pratschke J, Martins PN, Denecke C, Jurisch A, Kotsch K, Pascher A, Neuhaus P, Tullius SG: Ischemic preconditioning produces systemic protective and adoptively transferable effects. Kidney Int 2008;74:622–630.
76 Zager RA, Johnson A: Renal cortical cholesterol accumulation is an integral component of the systemic stress response. Kidney Int 2001;60:2299–2310.
77 Zager RA, Johnson AC, Lund S: 'Endotoxin tolerance': TNF-alpha hyper-reactivity and tubular cytoresistance in a renal cholesterol loading state. Kidney Int 2007;71:496–503.
78 Nath KA: Renal response to repeated exposure to endotoxin: implications for acute kidney injury. Kidney Int 2007;71:477–479.
79 Ratcliffe PJ: Understanding hypoxia signalling in cells – a new therapeutic opportunity? Clin Med 2006;6:573–578.
80 Bernhardt WM, Câmpean V, Kany S, et al: Preconditional activation of hypoxia-inducible factors ameliorates ischemic acute renal failure. J Am Soc Nephrol 2006;17:1970–1978.
81 Rosenberger C, Rosen S, Shina A, Frei U, Eckardt KU, Flippin LA, Arend M, Klaus SJ, Heyman SN: Activation of hypoxia inducible factors (HIF) ameliorates hypoxic distal tubular injury in the isolated perfused rat kidney. Nephrol Dial Transplant 2008;23: 3472–3478.
82 Weidemann A, Bernhardt WM, Klanke B, Daniel C, Buchholz B, Campean V,Amann K, Warnecke C, Wiesener MS, Eckardt KU, Willam C: HIF activation protects from acute kidney injury. J Am Soc Nephrol 2008;19:486–494.
83 Bernhardt WM, Gottmann U, Doyon F, Buchholz B, Campean V, Schödel J, Reisenbuechler A, Klaus S, Arend M, Flippin L, Willam C, Wiesener MS, Yard B, Warnecke C, Eckardt KU: Donor treatment with a PHD-inhibitor activating HIFs prevents graft injury and prolongs survival in an allogenic kidney transplant model. Proc Natl Acad Sci USA 2009;106:21276–21281.
84 Goldfarb M, Rosenberger C, Shina A, et al: A role for erythropoietin in the attenuation of radiocontrast-induced acute renal failure in rats. Ren Fail 2006;28:345–350.
85 Yokomaku Y, Sugimoto T, Kume S, et al: Asialoerythropoietin prevents contrast-induced nephropathy. J Am Soc Nephrol 2008;19:321–328.
86 Johnson DW, Pat B, Vesey DA, et al: Delayed administration of darbepoetin or erythropoietin protects against ischemic acute renal injury and failure. Kidney Int 2006;69:1806–1813.
87 Nath KA: Heme oxygenase-1: a provenance for cytoprotective pathways in the kidney and other tissues. Kidney Int 2006;70:432–443.

Seymour Rosen
Department of Pathology, Beth Israel Deaconess Medical Center
330 Brookline Avenue
Boston, MA 02215 (USA)
Tel. +1 6176674344, Fax +1 6179755620, E-Mail srosen@caregroup.harvard.edu

Herrera GA (ed): Experimental Models for Renal Diseases: Pathogenesis and Diagnosis.
Contrib Nephrol. Basel, Karger, 2011, vol 169, pp 297–310

Pathogenesis of Tubular Interstitial Nephritis

Tetsuhiro Tanaka[a,b] · Masaomi Nangaku[b]

[a]Division for Health Service Promotion, University of Tokyo, [b]Division of Nephrology and Endocrinology, University of Tokyo School of Medicine, Tokyo, Japan

Abstract

Injury to the interstitium of the kidney is regarded as a common pathway leading to end-stage renal insufficiency, regardless of etiology. Tubular interstitial nephritis is characterized histologically by inflammatory changes in the tubulointerstitial compartment, such as interstitial edema, leukocyte infiltration, accumulation of extracellular matrix proteins, tubular dilation and atrophy. Acute interstitial nephritis is often associated with use of drugs, such as β-lactam antibiotics and non-steroidal anti-inflammatory drugs, and is likely mediated through allergic mechanisms. On the other hand, chronic progressive tubular interstitial nephritis has a much more diverse etiology, ranging from infection and drugs to immune-mediated, hematologic, metabolic and hereditary disorders. Experimental studies in the past decade have focused mainly on common factors and mechanisms underlying chronic tubulointerstitial injury, such as activation of peritubular fibroblasts, leukocyte infiltration, release of inflammatory cytokines and growth factors at affected regions, epithelial-mesenchymal transition of tubular epithelium, and apoptosis. The execution of each is mediated by a number of local stimuli, such as filtered albumin, chronic hypoxia and oxidative stress, in addition to cytokines and growth factors. This chapter provides an overview of acute and chronic tubular interstitial nephritis, according to clinical manifestations of the disease. It also provides insight into common pathways underlying chronic tubular interstitial nephritis based on recent advances in translational and experimental research.

The tubulointerstitial compartment comprises as much as 80% of the entire renal mass and is made up of cellular and extracellular matrix (ECM) components. In chronic kidney diseases, the extent of injury to this compartment is known to correlate with the degree of residual renal function, irrespective of etiology [1]. Although this relationship has been reconfirmed by several groups independently and is now universally accepted, the basis of this relationship is not fully understood. Two of the most plausible explanations for loss of glomerular clearance are so-called tubuloglomerular feedback and the formation of atubular glomeruli. Tubular atrophy increases fluid delivery to the macula densa and triggers a decrease in the glomerular filtration rate. In addition, a histological study using serial sections across entire glomeruli in chronically damaged kidneys revealed that chronic tubulointerstitial damage ultimately leads to tubular atrophy and the generation of atubular glomeruli [2]. Therefore, tubular interstitium is responsible not only for the regulation of fluids, electrolytes and acid-base homeostasis, but is also critically important for clearance of unwanted uremic toxin out of the organic body. With this in mind, much effort has been devoted to elucidating the pathologic features of tubular interstitial

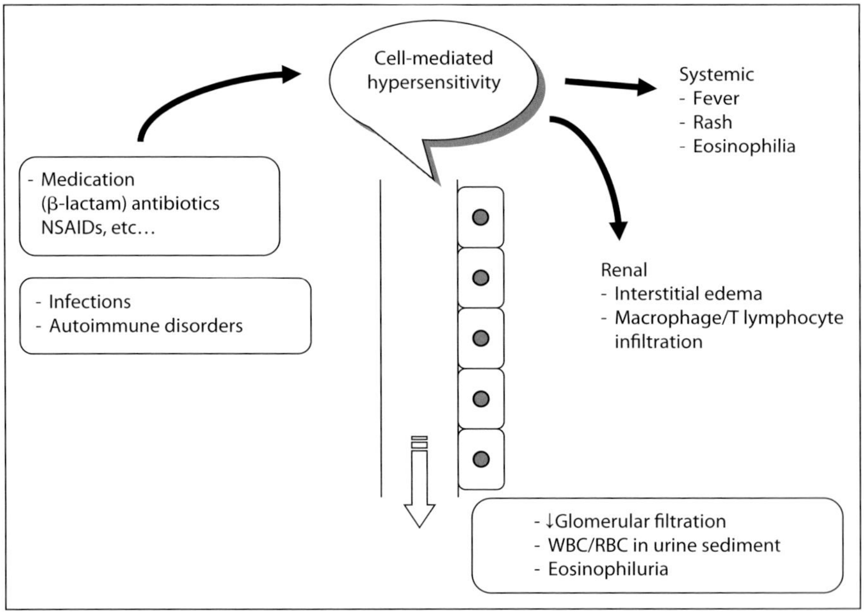

Fig. 1. Schematic diagram representing pathways of AIN.

nephritis in both human and animal studies. This chapter reviews the pathophysiology of tubular interstitial nephritis from the viewpoint of clinical practice and introduces some of the important molecular and cellular mechanisms that are common to varying etiologies and that characterize hypocellular tubulointerstitial fibrosis, as revealed by experimental research.

Renal disease can be classified as glomerular, vascular or tubulointerstitial in origin. Injury to the tubulointerstitial compartment may be primary or secondary to injury in other parts of the kidney. Pathological studies have clarified that the degree of injury to the tubulointerstitial compartment is a better indicator of residual renal function than that of other compartments, signifying its importance [3].

Although there is no generally accepted classification of tubular interstitial nephritis, it is practical to classify this entity according to etiology and clinical course. Tubular interstitial nephritis can be either acute or chronic. Acute tubular interstitial nephritis is often associated with acute renal failure, which develops over days to weeks because of a delayed hypersensitivity reaction to medication or other causes. On the other hand, chronic tubular interstitial nephritis develops over months to years and is associated with gradual loss of glomerular filtration rate as well as renal tubular dysfunction. The causes of chronic tubular interstitial nephritis range from infection and drugs to immune-mediated, hematologic, metabolic and hereditary diseases. However, common mechanisms as well as mediators appear to exist, such as epithelial-mesenchymal transition (EMT) and chronic hypoxia. Results from experimental studies suggest that these mechanisms are most likely to contribute to the development of hypocellular tubulointerstitial fibrosis, which will be discussed in the latter half of this chapter.

Acute Tubular Interstitial Nephritis

Acute tubulointerstitial nephritis (AIN) is frequently induced by medication and is associated

Table 1. Causes of chronic tubular interstitial nephritis

Category	Causes
Primary	
Infection	vesicoureteral reflux
Drugs	analgesics, lithium, calcineurin inhibitors, cisplatin, aristolochic acid (Chinese herb)
Toxins	lead, cadmium
Immune-mediated	sarcoidosis, Sjögren's syndrome, systemic lupus erythematosus, TIN with uveitis
Hematologic	multiple myeloma, lymphoproliferative disorders, light chain disease
Metabolic	hypokalemia, hypercalcemia, urate nephropathy
Genetic	cystinosis, Dent's disease, hyperoxaluria
Miscellaneous	Balkan nephropathy, radiation nephritis
Secondary	
Glomerular disease	
Vascular disease	
Obstructive nephropathy	

with autoimmune disorders and various infections, such as *Legionella,* leptospirosis and streptococcal organisms (fig. 1). Classically, AIN was most closely related to the use of β-lactam antibiotics, especially methicillin (no longer available in the United States). However, several other drugs can also cause AIN, including non-steroidal anti-inflammatory drugs (NSAIDs), antibiotics (penicillins, cephalosporins, ciprofloxacin and rifampicin), sulfonamides, furosemide and thiazide diuretics, cimetidine, allopurinol, and proton pump inhibitors (omeprazole and lansoprazole). By nature, drug-induced AIN is likely to involve cell-mediated hypersensitivity, and the pathological features resemble those seen in allograft rejection. On the other hand, the mechanisms involved in infection-induced AIN are not well understood.

There are no specific signs and symptoms of AIN. Rather, patients with AIN present with symptoms of acute renal dysfunction, and urine sediment reveals white and red blood cells, as well as white blood cell casts and eosinophiluria. Urinary protein excretion is usually less than 1 g/day. Although patients with drug-induced AIN may present with symptoms of allergic reactions, such as a rash, fever or eosinophilia, these findings are relatively less common [4]. Histological evaluation detected evidence of interstitial edema and a marked interstitial infiltrate consisting of macrophages and T lymphocytes. Renal biopsy remains the gold standard as a diagnostic procedure; however, it is less frequently performed since careful history taking and laboratory findings usually provide sufficient clues for diagnosis.

Treatment of patients with AIN includes cessation of suspect medications followed by careful observation of the clinical course. This approach results in spontaneous recovery in many patients [5]. Since an allergic mechanism is expected in drug-induced AIN, a corticosteroid is sometimes administered (e.g. prednisone, 1 mg/kg per day). Serum creatinine levels usually return to normal within 1–2 weeks.

Chronic Tubular Interstitial Nephritis

In contrast, chronic tubular interstitial nephritis is induced by a much broader spectrum of causes (table 1) and is characterized histologically by infiltration of leukocytes and interstitial edema, as seen in AIN, and tubular atrophy and

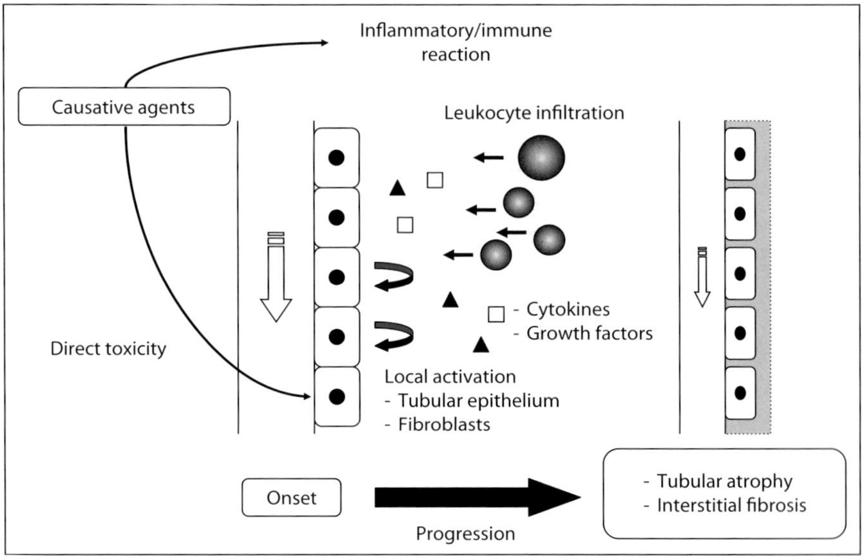

Fig. 2. Schematic diagram representing pathways of chronic tubular interstitial nephritis.

interstitial fibrosis at the advanced stages (fig. 2). In fact, secondary tubular interstitial nephritis is a final common pathway to end-stage kidney failure in a variety of glomerular disorders, as discussed below. Causative agents may either damage the tubulointerstitial compartment directly or influence it by inducing inflammatory or immune reactions in an indirect way. Direct nephrotoxicity is dependent on dose and duration of exposure, as can be seen in analgesic nephropathy, and often affects only a limited portion of the tubular interstitial structure. For example, necrosis in analgesic nephropathy is usually evident only in the renal papilla. Meanwhile, local activation of tubular and interstitial cells as well as infiltrating leukocytes, which are secondary to the initial insult, trigger the release of inflammatory cytokines and growth factors, such as transforming growth factor-β (TGF-β), which ultimately promotes fibrogenesis in the interstitium.

To provide an in-depth review of tubular interstitial nephritis, we first describe several representative disorders in detail and then explore several common mediators and underling mechanisms that characterize tubulointerstitial fibrosis, an end-stage feature of sustained tubular interstitial nephritis. The latter discussion is based on recent advances in experimental nephrology.

Reflux Nephropathy (Chronic Pyelonephritis)

Reflux nephropathy is often associated with vesicoureteral reflux and recurrent bacterial infection of the urinary tract. On the other hand, chronic pyelonephritis is defined as infection-related tubular interstitial nephritis without vesicoureteral reflux. Chronic inflammation of the urinary tract causes focal and segmental glomerulosclerosis and leads to nephrotic-range proteinuria.

Calcineurin Inhibitor Toxicity

Long-term use of calcineurin inhibitors is sometimes required in patients who have undergone kidney transplantation or in those with autoimmune disorders. These agents selectively constrict afferent arterioles and cause glomerular ischemia, which is associated with upregulation of the renin-angiotensin system and stimulation of TGF-β. Histologically, this scenario is associated with striped interstitial fibrosis in the renal cortex and medulla, afferent arteriolar hyalinosis, vacuolization of tubular epithelium and tubular atrophy. This stripe-like distribution is consistent with localized ischemic damage to nephrons, which results from an upstream occlusive afferent arteriolopathy [6].

Sarcoidosis

Sarcoidosis is a systemic disease of unknown etiology that is characterized by noncaseating granuloma in affected organs such as lung, skin and eyes. Pathologically, the kidney is characterized by interstitial granulomata formation and a disturbance in calcium homeostasis. An immune response in the center of the granulomata results in the accumulation of activated macrophages, capable of antigen presentation, which mainly recruit CD4-positive T lymphocytes and release inflammatory cytokines, such as interleukin-2 and interferon-γ. Although noncaseating granulomata in the interstitium are a unique pathological feature, this finding is rather uncommon.

Other Autoimmune Diseases

Sjögren's syndrome may be associated with chronic interstitial nephritis. Sjögren's syndrome is one of the most common causes of acquired distal (type I) renal tubular acidosis in adults, and the associated hypokalemia may be marked, resulting in a clinical presentation of severe weakness.

Systemic lupus erythematosus is the most common cause of interstitial nephritis with immune complexes defined by granular deposits of immunoglobulins and complement in the tubular basement membrane and/or interstitium. Interstitial involvement is present in half of the kidney biopsy specimens obtained from patients with lupus.

IgG_4-related sclerosing disease is a novel clinicopathological entity that has been proposed on histological and immunohistochemical examination of various organs of patients with autoimmune pancreatitis [7]. This systemic disease is characterized by extensive IgG_4-positive plasma cells and T lymphocyte infiltration of various organs, as well as tubular interstitial nephritis with dense infiltrations of IgG_4-positive mononuclear cells.

The response to corticosteroids at an early stage is generally favorable in these disorders.

Multiple Myeloma

Multiple myeloma is a B cell malignancy characterized by abnormal proliferation of plasma cells that are capable of producing monoclonal M protein. Involvement of the kidney occurs frequently (myeloma kidney) and is associated with light chain toxicity. The light chains may be nephrotoxic by themselves and/or exert a toxic effect by forming casts and obstructing tubular structure. Histologically, tubular casts surrounded by multinucleated giant cells, infiltration of plasma cells into the interstitial compartment, tubular atrophy and interstitial fibrosis have been observed.

Radiation Nephritis

Radiation nephritis is characterized histologically by glomerular capillary endothelial cell injury and swelling, which leads to vascular occlusion. Chronic radiation nephritis occurs in patients after receiving >1,500–2,500 rads to the kidney for more than 18 months. Clinical manifestations vary from progressive renal failure to massive proteinuria with preserved renal function, and hypertension with mild proteinuria.

Chronic Urate Nephropathy

Historically, chronic interstitial nephritis associated with chronic hyperuricemia was called

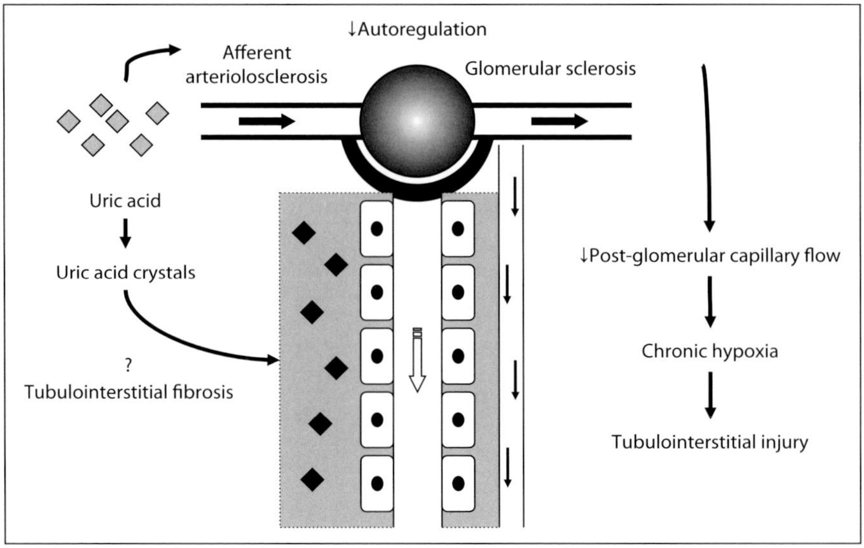

Fig. 3. Pathogenesis of chronic urate nephropathy.

'gouty nephropathy.' In the early 1980s, the concept of 'gouty nephropathy' as a disease was challenged because renal injury observed in subjects with gout was ascribed to co-existent hypertension, vascular disease or aging-associated renal injury. However, recent epidemiological studies have raised the interesting possibility that chronic hyperuricemia may be a true risk factor for the development and progression of chronic kidney disease (fig. 3).

The lesion is characterized by tubulointerstitial fibrosis, often with arteriolosclerosis and glomerulosclerosis. Within the kidney, precipitated uric acid crystals are often seen in the tubules and the interstitium. Although crystalline deposits of urate may account for some degree of renal injury, hyperuricemia may induce chronic renal injury independent of crystal formation. The mechanism appears to be the developing factor for pre-glomerular arteriolar disease that impairs the renal autoregulatory response, thereby causing glomerular hypertension.

Whether lowering uric acid is renoprotective in these patients remains controversial.

Analgesic Nephropathy

Historically, analgesic nephropathy resulted from the abuse of mixed analgesics containing phenacetin, aspirin and caffeine that were available as over-the-counter preparations in Europe, Australia and the United States. Histologically, the earliest changes in analgesic nephropathy are capillary sclerosis in the outer medulla [8], which can also be produced in experimental animals [9]. The pathological sequelae ultimately lead to papillary necrosis and interstitial fibrosis (fig. 4). Underlying molecular and cellular mechanisms are presently not completely understood, but may include prostaglandin inhibition that either impedes medullary blood flow or induces oxygen free and hydroxyl radicals [10]. Since medullary blood flow is critically dependent on local as well as systemic production of prostaglandins, inhibition of prostaglandin synthesis by analgesics causes regional ischemia, which along with its direct toxicity, eventually leads to papillary necrosis.

Due to restrictions limiting the sale of compound analgesics, especially those containing phenacetin, the number of cases of analgesic

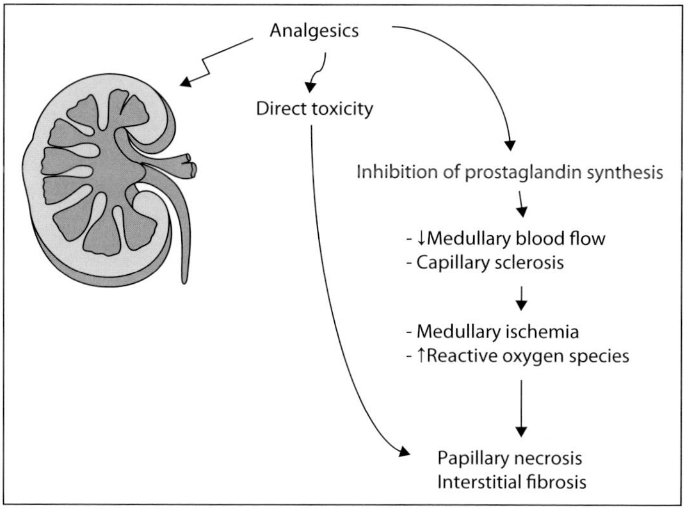

Fig. 4. Pathogenesis of analgesic nephropathy.

nephropathy has dramatically decreased. It is now extremely rare, and some consider this disease entity to be extinct [11]. A large study in the United States showed no association between use of current analgesic preparations and increased risk of renal dysfunction [12].

Translation from Experimental Research

Obstructive Nephropathy

Obstructive nephropathy is a major cause of renal failure in children. In animals, unilateral ureteral obstruction (UUO) is a well-studied model representing tubular interstitial nephritis [13]. Obstruction of the urinary tract leads to several molecular and cellular events that contribute (over the span of 1 week) to the development of tubular atrophy and interstitial fibrosis (fig. 5a).

Matrix Degradation
Interstitial fibrosis results from increased production of ECM components and insufficient degradation or turnover of such substrates. From the viewpoint of ECM accumulation during fibrogenesis, much attention has been paid to the metalloproteinases (MMPs) and the tissue inhibitor of MMP (TIMP) system. MMPs are a group of proteins that are capable of degrading ECM; therefore, it was first believed that they protect the interstitium from fibrogenesis. Contrary to the initial expectations, however, elimination of TIMP-1 alone did not alter the severity of interstitial fibrosis in the UUO model [14]. Furthermore, later studies identified MMPs as a critical factor in the migration of tubular epithelial cells into the interstitial space [15, 16]. MMP-2 and MMP-9, in fact, were found to contribute to disruption of tubular basement membrane, accelerate EMT and promote interstitial fibrosis.

Inflammatory Cell Infiltration
Previously, it was thought that infiltrating leukocytes, especially macrophages, played a critical role in the development of interstitial fibrosis. This idea was easy to envision based on the histological observation that the number of infiltrating macrophages, lymphocytes and fibroblasts accumulating in the interstitium parallels the degree of interstitial fibrosis. In fact, macrophages are an important source of cytokines and growth factors that play a critical role in the development of interstitial fibrosis. Results

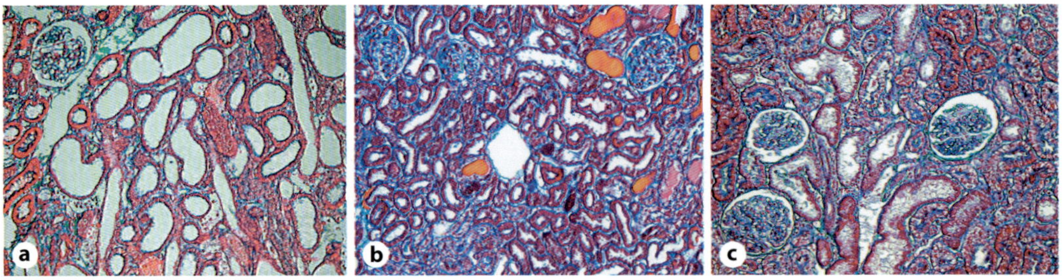

Fig. 5. Tubulointerstitial injury in experimental kidney injury models. Representative light microscope images showing tubulointerstitial injury in UUO (**a**), puromycin aminonucleoside nephropathy (**b**) and remnant kidney models in the rat (**c**). Varying degrees of tubular cell deformity together with leukocyte infiltration and interstitial scarring are observed.

of the experimental studies in animals with depleted macrophages or lacking genes that are critical for proper recruitment of macrophages also supported this view [17–21]. However, recent evidence suggests that this is not always the case. Macrophages of a certain category may be beneficial because of their ability to phagocytose ECM fragments [22].

Epithelial-Mesenchymal Transition

During the process of interstitial fibrosis, activated interstitial fibroblasts undoubtedly play a dominant role; however, the exact origin of such fibroblasts have long been a matter of debate. In addition to local fibroblasts, tubular epithelial cells are able to acquire a myofibroblast-like phenotype that is capable of producing extracellular matrix (EMT).

To address the question of the relative contribution of EMT during renal fibrogenesis, Iwano et al. [23] employed fibroblast-specific protein-1 as a marker of active fibroblasts and observed that a substantial number of interstitial fibroblasts, in fact, originated from neighboring tubular epithelial cells. Of the fibroblast-specific protein-1-positive fibroblast pool in the fibrotic kidney, as much as 36% co-expressed LacZ that had been tagged to renal proximal tubules, indicating that active interstitial fibroblasts can be generated from tubular epithelial cells under certain pathological conditions. In keeping with this observation, attempts to prevent EMT are known to conversely attenuate fibrosis, again signifying the contribution of EMT during interstitial fibrosis in the kidney.

The process of EMT is stepwise and composed of four distinct steps. The initial loss of epithelial cell adhesions allows cells to dissociate from their original position. Following de novo expression of α-SMA and actin reorganization, disruption of tubular basement membrane ensues and the cells finally acquire motility with which to migrate into the interstitial compartment. A number of mediators, such as TGF-β, are reported to facilitate EMT (table 2), and the underlying mechanisms are just beginning to be understood [24].

Among these underlying mechanisms, the TGF-β-mediated signaling pathway is the most intensively characterized. TGF-β induces phosphorylation of Smad-2 and -3 through its type I receptor and induces their association with the common partner, Smad-4. This complex translocates to the nucleus and controls the transcription of TGF-β target genes. TGF-β-induced EMT seems to mainly depend on Smad signaling because overexpression of inhibitory Smad-7 counteracts EMT [25, 26]. TGF-β also induces Smad-dependent expression of integrin-linked kinase (ILK), an intracellular serine/threonine protein kinase that interacts with cytoplasmic domains of β-integrins and other cytoskeleton-associated proteins. Overexpression of ILK in human proximal tubular cells suppresses the expression of E-cadherin and induces fibronectin and MMP-2, thus mediating EMT. Notably,

Table 2. Mediators facilitating EMT

Inducers/mediators
Transforming growth factor-β
Tumor necrosis factor-α
Fibroblast growth factor-2
Epidermal growth factor
MMP-2 and -9
Angiotensin II
Oxidative stress
Hypoxia
Inhibitors
BMP-7
Hepatocyte growth factor

introduction of a dominant-negative ILK mutant counteracts TGF-β-induced EMT, suggesting that ILK works downstream of TGF-β [27].

Mediators having a negative impact on EMT have also been reported. Bone morphogenic protein-7 inhibits TGF-β-induced EMT through Smad-mediated re-introduction of E-cadherin [28]. Hepatocyte growth factor is also reported to counteract TGF-β-mediated EMT through inhibition of ILK [29, 30].

Hypoxia and Oxidative Stress

Furthermore, tubulointerstitial hypoxia, oxidative stress and activation of the renin-angiotensin system also contribute to the process of EMT. Hypoxia in the tubulointerstitial compartment has been experimentally demonstrated in the past decade in a number of glomerular diseases (which will be discussed below). In fact, mice with defective hypoxia-inducible factor (HIF)-1α manifested much less frequent EMT following UUO, suggesting that HIF-1, a master regulator of hypoxia, is a key mediator of EMT in a hypoxic milieu [31]. Proposed mechanisms include abnormalities in the adherens, as well as the tight junction, in association with an increase in transcriptional repressors such as Snail. The Snail superfamily appears to have a key role as a transcriptional repressor of E-cadherin promoter [32, 33]. Other candidate HIF targets that potentially play a role in EMT are Twist [34] and lysyl oxidase [35].

The Renin-Angiotensin System and Plasminogen Activator Inhibitor-1

Activation of the renin-angiotensin system is also a key modulator of EMT. In cell culture experiments, angiotensin II per se fails to induce EMT, but induces TGF-β expression. Therefore, the fibrogenic effect of angiotensin II in vivo may largely depend on TGF-β signaling. However, a recent study identified a TGF-β-independent, plasminogen activator inhibitor-1-dependent pathway [36]. In β6-integrin-null mice, angiotensin II infusion into UUO mice led to a marked deposition of ECM, with lower levels of active TGF-β and Smad. These observations were in fact associated with local plasminogen activator inhibitor-1 expression. Results of this study clearly indicate that the renin-angiotensin system contributes to EMT in vivo through Smad-dependent as well as -independent pathways.

Serine proteases such as urokinase-type plasminogen activator and tissue-type plasminogen activator degrade matrix proteins by themselves. Furthermore, urokinase-type plasminogen activator and tissue-type plasminogen activator cleave plasminogen into the protease plasmin, which also enhances matrix degradation. Plasminogen activator inhibitor-1 inhibits these pathways and promotes fibrosis [37].

Cell Death

Hypocellularity of tubular epithelial cells is another important hallmark of hypocellular tubulointerstitial fibrosis. Apoptosis has been documented in both human renal diseases and animal models of renal injury [38–40]. In general, unregulated execution of tubular cell apoptosis in the pathological state is believed to contribute to a slowly progressive loss of resident renal cells, culminating in hypocellular scarring. To date, two representative pathways have been identified. An extrinsic pathway requires an external death signal, such as cytokines (e.g. TNF-α), and Fas-ligand (FasL, CD95L) to transduce death signaling. On binding of an extrinsic signal to its receptors [Fas (CD95) or TNF-R], a cascade of caspase activation ensues, in which active caspase-8 transduces a death signal to caspase-3, thus culminating in nucleosomal fragmentation. On the other hand, the intrinsic pathway is initiated by a broader array of stimuli, such as hypoxia, oxidative stress and DNA damage. It mediates death signaling through the release of cytochrome c from the mitochondria and caspase-9 activation. Although it is unclear to what degree each pathway is involved in the development of tubular hypocellularity, there is sufficient circumstantial evidence that both may be involved [41–43].

Chronic Urate Nephropathy

The hepatic enzyme, uricase, which degrades uric acid to allantoin, is present in most mammals; however, humans and the great apes lack this enzyme because of mutations that occurred during evolution. The existence of uricase in rodents hampered studies of hyperuricemia in animal models until Johnson and colleagues [44–46] utilized oxonic acid, an inhibitor of uricase, to develop rodents with hyperuricemia and elucidated involvement of endothelial dysfunction and oxidative stress in chronic urate nephropathy.

Tubulointerstitial Injury Secondary to Glomerular Diseases

Mechanisms of tubulointerstitial injury that ultimately lead to interstitial fibrosis have been studied extensively in animal models of glomerulopathies (fig. 5b, c). In comparison to non-glomerular tubular interstitial disease (such as UUO in experimental animals), major considerations in tubulointerstitial injury are: (1) direct tubular toxicity by filtered albumin and macromolecules, (2) activation by glomerulus-derived cytokines and growth factors, and (3) chronic hypoxia in the tubular interstitium, which most likely serves as a mediator to progressive renal insufficiency.

Proteinuria

Undoubtedly, one of the common mechanisms underlying tubulointerstitial damage secondary to glomerular injury is massive proteinuria [47, 48]. The relationship between massive proteinuria and progressive renal insufficiency has been established in several large-scale prospective studies, such as the Modification of Diet in Renal Disease (MDRD) study [49]. There is experimental evidence that filtered macromolecules elicit detrimental effects on tubular cells, such as lysosomal rupture and energy depletion [50, 51]. Putative toxic proteins include albumin, transferrin and lipoproteins. In addition, filtered complement components are activated by ammonia at the tubular epithelial brush border, leading to a cascade of events that culminates in cell injury [52–55].

On exposure of proximal tubular cells to albumin in vitro, the release of inflammatory cytokines, such as monocyte chemotactic protein-1 and platelet-derived growth factor, is evident. This could be a trigger of interstitial inflammation and scarring in vivo. However, it remains unclear whether sustained proteinuria itself is responsible for severe tubulointerstitial injury because the development of factors other than proteinuria in tubulointerstitial lesions cannot be excluded completely. In a rat model of podocyte injury with massive proteinuria, absence of any maneuvers known to induce disturbed renal hemodynamics resulted in interstitial inflammation without development of fibrosis [56]. Genetic components may be involved in renal fibrosis due to proteinuria, and some candidate genes were proposed by mapping of loci influencing proteinuria-induced collagen deposition in a certain strain of mice [57].

In primary glomerular injury, cytokines and growth factors originating from residual glomerular cells and infiltrating macrophages are reported to play an important role in tubulointerstitial injury [58]. Glomerular cytokines may reach tubular epithelial cells through the tubule lumen, via delivery through peritubular capillaries or by diffusion through the interstitial space [59, 60]. In fact, inflammatory cytokines and growth factors can be detected in the urine from patients and experimental animals with glomerular injury. Furthermore, histological analysis of interstitial leukocyte infiltration in experimental glomerular disease suggested that tubulointerstitial inflammation begins in the vicinity of glomeruli in which tubular epithelial cells are closest to the inflamed glomerular cells [61].

Hypoxia

Given the structural properties wherein peritubular capillary blood flow is followed by glomerular vascular beds, it is not surprising that initial injury to the glomerulus obstructs vascular beds and impairs post-glomerular capillary flow, thus causing tubulointerstitial hypoxia. Ischemia in the post-glomerular region then triggers several phenotypic changes in tubular cells, such as changes of proliferation rate, EMT and cell death. Hypoxic tubular cells, in turn, serve as a source of key mediators involved in macrophage infiltration and tubulointerstitial fibrosis, such as TGF-β. Interstitial accumulation of ECM further impairs local oxygenation by blocking oxygen diffusion, thus accelerating regional hypoxia. Tubulointerstitial hypoxia initiates a vicious cycle and serves as a critical mediator of end-stage kidney disease. This idea was first proposed by Fine et al. [62] about a decade ago, and subsequent studies involving animal nephritis models have proven that hypoxia really exists in the tubulointerstitial compartment [63].

On the other hand, a growing body of evidence indicates that individual cells adapt to hypoxia by transcriptionally upregulating a number of genes involved in angiogenesis, erythropoiesis and energy metabolism. HIF, especially HIF-1, plays a major role during this process [64]. HIF is a heterodimeric transcription factor belonging to the basic helix-loop-helix-per-arnt-sim (bHLH-PAS) family, composed of α- and β-subunits. Out of three HIF-α isoforms, HIF-1α and HIF-2α are relatively well characterized. In the hypoxic kidney, HIF-1α is expressed in tubular and glomerular epithelial cells as well as papillary interstitial cells. On the other hand, HIF-2α expression is restricted to endothelial and interstitial cells [65]. Published literature to date suggests a protective role of HIF in progressive glomerular diseases. In a heminephrectomized Thy-1.1 nephritis model, chemical activation of HIF using cobalt resulted in amelioration of tubulointerstitial injury, which was associated with a reduced number of cortical tubular cells exhibiting apoptosis [66]. Similar renoprotective action of HIF was observed in a rat remnant kidney model, in which administration of cobalt, and subsequent HIF accumulation, resulted in attenuation of tubulointerstitial injury. This impact of HIF was associated with better preservation of the peritubular capillary network as well as a marked decrease in the number of cortical tubular cells undergoing apoptosis [67]. A role for HIF in kidney protection has been suggested in several other models of glomerular diseases, such as type 2 diabetic nephropathy [68] and habu snake venom-induced nephropathy [69]. Studies to identify responsible mechanisms that represent amelioration of tubulointerstitial injury have recently begun [70, 71].

Conclusion

Tubular interstitial nephritis comprises a wide array of heterogeneous disorders of varying etiology. Given that there is substantial difference in pathophysiology and clinical course among disorders, studies over the past two decades have elucidated mechanisms and mediators specific to each disease entity. On the other hand, recent advances in molecular and cellular biology, together with animal experiments, have identified common mechanisms that characterize this disease entity as a whole and provoke hypocellular tubulointerstitial fibrosis. In particular, much progress has been made in discovering the molecular mechanisms that underlie EMT and apoptosis, in addition to the impact of chronic hypoxia in the diseased tubulointerstitium. A better understanding of these common mechanisms will lead to effective prevention and treatment of tubular interstitial nephritis.

References

1 Schainuck LI, Striker GE, Cutler RE, Benditt EP: Structural-functional correlations in renal disease. II. The correlations. Hum Pathol 1970;1:631–641.

2 Marcussen N: Atubular glomeruli and the structural basis for chronic renal failure. Lab Invest 1992;66:265–284.

3 Bohle A, Muller GA, Wehrmann M, Mackensen-Haen S, Xiao JC: Pathogenesis of chronic renal failure in the primary glomerulopathies, renal vasculopathies, and chronic interstitial nephritides. Kidney Int Suppl 1996;54:S2–S9.

4 Baker RJ, Pusey CD: The changing profile of acute tubulointerstitial nephritis. Nephrol Dial Transplant 2004;19:8–11.

5 Buysen JG, Houthoff HJ, Krediet RT, Arisz L: Acute interstitial nephritis: a clinical and morphological study in 27 patients. Nephrol Dial Transplant 1990;5:94–99.

6 Myers BD, Newton L: Cyclosporine-induced chronic nephropathy: an obliterative microvascular renal injury. J Am Soc Nephrol 1991;2:S45–S52.

7 Kamisawa T, Okamoto A: IgG4-related sclerosing disease. World J Gastroenterol 2008;14:3948–3955.

8 Schubert GE, Bethke BA: Capillarosclerosis of the renal pelvis: autopsy study for estimating the incidence of phenacetin and paracetamol abuse. Eur Urol 1986;12:327–330.

9 Burrell JH, Yong JL, Macdonald GJ: Experimental analgesic nephropathy: changes in renal structure and urinary concentrating ability in Fischer 344 rats given continuous low doses of aspirin and paracetamol. Pathology 1990;22:33–44.

10 Mohandas J, Duggin GG, Horvath JS, Tiller DJ: Metabolic oxidation of acetaminophen (paracetamol) mediated by cytochrome P-450 mixed-function oxidase and prostaglandin endoperoxide synthetase in rabbit kidney. Toxicol Appl Pharmacol 1981;61:252–259.

11 Mihatsch MJ, Khanlari B, Brunner FP: Obituary to analgesic nephropathy – an autopsy study. Nephrol Dial Transplant 2006;21:3139–3145.

12 Rexrode KM, Buring JE, Glynn RJ, Stampfer MJ, Youngman LD, Gaziano JM: Analgesic use and renal function in men. JAMA 2001;286:315–321.

13 Chevalier RL, Forbes MS, Thornhill BA: Ureteral obstruction as a model of renal interstitial fibrosis and obstructive nephropathy. Kidney Int 2009;75:1145–1152.

14 Kim H, Oda T, Lopez-Guisa J, Wing D, Edwards DR, Soloway PD, Eddy AA: TIMP-1 deficiency does not attenuate interstitial fibrosis in obstructive nephropathy. J Am Soc Nephrol 2001;12:736–748.

15 Yang J, Shultz RW, Mars WM, Wegner RE, Li Y, Dai C, Nejak K, Liu Y: Disruption of tissue-type plasminogen activator gene in mice reduces renal interstitial fibrosis in obstructive nephropathy. J Clin Invest 2002;110:1525–1538.

16 Cheng S, Lovett DH. Gelatinase A (MMP-2) is necessary and sufficient for renal tubular cell epithelial-mesenchymal transformation. Am J Pathol 2003;162:1937–1949.

17 Tesch GH, Schwarting A, Kinoshita K, Lan HY, Rollins BJ, Kelley VR: Monocyte chemoattractant protein-1 promotes macrophage-mediated tubular injury, but not glomerular injury, in nephrotoxic serum nephritis. J Clin Invest:1999;103:73–80.

18 Ophascharoensuk V, Giachelli CM, Gordon K, Hughes J, Pichler R, Brown P, Liaw L, Schmidt R, Shankland SJ, Alpers CE, Couser WG, Johnson RJ: Obstructive uropathy in the mouse: role of osteopontin in interstitial fibrosis and apoptosis. Kidney Int 1999;56:571–580.

19 Lange-Sperandio B, Cachat F, Thornhill BA, Chevalier RL: Selectins mediate macrophage infiltration in obstructive nephropathy in newborn mice. Kidney Int 2002;61:516–524.

20 Shimizu H, Maruyama S, Yuzawa Y, Kato T, Miki Y, Suzuki S, Sato W, Morita Y, Maruyama H, Egashira K, Matsuo S: Anti-monocyte chemoattractant protein-1 gene therapy attenuates renal injury induced by protein-overload proteinuria. J Am Soc Nephrol 2003;14:1496–1505.

21 Persy VP, Verhulst A, Ysebaert DK, De Greef KE, De Broe ME: Reduced postischemic macrophage infiltration and interstitial fibrosis in osteopontin knockout mice. Kidney Int 2003;63:543–553.

22 Nishida M, Fujinaka H, Matsusaka T, Price J, Kon V, Fogo AB, Davidson JM, Linton MF, Fazio S, Homma T, Yoshida H, Ichikawa I: Absence of angiotensin II type 1 receptor in bone marrow-derived cells is detrimental in the evolution of renal fibrosis. J Clin Invest 2002;110:1859–1868.

23 Iwano M, Plieth D, Danoff TM, Xue C, Okada H, Neilson EG: Evidence that fibroblasts derive from epithelium during tissue fibrosis. J Clin Invest 2002;110:341–350.

24 Zeisberg M, Kalluri R: The role of epithelial-to-mesenchymal transition in renal fibrosis. J Mol Med 2004;82:175–181.

25 Lan HY, Mu W, Tomita N, Huang XR, Li JH, Zhu HJ, Morishita R, Johnson RJ: Inhibition of renal fibrosis by gene transfer of inducible Smad7 using ultrasound-microbubble system in rat UUO model. J Am Soc Nephrol 2003;14:1535–1548.

26 Li JH, Zhu HJ, Huang XR, Lai KN, Johnson RJ, Lan HY: Smad7 inhibits fibrotic effect of TGF-Beta on renal tubular epithelial cells by blocking Smad2 activation. J Am Soc Nephrol 2002;13:1464–1472.

27 Li Y, Yang J, Dai C, Wu C, Liu Y: Role for integrin-linked kinase in mediating tubular epithelial to mesenchymal transition and renal interstitial fibrogenesis. J Clin Invest 2003;112:503–516.

28 Zeisberg M, Hanai J, Sugimoto H, Mammoto T, Charytan D, Strutz F, Kalluri R: BMP-7 counteracts TGF-beta1-induced epithelial-to-mesenchymal transition and reverses chronic renal injury. Nat Med 2003;9:964–968.

29 Kobayashi E, Sasamura H, Mifune M, Shimizu-Hirota R, Kuroda M, Hayashi M, Saruta T: Hepatocyte growth factor regulates proteoglycan synthesis in interstitial fibroblasts. Kidney Int 2003;64:1179–1188.

30 Yang J, Liu Y: Blockage of tubular epithelial to myofibroblast transition by hepatocyte growth factor prevents renal interstitial fibrosis. J Am Soc Nephrol 2002;13:96–107.

31 Higgins DF, Kimura K, Bernhardt WM, Shrimanker N, Akai Y, Hohenstein B, Saito Y, Johnson RS, Kretzler M, Cohen CD, Eckardt KU, Iwano M, Haase VH: Hypoxia promotes fibrogenesis in vivo via HIF-1 stimulation of epithelial-to-mesenchymal transition. J Clin Invest 2007;117:3810–3820.
32 Cano A, Perez-Moreno MA, Rodrigo I, Locascio A, Blanco MJ, del Barrio MG, Portillo F, Nieto MA: The transcription factor snail controls epithelial-mesenchymal transitions by repressing E-cadherin expression. Nat Cell Biol 2000;2:76–83.
33 Carver EA, Jiang R, Lan Y, Oram KF, Gridley T: The mouse snail gene encodes a key regulator of the epithelial-mesenchymal transition. Mol Cell Biol 2001;21:8184–8188.
34 Sun S, Ning X, Zhang Y, Lu Y, Nie Y, Han S, Liu L, Du R, Xia L, He L, Fan D: Hypoxia-inducible factor-1alpha induces Twist expression in tubular epithelial cells subjected to hypoxia, leading to epithelial-to-mesenchymal transition. Kidney Int 2009;75:1278–1287.
35 Erler JT, Bennewith KL, Nicolau M, Dornhofer N, Kong C, Le QT, Chi JT, Jeffrey SS, Giaccia AJ: Lysyl oxidase is essential for hypoxia-induced metastasis. Nature 2006;440:1222–1226.
36 Ma LJ, Yang H, Gaspert A, Carlesso G, Barty MM, Davidson JM, Sheppard D, Fogo AB: Transforming growth factor-beta-dependent and -independent pathways of induction of tubulointerstitial fibrosis in beta6(-/-) mice. Am J Pathol 2003;163:1261–1273.
37 Eddy AA, Fogo AB: Plasminogen activator inhibitor-1 in chronic kidney disease: evidence and mechanisms of action. J Am Soc Nephrol 2006;17:2999–3012.
38 Woo D: Apoptosis and loss of renal tissue in polycystic kidney diseases. N Engl J Med 1995;333:18–25.
39 Gobe GC, Axelsen RA: Genesis of renal tubular atrophy in experimental hydronephrosis in the rat. Role of apoptosis. Lab Invest 1987;56:273–281.
40 Thomas SE, Andoh TF, Pichler RH, Shankland SJ, Couser WG, Bennett WM, Johnson RJ: Accelerated apoptosis characterizes cyclosporine-associated interstitial fibrosis. Kidney Int 1998;53:897–908.
41 Schelling JR, Nkemere N, Kopp JB, Cleveland RP: Fas-dependent fratricidal apoptosis is a mechanism of tubular epithelial cell deletion in chronic renal failure. Lab Invest 1998;78:813–824.
42 Khan S, Cleveland RP, Koch CJ, Schelling JR: Hypoxia induces renal tubular epithelial cell apoptosis in chronic renal disease. Lab Invest 1999;79:1089–1099.
43 Saikumar P, Dong Z, Patel Y, Hall K, Hopfer U, Weinberg JM, Venkatachalam MA: Role of hypoxia-induced Bax translocation and cytochrome c release in reoxygenation injury. Oncogene 1998;17:3401–3415.
44 Sanchez-Lozada LG, Soto V, Tapia E, Avila-Casado C, Sautin YY, Nakagawa T, Franco M, Rodriguez-Iturbe B, Johnson RJ: Role of oxidative stress in the renal abnormalities induced by experimental hyperuricemia. Am J Physiol Renal Physiol 2008;295: F1134–F1141.
45 Mazzali M, Kanellis J, Han L, Feng L, Xia YY, Chen Q, Kang DH, Gordon KL, Watanabe S, Nakagawa T, Lan HY, Johnson RJ: Hyperuricemia induces a primary renal arteriolopathy in rats by a blood pressure-independent mechanism. Am J Physiol Renal Physiol 2002;282:F991–F997.
46 Kang DH, Nakagawa T, Feng L, Watanabe S, Han L, Mazzali M, Truong L, Harris R, Johnson RJ: A role for uric acid in the progression of renal disease. J Am Soc Nephrol 2002;13:2888–2897.
47 Remuzzi G, Bertani T: Pathophysiology of progressive nephropathies. N Engl J Med 1998;339:1448–1456.
48 Eddy AA Experimental insights into the tubulointerstitial disease accompanying primary glomerular lesions. J Am Soc Nephrol 1994;5:1273–1287.
49 Klahr S, Levey AS, Beck GJ, Caggiula AW, Hunsicker L, Kusek JW, Striker G: The effects of dietary protein restriction and blood-pressure control on the progression of chronic renal disease. Modification of Diet in Renal Disease Study Group. N Engl J Med 1994;330:877–884.
50 Harris KP, Burton C, Walls J: Proteinuria: a mediator of interstitial fibrosis? Contrib Nephrol 1996;118:173–179.
51 Burton C, Harris KP: The role of proteinuria in the progression of chronic renal failure. Am J Kidney Dis 1996;27:765–775.
52 Tang S, Lai KN, Sacks SH: Role of complement in tubulointerstitial injury from proteinuria. Kidney Blood Press Res 2002;25:120–126.
53 Nangaku M, Pippin J, Couser WG: C6 mediates chronic progression of tubulointerstitial damage in rats with remnant kidneys. J Am Soc Nephrol 2002;13:928–936.
54 Nangaku M, Pippin J, Couser WG: Complement membrane attack complex (C5b-9) mediates interstitial disease in experimental nephrotic syndrome. J Am Soc Nephrol 1999;10:2323–2331.
55 Matsuo S, Morita Y, Mizuno M, Nishikawa K, Yuzawa Y: Proteinuria and damage to tubular cells – is complement a culprit? Nephrol Dial Transplant 1998;13:2723–2726.
56 Kikuchi H, Kawachi H, Ito Y, Matsui K, Nosaka H, Saito A, Orikasa M, Arakawa M, Shimizu F: Severe proteinuria, sustained for 6 months, induces tubular epithelial cell injury and cell infiltration in rats but not progressive interstitial fibrosis. Nephrol Dial Transplant 2000;15:799–810.
57 Kato N, Watanabe Y, Ohno Y, Inoue T, Kanno Y, Suzuki H, Okada H: Mapping quantitative trait loci for proteinuria-induced renal collagen deposition. Kidney Int 2008; 73:1017–1023.
58 Healy E, Brady HR: Role of tubule epithelial cells in the pathogenesis of tubulointerstitial fibrosis induced by glomerular disease. Curr Opin Nephrol Hypertens 1998;7:525–530.
59 Schena FP, Gesualdo L, Grandaliano G, Montinaro V: Progression of renal damage in human glomerulonephritides: is there sleight of hand in winning the game? Kidney Int 1997;52:1439–1457.
60 Remuzzi G, Ruggenenti P, Benigni A: Understanding the nature of renal disease progression. Kidney Int 1997;51:2–15.
61 Lan HY, Paterson DJ, Atkins RC: Initiation and evolution of interstitial leukocytic infiltration in experimental glomerulonephritis. Kidney Int 1991;40:425–433.

62 Fine LG, Bandyopadhay D, Norman JT: Is there a common mechanism for the progression of different types of renal diseases other than proteinuria? Towards the unifying theme of chronic hypoxia. Kidney Int Suppl 2000;75:S22–S26.
63 Tanaka T, Miyata T, Inagi R, Fujita T, Nangaku M: Hypoxia in renal disease with proteinuria and/or glomerular hypertension. Am J Pathol 2004;165:1979–1992.
64 Schofield CJ, Ratcliffe PJ: Oxygen sensing by HIF hydroxylases. Nat Rev Mol Cell Biol 2004;5:343–354.
65 Rosenberger C, Mandriota S, Jurgensen JS, Wiesener MS, Horstrup JH, Frei U, Ratcliffe PJ, Maxwell PH, Bachmann S, Eckardt KU: Expression of hypoxia-inducible factor-1alpha and -2alpha in hypoxic and ischemic rat kidneys. J Am Soc Nephrol 2002;13:1721–1732.
66 Tanaka T, Matsumoto M, Inagi R, Miyata T, Kojima I, Ohse T, Fujita T, Nangaku M: Induction of protective genes by cobalt ameliorates tubulointerstitial injury in the progressive Thy1 nephritis. Kidney Int 2005;68: 2714–2725.
67 Tanaka T, Kojima I, Ohse T, Ingelfinger JR, Adler S, Fujita T, Nangaku M: Cobalt promotes angiogenesis via hypoxia-inducible factor and protects tubulointerstitium in the remnant kidney model. Lab Invest 2005;85:1292–1307.
68 Ohtomo S, Nangaku M, Izuhara Y, Takizawa S, Strihou CY, Miyata T: Cobalt ameliorates renal injury in an obese, hypertensive type 2 diabetes rat model. Nephrol Dial Transplant 2008;23:1166–1172.
69 Kudo Y, Kakinuma Y, Mori Y, Morimoto N, Karashima T, Furihata M, Sato T, Shuin T, Sugiura T: Hypoxia-inducible factor-1alpha is involved in the attenuation of experimentally induced rat glomerulonephritis. Nephron Exp Nephrol 2005;100:e95–e103.
70 Tanaka T, Nangaku M: Drug discovery for overcoming chronic kidney disease (CKD): prolyl-hydroxylase inhibitors to activate hypoxia-inducible factor (HIF) as a novel therapeutic approach in CKD. J Pharmacol Sci 2009;109:24–31.
71 Nangaku M, Eckardt KU: Hypoxia and the HIF system in kidney disease. J Mol Med 2007;85:1325–1330.

Dr. Masaomi Nangaku, MD, PhD
Division of Nephrology and Endocrinology,
University of Tokyo School of Medicine 7–3–1 Hongo,
Bunkyo-ku 113–8655 Tokyo (Japan)
Tel. +81 3 3815 5411, ext. 33128, Fax +81 3 5800 8806,
E-Mail mnangaku-tky@umin.ac.jp

Herrera GA (ed): Experimental Models for Renal Diseases: Pathogenesis and Diagnosis.
Contrib Nephrol. Basel, Karger, 2011, vol 169, pp 311–326

Obstructive Uropathy

Luan D. Truong[a–d] · Lillian Gaber[a,b] · Garabed Eknoyan[d]

[a]Department of Pathology, The Methodist Hospital, Houston, Tex., [b]Department of Pathology, Cornell University, New York, N.Y., and Departments of [c]Pathology and [d]Medicine, Baylor College of Medicine, Houston, Tex., USA

Abstract

Chronic tubulointerstitial nephritis (TIN), characterized by tubular atrophy, interstitial fibrosis and inflammation, is a major prognostic determinant of chronic kidney disease, regardless of the original cause of the kidney disease. Understanding the pathogenesis of TIN has been hampered by the lack of an adequate experimental model. However, the demonstration that the renal lesions of obstructive uropathy induced by experimental urinary obstruction (UO) has provided an excellent model to study the pathogenesis of TIN in general and especially congenital obstructive nephropathy, the most common cause of pediatric end-stage renal disease. Since relief of experimental UO is technically possible, this model is particularly useful for studying the potential reversibility of TIN. Experimental UO is usually created by the unilateral ligation of a ureter. This induces progressive tubular epithelial cell injury, including apoptosis, proliferation, loss of differentiation and atrophy; interstitial inflammatory cell infiltrates composed predominantly of macrophages and T cells; and interstitial fibrosis characterized by an increase and activation of interstitial fibroblasts, deposition of extracellular matrix proteins and loss of peritubular capillaries. These changes collectively lead to progressive scarring and the loss of renal parenchyma and kidney function. The glomeruli and large blood vessels remain either normal or show mild changes later in the course of the disease. In addition to TIN, congenital obstructive nephropathy causes marked derangement of renal and glomerular development. Relief of UO does not seem to reverse TIN. In fact, the renal lesions of obstructive uropathy not only persist, but also progress long after UO is relieved in both adult and neonatal rats. The pathogenesis of UO-induced TIN has been well studied, at least in part because of the ready application of this model to mice, in which genetic manipulation including gene deletion or transfection of putative pathogenic molecules is technically feasible. Experimental UO immediately induces mechanical stretching of tubular epithelial cells and activates the renin-angiotensin system, leading to profound changes of the cells, including neo-expression of a large number of molecules which control cell cycle (e.g. caspases, intrinsic and extrinsic death pathway molecules, inhibitors of cyclin-dependent kinases p27 and p21, reactive oxygen species, and catalase), hypoxic response (HIF-α), epithelial-mesenchymal transformation (e.g. hepatocyte growth factor, bone morphogenic protein 7 and nestin), and the upregulation of cytokines and growth factors (e.g. TGFβ-1, EGF, PDGF, VEGF and TNF-α) as well as chemokines (MCP-1, osteopontin, IL-1, ICAM-1, VCAM-1 and selectins). Inflammatory cells are recruited immediately after UO, probably under the effect of the renin-angiotensin system and later by tubular cell-derived chemokines. Several chemokines and their receptors are also expressed by the infiltrating inflammatory cells, thereby augmenting the recruitment of additional inflammatory cells through an autocrine loop. These molecules are probably also responsible for an increased number interstitial fibroblasts, which are derived not only from the proliferation of resident interstitial fibroblasts, but also from the renal homing of bone marrow-derived fibrocytes and the transformation of tubular epithelial cells, endothelial cells and pericytes into interstitial fibroblasts. Activated interstitial fibroblasts are responsible for the increased synthesis of extracellular matrix protein. This together with an impairment of various fibrolytic pathways leads to the increased deposition of extracellular matrix protein. In summary, typical features of TIN are regularly induced in the experimental model of obstructive uropathy. This versatile model has contributed much to elucidate the mechanism of TIN. The translation of this body of knowledge into TIN in general and its effective treatment of obstructive uropathy remains to be explored.

Urinary Obstruction Is an Excellent Model to Study the Pathogenesis of Chronic Tubulointerstitial Injury

Chronic tubulointerstitial nephritis (TIN), characterized by tubular atrophy and interstitial fibrosis and inflammation, is an almost constant component of chronic kidney disease (CKD) in general and of obstructive uropathy in particular. It is well established that the course and prognosis of CKD is determined to a great extent by the extent and severity of TIN, regardless of the original cause of the kidney disease [1]. Until recently, understanding the pathogenesis of TIN has been hampered by the lack of an adequate experimental model. Although classical TIN develops in traditional experimental models of CKD induced by exogenous administration of tubular antigen or ablation of a major portion of the renal parenchyma ('5/6 nephrectomy'), several considerations limit their usefulness for the study of TIN. The most important among these is the sporadic and variable nature of the renal changes observed among experimental animals of the same species, and the significant interspecies variation (mice are notoriously resistant to renal injury) [2–4] which is especially frustrating since genetic manipulation is most practical in mice, a technique that would be helpful in dissecting the mechanism of TIN.

The model of urinary obstruction (UO) obviates several of these limitations. Unilateral ligation of the ureter induces typical changes of TIN in the kidney [5] (see below). This model of TIN offers several advantages: (1) it is technically easy to induce, (2) there is constancy of the renal lesions that develop over a relatively short period of time, (3) it allows for the precise control of timing and severity of the renal lesions, (4) there is uniformity in the structural changes among different tested animals, (5) it is free of confounding factors including exogenous toxins or uremia since the unobstructed kidney maintains normal homeostasis and renal function, and (6) the renal lesions readily develop in all tested species, including mice, which allows for the use of gene transfection or deletion to dissect the pathogenesis of TIN [2]. Reversibility of TIN is an important therapeutic consideration. The UO model of TIN allows for the removal of the initiating cause (obstruction) at variable intervals and, therefore, is practically the only pertinent model for the study of reversibility of TIN [2, 6–9].

Additionally, congenital obstructive nephropathy is the most important cause of end-stage renal disease in children, and experimental UO in neonatal rats or mice is an excellent model for the study of this condition [6–9]. Since sudden and complete urine obstruction induced by ureteral ligation may not closely simulate the human situation in which UO is often partial and progressive, partial UO models created by embedding the ureter into the underlying psoas muscle or partial narrowing of the ureter have been used, albeit with variable success in creating TIN [2]. As a result of all these considerations, the experimental model of obstructive uropathy has emerged as the single most utilized approach for studying the pathogenesis of TIN, with almost 1,000 articles on the subject published over the past 6 years (data gleaned from PubMed).

Renal Lesions Induced by UO

UO induces tubular cell injury, recruitment of interstitial inflammatory cells, proliferation of interstitial fibroblast, loss of peritubular capillaries and accumulation of extracellular matrix [2, 5, 10]. These lesions are progressive, depending

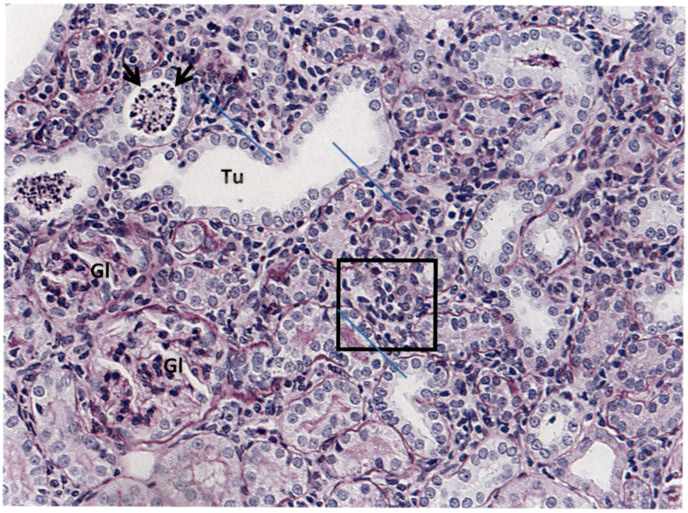

Fig. 1. Rat kidney 15 days after ureteral ligation showing tubular atrophy, tubular dilatation, interstitial mononuclear inflammatory cell infiltration and interstitial expansion square. The glomeruli, as shown here, do not show any significant changes (periodic acid-Schiff stain, original magnification ×100). Tu = Tubule; Gl = glomerulus.

on the duration of UO, and show little variation among the various species studied. The glomeruli and the large blood vessels are either normal or show mild changes later in the course of the disease (fig. 1, 2a–f).

Tubular Changes

The earliest changes observed following obstruction is tubular dilatation throughout the nephron, which is most pronounced in the collecting ducts [5, 10, 11]. This mechanical stretch is followed by a short period of marked tubular cell proliferation and subsequent progressive tubular cell apoptosis, accounting for the marked loss of renal parenchyma that follows [10, 11] (fig. 1, 2a and b). Tubular cell necrosis is noted in occasional tubules, but is never pronounced. The different tubular segments then lose their distinctive morphologic features and immunohistochemical profiles and become progressively atrophic [5, 10, 11]. Some tubular segments show progressive attenuation of epithelial cell markers (e.g. keratin, E-cadherin) simultaneously with the neo-expression of mesenchymal cell markers (e.g. vimentin, actin), which likely indicates the transformation of tubular epithelial cells into interstitial fibroblasts, a phenomenon termed epithelial-mesenchymal transformation [12]. In the model of congenital obstructive nephropathy, glomerulotubular disconnection is frequent, resulting in atubular glomeruli and aglomerular tubular segments [7].

These morphologic changes are accompanied by the neo-expression of a large variety of molecules implicated in mediating the tubular and interstitial changes described (partial listing in table 1), including those related to apoptosis and cell cycle regulation (e.g. caspases, intrinsic and extrinsic death pathway molecules, inhibitors of cyclin-dependent kinases p27 and p21, reactive oxygen species, and catalase) [13–39], the regulation of epithelial-mesenchymal transformation (hepatocyte growth factor, bone morphogenic protein 7, TGFβ-1, HIF-α and nestin) [40–43], hypoxic injury response (HIF-α) [44, 45], cytokines and growth factors (TGFβ-1, EGF, PDGF-C, VEGF, IGF-1, connective tissue growth factor and TNF-α) [3, 13, 14, 20, 22, 31, 36, 37, 46–49], and chemokines and chemoattractants (MCP-1, osteopontin, IL-1, ICAM-1, VCAM-1, angiotensin II and selectins) [13, 50–58]. This listing, impressive as it may appear, is by no means complete. In fact, the tissue mRNA array technique has helped reveal a markedly increased expression of several

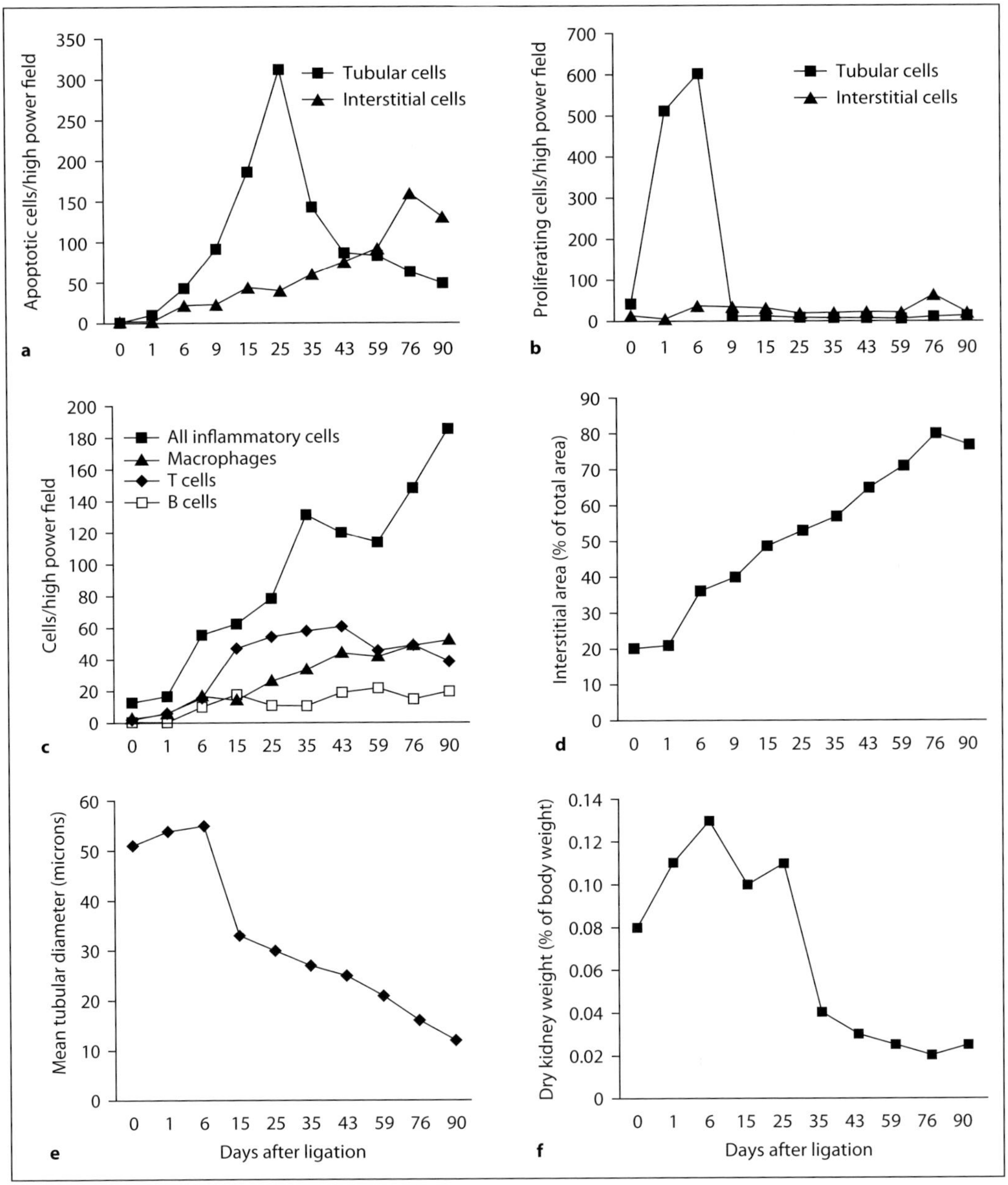

Fig. 2. Progressive tubulointerstitial changes induced by UO. **a** Tubular and interstitial cell apoptosis. **b** Tubular and interstitial cell proliferation. **c** Interstitial cell inflammatory infiltrates. **d** Interstitial area. **e** Tubular diameter. **f**. Progressive loss of renal weight (modified from [10]).

novel molecules, some of which carry no obvious or as yet known connection with the pathogenesis of TIN, such as PAX-2, PAX-8, GADD-45 and K9ca03.1 [19, 35, 59]. Continuous exploration and detection of candidate molecules offers new venues to study the pathogenesis of TIN in general. At the same time it indicates the extraordinary complexity and bewildering molecular interactions involved in the development of TIN.

Interstitial Inflammation

Recruitment of inflammatory cells occurs immediately after UO, leading to a several-fold increase in macrophages first noted in the corticomedullary junction within a few hours of obstruction [60]. This is followed by progressive interstitial inflammatory cellular infiltration, reaching a magnitude several times that at baseline in lesions of longer duration (weeks) of obstruction (fig. 1, 2c). The inflammatory cell infiltrates are composed principally of macrophages and activated T lymphocytes, with a smaller component of other inflammatory cells, such as B cells or plasma cells [60–62]. Neutrophils are absent or very scanty. This distinctive phenotype described in obstructive uropathy is also shared by other causes of TIN in both human and animal models [61, 62].

Apart from being the source of several pathogenic cytokines, the inflammatory cells may also express chemokines and chemokine receptors (e.g. CCR1, CCR5, MCP-1, CSF-1, macrophage-stimulating factor receptor, Toll-like receptor-2 and P2X7, see also table 1), implying both paracrine and autocrine loops for inflammatory cell recruitment [63–68]. Different macrophage phenotypes, either harmful or protective, have been identified [69]. Whether such a dual function is also true of the infiltrating T cells has not been examined, but is plausible.

Increased Interstitial Fibroblasts

Interstitial fibrosis develops in parallel with the tubular injury and interstitial inflammation observed in obstructive uropathy. This mainly consists of increased interstitial fibroblasts, interstitial expansion by the deposition of extracellular matrix, and loss of peritubular capillaries (fig. 1, 2d–f). Increased fibroblast density was traditionally attributed to local mitogenetic proliferation of resident fibroblasts [10, 70] (fig. 2b). Recent studies have shown that this mechanism accounts for only about half of these cells, with the rest due either to recruitment of bone marrow-derived fibrocytes from the circulation, or to local epithelial-mesenchymal transformation [12, 71, 72]. Collars of fibroblast-like cells surrounding dilated tubular segments have been described and may represent a morphologic manifestation of epithelial-mesenchymal transition [73, 74]. Pericytes and endothelial cells have most recently been implicated as additional sources of the interstitial fibroblasts [75, 76]. There is a temporal sequence for these different sources of interstitial fibrocytes, thus pericytic contribution develops immediately after UO, whereas epithelial-mesenchymal transformation starts several days later [71]. Aside from their increased density, the interstitial fibroblasts become 'activated', a phenomenon associated with the neo-expression of α-smooth muscle actin (myofibroblasts) and increased synthesis of extracellular matrix proteins [70, 71].

Extracellular Matrix Deposition

The interstitial fibrosis of obstructive uropathy is mainly due to the increased deposition of extracellular matrix proteins found in normal renal interstitium, including collagen types I, II, III, IV, V, VII and XV, laminin, fibronectin and heparan sulfate proteoglycans [5, 77]. Neo-expression of a novel type of extracellular matrix protein has not been described. This increased deposition of interstitial matrix is associated with loss of peritubular capillaries and endothelial cell apoptosis [78].

Table 1. Pathogenesis of the Renal Injury Caused by UO: Interventional Studies

Molecules/pathways	Function[a]	Interventional techniques	Results
		Tubular cell injury, apoptosis and proliferation	
Reactive oxygen species [2,3,13–17]	tubulotoxic	antagonist (exogenous antioxidants)	less apoptosis and fibrosis
Catalase [14, 15]	endogenous antioxidant in tubular cells	KO	more apoptosis
c-Jun amino-terminal kinase (JNK) 1 and 2 [18]	tubular cell apoptosis and renal fibrosis through the JNK/MAP kinase pathway	antagonist; KO	less fibrosis and apoptosis (antagonist); less apoptosis and unchanged fibrosis (KO)
Growth arrest and DNA damage-45 (GADD-45) δ [19]	nuclear protein regulating cell cycle, JNK pathway chemokines, and fibrosis	siRNA	less inflammation
TNF-α and its receptors [14, 20]	tubulotoxic	antagonist of receptor; KO of receptor genes	less tubular cell apoptosis and fibrosis
p53 [13, 21]	promotes apoptosis	KO	less tubular cell apoptosis
IGF-1 [14, 22]	cytoprotective	exogenous administration	less apoptosis and fibrosis
STAT-6 [30]	endogenous inhibitor of apoptosis	KO	more apoptosis
EGF [4, 13, 31]	tubular cell integrity	exogenous administration	less tubular cell apoptosis and fibrosis in rats; more renal injury in mice
Nitric oxide [13, 32]	cytoprotective; anti-apoptotic	antagonist; KO of nitric oxide synthetase gene	more tubular cell apoptosis, fibrosis and inflammation
Heat shock protein 72 [33]	tubular cell protection; inhibition of EMT	agonist	less apoptosis and fibrosis
Erythropoietin [34]	renoprotective	exogenous administration	less apoptosis, inflammation and fibrosis
PAX-2 [35]	renoprotective; anti-apoptosis	KO	more tubular cell apoptosis, more parenchymal loss
VEGF [36, 37]	survival factor for tubular and endothelial cells	exogenous administration	no effects
		Interstitial Inflammation	
CCR1 [63]	a chemokine receptor on macrophage and T cell membrane, chemoattractant	KO	less macrophage infiltration and fibrosis,
CCR5 [63]	a chemokine receptor on macrophage and T cell membrane, chemoattractant	KO	unchanged macrophage infiltration or renal injury

Table 1. Continued

Molecules/pathways	Function[a]	Interventional techniques	Results
MCP-1 [51]	chemoattractant of macrophages and other inflammatory cells	KO; antagonist	less macrophage infiltration
Selectins [52]	chemoattractant of macrophages and other inflammatory cells	KO; antagonist	less macrophage infiltration
Osteopontin [13, 14, 50, 54]	macrophage chemoattractant	KO	less macrophages; more apoptosis of tubular and interstitial cells
CD44 [55]	receptor for osteopontin	KO	less fibrosis and inflammation; more tubular cell injury
Colony-stimulating factor-1 [66]	macrophage chemoattractant	KO	less macrophages and tubular apoptosis
Toll-like receptor 2 [67]	Pro-inflammatory	KO	less inflammation and fibrosis; unchanged tubular cell injury
IL-1β and its receptor [64]	macrophage chemoattractant	KO of ligand or receptor gene	less fibrosis and apoptosis, unchanged macrophage infiltration
Macrophage colony-stimulating factor receptor c-fms [65]	receptor on macrophage cell membrane, macrophage recruitment	KO; antagonist	macrophage depletion; less tubular cell apoptosis; unchanged fibrosis
P2X7 [68]	receptors on inflammatory cell surface, inflammatory cell recruitment	KO	less inflammation
	Fibroblast Proliferation		
Renin-angiotensin system [4, 13, 56–58]	multiple functions (vasoconstriction, inflammatory cell chemoattractant, fibroblast proliferation, fibroblast activation, activation of TGF-β, TNF-α, transcription factor NF-κB)	KO of angiotensinogen, At_1 receptor or At_2 receptor genes; antagonists of angiotensin II; antagonist of NF-κB	attenuation of all lesions: less fibrosis, apoptosis and inflammation
TGFβ-1 and its signaling molecules Smads [2–4, 13, 46, 47]	multiple functions (reduced matrix metalloproteinase activity and tubular cell survival; increased matrix metalloproteinase inhibition, fibroblast proliferation, EMT, extracellular matrix synthesis and tubular cell apoptosis)	antagonists	less fibrosis, inflammation and apoptosis

Table 1. Continued

Molecules/pathways	Function[a]	Interventional techniques	Results
Smad3 [88, 89]	TGF signal pathway molecule, promoting TGFβ-1 activity	KO; antagonist	less TGF-β expression, inflammation, apoptosis and fibrosis
Smad7 [90]	TGF signal pathway molecule, inhibiting TGFβ-1 activity	KO	more TGF-β expression, inflammation, apoptosis and fibrosis
$\alpha_v\beta_6$ integrin [91]	activates TGFβ-1	KO	less fibrosis
Wnt/β catenin signaling pathway [84]	mesenchymal cell proliferation, renal development	antagonist; renal transfection of the Wnt inhibitory gene DKK-1	less fibrosis, renal catenin expression and fibroblast activation
Connective tissue growth factor [49]	fibrogenesis, effector of the TGF-β pathway	antisense treatment	less fibrosis
Intermediate-conductance Ca (2+)-activated K+ channel [K(Ca)3.1] [59]	mitogenic, fibroblast proliferation	KO; antagonist	less fibrosis and fibroblastic proliferation
Histone deacetylase [93, 104]	mitogenic, fibroblast proliferation (mediated by STAT3, a fibrogenic transcription factor)	antagonist	less fibroblast proliferation and activation; less tubular cell apoptosis
Integrin-linked kinase [94]	cell adhesion; cell survival; promotes EMT	antagonist	less fibrosis and fibroblast activation
Hypoxia induced factor (HIF)-α [44, 45]	fibrogenic; promotes EMT	antagonist; KO	less fibrosis
PDGF-C [48]	fibroblast mitogen	antagonist; KO	less fibrosis and inflammation
Decorin [95]	endogenous antagonist of TGFβ-1	KO	more fibrosis, inflammation and apoptosis
Death-associated protein kinase [96]	tubular cell endogenous promoter of apoptosis and inhibitor of fibrosis	KO	more fibrosis; unchanged apoptosis
Cyclin-dependent kinase inhibitor p27 and p21 [14, 38, 39]	inhibits cell proliferation	KO	more fibroblast proliferation and fibrosis
SCF-Skp2 [97]	ubiquitin ligase degrading p27	KO	less tubular cell proliferation, fibrosis and atrophy
Hepatocyte growth factor (HGF) [13, 14, 40, 41]	inhibits EMT (by Smads-mediated inhibition of TGF-β); cell survival; tissue regeneration	KO	more fibrosis and tubular cell apoptosis
		exogenous administration; renal transfection of HGF gene	less fibrosis, apoptosis, inflammation and atrophy

Table 1. Continued

Molecules/pathways	Function[a]	Interventional techniques	Results
HGF receptor Met [74]	receptor for hepatocyte growth factor	KO	more fibrosis and tubular cell apoptosis
Heme oxygenase-1 [98]	inhibits fibrogenesis and EMT	KO	more fibrosis, EMT and inflammation
Bone morphogenic protein-7 [42]	embryonic tubulogenesis, mature tubular cell integrity, inhibition of EMT and apoptosis	exogenous administration	less fibrosis and apoptosis
Bradykinin [99]	anti-fibrosis	KO of bradykinin receptor gene	more fibrosis
PPAR-λ [100]	transcription factor regulating inflammation, and cell growth	agonist	less fibrosis and inflammation,
Extracellular Matrix Deposition			
Plasminogen/plasmin [101]	extracellular matrix protein degradation	KO	less fibrosis (unexpected results)
Plasminogen activator [102]	converting plasminogen into plasmin	KO	less fibrosis (unexpected results, this KO probably also blocks EMT)
Tissue inhibitor of matrix metalloproteinase (TIMP)3 [103]	Inhibition of matrix metalloproteinase, extracellular matrix integrity	KO	more fibrosis and apoptosis

KO = Knockout mice; siRNA = small interfering RNA; EMT = epithelial-mesenchymal transformation; [a] Only known or putative functions pertinent to the pathogenesis of obstructive nephropathy are listed.

Glomerular and Vascular Changes

The glomeruli, arteries and veins show little or no changes early in the course of disease. In advanced stages, and against the background of very severe TIN, there may be mild, non-specific glomerular changes including capillary wall collapse or thickening, global sclerosis and vascular fibrous intimal thickening. This discrepancy in tubular and interstitial changes from the vascular and glomerular ones is also observed in human obstructive uropathy. The glomerular and vascular changes are collectively thought to reflect chronic ischemia common to advanced CKD of any cause and are not necessarily related per se or unique to obstructive uropathy. The relative glomerular and vascular sparing highlights the value of UO as a model to evaluate primary TIN, in which the tubulointerstitial injury is not pathogenetically due to a pre-existing glomerular or vascular disease. Of note and in contrast to the adult situation, glomerular changes are pronounced in neonatal UO and congenital obstructive nephropathy.

Kidney Function in Obstructive Uropathy

Changes in renal blood flow, glomerular filtration rate and tubular function develop immediately after experimental UO. The severity of these functional changes depends to a great extent on the degree of obstruction (partial vs. complete), its proximity to the kidney and the duration of obstruction. In general, the more complete the obstruction, the closer it is to the kidney, and the longer its duration, the more severe and less reversible the changes of kidney function become [79, 80].

The immediate response to UO is a brief period (min) of hyperemia and transient increase in renal blood flow, followed quickly by a progressive and gradually permanent vasoconstriction mediated by the increased generation of the vasoconstrictive agents mentioned above, such as angiotensin II, thromboxane A_2 and prostacyclines, and the diminished production of vasodilatory agents, such as nitric oxide, prostaglandins, platelet-activating factor and endothelin. The diminished renal blood flow is accompanied by a decline in glomerular filtration rate, initially due to the obstruction-induced increased intratubular hydrostatic pressure which opposes the normal filtration pressure gradient across the glomerular capillary ultrafiltration surface. This initial change in the physical forces governing filtration is soon magnified by the reduced glomerular capillary pressure induced by vasoconstriction, and thereafter by the diminution of available ultrafiltration surface caused by the decrease in number of functional nephrons [79–82].

The principal abnormalities in tubular function that develop consequent to the tubular epithelial cell injury are impaired reabsorption of solutes and water, and reduced capacity to concentrate and acidify the urine [42–85]. Most of these tubular abnormalities are especially evident after relief of obstruction and may persist as permanent abnormalities, depending on the duration of UO [80, 84, 86, 87].

Renal Lesions in Neonatal UO

Congenital obstructive nephropathy, the onset of which is often prenatal, is the most frequent cause of end-stage renal diseases in children. Treatment of this condition, including various considerations related to surgical relief of the obstruction, and the final renal outcome are predicated on understanding the pathogenesis and development of the renal lesions both before and after the relief of UO. Creation of prenatal UO is theoretically possible in larger animals such as sheep or monkeys, but is often impractical [6–9]. However, neonatal UO in rats or mice is suitable for studying the effect of prenatal obstructive uropathy in humans for the following reason. In humans, nephrogenesis is complete at the 34th gestational week, and is followed by glomerular maturation in subsequent months of gestation. In rats or mice, only about 10% of glomeruli are generated at birth, with the rest formed in the first 2 weeks of the neonatal period. Thus, neonatal UO in these species is expected to have the same impact on nephrogenesis as obstructive uropathy in the human fetus at the mid-trimester of gestation, a time of critical nephrogenesis [6–9].

UO in neonatal rats or mice induces the same spectrum of TIN as seen in adults [6–9]. In addition, UO during this critical period of embryogenesis irreversibly impairs renal and glomerular development and maturation. The magnitude of these impairments depends at least in part on the duration of UO, but may be pronounced even after a short period of UO. Thus, neonatal UO of a 4-day duration may reduce renal growth by 60% and the number of glomeruli by 50% [6–9]. There are also several glomerular changes including global sclerosis, immature glomeruli, collapsed glomerular capillaries, enlarged urinary space, prominent visceral or parietal epithelial cells, and atubular glomeruli due to a glomerulotubular disconnection at the tubular pole [6–9]. The same spectrum of glomerular changes is noted in kidney biopsies of children obtained during pyeloplasty to relieve congenital obstructive uropathy [87]. The renal lesions, including the glomerular changes, of congenital UO have a profound prognostic implication since their severity, which in turn reflects the duration of UO, is a major determinant of residual kidney function (stable, better or worse) and of renal outcome after UO is relieved (see below).

Reversibility of TIN

Since most renal diseases are diagnosed at a stage when TIN may already be well established, there is considerable clinical interest in the course of these lesions, especially whether they are reversible once the original kidney disease is treated. The UO model in which obstruction of the urine outflow can be relieved at various and precise time intervals is most suitable to address this question.

Chevalier et al. [8] have shown that relief of complete UO of a 5-day duration in neonatal rats, when the renal lesions are already severe, incurred no reversal of the lesions. There was less interstitial fibrosis and fibroblast proliferation 1 month after relief of UO; however, tubular cell proliferation and apoptosis, and elevated TGF-β, persisted, leading to progressive tissue injury and loss of kidney function (80% loss of glomerular filtration rate) and proteinuria 1 year later. In contrast, relief of partial UO of 5-day duration, when there was only mild tubular apoptosis and atrophy, was followed by progression of the injury, including fibrosis at day 21; however, a complete resolution of structure and function was noted at 42 days after the partial UO was relieved.

The same group compared the renal lesions after relief of complete UO of a 5-day duration induced in rats 1 day postnatally (when nephrogenesis is active) versus 19 days postnatally (when nephrogenesis is almost completed). They confirmed that 3 months after the relief of obstruction, renal lesions not only failed to reverse, but also progressed in both groups. The progression was more pronounced in the 19-day group [9].

Ito et al. [6] showed that relief of complete UO of a 3-day duration in adult rats restored renal blood flow and glomerular filtration rate; however, interstitial inflammation, interstitial fibrosis and tubular cell apoptosis persisted at 28 days after the relief.

These observations suggest that obstructive uropathy induced by UO in either neonatal immature or mature adult kidney shows little reversibility after the obstruction is relieved and that the renal injury progresses after release of the obstruction. The rate of progression may depend on the level of renal maturity. How much of this pessimistic sequence of events noted in experimental models is applicable to the fate of TIN in obstructive uropathy encountered in the clinical setting remains unknown.

Pathogenesis of UO-Induced TIN

The characteristic renal changes induced by UO, together with the identification of several molecules with known pathogenic effects have provided considerable insight into the pathogenesis of TIN. Recent proteomic techniques and tissue mRNA microarray have expanded the list of molecules expressed by the obstruction-injured kidney tissue. These findings provide a foundation for potential interventional studies using the exogenous administration of pharmacologic agonist or antagonist, and the genetic manipulation of these molecules to treat and further dissect the complex pathogenesis of UO-induced TIN [88–104] (table 1).

UO immediately activates the renin-angiotensin system and creates an imbalance of vasoconstrictors (e.g. angiotensin II) and vasodilators (e.g. prostaglandin, thromboxane), in favor of the former, leading to acute renal ischemia and tissue hypoxia [4, 13, 56–58]. UO also induces mechanical stretching of tubular cells [3, 13, 28, 29]. These two insults initiate the profound structural and functional changes of the tubular epithelial cells including the neo-expression of a large number of molecules implicated in the control of cell proliferation, apoptosis, renal hemodynamics, inflammatory cell recruitment and epithelial-mesenchymal transformation (table 1 and see also above). Several of these molecules are likely responsible for the tubular cell injury that ensues, including apoptosis, atrophy, loss of differentiation and transformation into interstitial fibroblasts.

Inflammatory cells, especially macrophages, are recruited immediately after UO, probably under the influence of the activated renin-angiotensin system [56–58, 60]. Additional and more robust inflammatory cell chemoattraction is probably provided by tubular cell-derived chemokines [13, 50–58]. Many of these same chemokines and their receptors are also expressed by the infiltrating inflammatory cells, thus magnifying inflammatory cell recruitment through an autocrine loop [63–68]. The inflammatory cell profile is distinctive for the predominance of macrophages and T cells, with a paucity of other cell types, likely reflecting a chemokine profile specific for the recruited inflammatory cells.

Cytokines and chemokines may also induce proliferation and apoptosis of the resident interstitial fibroblasts (table 1). The increased interstitial fibroblast density is to a great extent determined by the transformation of tubular epithelial cells into interstitial fibroblasts, a process under dual-stimulating (TFG-β, angiotensin II, HIF-α) [46, 47, 56–58] and inhibitory (hepatocyte growth factor, bone morphogenic protein 7, heme oxygenase-1, nestin) [40–43, 90] effects of various epithelial cell-derived molecules (table 1). Recently, interstitial homing of bone marrow-derived fibrocytes and transformation of endothelial cells or pericytes into interstitial fibroblasts have been shown to contribute to this pool. The molecular control of these novel processes is currently unknown [71–76]. Infiltrating interstitial fibroblasts are activated and responsible for the increased synthesis of extracellular matrix proteins [71–76]. This together with an impairment of different fibrolytic pathways [101–103] leads to the preferential increase of extracellular matrix deposition. These changes collectively result in renal tissue loss and scarring.

In the neonatal context, UO not only causes TIN, but also profoundly affects normal nephrogenesis. Several molecules known to regulate renal development have been shown to be altered by UO, including angiotensin II, bone morphemic protein 7, PAX-2, PAX-8 and molecules of the Wnt/catenin pathway [35, 42, 56, 57, 58, 92] (table 1). Their role in UO-mediated maldevelopment of the neonatal kidney remains to be determined.

Several considerations related to the above pathogenetic scheme deserve further comment. First, there is marked cross-talk, as well as overlapping and molecular redundancy of the various pathogenetic pathways in which the same molecule can exert variable effects, and conversely one type of change that can be mediated by a number of otherwise unrelated molecules. For example, TGFβ-1 can mediate apoptosis, fibrosis, epithelial-mesenchymal transformation and chemotaxis; tubular cell apoptosis is under the control of several apoptotic molecules, reactive oxygen species and angiotensin II; and manipulation of a molecule with a known function, e.g. anti-apoptotic, may attenuate not only cell death, but also inflammation. This extraordinary complexity and redundancy may account for the fact that no interventional manipulation aiming at a single molecule can normalize the renal changes; and the initial protective effect of an interventional manipulation tends to wear off during the course of longer periods of UO, indicating the emergence of compensatory or counter-regulatory pathogenetic pathways.

Secondly, UO induces not only harmful molecules, but also molecules with opposing protective effects. This dual control is noted for most major pathogenetic pathways. For example, epithelial-mesenchymal transformation is promoted by TGFβ-1 and angiotensin II, but is inhibited by hepatocyte growth factor, bone morphogenic protein 7 and nestin; tubular cell apoptosis is promoted by reactive oxygen species, but protected by endogenous catalase, STAT-6 and heat shock protein 72; and there is a dual population of infiltrating macrophages with either harmful or protective effects. Strategies to enhance these protective elements may have therapeutic value, but remain to be identified.

Thirdly, the same molecule may mediate a harmful effect in one cell type but is protective in another. For example, osteopontin is a macrophage chemoattractant, but is also anti-apoptotic.

Finally, the effect of a molecule may be species- or age-dependent. For example, exogenous EGF attenuates UO-induced renal injury in humans and adult rats, but enhances these changes in neonatal mice.

Conclusion

The classical features of obstructive uropathy are regularly induced in the experimental model of UO. This versatile model has contributed much to elucidating the mechanism of UO-induced TIN. Translation of this vast and complex body of knowledge into TIN in general and into effective treatment of obstructive uropathy remains the challenge of future studies.

References

1 Schainuck LI, Striker GE, Cutler RE, et al: Structural-functional correlations in renal disease. Part II: the correlations. Hum Pathol 1970;1:631–641.
2 Chevalier RL, Forbes MS, Thornhill BA: Ureteral obstruction as a model of renal interstitial fibrosis and obstructive nephropathy. Kidney Int 2009;75:1145–52.
3 Chevalier RL, Thornhill BA, Forbes MS, Kiley SC: Mechanisms of renal injury and progression of renal disease in congenital obstructive nephropathy. Pediatr Nephrol 2010;25:687–697.
4 Chevalier RL: Pathogenesis of renal injury in obstructive uropathy. Curr Opin Pediatr 2006;18:153–160.
5 Nagle RB, Bulger RE, Cutler RE, et al: Unilateral obstructive nephropathy in the rabbit: early morphologic, physiologic, and histochemical changes. Lab Invest 1973;28:456–467.
6 Ito K, Chen J, El Chaar M, Stern JM, Seshan SV, Khodadadian JJ, Richardson I, Hyman MJ, Vaughan ED Jr, Poppas DP, Felsen D: Renal damage progresses despite improvement of renal function after relief of unilateral ureteral obstruction in adult rats. Am J Physiol 2004;287:F1283–F1293.
7 Thornhill BA, Forbes MS, Marcinko ES, Chevalier RL: Glomerulotubular disconnection in neonatal mice after relief of partial ureteral obstruction. *Kidney Int* 2007;72:1103–1112.
8 Chevalier RL, Kim A, Thornhill BA, Wolstenholme JT: Recovery following relief of unilateral ureteral obstruction in the neonatal rat. Kidney Int 1999;55:793–807.
9 Chevalier RL, Thornhill BA, Chang AY, Cachat F, Lackey A: Recovery from release of ureteral obstruction in the rat: relationship to nephrogenesis. Kidney Int 2002;61:2033–2043.
10 Truong LD, Petrusevska G, Yang G, Gurpinar T, Shappel S, Lechago J, Rouse D, Suki WN. Cell apoptosis and proliferation in experimental chronic obstructive uropathy. *Kidney Int* 1996;50:200–207.
11 Gobe GC, Axelsen RA: Genesis of renal tubular atrophy in experimental hydronephrosis in the rat: role of apoptosis. Lab Invest 1987;56:273–281.
12 Iwano M, Plieth D, Danoff TM, et al: Evidence that fibroblasts derive from epithelium during tissue fibrosis. J Clin Invest 2002;110:341–350.
13 Bascands JL, Schanstra JP: Bradykinin and renal fibrosis: Have we ACE'd it? J Am Soc Nephrol 2004;15:2504–2506.
14 Docherty NG, O'Sullivan OE, Healy DA, Fitzpatrick JM, Watson RW: Evidence that inhibition of tubular cell apoptosis protects against renal damage and development of fibrosis following ureteric obstruction. Am J Physiol Renal Physiol 2006;290:F4–F13.
15 Sunami R, Sugiyama H, Wang DH, Kobayashi M, Maeshima Y, Yamasaki Y, Masuoka N, Ogawa N, Kira S, Makino H: Acatalasemia sensitizes renal tubular epithelial cells to apoptosis and exacerbates renal fibrosis after unilateral ureteral obstruction. *Am J Physiol Renal Physiol* 2004; 286:F1030–F1038.
16 Pat B, Yang T, Kong C, Watters D, Johnson DW, Gobe G: Activation of ERK in renal fibrosis after unilateral ureteral obstruction: modulation by antioxidants. Kidney Int 2005;67:931–943.
17 Jones EA, Shahed J, Shoskes DA: Modulation of apoptotic and inflammatory genes by bioflavonoids and angiotensin II inhibition in ureteral obstruction. Urology 2000;56:346–351.
18 Ma FY, Flanc RS, Tesch GH, Han Y, Atkins RC, Bennett BL, Friedman GC, Fan JH, Nikolic-Paterson DJ: A pathogenic role for c-Jun amino-terminal kinase signaling in renal fibrosis and tubular cell apoptosis. J Am Soc Nephrol 2007;18:472–484.
19 Shin GT, Kim DR, Lim JE, Yim H, Kim H: Upregulation and function of GADD45gamma in unilateral ureteral obstruction. Kidney Int 2008;73:1251–1265.
20 Guo G, Morrissey R, McCracken R, Tolley T, Klahr S: Role of TNFR1 and TNFR2 receptors in tubulointerstitial fibrosis of obstructive nephropathy. Am J Physiol Renal Physiol 1999;277:F766–F772.
21 Choi YJ, Mendoza L, Rha SJ, Sheikh-Hamad D, Baranowska-Daca E, Nguyen V, Smith CW, Nassar G, Suki WN, Truong LD: Role of p53-dependent activation of caspases in chronic obstructive uropathy: evidence from p53 null mutant mice. J Am Soc Nephrol 2001;12:983–992.

22 Chevalier RL, Goyal S, Kim A, Chang AY, Landau D, LeRoith D: Renal tubulointerstitial injury from ureteral obstruction in the neonatal rat is attenuated by IGF-1. Kidney Int 2000;57:882–890.
23 Misseri R, Meldrum DR, Dinarello CA, et al: TNH-alpha mediates obstruction-induced renal tubular cell apoptosis and proapoptotic signaling. Am J Physiol 2004;288:F406–F411.
24 Truong LD, Choi YJ, Tsao CC, Ayala G, Sheikh-Hamad D, Nassar G, Suki WN: Renal cell apoptosis in chronic obstructive uropathy: the roles of caspases. Kidney Int 2001;60:924–934.
25 Choi Y-J, Baranowska-Daca E, Nguyen V, Koji T, Ballantyne CM, Sheikh-Hamad D, Suki WN, Truong LD: Mechanism of chronic obstructive uropathy: increased expression of apoptosis-promoting molecules. Kidney Int 2000;58:1481–1491.
26 Lapis H, Yu H, Steinhardt GF: Cell proliferation, apoptosis, Bcl-2 and Bax expression in obstructed opossum early metanephroi. J Urol 2000;164:511–517.
27 Zhang G, Oldroyd SD, Huang LH, Yang B, Li Y, Ye R, El Nahas AM: Role of apoptosis and Bcl-2/Bax in the development of tubulointerstitial fibrosis during experimental obstructive nephropathy. Exp Nephrol 2001;9:71–80.
28 Ricardo SD, Ding G, Eufemio M, Diamond JR: Antioxidant expression in experimental hydronephrosis: role of mechanical stretch and growth factors. Am J Physiol Renal Physiol 1997;272:F789–F798.
29 Power RE, Doyle BT, Higgins D, Brady HR, Fitzpatrick JM, Watson RW: Mechanical deformation induced apoptosis in human proximal renal tubular epithelial cells is caspase dependent. J Urol 2004;171:457–461.
30 Yukawa K, Kishino M, Goda M, et al: STAT6 deficiency inhibits tubulointerstitial fibrosis in obstructive nephropathy. Int J Mol Med 2005;15:225–230.
31 Kennedy WA II, Buttyan R, Garcia-Montes E, D'Agati V, Olsson CA, Sawczuk IS: Epidermal growth factor suppresses renal tubular apoptosis following ureteral obstruction. Urology 1997;49:973–980.
32 Hochberg D, Johnson CW, Chen J, Cohen D, Stern J, Vaughan ED Jr., Poppas D, Felsen D: Interstitial fibrosis of unilateral ureteral obstruction is exacerbated in kidneys of mice lacking the gene for inducible nitric oxide synthase. Lab Invest 2000;80:1721–1728.
33 Mao H, Li Z, Zhou Y, Li Z, Zhuang S, An X, Zhang B, Chen W, Nie J, Wang Z, Borkan SC, Wang Y, Yu X: HSP72 attenuates renal tubular cell apoptosis and interstitial fibrosis in obstructive nephropathy. Am J Physiol Renal Physiol 2008;295:F202–F214.
34 Srisawat N, Manotham K, Eiam-Ong S, Katavetin P, Praditpornsilpa K, Eiam-Ong S: Erythropoietin and its non-erythropoietic derivative: do they ameliorate renal tubulointerstitial injury in ureteral obstruction? Int J Urol 2008;15:1011–1017.
35 Cohen T, Loutochin O, Amin M, Capolicchio JP, Goodyer P, Jednak R: PAX2 is reactivated in urinary tract obstruction and partially protects collecting duct cells from programmed cell death. Am J Physiol Renal Physiol 2007;292:F1267–F1273.
36 Burt LE, Forbes MS, Thornhill BA, Kiley SC, Chevalier RL: Renal vascular endothelial growth factor in neonatal obstructive nephropathy. I. Endogenous VEGF. Am J Physiol Renal Physiol 2007;292:F158–F167.
37 Burt LE, Forbes MS, Thornhill BA, Kiley SC, Chevalier RL: Renal vascular endothelial growth factor in neonatal obstructive nephropathy. II. Exogenous VEGF. Am J Physiol Renal Physiol 2007;292:F168–F174.
38 Hughes J, Brown P, Shankland SJ: Cyclin kinase inhibitor p21CIP1/WAF1 limits interstitial cell proliferation following ureteric obstruction. Am J Physiol 1999;277: F948–F956.
39 Ophascharoensuk V, Fero ML, Hughes J, et al: The cyclin-dependent kinase inhibitor p27Kip1 safeguards against inflammatory injury. Nat Med 1998;4:575–580.
40 Gao X, Mae H, Ayabe N, Takai T, Oshima K, Hattori M, Ueki T, Fujimoto J, Tanizawa T: Hepatocyte growth factor gene therapy retards the progression of chronic obstructive nephropathy. Kidney Int 2002;62:1238–1248.
41 Yang J, Liu Y: Delayed administration of hepatocyte growth factor reduces renal fibrosis in obstructive nephropathy. Am J Physiol Renal Physiol 2003;284:F349–F357.
42 Klahr S, Morrissey J: Obstructive nephropathy and renal fibrosis: the role of bone morphogenic protein-7 and hepatocyte growth factor. Kidney Int Suppl 2003;87:S105–S112.
43 Sakairi T, Hiromura K, Yamashita S, Takeuchi S, Tomioka M, Ideura H, Maeshima A, Kaneko Y, Kuroiwa T, Nangaku M, Takeuchi T, Nojima Y: Nestin expression in the kidney with an obstructed ureter. Kidney Int 2007;72:307–318.
44 Higgins DF, Kimura K, Bernhardt WM, Shrimanker N, Akai Y, Hohenstein B, Saito Y, Johnson RS, Kretzler M, Cohen CD, Eckardt KU, Iwano M, Haase VH: J Clin Invest 2007;117:3810–3820.
45 Kimura K, Iwano M, Higgins DF, Yamaguchi Y, Nakatani K, Harada K, Kubo A, Akai Y, Rankin EB, Neilson EG, Haase VH, Saito Y: Stable expression of HIF-1alpha in tubular epithelial cells promotes interstitial fibrosis. Am J Physiol Renal Physiol 2008;295: F1023–F1029.
46 Miyajima A, et al: Novel nuclear factor κB activation inhibitor prevents inflammatory injury in unilateral ureteral obstruction. J Urol 2003;169:1559–1563.
47 Miyajima A, Chen J, Lawrence C, Ledbetter S, Soslow RA, Stern J, Jha S, Pigato J, Lemer ML, Poppas DP, Vaughan ED, Felsen D: Antibody to transforming growth factor-β ameliorates tubular apoptosis in unilateral ureteral obstruction. Kidney Int 2000;58:2301–2313.
48 Eitner F, Bücher E, van Roeyen C, Kunter U, Rong S, Seikrit C, Villa L, Boor P, Fredriksson L, Bäckström G, Eriksson U, Ostman A, Floege J, Ostendorf T: PDGF-C is a proinflammatory cytokine that mediates renal interstitial fibrosis. J Am Soc Nephrol 2008;19:281–9.
49 Yokoi H, Mukoyama M, Nagae T, et al: Reduction in connective tissue growth factor by antisense treatment ameliorates renal tubulointerstitial fibrosis. J Am Soc Nephrol 2004;15:1430–1440.

50 Giachelli CM, Lombardi D, Johnson RJ, et al: Evidence for a role of osteopontin in macrophage infiltration in response to pathological stimuli in vivo. Am J Pathol 1998;152:353–358.
51 Wada T, Furuichi K, Sakai N, et al: Gene therapy via blockade of monocyte chemoattractant protein-1 for renal fibrosis. J Am Soc Nephrol 2004;15:940–948.
52 Lange-Sperandio B, Cachat F, Thornhill BA, Chevalier RL: Selectins mediate machrophage infiltration in obstructive nephropathy in newborn mice. Kidney Int 2002;61:516–524.
53 Shappell SB, Mendoza LH, Gurpinar T, et al: Expression of adhesion molecules in kidney with experimental chronic obstructive uropathy: The pathogenic role of ICAM-1 and VCAM-1. Nephron 2000;85:156–166.
54 Yoo KH, Thornhill BA, Forbes MS, Coleman CM, Marcinko ES, Liaw L, Chevalier RL: Osteopontin regulates renal apoptosis and interstitial fibrosis in neonatal chronic unilateral ureteral obstruction. Kidney Int 2006;70:1735–1741.
55 Rouschop KM, Sewnath ME, Claessen N, et al: CD44 deficiency increases tubular damage but reduces renal fibrosis in obstructive nephropathy. J Am Soc Nephrol 2004;15:674–686.
56 Fan YY, Nishiyama A, Fujisawa Y, Kobori H, Nakano D, Matsuura J, Hase N, Hitomi H, Kiyomoto H, Urata H, Kohno M: Contribution of chymase-dependent angiotensin II formation to the progression of tubulointerstitial fibrosis in obstructed kidneys in hamsters. J Pharmacol Sci 2009;111:82–90.
57 Satoh M, Kashihara N, Yamasaki Y, et al: Renal interstitial fibrosis is reduced in angiotensin II type 1a receptor-deficient mice. J Am Soc Nephrol 2001;12:317–325.
58 Esteban V, Lorenzo O, Rupérez M, et al: Angiotensin II, via AT1 and AT2 receptors and NF-kappaβ pathway regulates the inflammatory response in unilateral ureteral obstruction. J Am Soc Nephrol 2004;15:1514–1529.
59 Grgic I, Kiss E, Kaistha BP, Busch C, Kloss M, Sautter J, Müller A, Kaistha A, Schmidt C, Raman G, Wulff H, Strutz F, Gröne HJ, Köhler R, Hoyer: Renal fibrosis is attenuated by targeted disruption of KCa3.1 potassium channels. Proc Natl Acad Sci USA 2009;106:14518–14523.
60 Schreiner GF, Harris KPG, Purkerson ML, Klahr S: Immunological aspects of acute ureteral obstruction: immune cell infiltrate in the kidney. Kidney Int 1988;34:487–493.
61 Truong LD, Farhood A, Tasby J, Gillum D: Experimental chronic renal ischemia: morphologic and immunologic studies. Kidney Int 1992;41:1676–1689.
62 Stachura I, Si L, Madan E, Whiteside T: Mononuclear cell subsets in human renal disease. Enumeration in tissue sections with monoclonal antibodies. Clin Immunol Immunopathol 1984;30:362–373.
63 Eis V, Luckow B, Vielhauer V, et al: Chemokine receptor CCR1 but not CCR5 mediates leukocyte recruitment and subsequent renal fibrosis after unilateral ureteral obstruction. J Am Soc Nephrol 2004;15:337–347.
64 Jones LK, O'Sullivan KM, Semple T, Kuligowski MP, Fukami K, Ma FY, Nikolic-Paterson DJ, Holdsworth SR, Kitching AR: IL-1RI deficiency ameliorates early experimental renal interstitial fibrosis. Nephrol Dial Transplant 2009;24:3024–3032.
65 Ma FY, Liu J, Kitching AR, Manthey CL, Nikolic-Paterson DJ: Targeting renal macrophage accumulation via cfms kinase reduces tubular apoptosis but fails to modify progressive fibrosis in the obstructed rat kidney. Am J Physiol Renal Physiol 2009;296: F177–F185.
66 Lenda DM, Kikawada E, Stanley ER, Kelley VR: Reduced macrophage recruitment, proliferation and activation in colony-stimulating factor-1-deficient mice results in decreased tubular apoptosis during renal inflammation. J Immunol 2003;170:3254–3262.
67 Leemans JC, Butter LM, Pulskens WP, Teske GJ, Claessen N, van der Poll T, Florquin S: The role of Toll-like receptor 2 in inflammation and fibrosis during progressive renal injury. PLoS One 2009;4:e5704.
68 Gonçalves RG, Gabrich L, Rosário A Jr, Takiya CM, Ferreira ML, Chiarini LB, Persechini PM, Coutinho-Silva R, Leite M Jr: The role of purinergic P2X7 receptors in the inflammation and fibrosis of unilateral ureteral obstruction in mice. Kidney Int 2006; 70:1599–1606.
69 Nishida M, Hamaoka K: Macrophage phenotype and renal fibrosis in obstructive nephropathy. Nephron Exp Nephrol 2008;110:e31–e36.
70 Picard N, Baum O, Vogetseder A, Kaissling B, Le Hir M: Origin of renal myofibroblasts in the model of unilateral ureter obstruction in the rat. Histochem Cell Biol 2008;130:141–155.
71 Grande MT, López-Novoa JM: Fibroblast activation and myofibroblast generation in obstructive nephropathy. Nat Rev Nephrol 2009;5:319–328.
72 Wada T, Sakai N, Matsushima K, Kaneko S: Fibrocytes: a new insight into kidney fibrosis. Kidney Int 2007;72:269–273.
73 Butt MJ, Tarantal AF, Jimenez DF, Matsell DG: Collecting duct epithelial-mesenchymal transition in fetal urinary tract obstruction. Kidney Int 2007;72:936–944.
74 Ma H, Saenko M, Opuko A, Togawa A, Soda K, Marlier A, Moeckel GW, Cantley LG, Ishibe S: Deletion of the Met receptor in the collecting duct decreases renal repair following ureteral obstruction. Kidney Int 2009;76: 868–876.
75 Zeisberg EM, Potenta SE, Sugimoto H, et al: Fibroblasts in kidney fibrosis emerge via endothelial-to-mesenchymal transition. J Am Soc Nephrol 2008;19:2282–2287.
76 Lin SL, Kisseleva T, Brenner DA, Duffield JS: Pericytes and perivascular fibroblasts are the primary source of collagen-producing cells in obstructive fibrosis of the kidney. Am J Pathol 2008;173:1617–1627.
77 Sharma AK, Mauer SM, Kim Y, et al: Interstitial fibrosis in obstructive nephropathy. Kidney Int 1993;44:774–788.
78 Ohashi R, Shimizu A, Masuda Y, Kitamura H, Ishizaki M, Sugisaki Y, Yamanaka N: Peritubular capillary regression during the progression of experimental obstructive nephropathy. J Am Soc Nephrol 2002;13:1795–1805.
79 Klahr S, Morrisey J: Obstructive nephropathy and renal fibrosis. Am J Physiol Renal Physiol 2002;283:F861–F875.
80 Klahr S: Urinary tract obstruction. Semin Nephrol 2001;21:137–145.

81 Suki W, Eknoyan G, Rector FC Jr, Seldin DW: Patterns of nephron perfusion in acute and chronic hydropnephrosis. J Clin Invest 1966;45:122–131.
82 Pedersen TS, Hvistendahl JJ, Djurhuus JC, Frokiaer J: Renal water and sodium handling during graded unilateral ureter obstruction. Scand J Urol Nephrol 2002;36:163–172.
83 Eknoyan G, Suki WN, Martinez-Maldonado M, Anhalt M: Chronic hydronephrosis. Observations on the mechanism of the defect in urine concentration. Proc Soc Exp Biol Med 1970;134:634–639.
84 Li C, Wang W, Kwon T, Knepper MA, Nielsen S, Frokiaer J: Altered expression of major renal Na transporters in rats with bilateral ureteral obstruction and release of obstruction. Am J Physiol Renal Physiol 2003;385:F889–F901.
85 Wang G, Ring T, Li C, Kim SW, Wen J, Djurhuus JC, Nielsen S, Frokiaer J: Unilateral ureteral obstruction alters expression of acid-base transporters in rat kidney. J Urol 2009;182:2964–2973.
86 Shi Y, Pedersen M, Li C, Wen JC, Thomsen K, Jorgensen H, Jorgensen TM, Knepper MA, Nielsen S, Djurhuus JC, Frokaier J: Early release of neonatal ureteral obstruction preserves renal function. Am J Physiol Renal Physiol 2004;286:F1087–F1099.
87 Huang W-Y, Peters CA, Zurakowski D, Borer JG, Diamond DA, Bauer SB, McLellan DL, Rosen S: Renal biopsy in congenital ureteropelvic junction obstruction: Evidence for parenchymal maldevelopment. Kidney Int 2006;69:137–143.
88 Moon JA, Kim HT, Cho IS, et al: IN-1130, a novel TGF-beta type 1 receptor kinase (ALKS) inhibitor, suppresses renal fibrosis in obstructive nephropathy. Kidney Int 2006;70: 1234–1243.
89 Ihazaki K, Kanamaru Y, Kojima Y, et al: Smad3 deficiency attenuates renal fibrosis, inflammation and apoptosis after unilateral ureteral obstruction. Kidney Int 2004;66:597–604.
90 Chung AC, Huang XR, Zhou L, Heuchel R, Lai KN, Lan HY: Disruption of the Smad7 gene promotes renal fibrosis and inflammation in unilateral ureteral obstruction (UUO) in mice. Nephrol Dial Transplant 2009;24:1443–1154.
91 Ma LJ, Yang H, Gaspert A, et al: Transforming growth factor-beta-dependent and -independent pathways of induction of tubulointerstitial fibrosis in beta 6(-/-) mice. Am J Pathol 2003;163:1261–1273.
92 He W, Dai C, Li Y, Zeng G, Monga SP, Liu Y: Wnt/beta-catenin signaling promotes renal interstitial fibrosis. J Am Soc Nephrol 2009;20:765–776.
93 Pang M, Kothapally J, Mao H, Tolbert E, Ponnusamy M, Chin YE, Zhuang S: Inhibition of histone deacetylase activity attenuates renal fibroblast activation and interstitial fibrosis in obstructive nephropathy. Am J Physiol Renal Physiol 2009;297: F996–F1005.
94 Li Y, Tan X, Dai C, Stolz DB, Wang D, Liu Y: Inhibition of integrin-linked kinase attenuates renal interstitial fibrosis. J Am Soc Nephrol 2009;20:1907–1918.
95 Schaeffer L, Macakova K, Raslik I, et al: Absence of decorin adversely influences tubulointerstitial fibrosis of the obstructed kidney by enhanced apoptosis and increased inflammatory reaction. Am J Pathol 2002; 160:1181–1197.
96 Yukawa K, Shirasawa N, Ohshima A, Mune M, Kimura A, Bai T, Tsubota Y, Owada-Makabe K, Tanaka T, Kishino M, Tsuruo Y, Umesaki N, Maeda M: Death-associated protein kinase localization to human renal tubule cells and increased expression of chronic obstructive uropathy in rats. J Nephrol 2004;17:26–33.
97 Suzuki S, Fukasawa H, Kitagawa K, Uchida C, Hattori T, Isobe T, Oda T, Misaki T, Ohashi N, Nakayama K, Nakavama KI, Hishida A, Yamamoto T, Kitagawa M: Renal damage in obstructive nephropathy is decreased in Skp2-deficient mice. Am J Pathol 2007;171:473–483.
98 Kie JH, Kapturczak MH, Traylor A, Agarwal A, Hill-Kapturczak N: Heme oxygenase-1 deficiency promotes epithelial-mesenchymal transition and renal fibrosis. J Am Soc Nephrol 2008;19:1681–1691.
99 Schanstra JP, Duchene J, Desmond L, et al: The protective effect of angiotensin converting enzyme inhibition in experimental renal fibrosis in mice is not mediated by bradykinin B2 receptor activation. Thromb Haemost 2003;89:735–740.
100 Kawai T, Masaki T, Doi S, Arakawa T, Yokovama Y, Doi T, Kohno N, Yorioka N: PPAR-gamma agonist attenuates renal interstitial fibrosis and inflammation through reduction of TGF-beta. Lab Invest 2009;89:47–58.
101 Edgtton KL, Gow RM, Kelly DJ, et al: Plasmin is not protective in experimental renal interstitial fibrosis. Kidney Int 2004;66:68–76.
102 Yang J, Shultz RW, Mars WM, et al: Disruption of tissue-type plasminogen activator gene in mice reduces renal interstitial fibrosis in obstructive nephropathy. J Clin Invest 2002;110:1525–1538.
103 Kassiri Z, Oudit GY, Kandalam V, Awad A, Wang X, Ziou X, Maeda N, Herzenberg AM, Scholey JW: Loss of TIMP3 enhances interstitial nephritis and fibrosis. J Am Soc Nephrol 2009;20:1223–1235.
104 Kuratsune M, Masaki T, Hirai T, Kiribayashi K, Yokoyama Y, Arakawa T, Yorioka N, Kohno N: Signal transducer and activator of transcription 3 involvement in the development of renal interstitial fibrosis after unilateral ureteral obstruction. Nephrology 2007;12:565–571.

Luan D. Truong, MD
Department of Pathology, MS 205, The Methodist Hospital
6565 Fannin Street
Houston, TX 77030 (USA)
Tel. +1 713 441 6481, Fax +1 713 793 1603, E-Mail ltruong@tmhs.org

Herrera GA (ed): Experimental Models for Renal Diseases: Pathogenesis and Diagnosis.
Contrib Nephrol. Basel, Karger, 2011, vol 169, pp 327–336

Renal Vasculature in Essential Hypertension: A Review of Some Contrarian Evidence

Richard E. Tracy

Department of Pathology, Louisiana State University Health Sciences Center, New Orleans, La., USA

Abstract

This chapter reviews a body of evidence concerning the anatomic pathology, pathogenesis, epidemiology and possible etiologic agents of benign essential hypertension in human populations. A core finding serves as the starting point for further reasoning: intimal fibroplasia of renal interlobular arteries (arteriosclerosis) increases with age at varying rates in all populations around the world, and the rise of mean arterial pressure (MAP) with age is closely tied to this process. The weight of evidence favors the view that fibroplasia progresses for wholly unknown reasons, is not accelerated by elevations in MAP, and that it raises MAP in proportion to its severity by creating nephron heterogeneity that initiates Goldblatt mechanisms. This form of hypertension has been designated type 2 to distinguish it from a less common form, called type 1, where the fibroplasia is of mild or minimal degree. This chapter reviews evidence that indicates type 1 does not evolve into type 2 because hypertension is not accelerating the process. Experimental models of hypertension include the Goldblatt model, which resembles some aspects of type 2, and spontaneous genetic models, which mimic some aspects of type 1 hypertension. There is currently no persuasive evidence that type 2 hypertension, as it naturally develops in the human during aging of 50 years or more, can be reproduced in laboratory animals. Clues to the etiology of the arterial fibroplasia that appears to underlie most instances of essential hypertension would best be sought in the study of variations among human populations around the world, and especially of their migrants.

The features of the renal arterial tree in human hypertension have gathered no consensus descriptions of their anatomy or of their relationships to blood pressure. The review here considers some of the contrarian evidence that emphasizes perspectives outside of most mainstream viewpoints.

The disease entity, hypertension, is named for a clinical sign, not for an etiologic agent or pathognomonic lesion, or even a symptom. This odd circumstance is at least in part due to the lack of clear understanding of the anatomic or biochemical abnormalities that may properly belong to this entity.

Much is said about medial hypertrophy in the arterial tree. This usage is sometimes correct, but is seriously misdirected in other circumstances. Patients with long-standing hypertension followed at autopsy show a pattern of arterial features summarized in figure 1a [1]. Medial hypertrophy, attributable chiefly to smooth muscle cell (SMC) enlargement, typifies the arteries that serve mainly as a conduit function (i.e. generally those with diameters greater than 0.3–0.5 mm). Resistance level arteries, however, with diameters less than 0.3 mm, display medial hypertrophy rarely and only in special circumstances. These typically undergo remodeling to a smaller caliber while preserving an unchanged medial muscle mass, thereby increasing the wall thickness to diameter ratio (eutrophic inward remodeling [2]).

This pattern affects the arteries within the bulk tissues of the body, skin, muscle, fat, bone, etc., and therefore accounts for the increased peripheral resistance that persists in the presence of normal or slightly increased cardiac output. This increased resistance serves to raise the sheer stress needed to disperse the energy of high pressure as heat so that locally autoregulated capillary flow is maintained at normal pressures. Some data have suggested that the afferent arterioles of hypertensive kidneys also participate in this remodeling to smaller calibers in humans [3] and, from indirect evidence, in rats with spontaneous hypertension [2, 4].

Arteries within the renal cortex are essentially all in the diameter range less than 0.3 mm, and therefore do not manifest medial hypertrophy. Medial SMC hypertrophy can sometimes be seen in glomerulonephritis, but in no other circumstances [5]. This observation supports that the hypertrophy can be used as a helpful diagnostic criterion for recognizing glomerular disorders. This principle might explain the medial hypertrophy that often accompanies glomerulosclerosis in spontaneously hypertensive rats [6]. In benign essential hypertension, alterations of arteries and arterioles are mainly of the kinds illustrated in figures 1b–e.

Of major concern to the topic of benign essential hypertension is the structural alteration illustrated in figure 1b. This has been called arteriosclerosis, arteriolosclerosis, arterial intimal fibroplasia (fibroplasia for short), and other names. Fishberg [7] noted in 1925, 'As far back as 1868, Johnson described hypertrophy of the media as a characteristic in granular kidney and most current textbooks contain the same statement, but, as a matter of fact, one finds in arteriosclerosis a gradual and progressive degenerative atrophy of the medial muscle with replacement fibrosis'. More than 80 years later, this statement is still valid today. This arteriosclerosis is rarely seen in more than a minimal degree in the bulk tissues of the body; it is most severe and extensive in the kidney, although it is often found in the pancreas, spleen, liver and other abdominal organs as well, but generally not the gastrointestinal tract. Related kinds of change sometimes affect small arteries of the brain and heart.

The condition shown in figure 1b is a kind of metaplasia, wherein medial tissue withers and disappears, and is replaced by a newly formed intima molded of cell-poor, densely compacted collagen and associated matrix. It should be noted that in comparison with a normal artery, the span from positions 1 to 2 in the two examples display equally about 8 or 9 SMC nuclei, as if the intimal cells in the fibroplasia are the pre-existing medial SMCs that have altered size, orientation, shape and inter-relationships. In figure 1c, the features suggest that a collagenous zone has formed within the mid-depth of media displacing some of it inward (z), retaining the original internal elastica (x) and forming a new one for the remnant outer media (y). If this is the usual pattern of progression, then the inner remnant of media nearly always soon vanishes, perhaps by transformation into additional intima. In early stages, the fibrotic media can have increased thickness [8], which can deceptively be mistaken for hypertrophy (or even for hyperplasia). There is clearly no indication of hyperplasia, although a quantitative test of this question has never been reported. For purposes of measurement, the best results are obtained from measuring intimal thickness rather than wall thickness because the thickness of a shrinking media countervails the process of interest and dilutes useful information. Intima as a percentage of outer diameter is called R, R_C applying to diameters of 150–300 μm and R_R to those of 80–149 μm.

Something very different occurs in malignant hypertension. A major diagnostic finding is the mucinous edema of an expanding intima affected by SMC proliferation, lymphocytic and PMN infiltration, and other features (fig. 1d) [9–11]. It is easy to confuse this change with the benign condition, which might explain the faulty use of the term hyperplasia to describe the benign

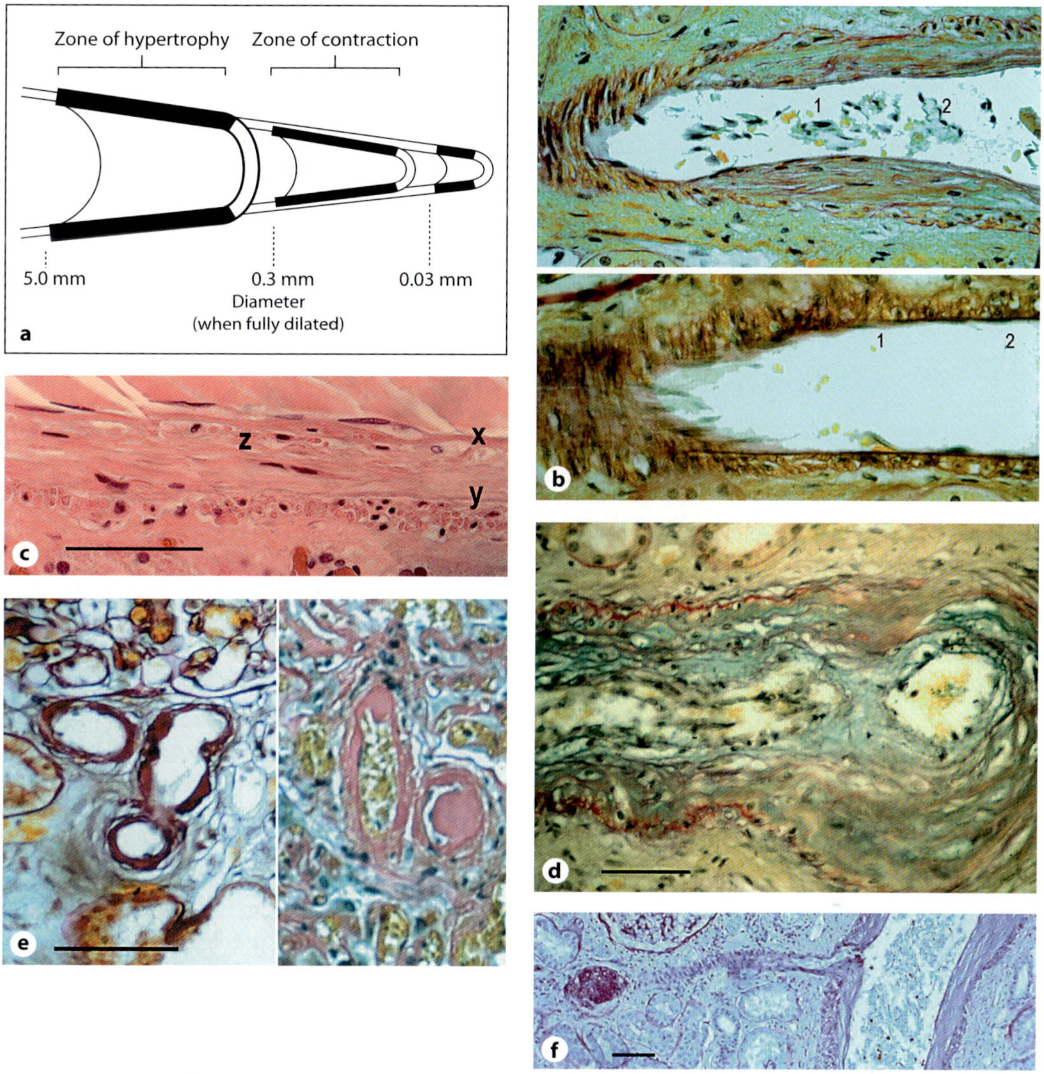

Fig. 1. **a** Arterial features of patients with long-standing hypertension [1]. **b** Fibroplastic (upper) and normal (lower) interlobular arteries. **c** Interlobular artery with benign features. **d** Interlobular artery with malignant features. **e** Pressure-fixed and immersion-fixed afferent arterioles with hyalinization; glomeruli at top of fields. **f** Obsolete glomerulus with arteriolar constriction at the ostium of its afferent arteriole. Frames **b**, **d**, **e** and **f**: PAS-Alcian blue. Frame **c**: H&E. Bars = 60 μm.

condition. Autoregulation of renal blood flow can fail when MAP exceeds 160 mm Hg, severely increasing flow rates, and the consequent excessive sheer stress may be the cause of these malignant structural changes.

The hyalinization of afferent and efferent arterioles [12, 13] and of terminal arteries (diameter <80 μm) is almost universally described as constricting or occluding the lumen, causing nephron atrophy and nephrosclerosis. However, when formalin is perfused under pressure into the renal artery, these deposits are pressed outward to restore a fully intact lumen (fig. 1e) [3]. When lacking pressure, the hyaline deposits

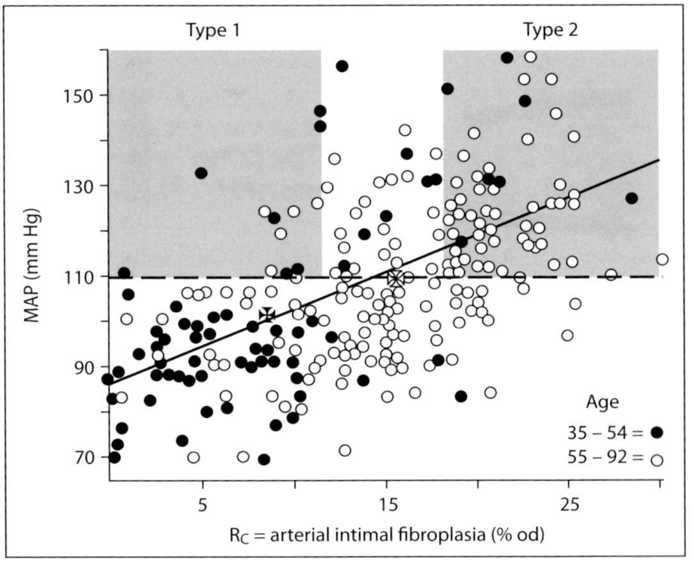

Fig. 2. Each dot represents a case and the Maltese crosses are the two-dimensional means. Reproduced from [16].

take a rounded form like that of liquid droplets. When pressure is restored, they reform back into flattened components of the vessel wall with outward bulging, reminiscent of the outward remodeling at sites of atherosclerotic plaques. Rather than hyalin as a cause of nephron atrophy, it may be that the arteriole is constricted as it originates through the fibroplastic intima of its parent artery, a suggestion that has some observational support [14] (fig. 1f). Hence, it may be that the fibroplasia fosters fibrous replacement of atrophied nephrons (nephrosclerosis) [15] because of these strictures at ostia of afferent arterioles. Features of atherosclerosis can occur in arteries as small as 500 μm (arcuates), but never at sizes <300 μm (interlobulars). Atherosclerotic materials in arteries of these sizes are attributed to emboli from a diseased aorta.

Some years ago, a series of autopsies was assembled by reviewing outpatient clinical records of patients at autopsy [16]. Those records sometimes provided at least 3 years of outpatient blood pressures [MAP (mean arterial pressure)= (2 diastolic + systolic)/3] and at least 10 readings, and these cases were retained for measurement of fibroplasia in interlobular arteries. A transition from low to high or from high to low MAP (at least to some minimal degree, but often of a large amount), was identified in all records. It was subsequently found that only the levels after transition were of interest, and the earlier records offered no useful additional information. Figure 2 summarizes the pertinent findings for the present purposes.

MAP correlates with R_C in accordance with a linear regression that applies to men and women of all ages in the range 35–74 years (r = 0.6). Taking an arbitrary MAP cutoff of 110 mm Hg for hypertension, it is possible to recognize two types of hypertension. Type 2 occurs in association with severe degrees of fibroplasia, while type 1 reveals only mild or minimal degrees. This range of variation has been recognized since the early 20th century [7]. An especially influential demonstration of this spectrum was provided by Sommers et al. [17]. While performing lumbar sympathectomy in hopes of reversing hypertension, they obtained tiny superficial kidney biopsies to evaluate for arterial and arteriolar abnormalities. The authors were especially impressed by the finding of some hypertensives with mild or minimal abnormalities. Although they did not use the terms type 1 and type 2, these terms could

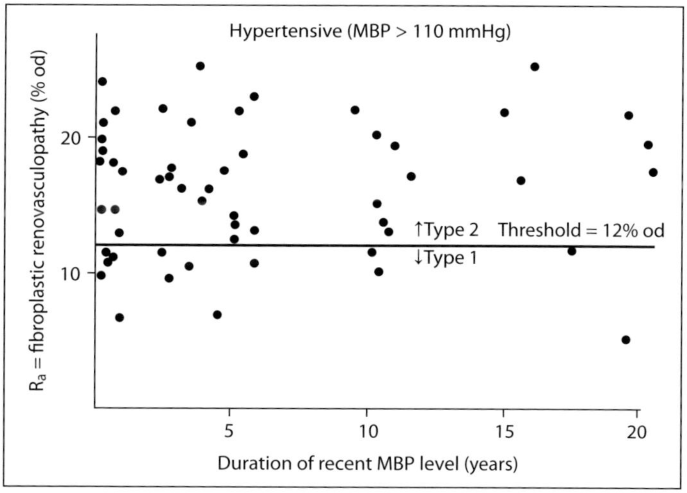

Fig. 3. Each dot represents a case. Reproduced from [16].

have been introduced from their findings. In their conclusions we read, 'Although two thirds of the cases revealed advanced renal arteriolar sclerosis, the presence of even a few individuals with apparently normal renal vessels is consistent with the view that hypertension precedes structural changes in the kidney vasculature.' However, what was left unsaid is that these findings are also consistent with the possibility that the type 1 hypertensives will remain type 1 and experience no more than average rates of fibroplasia progression, or perhaps less, as they age.

Whether type 1 hypertension will fail to convert into type 2 over time (by analogy with types 1 and 2 diabetes) could best be determined with a prospective epidemiological study. This is not practical, however, because a biopsy at the beginning of such a study would be unethical. Biopsies available to Sommers et al. [17] were unique to those circumstances and have not been reproduced; those studies did not include later follow-up with autopsies. Figure 3, however, offers an approximation to the optimal study. Subjects whose lengthy records of MAP documented the onset of hypertension provided subsequent MAP readings with sustained hypertension for intervals ranging from 1–19 years [16]. At autopsy, the fibroplasia did not correlate with the duration of the persistently elevated MAP. Nine of these subjects with observed onset of hypertension were classified as type 1 after endurance of the hypertension for up to 19 years. This outcome offers the possibility that some type 1 subjects will remain type 1, implying that hypertension does not accelerate fibroplasia. The findings in figure 3 provide evidence for the feasibility of such a study.

In recent times it has become possible to measure renal blood flow using Doppler ultrasound methods [18]. These methods have consistently identified what might be called type 1 and type 2 hypertensives on the basis of normal or diminished flow rates. The question of the progression of fibroplasia in type 1 hypertensives is therefore accessible to an answer with the use of autopsy follow-up. This would not be an ideal study because many of the hypertensives with normal renal blood flow are likely to have nephron heterogeneity based on fibroplasia unevenly distributed among the nephrons [19–21]. The outcome of this proposed study would therefore rely on whether or not a subset of hypertensives with persistent type 1 hypertension would be found at autopsy, after experiencing many years of sustained hypertension, thus reproducing the provisional conclusion of figure 3.

Sealy et al. [21] have explained how such heterogeneity is uniquely structured to induce both high and low renin hypertension even in the

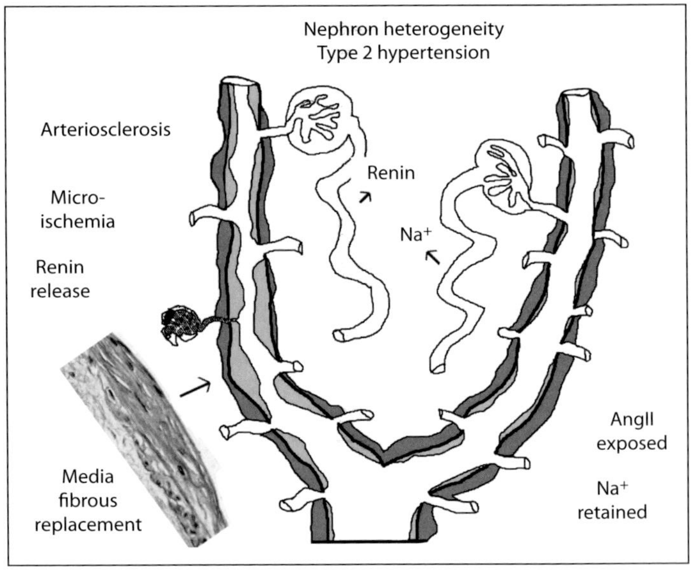

Fig. 4. A simplified summation of the proposed thesis. Arteriosclerosis is unevenly scattered and generates one subset of nephrons that are ischemic and releasing renin, and another subset that are hyperemic and retaining sodium in response to the renin. The mean rate of function of all nephrons averaged can be within the normal range and clearance studies can fail to detect that kidneys are abnormal.

presence of normal overall flow rates. Figure 4 outlines the main features of this theory whereby ischemic nephrons supplied by sclerotic arteries release renin which then acts on the hyperemic nephrons supplied by competent arteries to cause sodium retention. Renin will be high if sodium is deficient and low if sodium space expansion is sustaining the hypertension.

Washout curves for radioactive xenon [19, 20, 22] have been used to partition renal blood flow into a fast and a slow component. It is customary to call these components cortical and juxtaglomerular, respectively, but Ladefoged [20] has noted, 'It is possible that areas in the cortex with low perfusion may be represented in the second component of the wash-out curve. . .'. Data consistent with this possibility have been repeatedly reported by multiple observers [19, 22]: the slow component acquires an increasingly greater share of renal flow, and may sometimes show increases in its total amount, while the fast component declines dramatically with progressive disease. This outcome affirms an important prediction of the nephron heterogeneity hypothesis. This washout method has been challenged as lacking credibility [23].

After assembling the autopsy series in figure 2, the series was expanded by adding subjects from New Orleans, Bogalusa, Greenland and Bombay (Mumbai), India. Two regression equations, one relating MAP to R_C and the other to R_R, were suitable to all populations equally well [24]. Using those equations, MAP was estimated from renal fibroplasia measured in autopsy specimens from New Orleans, Bombay and Tokyo [16] (fig. 5). Figure 5 also displays the trends of MAP revealed in national survey data. The patterns of MAP rising with age are nearly identical to those of renal arterial fibroplasia in these populations.

The patterns in figure 5 might seem equally consistent with the proposals that fibroplasia is the cause or the consequence of sustained hypertension. It should be noted, however, that fibroplasia begins at least as early as 20 years of age, at a time when hypertension is extremely rare in these populations, and accelerates most rapidly around 40 years of age when hypertension is still uncommon. Into old age, as the prevalence of hypertension becomes widespread, the progression of fibroplasia abates. These details argue against hypertension as a cause of fibroplasia.

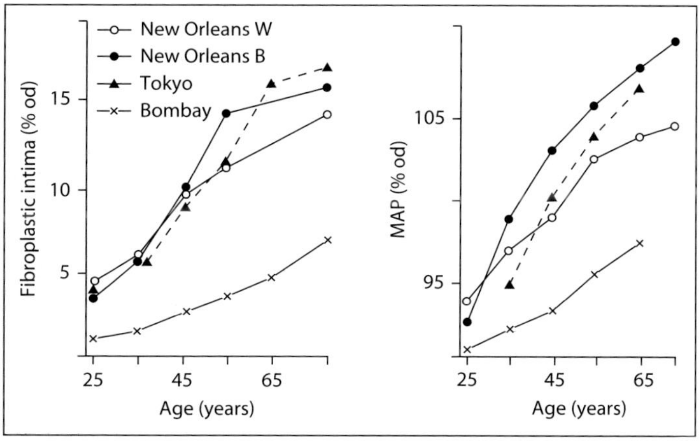

Fig. 5. Each dot represents the mean of a population-specific 10-year age group. Reproduced from [16].

Figure 3 offers additional support for the view that hypertension does not accelerate fibroplasia because some subjects with many years of sustained hypertension can still be found at autopsy to have mild or minimal fibroplasia. It should be considered that the vascular features of malignant hypertension are reversible with therapeutic correction of the high blood pressure, but that the forerunning fibroplasia later revealed at autopsy is not [9]. This observation indicates that the effect of hypertension is to accelerate the malignant form of renal vasculopathy, but not the benign form. Malignant hypertension is a vicious cycle with hypertension accelerating vasculopathy, which then accelerates the hypertension. This closed cycle, if left untreated as in the past, goes rapidly to its fatal completion. If the benign condition were also a vicious cycle, then all hypertension would be malignant. It seems easier to accept that the cycle is not closed in the benign condition because hypertension does not accelerate the fibroplasia. The correlations seen in figures 2 and 5 are therefore most readily explained by nephron ischemia scattered heterogeneously through the renal cortex and raising blood pressure by Goldblatt mechanisms [21].

The Intersalt Study was a large collaborative study carried out in 52 populations around the world [25]. Its major conclusion was that urinary sodium excretion, a surrogate for dietary salt intake, was correlated between populations with the rise with age of systolic and diastolic pressures, but that the actual attained levels had no such correlation (omitting four exceptional primitive groups in Brazil, New Guinea and Kenya). The authors did not ask how the age-related rise correlated with the initial value, but their tabulations provide the data needed to assess the correlation [26]. The association showed no significant correlation, but rather revealed that the initial average systolic pressure in a population had no perceptible bearing on the average level attained at 55 years of age. This outcome offers evidence that higher pressure levels early in life do not affect the progression toward hypertension later in life.

It should also be considered that resistance level arteries in the human foot are among those most resistant to intimal fibroplasia, even though they are exposed to systolic pressures in the range of 300 mm Hg for many hours each day while standing. Why hypertension should only have a harmful effect on arteries in a few organs, and especially the kidney, has rarely been discussed by advocates of the proposal that hypertension causes these arterial changes.

A logical consequence of the foregoing analysis is that blood pressure itself may not be a controlling influence upon the progression of intimal fibroplasia in renal interlobular arteries or upon its associated rise of MAP with age. Hence, if a risk factor is correlated with MAP, then it is unlikely to be a cause of fibroplasia. Salt intake, for instance, correlates with systolic and diastolic pressure, and the effect of salt increases progressively with rising levels of MAP [26]. That increment added by salt to the elevated MAP, like all contributors to MAP, can be viewed within the foregoing analysis as failing to accelerate the disease progression. Greater intakes of salt produce increments of elevated MAP at all ages from 20–60 years [27]. Hence, subjects with lifelong excesses in salt consumption should be exposed to several decades of elevated pressures. Yet, when at 50 years of age the salt is removed from their diets, the MAP quickly falls back to the levels of subjects never exposed to the salt effect [27]. Those many years of salt-induced MAP elevations leave no record within the relevant tissues of the body because removal of the salt obliterates its effect. The Intersalt finding that urinary sodium correlates across populations with the rise in systolic pressure on age can easily be explained by the exaggerated effect of salt at ever increasing levels of MAP. That is to say, populations that experience rising pressures are the populations with the most subjects who are salt-sensitive at 55 years of age. Presumably, removal of the salt from the diets of these subjects should obliterate the correlation found by Intersalt [27].

A similar logic applies to obesity. When obese subjects succeed in dietary weight loss, their blood pressures decline by the amount attributable to the weight change [26]. Again, many decades of pressure elevations induced by excessive body weight leave no record in the relevant body tissues because weight loss through dieting (and perhaps by bariatric surgery, which needs to be examined for this effect) fully returns pressures to levels seen in subjects who were never obese. Any other factor that acts to raise blood pressure can be seen in the same way. MAP itself is not apt to be a cause for fibroplasia (other than the malignant form) or the rise of MAP with age, therefore the increment of MAP induced by the factor in question is among those not having an effect on microscopic arteriosclerosis. The search for the etiology of benign essential hypertension should put less emphasis on the factors known to elevate MAP, and look elsewhere instead [28].

Analyses like those in figures 2 and 5 were also carried out for arteriolar hyalinization. The hyalinization was found to increase with age up to about age 40 years and then to level off at a constant range into old age [29]. Hence, the process is accelerated at ages when hypertension is rare (20–39 years), and fails to progress further at ages when hypertension is commonplace (ages 40–74 years). Also, the correlation across populations was not significant when relating hyalinization to MAP [16]. The correlation of arteriolar hyalinization with MAP across individuals is generally weak (in the range of $r = 0.4$), and could be explained by an indirect correlation of hyaline in some way associated with accelerating arterial fibroplasia [12]. In a multivariate setting, fibroplasia correlates with MAP and hyalin loses its correlation with MAP, carrying nothing of additional significance. These findings favor the view that high MAP is not the driving force for arteriolar hyalinization, and that hyalinization does not induce sizable increases in blood pressure.

The curves shown in figure 5 can be described as fitting the equation $R_C = \beta A^2(1 - 0.1A)$ [28]. This equation was accepted without statistically significant rejection in men and women of every population so far examined. Hence, the diversity in fibroplasia rates is governed by a single variable parameter, β. The values of β so far observed around the world range from 0.0064 for men in Mexico to 0.0119 for women in Greenland. A similar equation for arteriolar hyalinization is $Hy = \gamma A^2(1 - .0117A)$. The global range is from $\gamma = 0.00098$ for women in Peru to $\gamma = 0.00485$ for men in Greenland. The etiologic agents that account for these global variations are wholly unknown. Recognized agents such as salt, obesity and others are likely to contribute nothing of significance to this process (see above). These global variations could therefore be rich sources for clues to the mysterious etiologic agents, especially in migrant populations [28]. The finding that R_C correlates with a related kind of intimal fibroplasia in coronary arteries might be of some help for this purpose [30].

In many various kinds of experimental models for hypertension, beginning historically with renal artery strictures, irritant wrapping of the kidneys or aldosterone-salt models and extended to later models [8, 11], the interlobular arteries readily develop the features of malignant hypertension. These include the hyperplastic mucinous edema of arteries (fig. 1d) and fibrinoid necrosis of arterioles (not illustrated here) along with characteristic changes in glomeruli, tubules and interstitium. Whether the characteristic fibroplasia of the benign human condition can also develop in these models is not clear. Most authors omit this feature from their descriptions of renal histology. Rare claims of finding appearances like those in figure 1b raise the possibility that lesions of malignant hypertension might resolve and leave behind some alterations that resemble the human condition without actually replicating it in its natural form.

Textbooks of veterinary pathology put little emphasis on arteriosclerosis of any sort. Jubb and Kennedy [31], for instance, offer one short paragraph on sclerosis of visible arteries, and no mention of microscopic arteries or arterioles (other than of rare instances of hyalinization). The full range of domesticated animals apparently fail to manifest any significant degree of arteriosclerosis or arteriolosclerosis, and these processes are outside the range of interest of veterinarians. In a study of renal interlobular arteries in various mammalian species [32], it was found that the 3-year-old rat had arteries that resemble those of 3-year-old humans, 6-year-old ferrets resemble 6-year-old humans, 12-year-old cats and dogs resemble 12-year-old humans, and 20-year-old horses and rhesus monkeys resemble 20-year-old humans. None of those species displayed anything resembling type 2 hypertension. If that type of hypertension can occur in animals, then it has to be experimentally induced, and the possibility of doing this has not been clearly demonstrated. The relationship of fibroplasia progression to age appears to be governed by the calendar and not by the longevity of each species.

Arteriolar hyalinization of afferent arterioles is readily induced in rats by administering antagonists of nitric oxide synthase such as L-NAME [33]. The effect of blood pressure on hyalinization, as distinct from glomerulosclerosis, however, has not been clarified.

Conclusions

The weight of evidence favors the view that type 1 hypertension does not evolve over time into type 2, indicating that elevated MAP does not accelerate the intimal fibroplasia of interlobular arteries in those individuals. Similarly, the intimal fibroplasia of interlobular arteries that is diagnostic of type 2 hypertension is most easily seen as a cause and not a consequence of elevated MAP. Agents such as salt and obesity can elevate MAP as long as the agents are actively in place, but their effects are readily reversible upon withdrawal of the agents. Hence, these agents do not appear to cause the rise of MAP with age, which introduces the high prevalence of type 2 hypertension in some aging populations. The search for etiology of benign essential hypertension should emphasize agents that are not on the usual lists of suspects.

References

1 Short D: The vascular fault in chronic hypertension. Lancet 1966;1:1302–1304.

2 Mulvany MJ, Baumbach GL, Aalkjaer C, Heagerty AM, Korsgaard N, Schiffrin EL, Heistad DD: Vascular remodeling. Hypertension 1996;28:505–506.

3 Tracy RE, Overll EO: Arterioles of perfusion-fixed hypertensive and aged kidneys. Arch Path 1966:82:526–534.

4 Folkow B, Gothberg G, Lundin S, Richsten SE: Structural 'resetting' of the renal vascular bed in spontaneously hypertensive rats (SHR). Acta Physiol Scand 1977;100:270–272.
5 Fishberg AM: The arteriolar lesions of glomerulonephritis. Arch Int Med 1927;40:80–97.
6 Geiger H, Fierlbeck W, Mai M, Ruchti H, Schönlfeld, Dämmrich J, Hugo C, Neumayer HH: Effects of early and late antihypertensive treatment on extracellular matrix proteins and mononuclear cells in uninephrectomized SHR. Kidney Int 1997;51:750–761.
7 Fishberg AM: Anatomic findings in essential hypertension. Arch Int Med 1925;35:659–668.
8 Stoddard LD, Puchtler H: Human renal vascular lesions and hypertension. Pathol Annu 1969;4:253–268.
9 McCormack LJ, Beland JE, Schneckloth RE, Corcoran AC: Effects of antihypertensive treatment on the evolution of the renal lesions in malignant nephrosclerosis. Am J Pathol 1958;34:1022–1018.
10 Jespersen B, Eiskjaer H, Christiansen NO, Olsen S, Pedersen EB: Malignant hypertension; relationship between blood pressure control and renal function during long-term observations of patients with malignant nephrosclerosis. J Clin Hypertens 1987;3:409–418.
11 Helmchen U, Bohle RM, Kneissler U, Groene HJ: Intrarenal arteries in rats with early two-kidney, one clip hypertension. Hypertension 1984;6:III87–III92.
12 Tracy RE, Parra D, Eisaguirre W, Balanza AT: Influence of arteriolar hyalinization on arterial intimal fibroplasia in the renal cortex of subjects in the United States, Peru, and Bolivia, applicable also to other populations. Am J Hypertens 2002;15:1064–1073.
13 Tracy RE: Age trends of renal arteriolar hyalinization explored with the aid of serial sections. Nephron Clin Pract 2007;105: c171–c177.
14 Ljungqvist A: The intrarenal arterial pattern in essential hypertension; a micro-angiographic and histological study. J Path Bact 1962;84:313–325.
15 Tracy RE, Ishii T: What is 'nephrosclerosis'? Lessons from Japan and Mexico. Nephrol Dial Transplant 2000;15:1357–1366.
16 Tracy RE: The heterogeneity of vascular findings in the kidneys of patients with benign essential hypertension. Nephrol Dial Transplant 1999;14:1634–1639.
17 Sommers SC, Relman AS, Smithwick RH: Histologic studies of kidney biopsy specimens from patients with hypertension. Am J Pathol 1958;34:685–702.
18 Pontremoli RP, Viazzi F, Martinoli C, Ravera M, Nicolella C, Berruti V, Leoncini G, Ruello N, Zagami P, Bezante GP, Derchi LE, Deferrari G: Increased renal resistive index in patients with essential hypertension: a marker of target organ damage. Nephrol Dial Transplant 1999;14:360–365.
19 Hollenberg NK, Epstein M, Basch RI, Merrill JP: 'No man's land' of renal vasculature; an arteriographic and hemodynamic assessment of the interlobar and arcuate arteries in essential and accelerated hypertension. Am J Med 1969;47:845–854.
20 Ladefoged J: Measurements of the renal flood flow in man with the ^{133}xenon wash-out technique. Scand J clin Lab Invest 1966;18:299–315.
21 Sealy JE, Blumenfeld JD, Bell GM, Pecker MS, Sommers SC, Laragh JH: On the renal basis for essential hypertension: nephron heterogeneity with discordant renin secretion and sodium excretion causing a hypertensive vasoconstriction-volume relationship. J Hypertens 1988;6:763–777.
22 Gatta A, Merkel C, Pessina AC, Milani L, Sacerdoti D, Zuin R: Renal hemodynamics in essential hypertension assessed by 133-xenon washout and selective renal angiography. Angiology 1982;33:818–824.
23 Aukland K: Intrarenal distribution of blood flow. Are reliable methods available for measurements in man? Scan J Clin Lab Invest 1975;35:481–486.
24 Tracy RE, Lanjewar DN, Ghorpade KG, Valand AG, Raghuwanshi SR: Renovasculopathies of hypertension in Bombay, India. Geriat Nephrol Urol 1997;7:101–109.
25 Intersalt Cooperative Research Group: Intersalt: an international study of electrolyte excretion and blood pressure. Results for 24 hour urinary sodium and potassium excretion. BMJ 1988;297:319–328.
26 Tracy RE: Salt, obesity, and alcohol fail to induce a lasting rise of blood pressure with age, and may be independent of renocortical vasculopathy. QJM 1999;92:601–607.
27 Law MR, Frost CD, Wald NJ: By how much does salt reduction lower blood pressure? III – analysis of data from trials of salt reduction. BMJ 1991;302:819–823.
28 Tracy RE, Guileyardo JM: Renovasculopathies of hypertension in Hispanics of Dallas, Texas. Arch Med Res 1999;30:40–48.
29 Tracy RE, Parra D, Eisaguirre W, Balanza AT: Influence of arteriolar hyalinization on arterial intimal fibroplasia in the renal cortex of subjects in the United States, Peru, and Bolivia, applicable also to other populations. Am J Hypertens 2002;15:1064–1073.
30 Tracy RE: Introducing the sequestration hypothesis; in Tracy RE (ed): The Role of Aging in Atherosclerosis. Dordrecht, Kluwer Academic Publishers, 2003, pp 1–10.
31 Jubb KVF, Kennedy PC: Pathology of domestic animals. New York, Academic Press, 1970.
32 Tracy RE, Johnson LK: Aging of a class of arteries in various mammalian species in relation to life span. Gerontology 1994;40:291–297.
33 Ono H, Saitoh M, Ono Y, Ishimitu T, Matsuoka H: Imidapril improves L-NAME-exacerbated nephrosclerosis with TGF-β1 inhibition in spontaneously hypertensive rats. J Hypertens 2004;22:1389–1395.

Richard E. Tracy, MD, PhD, Professor Emeritus
Department of Pathology, Louisiana State University Health Sciences Center
1901 Perdido St, 1P5
New Orleans, LA 70112 (USA)
Tel. +1 504 568 6072, Fax +1 504 568 6037, E-Mail rtracy@lsuhsc.edu

Herrera GA (ed): Experimental Models for Renal Diseases: Pathogenesis and Diagnosis.
Contrib Nephrol. Basel, Karger, 2011, vol 169, pp 337–350

Thrombotic Microangiopathies: From Animal Models to Human Disease and Cure

Jessica Caprioli[a] · Giuseppe Remuzzi[a,b] · Marina Noris[a]

[a]Mario Negri Institute for Pharmacological Research, Clinical Research Center for Rare Diseases 'Aldo e Cele Daccò', Villa Camozzi, Ranica, and [b]Department of Nephrology and Dialysis, Azienda Ospedaliera, Ospedali Riuniti di Bergamo, Bergamo, Italy

Abstract

Thrombotic microangiopathies are a group of microvascular disorders, with reduced organ perfusion and hemolytic anemia. The two most relevant conditions characterized by thrombotic microangiopathic anemia (TMA) are thrombotic thrombocytopenic purpura (TTP) and hemolytic uremic syndrome (HUS). In TTP, systemic microvascular aggregation of platelets causes ischemia in the brain and other organs. In HUS, platelet-fibrin thrombi predominantly occlude the renal circulation. TTP can be inherited due to deficiencies in the activity of von Willebrand factor cleaving protease (ADAMTS13) or acquired due to the presence of autoantibodies directed against ADAMTS13. The majority of HUS cases are secondary to infections by strains of *Escherichia coli* that produce Shiga-like toxins (Stx-HUS), while about 5–10% of all cases are classified as atypical HUS (aHUS). Genetically derived impaired regulation of the complement system is associated with aHUS. Infusion or the exchange of fresh frozen plasma have ameliorated the prognosis of TMA; however, no specific therapies aimed at preventing or limiting the microangiopathic process have been proven to affect the course of TMA. Large mammals, small animal models, knockout and transgenic mouse models of TTP and both Stx-HUS and aHUS have been developed and have provided outstanding contributions to nearly all areas of TMA research. A better understanding of the key clinical features of the diseases and of the importance of genetic and/or environmental factors involved in the pathogenesis of the diseases have been obtained. These animal models have also allowed the set up of protocols aimed at ameliorating the clinical approach to patients and for the development of new drugs and vaccines.

Thrombotic microangiopathies are microvascular disorders characterized by thrombocytopenia due to increased platelet aggregation and thrombus formation, eventually leading to disturbed microcirculation with reduced organ perfusion and hemolytic anemia with erythrocyte fragmentation. Endothelial dysfunction appears to be an important factor in the sequence of events leading to microvascular thrombosis [1–3].

Although several disease states may manifest as thrombotic microangiopathy (TMA), the two most relevant conditions associated with TMA are thrombotic thrombocytopenic purpura (TTP) and hemolytic uremic syndrome (HUS) [4]. In TTP, systemic microvascular aggregation of platelets causes ischemia in the brain and other organs. In HUS, platelet-fibrin thrombi predominantly occlude the renal circulation [2].

Various forms of both TTP and HUS exist, and these can now be defined by specific etiological factors. As for TTP, both acquired [5] and inherited deficiencies in the activity of

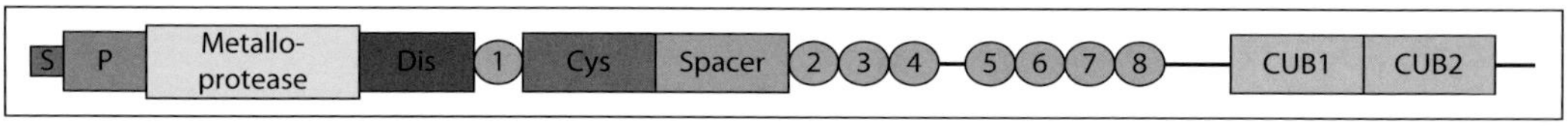

Fig. 1. Protein structure of ADAMTS13. ADAMTS13 consists of a signal peptide (S), a propeptide (P), a metalloprotease domain, a disintegrin-like domain (Dis), thrombospondin type 1 repeats, (numbered 1–8) a Cys-rich domain (Cys), a spacer domain, and 2 complement Clr/Cls sea urchin epidermal growth and bone morphogenic protein 1 or CUB domains (CUB1 and CUB2).

von Willebrand factor (vWF) cleaving protease (ADAMTS13) [6] have been described. About 90% of HUS cases are secondary to infections by *Escherichia coli* strains which produce Shiga-like toxins (Stx-HUS) [7–9]. Approximately 5–10% of all cases of HUS are unrelated to infections by Shiga toxin-producing *E. coli* and are classified as atypical [10]. According to a recent U.S. study, the incidence of atypical HUS (aHUS) is about one tenth that of Stx-HUS [11] and corresponds to approximately 2 cases/year per 1,000,000 total population. The prognosis is poor, with a mortality rate of 60–90%. Genetically derived impaired regulation of the complement system is associated with aHUS [10, 12, 13].

Differentiation of the several primary forms of TMA is crucial to predict disease outcome and to establish the most appropriate therapeutic approach. Infusion or exchange of fresh frozen plasma have dramatically changed the outcome of TMA, which in the 1960s was almost invariably fatal [14–16]. Plasma infusion has been extensively used for TTP, and remission has been documented in most cases [17, 18]. Stx-HUS, mostly due to *E. coli* O157:H7 infection, is associated with prodromal diarrhea followed by acute renal failure, and considered a disease with a good outcome [19]. Antibiotics are not necessary in most instances [20] and antimotility agents are contraindicated [21]. No specific therapy aimed at preventing or limiting the microangiopathic process has shown an affect on the course of the disease in children [19]. aHUS is often associated with progressive renal function deterioration and is characterized by a high mortality rate [11]. Supportive therapy and plasma infusion or exchange are normally given to these patients [14, 22].

Animal models of TTP and HUS have been developed that tend to recapitulate the key clinical features of the diseases and have been instrumental in elucidating the importance of genetic and/or environmental factors in the pathophysiology of the diseases. Moreover, these models have allowed for the creation of protocols aimed at ameliorating the clinical approach to patients and for the development of new drugs and vaccines.

Thrombotic Thrombocytopenic Purpura

A major contribution to the understanding of the pathogenesis of TTP has been provided by Furlan et al. [23] and Tsai and Lian [5] who partially purified and characterized a plasma metalloprotease that physiologically reduces UL-vWF multimers by cleaving vWF at the peptide bond between the 842Y and 843M residues in the A2 domain of the subunit. The activity of this protease was found deficient in the majority of patients with TTP leading to accumulation of UL-vWF multimers that are highly reactive with platelets. In 2001, the protease was identified as ADAMTS13, the thirteenth member of the ADAMTS (a disintegrin and metalloprotease with thrombospondin type 1 domains) family [24] (fig. 1).

Two primary mechanisms for deficiency of the ADAMTS13 activity have been identified in TTP patients. In 70–80% of patients, ADAMTS13 deficiency is acquired and caused by the presence of circulating autoantibodies that develop transiently and tend to disappear during remission [5, 23]. In rare cases (about 10%), mutations in the *ADAMTS13* cause a congenital deficiency of the protease [25, 26] and result in a familial recessive form of TTP. More than 80 different mutations have been identified in families with hereditary TTP. Studies on secretion and activity of the mutant forms of the protease have shown that most of these mutations lead to an impaired secretion, and when the mutant protein is secreted, the proteolytic activity is greatly reduced [27–30].

Approximately 60% of the mutations are missense causing single amino acid substitutions, and the remaining are nonsense, deletions or insertions causing frame-shifts, or splice site mutants leading to a truncated protein. Although the mutations that result in a truncated protein are equally distributed throughout the molecule, more than 75% of the missense mutations cluster in the first half of the protein.

Most patients are carriers of compound heterozygous mutations. Emerging data suggest that the type and location of *ADAMTS13* mutations may influence the age of onset of TTP and the penetrance of the disease in mutation carriers. It was proposed that deficiency of ADAMTS13 might predispose to microvascular thrombosis and environmental factors may contribute to induce full-blown manifestation of the disease [31–33]. According to this model, a triggering event, such as infections, pregnancy, surgery, bone marrow transplant and certain medications, activates microvascular endothelial cells and causes the secretion of UL-vWF multimers and P-selectin expression. Platelets in flowing blood adhere transiently to secreted vWF anchored on P-selectin expressed by activated endothelium [34]. In normal subjects, ADAMTS13 cleaves vWF multimers, thus limiting the thrombus growth [34]. Instead, in the absence of ADAMTS13, long strings of secreted UL-vWF and platelets remain bound to endothelial cells, thus initiating the formation of thrombi.

The identification of *ADAMTS13* as the gene responsible for familial TTP triggered efforts to develop a 'knockout' mouse model of the disease. Two groups generated ADAMTS13-deficient mice by gene targeting [35, 36]. The first *Adamts13*$^{-/-}$ mouse model [35] was obtained in the mixed C57BL/6x129x1/Sv background. vWF-cleaving activity in the *Adamts13*$^{-/-}$ mice was completely lost with normal activity observed in the *Adamts13*$^{+/+}$ controls (and 50% activity in the heterozygotes). However, findings consistent with TTP were not present in *Adamts13*$^{-/-}$ mice. In fact, peripheral blood counts and blood smears were normal without signs of hemolytic anemia or thrombocytopenia, and there was no evidence of hyaline thrombus formation on histological examination of numerous TTP target organs. These observations in mice suggest that loss of ADAMTS13 activity may not be sufficient for the induction of TTP and may imply the need for other key genetic modifying factors and/or environmental triggers. These observations may reflect the situation in humans, where many familial TTP cases are not diagnosed until early adulthood. Some rare individuals carrying ADAMTS13 disease-causing mutations have been reported to remain free of TTP through the third decade of life [30].

Of note, the vWF multimer profile in *Adamts13*$^{-/-}$ mice was not different from wild-type animals. Furthermore, vWF multimers from both *Adamts13*$^{-/-}$ and wild-type mice appeared similar in size to UL-vWF found in plasma from TTP patients with ADAMTS13 deficiency. Thus, mice with the C57BL/6x129x1/Sv genetic background exhibited UL-vWF regardless of whether or not they expressed ADAMTS13. C57BL/6x129x1/Sv wild-type mice express a form of ADAMTS13 that lacks the CUB domains that are critical for vWF cleaving activity in humans, which could potentially account for the UL-vWF multimers observed in this particular strain of mice.

The second *Adamts13*$^{-/-}$ mouse model was produced on a pure 129x1/Sv genetic background [37]. These mice also failed to demonstrate any findings consistent with TTP. However, UL-vWF was present in the *Adamts13*$^{-/-}$ mice, while the *Adamts13*$^{+/+}$ controls demonstrated a normal multimeric pattern. Thus, on the pure 129x1/Sv genetic background, ADAMTS13 efficiently cleaves UL-vWF multimers and ADAMTS13 deficiency resulted in the appearance of UL-vWF, but without any evidence of TTP.

In both C57BL/6x129x1/Sv and 129x1/Sv mouse strains, hematologic and coagulation parameters, including platelet count, were comparable between ADAMTS13-deficient and wild-type mice, although some subtle phenotypic differences were evidenced. Using intravital microscopy and infusion of fluorescent platelets, it was observed that vWF-mediated interactions between platelets and the vascular endothelium were significantly prolonged in the C57BL/6x129x1/Sv *Adamts13*$^{-/-}$ mice. In the other strain (129x1/Sv), thrombocytopenia was induced when *Adamts13*$^{-/-}$ mice were challenged with the platelet and endothelial agonists collagen and epinephrine, thus indicating that a complete ADAMTS13 deficiency in mice caused a prothrombotic state, although not TTP.

Of interest, the C57BL/6x129x1/Sv *Adamts13*$^{-/-}$ mice were used in gene transfer experiments aimed at correcting ADAMTS13 genetic deficiency in the circulation. Intravenous administration of Ad-ADAMTS13 (adenoviral-mediated gene transfer) restored the release into the circulation of a functionally active protease for at least 12 weeks at levels consistently above 5–10% [38]. Of relevance, such levels in humans are sufficient to protect patients from TTP recurrences. In the same strain of mice, either by single in utero injection of a lentiviral vector encoding human *Adamts13* gene [39] or by autologous transplantation of hematopoietic progenitor cells transduced ex vivo with a lentiviral vector encoding murine *Adamts13* [40], long-term expression of ADAMTS13 and reduction of plasma UL-vWF multimer size was obtained, which was associated with systemic protection against ferric chloride-induced arterial thrombosis. Altogether the above data represent the proof of principle that gene therapy can restore ADAMTS13 plasma levels and open new therapeutical perspectives for patients affected by hereditary TTP.

In both humans and mice, genetic factors may exist that regulate vWF levels which account for the high degree of vWF level variation. The CASA/Rk mouse strain exhibits plasma vWF levels significantly higher than C57BL/6 and 129x1/Sv mice [41]. In the attempt to obtain a mouse model that faithfully recapitulates the key clinical findings of human TTP, the CASA/Rk strain was backcrossed to the C57BL/6–129x1/Sv *Adamts13*$^{-/-}$ mice. A subgroup of the *CASA/Adamts13*$^{-/-}$ mice had severe thrombocytopenia and the presence of UL-vWF in blood. Moreover, spontaneous thrombocytopenia and microangiopathic hemolytic anemia on peripheral blood smear and vWF-rich/fibrin-poor thrombi in multiple organs, including the brain, heart and kidneys, were observed. These mice also demonstrated a significantly increased mortality rate as compared to wild-type CASA mice.

When Shiga toxin was infused intravenously in these mice, TTP-like symptoms were observed, including variable degrees of thrombocytopenia, microangiopathic hemolytic anemia and vWF-rich/fibrin-poor hyaline thrombi in the brain, heart and kidneys. On the other hand, TTP-like symptoms were not observed with Stx challenge in the original ADAMTS13-deficient mice on the mixed C57BL/6–129x1/Sv background. Thus, the introduction of the CASA/Rk genetic background appeared to favor susceptibility to TTP in the setting of ADAMTS13 deficiency [42]. No correlation was observed between plasmatic levels of vWF and severity of TTP, implying the existence of TTP modifying genes distinct from vWF in CASA/Rk mice [42]. However, when the vWF null allele was crossed onto the *CASA/Adamts13*$^{-/-}$ background thus generating *Vwf*$^{-/-}$*CASA/Adamts13*$^{-/-}$ mice, Stx did not induce thrombocytopenia nor TMA-related mortality in these animals. It is clear that vWF is required for induction of Stx-induced thrombocytopenia (and therefore TTP-like symptoms) [42].

Endothelial injury has long been postulated to be a trigger for TTP. The use of Stx exposed the CASA ADAMTS13-deficient mice to an agent known to be toxic to endothelial cells [43]. It is possible that other substances similarly able to affect endothelial cell activation may be potential triggers for TTP. Such compounds might include known endothelial agonists such as ddAVP, histamine, epinephrine, thrombin and certain cytokines and their activity as TTP triggers could be tested in the ADAMTS13 deficient model.

The ADAMTS13-deficient mice may prove helpful in identifying potential genetic modifying factors that account for the differences in TTP-susceptibility (TTP-resistant C57BL/6 and 129x1/Sv mice as compared to the TTP-sensitive *Adamts13/CASA* mice). The possibility to generate 'double knockouts' (by crossing *Adamts13/CASA*$^{-/-}$ with animals that are deficient for complement genes or with those harboring the coagulation factor V Leiden) might allow for investigating the potential contribution of these genes to TTP susceptibility and pathogenesis.

Our discussion thus far has focused on mouse models of congenital *Adamts13* deficiency. Other studies have also used mice to investigate the functional consequences of anti-ADAMTS13 antibodies [44]. Using intravital microscopy, C57BL/6J/129Sv *Adamts13*$^{+/+}$ mice receiving an injection of polyclonal rabbit anti-human ADAMTS13 antisera demonstrated significantly prolonged vWF-mediated platelet-endothelial interactions, similar to control *Adamts13*$^{-/-}$ mice not receiving antisera. Thus the anti-ADAMTS13 antisera was able to induce functional ADAMTS13 deficiency, possibly in a manner similar to humans with acquired TTP and anti-ADAMTS13 antibodies. Although the antibodies used in these experiments were not isolated from patients with acquired TTP, these experiments demonstrated that antibody-mediated inhibition of ADAMTS13 in the mouse could possibly be used to model this form of TTP.

Stx-HUS

Shiga toxin-producing *E. coli* (STEC) is the principal etiologic agent of typical diarrhea-associated HUS. After exposure to Stx-*E. coli*, 38–61% of individuals develop hemorrhagic colitis and 3–9% (in sporadic infections) to 20% (in epidemic forms) progress to overt HUS [45, 46]. The disease is characterized by prodromal diarrhea followed by acute renal failure. Seventy percent of patients who develop HUS require red blood cell transfusions, 50% need dialysis and 25% have neurologic involvement, including stroke, seizure and coma [46–48]. Although mortality for infants and young children in industrialized countries decreased when dialysis became available, as well as after the introduction of intensive care facilities, 3–5% of patients die during the acute phase of HUS [47].

When STECs colonize the large intestine, they tightly adhere to the intestinal epithelial cell via a distinctive 'attaching and effacing' lesion [49]. Stxs, a family of powerful exotoxins, are released by STEC into the intestine, translocate across the gut epithelium into the circulation, and are transported to capillary endothelial cells in renal glomeruli and other organs [50]. Stxs then bind to globotetraosylceramide (Gb4) and neutral glycolipid (Gb3) receptors expressed on glomerular endothelial cells and tubular epithelial cells. They are internalized via receptor-mediated endocytosis and routed to the endoplasmic reticulum [51]. The toxins enzymatically inactivate the ribosome by depurination of the 28s rRNA component, resulting in a blockade of peptide elongation and inhibition of protein synthesis [52]. Moreover, Stxs trigger a cascade of signaling events in endothelial cells, which culminates in the expression of chemoattractants and adhesion molecules leading to leukocyte and platelet adhesion. The resulting injured endothelium changes its normal thromboresistant phenotype and becomes thrombogenic, which can explain the microvascular thrombus formation in Stx-HUS [53, 54].

Various attempts have been made since the late 1980s to obtain an animal model with which to explore the pathogenesis of Stx-induced HUS. Rabbits, mice, piglets and baboons have been evaluated as potential experimental models by simply administering Stxs or STEC to animals [55].

Intravenous administration of purified Stx1 in 2-kg male rabbits induced non-bloody diarrhea and a progressive flaccid paresis, usually culminating in animal death [55]. The histopathology was characterized by edema and hemorrhage in the mucosa and submucosa of the cecum and edema, hemorrhage and neuronal necrosis in the brain and grey matter of the spinal cord. Thrombotic microangiopathy was found to be the underlying lesion in the gastrointestinal trait and in the central nervous system. However, no renal involvement was observed. Indeed the kidneys appeared normal macroscopically, the glomeruli were of normal size and cellularity, and there was no evidence of capillary thrombosis or endothelial cell damage.

The gastrointestinal and central nervous system microvascular lesions were similar to those seen in the kidneys and other organs in patients with STEC-associated HUS. The different organ distribution of the microvascular lesions as compared to those observed in humans very likely reflects the different localization of Stx1 receptors that in the rabbit are expressed in the spinal cord, brain, cecum, colon and small bowel, but not in the kidney.

On the other hand, the sites and quantities of Gb3 in human and baboon kidney tissues are similar [56]. For this reason the baboon was chosen to characterize the responses to intravenous infusion of Stx1 [57]. Depending on the amount of toxin infused, Stx1 produced a variety of responses, including acute renal failure, hematuria, proteinuria, thrombocytopenia, schistocytosis, anemia and melena. Light and electron microscopy showed organelle disintegration and necrosis of the renal proximal tubular epithelium and intestinal mucosal epithelium, both of which were previously shown to bear Gb3 receptors. The renal proximal tubular epithelial changes were accompanied by swelling of visceral epithelial cells (podocytes) and by swelling and detachment of endothelial cells of the glomerular capillaries. In addition, microvascular fibrin deposition and thrombosis in renal glomerular and peritubular capillaries were present, though scanty.

Differences in Stx1 administration modulated disease severity: animals that received the Stx1 as a single dose (100 ng/kg i.v. bolus) developed thrombocytopenia, schistocytosis and acute renal failure [58], whereas animals that received four divided doses of Stx1 (25-ng/kg doses at 12-hour intervals) developed no clinical or histologic features of HUS. The baboon model was then used to study the response to small intravenous doses of Stx1, with and without the co-administration of lipopolysaccharide (LPS) [59]. The co-administration of two doses of LPS and four doses of Stx1 resulted in HUS, but the administration of either toxin separately did not. It appears that LPS is not required for disease expression, but that it can augment the response to otherwise subtoxic amounts of Stx1. At variance with Stx1, low doses of Stx2 given alone to baboons resulted in progressive thrombocytopenia, hemolytic anemia, azotemia, high urinary interleukin-6 levels and glomerular TMA at necropsy [60]. This finding is consistent with epidemiological studies showing that Stx2-producing *E. coli* are more likely to cause HUS than strains that produce only Stx1 [61, 62].

A canine model of typical HUS derives from a naturally occurring epidemic disease of kenneled greyhound dogs, associated with exposure to STEC-contaminated food [63, 64]. An idiopathic disease involving skin and occasionally kidneys was seen. Hemorrhages, fibrinoid arteritis, thrombosis and infarction with deep, slowly healing ulcers characterized the skin lesions. Pre-acute renal glomerular necrosis, particularly involving afferent arterioles with intravascular coagulation in glomerular capillaries were the distinctive findings in affected kidneys. Severely affected dogs developed precipitous thrombocytopenia and microangiopathic hemolytic anemia followed by acute renal failure and death from uremia. Glomerular ultrastructural changes resembled those in human HUS. A complex pathogenesis of this disorder was suspected in which genetic predisposition plays a prominent role.

Potential advantages of greyhound dogs over the baboons are that dogs do not have to be anesthetized for the duration of the study, are less expensive, easier to work with and evoke a lesser degree of ethical objection. Unlike smaller animals, dogs are large enough to allow repeated blood sampling without significant physiological impact. However these large animal models are expensive and impractical for the common researcher.

In mice, either the inoculation of STEC [65–67] or the injection of purified Stxs alone [35, 61, 68–70] never induced HUS. On the other hand, a mouse model of HUS has been obtained by concurrent intraperitoneal injection of both Stx2 and bacterial LPS [71, 72]. The idea derived from the observation that renal disease is probably due to the combined action of Stxs and LPS on the renal glomeruli and tubules. The Stx2 plus LPS mouse model of HUS recapitulates the human disease in both its signs and symptoms, including the clinical diagnostic triad of renal failure, thrombocytopenia and hemolytic anemia. Increased serum creatinine and BUN in these mice demonstrate kidney dysfunction. Low platelet counts and evidence of hemolytic anemia were found in the model. The mice also exhibited the neutrophilia and monocytosis that is found clinically [73], and fibrin and red cell staining of kidneys showed thrombus formation in the microvasculature, as seen in patients with HUS. Furthermore, electron microscopy showed ultrastructural changes in the glomeruli, including irregular endothelial cell fenestrations and partial detachment from the glomerular basement membrane (GBM), that are consistent with HUS. On the other hand, neither Stx2 nor LPS alone were able to mimic HUS in the mouse. Of note, anti-O157 LPS antibodies were found in the large majority of patients with STEC infection, suggesting that LPS may contribute to Stx-induced HUS in humans as well [45, 74].

Doubts have been expressed [75, 76] on the above mentioned model as a good representation of human HUS, as other authors failed to find glomerular alterations in mice injected with purified Stxs and LPS [71, 77–81]. The development of tubular injury in the absence of glomerular damage was explained by the localization of the Stx receptors in the mouse kidney as it was found that renal tubuli, but not glomeruli, express Gb3, the receptor for Stx [61, 68, 69, 82]. However, it is possible that LPS and inflammatory cytokines, such as TNF-α, may upregulate glomerular Stx receptor in mice, as it was reported to occur in vivo in non-human primates [83] and in vitro in human glomerular endothelial cells [84].

Another area of research is related to the use of experimental animal models of STEC infection that do not develop HUS. These animals are useful for testing the advantages/disadvantages of various therapies aimed at preventing or limiting STEC infection or the exposure to Stxs. For example, models of STEC infection have been used to assess the effects of various antibiotics, i.e. to disclose whether they should be administered or not to treat STEC infection. One study compared the efficacy of levofloxacin and fosfomycin [85] in mice inoculated with two different strains of *E. coli* O 157 (EDL931 and #7). Levofloxacin exhibited higher therapeutic efficacy than fosfomycin: a significant decrease in overall mortality was observed after treatment with levofloxacin, with STEC disappearing immediately from the feces of mice. Fosfomycin had only a partial effect on mortality and the Stx level decreased gradually. In another study [86], piglets were treated with either ciprofloxacin or azithromycin. After treatment with ciprofloxacin, infected piglets had diarrhea and the severe fatal neurological symptoms associated with Stx2 intoxication. Characteristic petechial hemorrhages in the cerebellum were more severe in ciprofloxacin-treated animals than in control animals. In contrast, azithromycin-treated piglets survived the infection and had little or no brain hemorrhaging.

In recent years, other therapeutic strategies to fight STEC infection have been developed and tested in animal models, such as new vaccines against STEC. A fusion protein carrying the immunogenic B subunits derived from Stx1 and 2, named Stx2B-Stx1B (2S for short) was produced and used to immunize mice [87]. This protein induced high levels of IgG in mice that inhibited cytotoxicity of the toxins and protected mice challenged with a lethal dose of lysed STEC. In another study [88], a vaccine against STEC was produced. EspA (*E. coli*-secreted protein A), intimin and Stx2 (Shiga toxin 2) are three predominant virulence factors of STEC, and each of them has proved to be capable of inducing partial protective immunity. A trivalent recombinant protein designated EIS that is composed of EspA (E), the C-terminal 300 amino acids of intimin (I) and the B subunit of Stx2 (S) was produced and tested as a vaccine using a mouse model. Immunization with EIS induced a strong humoral response to EspA, intimin and Stx2 and protected mice against the challenges with live STEC or STEC-sonicated lysate. Moreover, EIS enhanced clearance of intestinally colonized bacteria. These studies demonstrated the efficacy of vaccines against STEC infection in animal models, suggesting that they could be promising agents in clinical trials [88].

Atypical HUS

Research over the last 10 years has linked aHUS to hyperactivation of the complement cascade [89]. In 1998, Warwicker et al. [90] studied three families with aHUS and established linkage in the affected individuals to the regulator of complement activation gene cluster (RCA) on human chromosome 1q32, which encodes for several complement regulatory proteins. The first examined candidate gene in this region was factor H *(CFH)* because an association between familial aHUS and CFH abnormalities had been described previously [91–93]. Following this seminal article [90], more than 120 different mutations have been reported in patients with aHUS in *CFH* and in two other regulators of the complement system, *MCP* and *CFI* (www.FH-HUS.org). More recently gain of function mutations in key proteins of the alternative pathway cascade, *CFB* and *C3*, and in thrombomodulin, an antithrombotic protein with complement inhibitory activity, have also been reported [94, 95]. In addition, a form of aHUS associated with anti-CFH autoantibodies has been described, mainly in children [96].

CFH plays a key role in the regulation of the alternative pathway of complement (fig. 2). It serves as a co-factor for the C3b-cleaving enzyme, CFI, in the degradation of newly formed C3b molecules and controls decay, formation and stability of the C3b convertase C3bBb [97]. CFH consists of 20 homologous short consensus repeats (SCRs). The complement regulatory domains needed to prevent fluid phase alternative pathway amplification have been localized within the N-terminal SCR1–4 [98]. The inactivation of surface-bound C3b is dependent on the binding of the C-terminal domain of CFH to polyanionic molecules, which increases CFH affinity for C3b and exposes its complement regulatory N-terminal domain. *CFH* mutations in aHUS patients are generally heterozygous and cause either single amino acid changes or premature translation interruptions, mainly clustering in the C-terminus domains, and are commonly associated with normal CFH plasma levels [22].

Expression and functional studies have demonstrated that CFH proteins carrying HUS-associated mutations have a severely reduced capability to interact with polyanions and surface-bound C3b [99–101], which results in a lower density of mutant CFH molecules bound to the endothelial cell surface and a diminished complement regulatory activity on the cell membrane [101]. In contrast, these mutants have a normal capacity to control activation of the complement in plasma, as indicated by data that they retain a normal cofactor activity in the proteolysis of fluid-phase C3b [101].

Impaired control of the complement system activation due to mutations in *CFH* has also been described in membranoproliferative glomerulonephitis type II (MPGN2). MPGN2 is a rare cause of chronic nephritis characterized by the presence of dense deposits within the GBM, capillary wall thickening, mesangial cell proliferation and glomerular fibrosis [102]. The disease is associated with multiple complement abnormalities, including intense accumulation of C3c in GBM deposits and persistent reduction of C3 serum levels [102, 103]. Seven mutations in *CFH* in MPGN2 patients that determine undetectable or markedly reduced protein levels have been described [104, 105]. Out of 8 patients, 7 are homozygous or compound heterozygous. Four mutations affect framework cysteine residues and block CFH secretion and intracellular catabolism [106, 107].

The histopathological differences observed in MPGN2 and aHUS and the genetic data

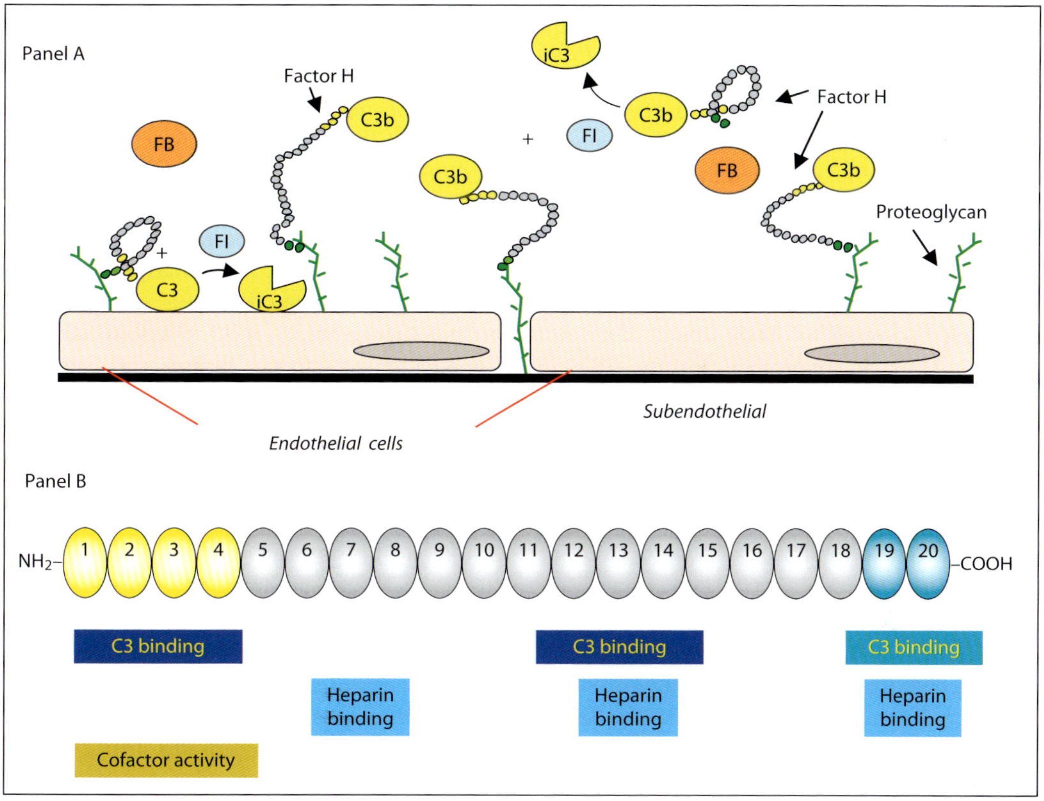

Fig. 2. a Proposed model of CFH action. In the presence of normal CFH, C3b is rapidly inactivated to iC3b. CFH in the circulation binds to C3 and favors its degradation by CFI. In addition, it binds to polyanionic proteoglycans that are present on the endothelial surface and in the subendothelial matrix, where, because of its high affinity for C3b, it entraps fluid-phase C3b, thus preventing its deposition on the host surface and its binding to CFB to form the convertase complex (C3bBb). **b** Protein structure of CFH and localization of its functional domains.

discussed above support the hypothesis that distinct functional alterations in CFH are critical in conferring the specific phenotype of aHUS and MPGN2.

Studies in animal models of MPGN2 and aHUS have provided precious information to clarify the mechanisms that cause the disparate phenotype underlying *CFH* genetic defect. Lethal MPGN has been described in a strain of pigs [108, 109] that were deficient in CFH [110, 111] due to a mutation in SCR20 which prevented secretion of the mutant protein [112]. The disease was 100% penetrant in homozygous CFH-deficient pigs, while no heterozygous carrier developed the disease. This strain has now been eradicated [113], but its elegant characterization has provided important insights into the pathological changes that occur in MPGN associated with CFH deficiency.

Further insights into the pathogenesis of MPGN associated with CFH deficiency derived from studies in Cfh-deficient mice [114]. These animals were developed by targeted gene deletion of murine Cfh. Similar to homozygous CFH-deficient humans, $Cfh^{-/-}$ mice had extremely low levels of plasma C3 as a result of uncontrolled alternative pathway activation. At 8 months, all the homozygous $Cfh^{-/-}$ mice had developed light microscopic features of MPGN [114], while no evidence of spontaneous renal disease was observed in heterozygous Cfh-deficient mice. Sequential analysis of the renal histopathology in these mice showed that glomerular C3 deposition preceded the appearance of GBM-dense deposits which, in turn, preceded the histological light microscopic appearances of MPGN [114].

$Cfh^{-/-}$ mice were also crossed with $Bf^{-/-}$ and $C5^{-/-}$ mice, in order to obtain double knockout animals [114]. No abnormalities in the GBM in the $Cfh^{-/-}Bf^{-/-}$ mice were observed. Hence, preventing alternative pathway activation completely prevented the development of MPGN. $Cfh^{-/-}C5^{-/-}$ mice still developed GBM deposits, but serum creatinine levels, neutrophil numbers, glomerular cellularity and mortality were reduced in comparison with $Cfh^{-/-}$ mice. This observation suggests that chronic C5 activation may be involved in glomerular inflammatory response in MPGN associated with factor H dysfunction.

The only available model of aHUS is represented by $Cfh^{-/-}$ mice transgenically expressing a mouse factor H (FHΔ16–20) that lacked the terminal five SCR domains, the equivalent mouse location of the majority of aHUS-associated CFH human mutations [115]. FHΔ16–20 transgenic mice were intercrossed with $Cfh^{-/-}$ mice to generate mice expressing the mutant protein ($Cfh^{-/-}$FHΔ16–20). $Cfh^{-/-}$.FHΔ16–20 mice were viable, and FHΔ16–20 was detectable in plasma. Analogous to aHUS-associated CFH human mutants, FHΔ16–20 retained plasma C3 regulation but showed impaired binding to heparin and human umbilical vein endothelial cells (HUVECs) in vitro. In fact endothelial damage characteristic of TMA was evident on renal ultrastructural examination of $Cfh^{-/-}$.FHΔ16–20. Importantly, electron-dense GBM deposits, an ultrastructural feature of MPGN2, were absent. Renal TMA in $Cfh^{-/-}$.FHΔ16–20 mice was associated with renal failure, red cell fragmentation and thrombocytopenia, all cardinal features of aHUS. Immunofluorescence studies in the $Cfh^{-/-}$.FHΔ16–20 mice with hematuria showed C3 deposition along the endothelium and within the smooth muscle of renal arteries, in addition to abnormal deposition within the glomerular mesangium and capillary walls. The spontaneous pathology in the $Cfh^{-/-}$.FHΔ16–20 mice, like that of humans with functionally similar CFH mutations, targeted the renal vasculature, suggesting that there are unique anatomical and/or physiological properties of this endothelial bed that render it particularly sensitive to complement-mediated damage.

In other words, aHUS develops when a combination of effective plasma C3 regulation and defective regulation on renal endothelium are present simultaneously. This observation indicates that new therapeutic agents that can restore C3 regulation on cell surfaces should be developed. In fact, the restoration of fluid phase regulation alone may be deleterious by increasing circulating plasma levels of C3.

Further investigation on $Cfh^{-/-}$.FHΔ16–20 mice revealed that the absence of C5 prevented the development of spontaneous aHUS in these animals, demonstrating that aHUS is dependent on C5 activation in this model [116]. $Cfh^{-/-}$.FHΔ16–20 mice that were also deficient in C3 did not develop spontaneous renal injury until 8 months of age, demonstrating that C3 activation is an absolute requirement for spontaneous renal damage in these mice. These data indicate that C3-dependent C5 activation is required for the development of aHUS in $Cfh^{-/-}$.FHΔ16–20 mice.

On the other hand, heterozygous $Cfh^{+/-}$.FHΔ16–20 mice do not develop aHUS. This is at variance with aHUS in humans, which is generally associated with *CFH* heterozygous mutations. There is evidence that the combination of genetic defects and/or conditions that trigger complement activation, either directly (bacterial and viral infections) or indirectly by causing endothelial insult (drugs, certain systemic diseases), are required to induce the aHUS phenotype in humans [22]. Whether heterozygous $Cfh^{+/-}$.FHΔ16–20 mice develop the disease after environmental insults requires further investigation.

Conclusions

Outstanding advances have been obtained in recent years in the comprehension of pathogenetic and clinical features of TMA; however, knowledge about these diseases is still incomplete.

The development of proper animal models has the potential to contribute to nearly all areas of TMA research, including the direct study of the pathophysiology, the ability to serve as important tools for the screening of potential therapeutics and recombinant products, and the further identification of potential environmental triggers and genetic modifying factors.

The therapeutics now available are inadequate. This provides the impetus for the development of new approaches using animal models to better understand the pathogenesis and obtain new treatments, which is the ultimate translational objective of animal modeling.

References

1 Copelovitch L, Kaplan BS: The thrombotic microangiopathies. Pediatr Nephrol 2008;23:1761–1767.
2 Ruggenenti P, Noris M, Remuzzi G: Thrombotic microangiopathy, hemolytic uremic syndrome, and thrombotic thrombocytopenic purpura. Kidney Int 2001;60:831–846.
3 Kaplan BS, Meyers KE, Schulman SL: The pathogenesis and treatment of hemolytic uremic syndrome. J Am Soc Nephrol 1998;9:1126–1133.
4 Besbas N, et al: A classification of hemolytic uremic syndrome and thrombotic thrombocytopenic purpura and related disorders. Kidney Int 2006;70:423–431.
5 Tsai HM, Lian EC: Antibodies to von Willebrand factor-cleaving protease in acute thrombotic thrombocytopenic purpura. N Engl J Med 1998;339:1585–1594.
6 Galbusera M, Noris M, Remuzzi G: Inherited thrombotic thrombocytopenic purpura. Haematologica 2009;94:166–170.
7 Thorpe CM: Shiga toxin-producing *Escherichia coli* infection. Clin Infect Dis 2004;38:1298–1303.
8 Karmali MA, et al: Sporadic cases of haemolytic-uraemic syndrome associated with faecal cytotoxin and cytotoxin-producing *Escherichia coli* in stools. Lancet 1983;1:619–20.
9 Riley LW, et al: Hemorrhagic colitis associated with a rare *Escherichia coli* serotype. N Engl J Med 1983;308:681–685.
10 Noris M, Remuzzi G: Hemolytic uremic syndrome. J Am Soc Nephrol 2005;16:1035–1050.
11 Constantinescu AR, et al: Non-enteropathic hemolytic uremic syndrome: causes and short-term course. Am J Kidney Dis 2004;43:976–982.
12 Noris M, Remuzzi G: Atypical hemolytic uremic syndrome. New Engl J Med 2009;361:1676–1687.
13 Jokiranta TS, et al: Where next with atypical hemolytic uremic syndrome? Mol Immunol 2007;44:3889–3900.
14 Bresin E, et al: Outcome of renal transplantation in patients with non-Shiga toxin-associated haemolytic uremic syndrome: prognostic significance of genetic background. Clin J Am Soc Nephrol 2006;1:88–99.
15 Loirat C, Noris M, Fremeaux-Bacchi V: Complement and the atypical hemolytic uremic syndrome in children. Pediatr Nephrol 2008;23:1957–1972.
16 Ariceta B, et al: Guideline for the investigation and initial therapy of diarrhea-negative hemolytic uremic syndrome. Pediatr Nephrol 2009;24:687–696.
17 Vesely SK, et al: ADAMTS13 activity in thrombotic thrombocytopenic purpura-hemolytic uremic syndrome: relation to presenting features and clinical outcomes in a prospective cohort of 142 patients. Blood 2003;102:60–68.
18 Zheng XL, et al: Effect of plasma exchange on plasma ADAMTS13 metalloprotease activity, inhibitor level, and clinical outcome in patients with idiopathic and nonidiopathic thrombotic thrombocytopenic purpura. Blood 2004;103:4043–4049.
19 Scheiring J, Andreoli SP, Zimmerhackl LB: Treatment and outcome of Shiga-toxin-associated hemolytic uremic syndrome (HUS). Pediatr Nephrol 2008;23:1749–1760.
20 Wong CS, et al: The risk of the hemolytic-uremic syndrome after antibiotic treatment of *Escherichia coli* O157:H7 infections. N Engl J Med 2000;342:1930–1936.
21 Tarr PI, Gordon CA, Chandler WL: Shiga-toxin-producing *Escherichia coli* and haemolytic uraemic syndrome. Lancet 2005;365:1073–1086.
22 Caprioli J, et al: Genetics of HUS: the impact of MCP, CFH, and IF mutations on clinical presentation, response to treatment, and outcome. Blood 2006;108:1267–1279.
23 Furlan M, et al: von Willebrand factor-cleaving protease in thrombotic thrombocytopenic purpura and the hemolytic-uremic syndrome. N Engl J Med 1998;339:1578–1584.
24 Soejima K, Nakagaki T: Interplay between ADAMTS13 and von Willebrand factor in inherited and acquired thrombotic microangiopathies. Semin Hematol 2005;42:56–62.
25 Upshaw JD Jr: Congenital deficiency of a factor in normal plasma that reverses microangiopathic hemolysis and thrombocytopenia. N Engl J Med 1978;298:1350–1352.
26 Moake JL: Thrombotic microangiopathies. N Engl J Med 2002;347:589–600.
27 Kokame K, et al: Mutations and common polymorphisms in ADAMTS13 gene responsible for von Willebrand factor-cleaving protease activity. Proc Natl Acad Sci USA 2002;99:11902–11907.

28 Matsumoto M, et al: Molecular characterization of ADAMTS13 gene mutations in Japanese patients with Upshaw-Schulman syndrome. Blood 2004;103:1305–1310.
29 Uchida T, et al: Identification of novel mutations in ADAMTS13 in an adult patient with congenital thrombotic thrombocytopenic purpura. Blood 2004;104:2081–2083.
30 Donadelli R, et al: In-vitro and in-vivo consequences of mutations in the von Willebrand factor cleaving protease ADAMTS13 in thrombotic thrombocytopenic purpura. Thromb Haemost 2006;96:454–464.
31 Lotta LA GI, Cairo A, Klaassen R, Metin A, Gurgey A, et al: Genotype-phenotype correlation in congenital ADAMTS13 deficient patients. Blood 2008; ASH Annual Meeting Abstracts (112):273.
32 Schneppenheim R, et al: A common origin of the 4143insA ADAMTS13 mutation. Thromb Haemost 2006;96:3–6.
33 Garagiola I, et al: Nonsense-mediated mRNA decay in the ADAMTS13 gene caused by a 29-nucleotide deletion. Haematologica 2008;93:1678–1685.
34 Dong JF, et al: ADAMTS-13 rapidly cleaves newly secreted ultralarge von Willebrand factor multimers on the endothelial surface under flowing conditions. Blood 2002;100:4033–4039.
35 Motto DG, et al: Shigatoxin triggers thrombotic thrombocytopenic purpura in genetically susceptible ADAMTS13-deficient mice. J Clin Invest 2005;115:2752–2761.
36 Banno F, et al: Identification of strain-specific variants of mouse Adamts13 gene encoding von Willebrand factor-cleaving protease. J Biol Chem 2004;279:30896–30903.
37 Banno F, et al: Complete deficiency in ADAMTS13 is prothrombotic, but it alone is not sufficient to cause thrombotic thrombocytopenic purpura. Blood 2006;107:3161–3166.
38 Trionfini P, Tomasoni S, Galbusera M, et al: Adenoviral-mediated gene transfer restores plasma ADAMTS13 antigen activity in ADAMTS13 knockout mice. Gene Ther 2009;16:1373–1379.
39 Niiya M, et al: Correction of ADAMTS13 deficiency by in utero gene transfer of lentiviral vector encoding ADAMTS13 genes. Mol Ther 2009;17:34–41.
40 Laje P, et al: Correction of murine ADAMTS13 deficiency by hematopoietic progenitor cell-mediated gene therapy. Blood 2009;113:2172–2180.
41 Mohlke KL, et al: Mvwf, a dominant modifier of murine von Willebrand factor, results from altered lineage-specific expression of a glycosyltransferase. Cell 1999;96:111–120.
42 Chauhan AK, et al: The combined roles of ADAMTS13 and VWF in murine models of TTP, endotoxemia, and thrombosis. Blood 2008;111:3452–3457.
43 Burns ER, Zucker-Franklin D: Pathologic effects of plasma from patients with thrombotic thrombocytopenic purpura on platelets and cultured vascular endothelial cells. Blood 1982;60:1030–1037.
44 Chauhan AK, et al: Systemic antithrombotic effects of ADAMTS13. J Exp Med 2006;203:767–776.
45 Banatvala N, et al: The United States National Prospective Hemolytic Uremic Syndrome Study: microbiologic, serologic, clinical and epidemiologic findings. J Infect Dis 2001;183:1063–1070.
46 Mead PS, Griffin PM: *Escherichia coli* O157:H7. Lancet 1998;352:1207–1212.
47 Milford D: The hemolytic uremic syndromes in the United Kingdom; in Kaplan BS, Trompeter RS, Moake JL (eds): Hemolytic Uremic Syndrome and Thrombotic Thrombocytopenic Purpura. New York, Marcel Dekker Inc., 1992, pp 39–59.
48 Garg AX, et al: Long-term renal prognosis of diarrhea-associated hemolytic uremic syndrome: a systematic review, meta-analysis, and meta-regression. JAMA 2003;290:1360–1370.
49 Donnenberg MS, et al: Role of the eaeA gene in experimental enteropathogenic *Escherichia coli* infection. J Clin Invest 1993;92:1412–1417.
50 Hurley BP, Thorpe CM, Acheson DW: Shiga toxin translocation across intestinal epithelial cells is enhanced by neutrophil transmigration. Infect Immun 2001;69:6148–6155.
51 Nakajima H, et al: Kinetic analysis of binding between Shiga toxin and receptor glycolipid Gb3Cer by surface plasmon resonance. J Biol Chem 2001;276:42915–42922.
52 Obrig TG, Moran TP, Brown JE: The mode of action of Shiga toxin on peptide elongation of eukaryotic protein synthesis. Biochem J 1987;244:287–294.
53 Morigi M, et al: Verotoxin-1-induced up-regulation of adhesive molecules renders microvascular endothelial cells thrombogenic at high shear stress. Blood 2001;98:1828–1835.
54 Ruggeri Z: Endothelial cells: they only look all alike. Blood 2001;98:1644.
55 Richardson SE, et al: Experimental verocytotoxemia in rabbits. Infect Immun 1992;60:4154–4167.
56 Tesh VL, Samuel JE, Burris JA, Owens JW, Taylor FBJ, Siegler RL: Quantitation and localization of shiga toxin/shiga-like toxin-binding glycolipid receptors in human and babbon tissues. in: Recent Advances in Verocytotoxin-Producing Escherichia Coli Infections, edited by Karmali MA, Goglio AG, Bergamo, Italy, Elsevier Science BV, 1994, pp 189–192.
57 Taylor CM, et al: A laboratory model of toxin-induced hemolytic uremic syndrome. Kidney Int 1999;55:1367–1374.
58 Siegler RL, et al: Response to single and divided doses of Shiga toxin-1 in a primate model of hemolytic uremic syndrome. J Am Soc Nephrol 2001;12:1458–1467.
59 Siegler RL, et al: Response to Shiga toxin-1, with and without lipopolysaccharide, in a primate model of hemolytic uremic syndrome. Am J Nephrol 2001;21:420–425.
60 Siegler RL, et al: Response to Shiga toxin 1 and 2 in a baboon model of hemolytic uremic syndrome. Pediatr Nephrol 2003;18:92–96.
61 Tesh VL, et al: Comparison of the relative toxicities of Shiga-like toxins type I and type II for mice. Infect Immun 1993;61:3392–3402.
62 O'Brien AD, et al: Shiga toxin: biochemistry, genetics, mode of action, and role in pathogenesis. Curr Top Microbiol Immunol 1992;180:65–94.
63 Carpenter JL, et al: Idiopathic cutaneous and renal glomerular vasculopathy of greyhounds. Vet Pathol 1988;25:401–407.
64 Hertzke DM, et al: Glomerular ultrastructural lesions of idiopathic cutaneous and renal glomerular vasculopathy of greyhounds. Vet Pathol 1995;32:451–459.

65 Wadolkowski EA, et al: Acute renal tubular necrosis and death of mice orally infected with *Escherichia coli* strains that produce Shiga-like toxin type II. Infect Immun 1990;58:3959–3965.
66 Karpman D, et al: The role of lipopolysaccharide and Shiga-like toxin in a mouse model of *Escherichia coli* O157:H7 infection. J Infect Dis 1997;175:611–620.
67 Wadolkowski EA, Burris JA, and O'Brien AD, Mouse model for colonization and disease caused by enterohemorrhagic *Escherichia coli* O157:H7. Infect Immun 1990;58:2438–2445.
68 Rutjes NW, et al: Differential tissue targeting and pathogenesis of verotoxins 1 and 2 in the mouse animal model. Kidney Int 2002;62:832–845.
69 Wolski VM, Soltyk AM, Brunton JL: Tumour necrosis factor alpha is not an essential component of verotoxin 1-induced toxicity in mice. Microb Pathog 2002;32:263–271.
70 Dran GI, et al: Protective role of nitric oxide in mice with Shiga toxin-induced hemolytic uremic syndrome. Kidney Int 2002;62:1338–1348.
71 Keepers TR, et al: A murine model of HUS: Shiga toxin with lipopolysaccharide mimics the renal damage and physiologic response of human disease. J Am Soc Nephrol 2006;17: 3404–3414.
72 Zanchi C, et al: Fractalkine and CX3CR1 mediate leukocyte capture by endothelium in response to Shiga toxin. J Immunol 2008;181:1460–1469.
73 Pavia AT, et al: Hemolytic-uremic syndrome during an outbreak of *Escherichia coli* O157:H7 infections in institutions for mentally retarded persons: clinical and epidemiologic observations. J Pediatr 1990;116: 544–551.
74 Chaisri U, et al: Localization of Shiga toxins of enterohaemorrhagic *Escherichia coli* in kidneys of paediatric and geriatric patients with fatal haemolytic uraemic syndrome. Microb Pathog 2001;31:59–67.
75 Psotka MA, et al: Shiga toxin 2 targets the murine renal collecting duct epithelium. Infect Immun 2009;77:959–969.
76 Paixao-Cavalcante D, et al: Shiga toxin-2 results in renal tubular injury but not thrombotic microangiopathy in heterozygous factor H-deficient mice. Clin Exp Immunol 2009;155:339–347.
77 Palermo MS, et al: Depletion of liver and splenic macrophages reduces the lethality of Shiga toxin-2 in a mouse model. Clin Exp Immunol 1999;116:462–467.
78 Palermo M, et al: Pretreatment of mice with lipopolysaccharide (LPS) or IL-1beta exerts dose-dependent opposite effects on Shiga toxin-2 lethality. Clin Exp Immunol 2000; 119:77–83.
79 Ikeda M, Ito S, Honda M: Hemolytic uremic syndrome induced by lipopolysaccharide and Shiga-like toxin. Pediatr Nephrol 2004;19:485–489.
80 Roche JK, et al: CXCL1/KC and CXCL2/MIP-2 are critical effectors and potential targets for therapy of *Escherichia coli* O157:H7-associated renal inflammation. Am J Pathol 2007;170:526–537.
81 Harel Y, et al: A reporter transgene indicates renal-specific induction of tumor necrosis factor (TNF) by Shiga-like toxin. Possible involvement of TNF in hemolytic uremic syndrome. J Clin Invest 1993;92:2110–2116.
82 Liu XH, Lingwood CA, Ray PE: Recruitment of renal tubular epithelial cells expressing verotoxin-1 (Stx1) receptors in HIV-1 transgenic mice with renal disease. Kidney Int 1999;55:554–561.
83 Clayton F, et al: Lipopolysaccharide upregulates renal Shiga toxin receptors in a primate model of hemolytic uremic syndrome. Am J Nephrol 2005;25:536–540.
84 van Setten PA, et al: Effects of TNF alpha on verocytotoxin cytotoxicity in purified human glomerular microvascular endothelial cells. Kidney Int 1997;51:1245–1256.
85 Isogai E, et al: Effect of antibiotics, levofloxacin and fosfomycin, on a mouse model with *Escherichia coli* O157 infection. Microbiol Immunol 2000;44:89–95.
86 Zhang Q, et al: Gnotobiotic piglet infection model for evaluating the safe use of antibiotics against *Escherichia coli* O157:H7 infection. J Infect Dis 2009;199:486–493.
87 Gao X, et al: Immunogenicity of a novel Stx2B-Stx1B fusion protein in a mice model of Enterohemorrhagic *Escherichia coli* O157:H7 infection. Vaccine 2009;27:2070–2076.
88 Gu J, et al: Enterohemorrhagic *Escherichia coli* trivalent recombinant vaccine containing EspA, intimin and Stx2 induces strong humoral immune response and confers protection in mice. Microbes Infect 2009;11:835–841.
89 Noris M, et al: Hypocomplementemia discloses genetic predisposition to hemolytic uremic syndrome and thrombotic thrombocytopenic purpura: role of factor H abnormalities. Italian Registry of Familial and Recurrent Hemolytic Uremic Syndrome/Thrombotic Thrombocytopenic Purpura. J Am Soc Nephrol 1999;10:281–293.
90 Warwicker P, et al: Genetic studies into inherited and sporadic hemolytic uremic syndrome. Kidney Int 1998;53:836–844.
91 Thompson RA, Winterborn MH: Hypocomplementaemia due to a genetic deficiency of beta 1H globulin. Clin Exp Immunol 1981;46:110–119.
92 Pichette V, et al: Familial hemolytic-uremic syndrome and homozygous factor H deficiency. Am J Kidney Dis 1994;24:936–941.
93 Rougier N, et al: Human complement factor H deficiency associated with hemolytic uremic syndrome. J Am Soc Nephrol 1998;9:2318–2326.
94 Delvaeye M, et al: Thrombomodulin mutations in atypical hemolytic uremic syndrome. New Engl J Med 2009;361:345–357.
95 Fremeaux-Bacchi V, et al: Mutations in complement C3 predispose to development of atypical hemolytic uremic syndrome. Blood 2008;112:4948–4952.
96 Dragon-Durey MA, et al: Anti-Factor H autoantibodies associated with atypical hemolytic uremic syndrome. J Am Soc Nephrol 2005;16:555–563.
97 Zipfel PF, Skerka C: Complement factor H and related proteins: an expanding family of complement-regulatory proteins? Immunol Today 1994;15:121–126.
98 Rodriguez de Cordoba S, et al: The human complement factor H: functional roles, genetic variations and disease associations. Mol Immunol 2004;41:355–367.

99 Manuelian T, et al: Mutations in factor H reduce binding affinity to C3b and heparin and surface attachment to endothelial cells in hemolytic uremic syndrome. J Clin Invest 2003;111:1181–1190.
100 Sanchez-Corral P, et al: Functional analysis in serum from atypical hemolytic uremic syndrome patients reveals impaired protection of host cells associated with mutations in factor H. Mol Immunol 2004;41:81–84.
101 Jozsi M, et al: Attachment of the soluble complement regulator factor H to cell and tissue surfaces: relevance for pathology. Histol Histopathol 2004;19:251–258.
102 Appel GB, et al: Membranoproliferative glomerulonephritis type II (dense deposit disease): an update. J Am Soc Nephrol 2005; 16:1392–1403.
103 Mullins RF, Aptsiauri N, Hageman GS: Structure and composition of drusen associated with glomerulonephritis: implications for the role of complement activation in drusen biogenesis. Eye (Lond) 2001;15:390–395.
104 Zipfel PF, et al: Complement and diseases: defective alternative pathway control results in kidney and eye diseases. Mol Immunol 2006;43:97–106.
105 Montes T, et al: Genetic deficiency of complement factor H in a patient with age-related macular degeneration and membranoproliferative glomerulonephritis. Mol Immunol 2008;45:2897–2904.
106 Ault BH, et al: Human factor H deficiency. Mutations in framework cysteine residues and block in H protein secretion and intracellular catabolism. J Biol Chem 1997;272: 25168–25175.
107 Dragon-Durey MA, et al: Heterozygous and homozygous factor H deficiencies associated with hemolytic uremic syndrome or membranoproliferative glomerulonephritis: report and genetic analysis of 16 cases. J Am Soc Nephrol, 2004;15:787–795.
108 Jansen JH, Hogasen K, Mollnes TE: Extensive complement activation in hereditary porcine membranoproliferative glomerulonephritis type II (porcine dense deposit disease). Am J Pathol 1993;143:1356–1365.
109 Jansen JH, et al: In situ complement activation in porcine membranoproliferative glomerulonephritis type II. Kidney Int 1998;53:331–349.
110 Hogasen K, et al: Hereditary porcine membranoproliferative glomerulonephritis type II is caused by factor H deficiency. J Clin Invest 1995;95:1054–1061.
111 Jansen JH, Hogasen K, Grondahl AM: Porcine membranoproliferative glomerulonephritis type II: an autosomal recessive deficiency of factor H. Vet Rec 1995;137:240–244.
112 Hegasy GA, et al: The molecular basis for hereditary porcine membranoproliferative glomerulonephritis type II: point mutations in the factor H coding sequence block protein secretion. Am J Pathol 2002;161:2027–2034.
113 Hogasen K, Jansen JH, Harboe M: Eradication of porcine factor H deficiency in Norway. Vet Rec 1997;140:392–395.
114 Pickering MC, et al: Uncontrolled C3 activation causes membranoproliferative glomerulonephritis in mice deficient in complement factor H. Nat Genet 2002;31:424–428.
115 Pickering MC, et al: Spontaneous hemolytic uremic syndrome triggered by complement factor H lacking surface recognition domains. J Exp Med 2007;204:1249–1256.
116 Jorge E, Cook HT, Botto M, Pickering MC: Insights into the role of complement dysregulation in atypical haemolytic uremic syndrome. Mol Immunol 2009;46:2851.

Dr. Marina Noris
Mario Negri Institute for Pharmacological Research
Clinical Research Center for Rare Diseases 'Aldo e Cele Daccò'
Via Camozzi, 3
24020 Ranica (BG) (Italy)
Tel. +39 035 4535362, Fax +39 035 4535377, E-Mail marina.noris@marionegri.it

Herrera GA (ed): Experimental Models for Renal Diseases: Pathogenesis and Diagnosis.
Contrib Nephrol. Basel, Karger, 2011, vol 169, pp 351–362

Glomerular Repair: Present Status and Future Expectations

Guillermo A. Herrera · Elba A. Turbat-Herrera · Jiamin Teng

Nephrocor, Tempe, Ariz., USA

Abstract

This book covers much on mechanisms involved in glomerular damage which may lead to irreversible changes and loss of nephron function. The fact that these mechanisms have been elucidated is very important for the design of therapeutic options aiming at controlling and ameliorating tissue damage, thus delaying and, in some instances, completely stopping progression to end-stage renal disease, and facilitating repair. The ability of the body to maintain renal homeostasis indicates that a reservoir of cells should exist somewhere in the body to support normal turnover of glomerular cells but insufficient to adequately repair the damage in major glomerular damage. Although tubular cells can repopulate damaged tubules spontaneously after major injury, that is not the case with glomerular cells. Chronic progressive renal disease is characterized by glomerulosclerosis, interstitial inflammation, tubular damage and interstitial fibrosis. If the adaptive capacity of the cells present in a particular renal compartment is exceeded, the injurious agent/reaction produces irreversible damage leading to cell death which may be by apoptosis or necrosis. Lost cells need to be restored and the extracellular matrix scaffold must be remodeled to its original form. This chapter will recapitulate what is known about glomerular healing and repair. The field is evolving and changing rapidly.

Chronic renal disease continues increasing at a rate of 6–8% per annum. It is believed that presently about 10–20% of the total adult population has some degree of chronic renal disease. Dialysis and transplantation are the only viable options for treatment of patients with end-stage renal disease. In a significant number of the cases, dialysis is primarily viewed as a temporary treatment modality to make preparations towards renal transplantation. Approximately 0.1% of the adult population is on dialysis currently and this consumes about 2% of the total health expenditure in many developed countries [1]. While renal transplantation offers a continuously improving mode of therapy for end-stage renal disease, it is expensive and creates challenges that for some patients are extremely difficult to deal with. In essence, chronic renal disease represents a significant socioeconomic, healthcare and financial burden.

In general, the kidney has been regarded as an organ with minimal cellular turnover and limited capacity for repair. The overall dogma has been that the kidney reaches its maximal number of nephrons at a certain point and from then nephron loss is expected to occur as part of the process of aging [2]. This can be accelerated by renal diseases targeting the various renal compartments. While true renal regeneration does not occur, the kidneys maintain a significant capacity to undergo repair after acute damage, and this is best illustrated by the capacity of tubular cells to replace damaged ones. The same, though, does not occur with glomeruli; their capacity for repair is rather limited and once the damage has reached a certain point, the ability to recuperate from the injury may primarily depend on extrarenal cellular sources. A critical concept in chronic kidney disease is that sometimes in the course of the disease, renal tissue injury reaches a point of no-return beyond which the mechanisms associated with progressive tissue damage continue even when the initial insult(s) have disappeared [3].

The argument has been made that nephron survival after damage is dependent on the maintenance of glomerular viability rather than that of the surrounding tubular interstitial compartment. This notion is supported by the theory that maintains that one of the main reservoirs of renal progenitor cells is in the Bowman's capsule. Using a transgenic animal model, Appel et al. [4] showed that within the Bowman's capsule, there are podocyte progenitor cells that ensure a constant supply of podocytes and which may even replace damaged tubular cells [5]. Ronconi et al. [6] isolated a CD133 and CD24+ population of cells from the Bowman's capsule of human kidneys, and upon introducing them into an immunodeficient animal, these cells participated in the replacement of podocytes and tubular epithelium, suggesting that they represent stem cells. This brings up whether the appropriate treatment would involve stem cells and supporting cellular elements that together will provide cytokines and growth factors to facilitate survival of existing cells and repair of the particular compartment(s) that has been damaged with emphasis on glomerular rescue.

Transforming growth factor-β (TGF-β) has been recognized as a crucial effector of glomerulosclerosis and interstitial fibrosis [7–10], but is also the key player in mesangial repair by stimulating mesangial cells to produce matrix material and inhibiting platelet-derived growth factor-β, which is an inducer of mesangial cell proliferation. Controlling the activity of this growth factor is essential to avoid progressive nephron damage. Molecules such as hepatocyte growth factor and BMP-7 counteract the actions of TGF-β [11–15]. Decorin, an extracellular matrix protein has also been proposed as an inhibitor of TGF-β [16, 17]. These factors have been shown to decrease progression of renal damage in experimental situations, but not in humans [11–17].

An additional factor that makes glomerular repair difficult is the need for initial destruction of the sclerotic matrix that may have accumulated during pathologic remodeling. This matrix may contain very difficult to digest extracellular matrix proteins produced by various cell types. These are predominantly mesangial cells that have converted to a myofibroblastic phenotype essentially behaving like fibroblasts and producing fibrillary collagen among other extraneous extracellular matrix proteins, such as tenascin present in small amounts in the normal glomeruli, but increased significantly in nodular glomerulosclerosis [18–20]. Distortion of the balance between matrix synthesis and turnover also results in altered matrix composition [20]. The abnormal matrix that accumulates in glomerulosclerosis requires replacement prior to repair. The usual machinery present in the glomerulus for maintaining homeostasis and remodeling tissues may not be prepared to deal with the sclerotic matrix that has been laid down in some circumstances because of its unusual composition. Mesangial cells are integral elements in maintaining glomerular homeostasis and also participate actively in the development of sclerotic matrices [21, 22]. If new cells are delivered to a sclerotic environment, these may be capable of destroying and cleaning up prior to rebuilding the damaged matrix or may provide help to remaining cells to accomplish the same.

Furthermore, diseased glomeruli secrete substances that do not permit mesangial cell survival and potentiate apoptosis. Mesangial cell phenotype and survival are tightly regulated by changes in the matrix. It is well recognized that in addition to their structural functions, the components of the extracellular matrix serve as modulators of cell growth and tissue differentiation.

One major limit to the ability of glomeruli to heal is the requirement for the presence of a critical number of surviving cells to restore structural integrity and function. Glomerular and tubular function and survival are very closely related. It has been debated whether nephron loss is a result of glomerular injury itself or a consequence of associated tubular interstitial damage or possibly both [23, 24].

Bone marrow contains two populations of stem cells: hematopoietic stem cells and mesenchymal stromal cells which provide support for hematopoietic stem cells [25]. It has been proposed that bone marrow stem cells are able to participate in the repair of a variety of organs, including kidney. There is also a pool of stem cells located in the kidney itself that may participate in the process of regeneration/repair [26].

Stem cells must exhibit a number of unique characteristics, which include the ability to self maintain, indefinite proliferative potential and the ability to develop into different differentiated cell types, and when delivered to a location where damage has occurred, they can replace injured cells and re-establish homeostasis.

The bone marrow cells have been shown to differentiate into mesangial and vascular endothelial cells and eventually perform normal functions of these cells [27, 28]. Intra-renal administration of bone marrow-derived mesenchymal stem cells has been used in anti-Thy 1.1-mediated cell model of antibody-mediated mesangiolysis. Kunter et al. [29] emphasized the importance of paracrine effects in the restoration of normal glomerular function. A mix of several appropriate cytokines is a must for repair to take place. Some authors have referred to this as a 'trophic' effect [30] and emphasized that mesenchymal stem cells do not always fully differentiate into mature cells to produce the desired healing effect, as the secretion of growth factors and cytokines by the stem cells is enough to positively influence repair of damaged tissues.

The signals needed to accomplish site-specific differentiation are not well known. A proper tissue microenvironment is needed to support stem cell differentiation and needs to include the establishment of numerous signaling mechanisms including those that link proteins with growth factors, cell to cell, and cell to matrix.

The delivery of bone marrow stem cells to the kidney increases in the presence of damaged renal tissues, as is the case with other damaged organs [31]. Several studies have shown that bone marrow treatment may lead to improvement of renal function in patients with several disease, including IgA nephropathy and Alport's syndrome [32–35]. However, some have expressed skepticism as to whether administration of mesenchymal stem cells can delay renal failure in chronic diseases [36].

Glomerular Remodeling/Repair

Glomerular remodeling in health and disease involves the participation of the sublethally injured mesangial cells aided by other cellular elements which may originate in the bone marrow (bone marrow-derived cells), from stem cell pools located in certain areas in the kidneys, or be delivered to the glomerulus from exogenous sources [30, 37–46]. Glomerular cells tend to regress to a primitive embryonic mesenchymal phenotype in response to injury, which is crucial for remodeling to take place, as this endows them with a better ability to deal with the damaged environment and adapt to the remodeled matrix [21]. As our understanding of the events associated with cellular behavior in response to renal injury increases, the design of innovative therapeutic maneuvers aiming at promoting healing and stopping further damage will benefit considerably.

The specific mechanisms involved in glomerular repair remain unknown. Mechanisms involved in stem cell-driven tubular repair may provide some information that may apply to glomerular repair. Herrera et al. [47] suggested that bone marrow-derived cells contribute directly to regeneration of renal tubular epithelium. Li et al. [48] showed that a minority of bone marrow-derived cells replace damaged tubular cells. Once engrafted, these cells acquire an epithelial phenotype. The same occurs with glomerular and more specifically, mesangial cells. Morigi et al. [37] has demonstrated that bone marrow cells can differentiate into mesangial cells in murine recipients in a model of anti-glomerular basement membrane crescentic glomerulonephritis. A paracrine effect induced by these bone marrow cells appears to be responsible for some crucial, if not the main, reparative effects.

Migration towards sites of injury is the first step for stem cell-mediated regeneration to occur. Homing of mesenchymal stem cells in injured glomeruli relies on their ability to locate there and effectively engage in interaction with surrounding matrix permitting efficient integration into the anatomic and functional milieu. There is a need to find ligands that facilitate and make this process as efficient as possible. Very little is known about mechanisms involved in mesangial stem cell recruitment and homing in glomeruli.

The proper microenvironment to support stem cell differentiation requires a multitude of signals, matrix proteins, soluble growth factors, cytokines, cell to cell interactions and mechanical forces.

When rat kidneys are decellularized, retaining matrix signals for differentiating pluripotent cells, and embryonic stem cells are seeded in an antegrade fashion using the renal artery as a conduit for delivery or in a retrograde manner through the ureter, they proliferate in glomerular, interstitial and vascular compartments, expressing markers indicating differentiation along various cell types depending on location [49]. This indicates that the process of seeding stem cells in various compartments leads to differentiation and transformation into specific cell types as long as the microenviroment is there to provide direction and nourishment. The proper matrix and, sometimes, key structures need to be there. For example, the absence of tubular basement membrane impede proper differentiation of tubular cells when these are delivered to the interstitium and the stem cells become apoptotic instead [49].

In essence, repair and regeneration eventually require cellular proliferation and differentiation of stem precursor cells and, only when the appropriate extracellular matrix and a proper microenvironment are there, can this occur in an orderly and adequate manner.

Participants in the Process of Glomerular Repair/Regeneration

Players in the process of glomerular repair include not only sublethally injured mesangial cells, but also stem cells that may be derived from a variety of sources. A pool of stem cells appears to be located in the kidney [38–40]. The renal papilla has been suggested as a possible niche for renal stem cells. Oliver et al. [40] have identified the upper papilla as a site of slowly cycling cells which represents a characteristic of stem cells. These slow recycling cells are defined as cellular elements that retain the DNA analogue 5-bromo-2 deoxyuridine (BrdU) after a short pulse followed by a long chase. Due to the fact that subsequent cell divisions in the absence of labeling dilute the incorporated BrdU, only those cells with the lowest replication profile are detected and called labeled retaining cells. This approach has been used in various organs to identify stem cells. One problem with this hypothesis is that while the authors provided some evidence that these stem cells may be recruited to sites of tubular cell damage, it has never been shown that they may participate in the healing of the mesangium.

Another group of investigators [50] have determined that there are stem cells in the tubular epithelium and interstitium, but again they have been shown to replace tubular, not glomerular, cells. Some investigators have used the approach of looking for cells expressing CD133, a stem cell marker. Using this approach, Bussolati et al. [46] identified a pool of multipotential stem cells which express epithelial markers. Denkel et al. [51], using the marker Sca1 plus the absence of hematopoietic markers to identify renal stem cells, showed a subset a multipotential stem cell population that displays a tubular phenotype when delivery to the kidneys occurs. Furthermore, Sagrinati et al. [5] demonstrated a subgroup of parietal epithelial cells (including in humans) co-expressing CD24 and 133, as well as other stem cell-specific transcription factors such as BmI-1, Oct-4 and Nanog which may represent stem cells. Kitamura et al. [52] dissected portions of the rat kidney which would give rise to clones with potential to form epithelium in response to Wnt4. Unfortunately, these studies did not test whether these progenitor stem cells can aid in glomerular repair and serve as a reservoir glomerular cells if injured beyond repair.

While some studies have proposed that bone marrow-derived cells transdifferentiate into specific renal cell types, including glomerular cells, this does not appear to be a common event. However, there is some experimental data that supports this role for the bone marrow-derived stem cells. For example, bone marrow cells were shown to differentiate into mesangial cells in murine recipients when the glomerulus was damaged by an antibody-mediated glomerulonephritis [53]. Other pertinent studies have highlighted how bone marrow cells contribute to the maintenance and repair of the endothelial cells, including those in glomeruli [28].

Furthermore, although initial studies suggested that mesenchymal stem cells contributed significantly to tubular regeneration [47] or nephron formation in whole embryo systems [54], it is currently believed that their contribution to these is small. Cell fusion may explain some of the initial data that was interpreted as direct replacement of damaged cells by mesenchymal stem cells.

In vitro Platforms of Matrix Injury by Glomerulopathic Light Chains: Contributions to the Understanding of Mesangial Repair by Mesenchymal Stem Cells

In vitro and in vivo experimental platforms of mesangial injury by glomerulopathic light chains have provided a unique opportunity to test whether mesenchymal stem cells may play a role in glomerular healing. Mesangial cells grown on Matrigel (matrix that mimics normal mesangial matrix) are incubated with glomerulopathic light chains purified from the urine of patients with renal biopsy-proven light chain deposition disease and AL-amyloidosis. In the first condition, mesangial sclerosis occurs, while in amyloidosis, matrix destruction is the typical finding. In both, matrix damage is a key finding.

Mesangial cell deletion by apoptosis (fig. 1) and matrix damage are integral components of these experimental systems [55, 56]. One of the most significant aspects of these experimental platforms is that glomerulopathic light chains produce two opposite effects on the mesangial matrix: some light chains potentiate mesangial matrix increase or sclerosis (fig. 2a,b), while others participate in the formation of amyloid, resulting in activation of metalloproteinases and eventual replacement of the native mesangial matrix [57, 58] (fig. 2c). Because the cascade of events that are responsible for the matrix changes can be maneuvered effectively in these research platforms by changing a number of variables, a rather comprehensive understanding of the sequence of events involved in the participation of mesenchymal stem cells in replenishing damaged mesangial cells and repairing matrix has been achieved.

The first striking finding in these models is that after mesangial cells and matrix have been damaged by glomerulopathic light chains and when the mesenchymal stem cells are delivered, they are immediately attracted to and converge at the sites where damage has occurred (fig. 3). Then, they proceed to clean up the remaining damaged mesangial cells and matrix (fig. 4a and b). Once this is accomplished, the mesenchymal stem cells differentiate and acquireing cytoskeletal characteristics, making them appear like mature mesangial cells and become functional cellular elements. The differentiating mesenchymal stem cells acquire cytoplasmic projections and contractile functional elements.

The cytoplasmic processes allow these cells to contact neighboring mesenchymal cells, finally

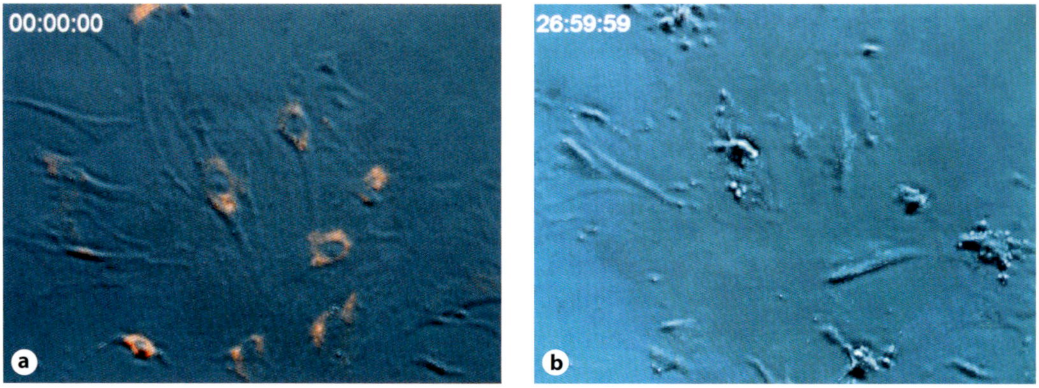

Fig. 1. Rat mesangial cells growing on Matrigel incubated with glomerulopathic light chains. Six-dimensional live cell system. Phase contrast microscopy. Magic Red (cresyl violet) real-time protease detection kit for caspases 3 and 7. Glomerulopathic light chains induce apoptosis of mesangial cells, as shown by the Magic Red cytoplasmic labeling of mesangial cells, which occurs immediately after incubation of mesangial cells with glomerulopathic light chains (time 00:00:00; **a** Fragments of these mesangial cells are seen detaching from the main cell bodies as part of the apoptotic process (time 26:59:59; **b** The debris generated by the apoptotic cellular process must be cleared before the cell replacement and matrix repair can occur.

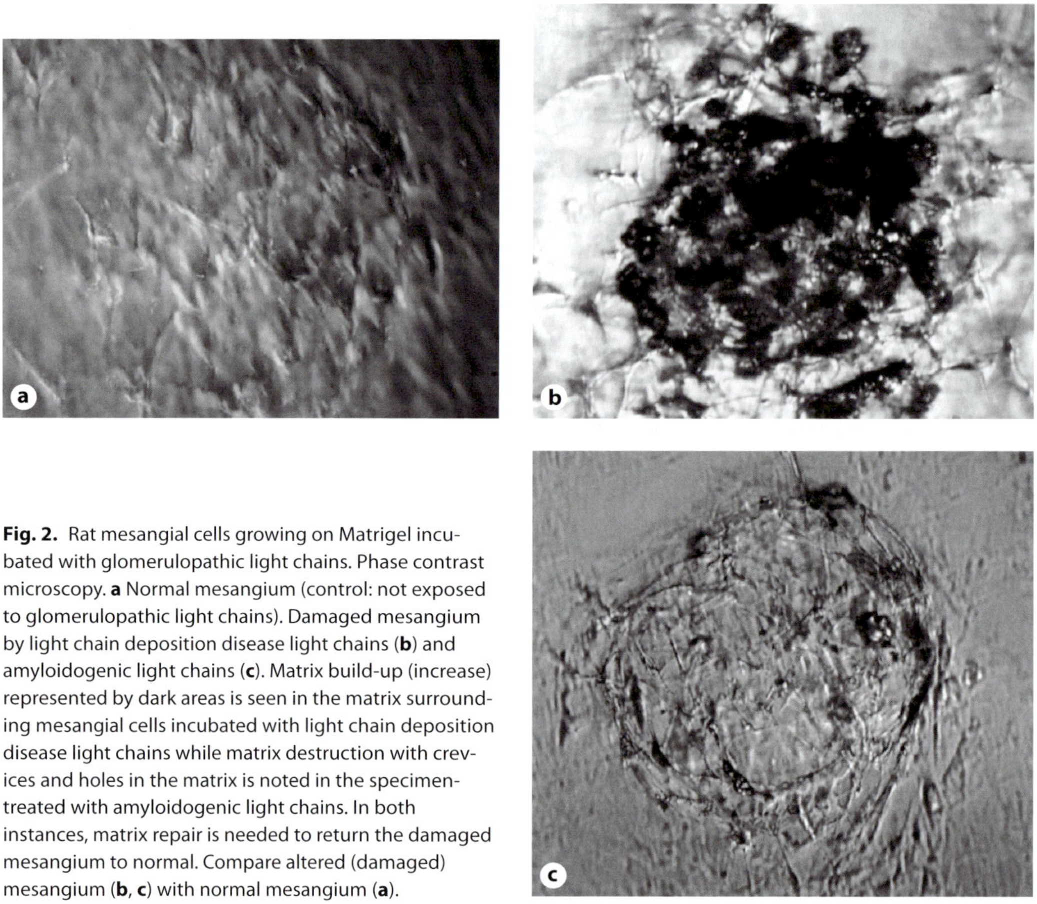

Fig. 2. Rat mesangial cells growing on Matrigel incubated with glomerulopathic light chains. Phase contrast microscopy. **a** Normal mesangium (control: not exposed to glomerulopathic light chains). Damaged mesangium by light chain deposition disease light chains (**b**) and amyloidogenic light chains (**c**). Matrix build-up (increase) represented by dark areas is seen in the matrix surrounding mesangial cells incubated with light chain deposition disease light chains while matrix destruction with crevices and holes in the matrix is noted in the specimen-treated with amyloidogenic light chains. In both instances, matrix repair is needed to return the damaged mesangium to normal. Compare altered (damaged) mesangium (**b**, **c**) with normal mesangium (**a**).

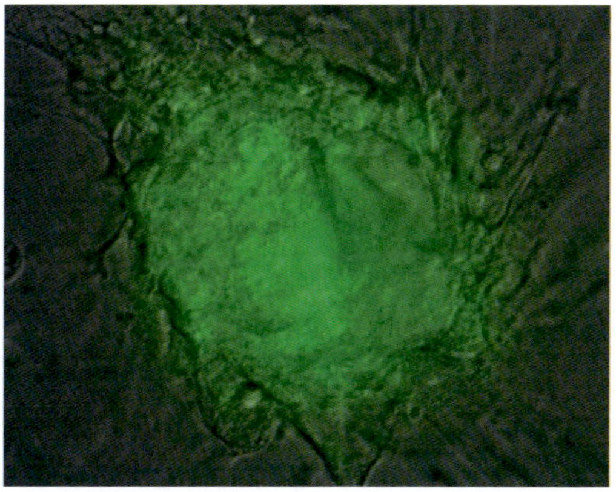

Fig. 3. Rat mesangial cells growing on Matrigel incubated with amyloidogenic light chains and PKH2GL (green)-labeled rat mesenchymal stem cells added to the culture. Six-dimensional live cell system. Phase contrast microscopy. PKH2GL-tagged mesenchymal stem cells converge at the site of matrix injury quickly after their introduction into the cell culture platform.

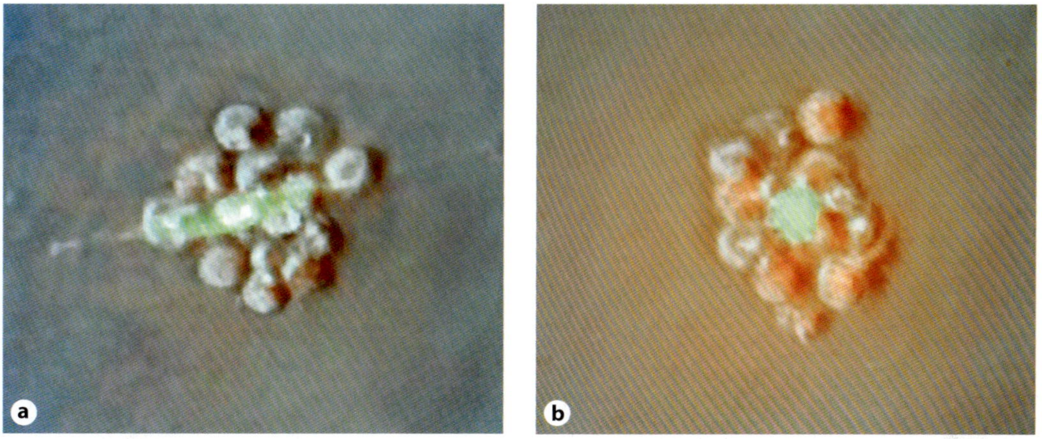

Fig. 4. a, b Rat mesangial cells growing on Matrigel (tagged with PKH2GL-green) incubated with glomerulopathic (light chain deposition disease) light chains and then Magic Red (cresyl violet)-tagged rat mesenchymal stem cells added to the culture. Six-dimensional live cell system. Phase contrast microscopy. Rat mesenchymal stem cells converge where apoptotic mesangial cells are located to dispose of the remains of these cells and surrounding debris prior to beginning the process of mesangial repair. Note the mesenchymal stem cells cleaning action and compare **a** with **b**, noting that the cells/debris are being destroyed. Apoptotic mesangial cells and associated debris are eventually completely destroyed and cleared.

resulting in the formation of a network that is typical of normal mesangial cells (fig. 5a, b). At the same time, they start laying down matrix and restoring the extracellular domain to its previous undamaged appearance and functional status. These models, whose most sophisticated expression is represented by a six-dimensional living cell system, have provided a unique way to monitor events that take place and understand how mesenchymal stem cells repopulate and reconstitute a damaged mesangial area [59, 60]. Unfortunately, these experiments do not address glomerular healing as a whole, but preliminary work has been done using similar platforms in

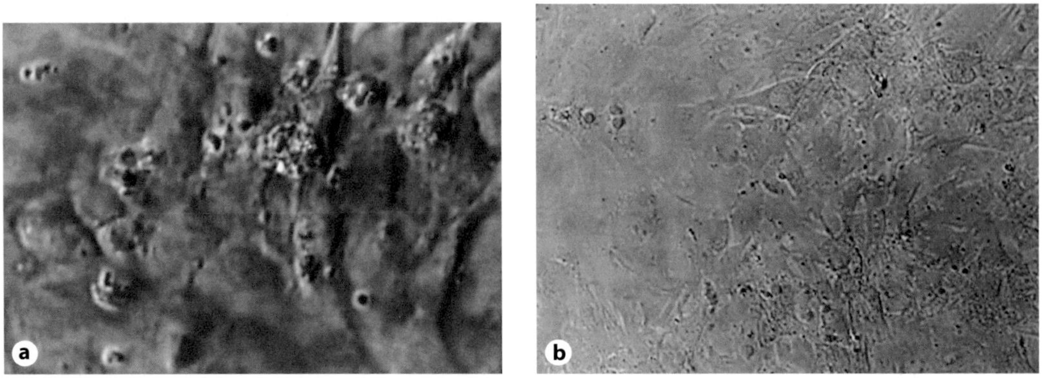

Fig. 5. **a**, **b** Rat mesangial cells growing on Matrigel incubated with glomerulopathic light chains and then rat mesenchymal stem cells added to the culture. Six-dimensional live cell system. Phase contrast microscopy. In **a**, note that a significant portion of the mesangial matrix has been repaired, but there are some areas where damaged matrix still remains. In **b**, note essentially complete mesangial repair has taken place. Rat mesenchymal stem cells have differentiated into larger mature mesangial cells with well-defined cell processes, contacts between cells have been established, and surrounding matrix now appears normal. No areas of matrix build-up (sclerosis) or crevices indicating matrix destruction are identifiable.

animal models providing a unified understanding of what is involved in glomerular healing (unpubl. observations).

Delivery of Stem Cells into the Renal Parenchyma

Involving resident cells seems to be the easiest way to aid repair, but in some circumstances these cells are no longer available for recruitment and an exogenous source for stem cells appears to be the only viable alternative.

The delivery of exogenous stem cells to the human kidney represents an additional challenge in using these cells as a therapeutic modality. If they are delivered at random to the renal parenchyma, they may never reach the areas where they are needed and these cells may not be able to become globally integrated into a functional renal unit. In experimental models, these stem cells can be delivered either through the bloodstream (i.e. renal artery) or ureter, and they find a way to home in the damaged renal compartment(s). If delivery is to occur through the circulation, entrapment of stem cells in the microvasculature must be avoided. To avoid this, the stem cells should be delivered as close as possible to the organ to be targeted to maximize the effects. These are feasible ways that should be pursued as avenues for therapeutic intervention, albeit with a clear understanding that there are obstacles and possible complications that need to be dealt with, some of which are addressed below.

Because mesangial cells, along with most mammalian cells types, are anchorage-dependent [61], adhesion to matrices, even when damaged, is essential for cell survival. It should be explored whether providing cell adhesion peptides/substrates together admixed with stem cells would facilitate and speed up the process of repair and regeneration.

One fundamental issue is how to deliver a precise dose of primed mesenchymal stem cells to the desired site [30]. There is a need to find ways to deliver stem cells directly to the kidney, as has been shown for example using CD44 and its major ligand, hyaluronic acid, to recruit exogenous mesenchymal stem cells to renal tubules to repair acute tubular injury in the glycerol-induced model of acute renal failure [62–64]. How much to deliver (specific quantities of stem cells) represents an additional vexing question to which no good answer exists today.

It is also conceivable that these stem cells and supporting growth factors/cytokines could be delivered as intravenous 'cocktails' combined with organ or compartment-specific ligands that would make them home in to the desired locations, much similar to what is currently done in the treatment of malignancies where the aim is to deliver therapeutic agents exclusively to the neoplastic cells, thus protecting normal (non-neoplastic) cellular elements.

Conclusions

There is no question that organ repair must be an important quest for future research endeavors. Regenerative medicine has a number of unique applications for the treatment of degenerative diseases, as well as those diseases with tissue damage that require replacement of injured cellular elements to regain organ function.

Repair of renal tissues is a tightly controlled process. The nature, intensity and duration of various complicated steps must be coordinated well to accomplish the desired results. Otherwise, the expected and required tissue homeostasis may not be successfully obtained. These processes involve: (1) breaking down of sclerotic tissue if present, i.e. the demolition phase; (2) disposal of unwanted material that may have been deposited and resolution of inflammatory and cellular debris, i.e. the debridement or preparatory phase; (3) delivery of stem cells and appropriate cytokines and growth factors to facilitate repair, i.e. the delivery/homing phase; and (4) cellular proliferation by remaining (local) and exogenously administered cells, and laying down of a new extracellular matrix, i.e. the recruitment and execution phase.

Preliminary data in the laboratory shows that stem cells in the proper environment can be programmed to accomplish what is needed to restore renal homeostasis [37, 65]. Combining the techniques used in tissue engineering over the last decades with stem cell technology could lead to novel sources of replacement organs, potentially extending the value of stem cells to the next level, by becoming participants in the replacement of entire damaged organs [61].

A number of phase I and II clinical trials using mesenchymal stem cells to treat a variety of renal diseases, including some with their main pathologic effects involving glomeruli, such as refractory systemic lupus erythematosus, are in progress. Other similar trials explore feasibility in renal transplantation and acute tubular necrosis [66]. Aside from therapies based on hematopoietic stem cells, mesenchymal stem cell-based therapies currently represent the most advanced stem cell-based interventions in terms of clinical testing [67].

From a practical point of view, attempts at therapy using stem cells to repair the damaged glomerular compartment will likely increase in the near future in humans, but a full understanding of applications, realistic expectations, pitfalls and risks associated with this therapeutic approach will take much time and effort to be defined. The diversity of pathology present in patients with chronic renal disease may mean that no one specific cellular therapy may be applicable to all conditions [66] and innovative approaches to different situations may be needed.

Recent progress in regenerative biology suggests that delivery of stem cells represents a viable therapeutic option for at least a subset of patients with various stages of renal disease. The opportunity to improve therapy by promoting recovery for selected renal disorders is here and future findings will obviously provide new crucial insights into how stem cell technology can be taken from the research laboratory to the patients afflicted with glomerulopathies. Translating information from the research laboratory to the patients' bedside will be crucial for the advance of this new exciting field. Additional clinical trials will further define the capabilities of stem cell therapies.

Regenerative treatments using stem cells may evolve into realistic therapeutic options in the near future, especially if used early in the course of renal insufficiency before end-stage renal disease is reached. Stem cell therapy may also become a viable treatment for at least a subset of patients with end-stage renal damage, where repair becomes a real challenge due the advanced stage of tissue damage in all renal compartments.

References

1 De Vecchi AF, Dratwa M, Wiedemann ME: Healthcare systems and end-stage renal disease (ESRD) therapies – an international review: costs and reimbursements/funding of ESRD therapies. Nephrol Dial Transplant 1999;14(Suppl 6):31–34.

2 Saxen L, Sariola H: Early organogenesis of the kidney. Pediatr Nephrol 1987;1:385–392.

3 Garcia-Sanchez O, Lopez-Hernandez FJ, Lopez-Novo JM: An integrative view on the role of TGF-β in the progressive tubular deletion associated with chronic renal disease. Kidney Int 2010;77:950–955.

4 Appel D, Kershaw D, Smeets B, Yuan G, Fuss A, Frye, B Elger, M Kriz, W Floege J, Moeller MJ: Derivation of podocytes from the parietal cell compartment. J Am Soc Nephrol 2009;20:333–343.

5 Sagrinati C, Netti GS, Mazzinghi B, Lazeri E, Liotta F, Frosali F, Ronconi E, Meini C, Gacci M, Squecco R, Carini M Gesualdo L, Francini F, Maggi E, Anunziato F, Lasagni L, Serio M, Romagnani S, Romagnani P: Isolation and characterization of multipotent progenitor cells from the Bowman's capsule of adult human kidneys. J Am Soc Nephrol 2006;17:2443–2456.

6 Ronconi E, Sagrinati C, Angelotti ML, Lazzer E, Mazzinghi B, Ballerini L, Prente E, Bechenucci F, Gacci M, Carini M, Maggi E, Serio M, Vannelli GB, Lasagni L, Romagnani S, Romagnani P: Regeneration of glomerular podocytes by human renal progenitors. J Am Soc Neprol 2009;20:322–332.

7 Border WA, Okuda S, Languino LR, Ruoslahti E: Transforming growth factor-β regulates production of proteoglycans by mesangial cells. Kidney Int 1990;37:689–695.

8 Yamamoto T, Noble NA, Miller DE, Border WA: Sustained expression of TGF-β underlies development of progressive kidney fibrosis. Kidney Int 1994;45:916–927.

9 Border WA, Noble NA: TGF-beta in kidney fibrosis: a target for gene therapy. Kidney Int 1987;51:1388–1396.

10 Tamaki K, Okuda S: Role of TGF-beta in the progression of renal fibrosis. Contrib Nephol 2003;139:44–65.

11 Zeisberg M, Hanai J, Sugimoto H, et al: BMP-7 counteracts TGF-beta 1-induced epithelial to mesenchymal transition and reverses chronic renal injury. Nat Med 2003;9:964–968.

12 Wang S, de Caestecker M, Kopp J, Mitu G, LaPage, J, Hirschberg, R: Renal bone morphogenic protein-7 protects against diabetic nephropathy. J Am Soc Nephrol 2006;17:2504–2512.

13 Zeisberg M, Kalluri R: Reversal of experimental renal fibrosis by BMP 7 provides insights into therapeutic strategies for chronic kidney disease. Pedriatr Nephrol 2008;23:1395–1398.

14 Herrero-Fresneda I, Torras J, Franquesa M, Vidal A, Cruzado JM, Loberas N, Fillat C, Grinyo JM: HGF gene therapy attenuates renal allograft scarring by preventing profibrotic inflammatory-induced mechanisms. Kidney Int 2006;70:265–274.

15 Yang J, Dai C, Liu Y: A novel mechanism by which hepatocyte growth factor blocks tubular epithelial to mesenchymal transition. J Am Soc Nephrol 2005;16:68–78.

16 Border WA, Noble NA, Yamamoto T, Harper JR, Yamaguchi Y, Pierschbacher MD, Ruoslahti E: Natural inhibitor of transforming growth factor-β protects against scarring in experimental kidney disease. Nature 2002;360:361–364.

17 Schaefer L, Macakova K, Raslik I, Mocegova M, Grone H-J, Schonherr E, Robenek H, Echtermeyer FG, Grassel S, Bruckner P, Schaefer RM, Iozzo R, Kresse H: Absence of decorin adversely influences tubulointerstitial fibrosis of the obstructed kidney by enhanced apoptosis and increased inflammatory reaction. Am J Pathol 2002;160:1181–1191.

18 Morel-Maroger SL, Killen PD, Chi E, Striker, GE: The composition of glomerulosclerosis: I. Studies in focal sclerosis, crescentic glomerulonephritis, and membranoproliferative glomerulonephritis. Lab Invest 1984;51:181–192.

19 Truong LD, Pundur J, Barrios R, D'Agatti, V, Lechago J, Suki W, Majesky M: Tenascin is an important component of the extracellular matrix in normal and pathologic conditions. Kidney Int 1994;45:201–210.

20 Stetler-Stevenson WG: Dynamic of matrix turnover during pathologic remodeling of the extracellular matrix. Am J Pathol 1996;148:1345–1350.

21 El Nahas AM: Plasticity of mesangial cells: role in kidney remodeling and scarring. Kidney Int 2003;64:1553–1563.

22 Herrera GA: Plasticity of mesangial cells: a basis for understanding pathological alterations. Ultrastruct Pathol 2006;30:471–479.

23 Schlondorff DO: Overview of factors contributing to the pathophysiology of progressive renal disease. Kidney Int 2008;74:860–866.

24 Nath KA: Tubulointerstitial changes as a major determinant in the progression of renal damage. Am J Kidney Dis 1992;20:1–17.

25 Bryder D, Rossi DJ, Weissman IL: Hematopoeitic stem cells. The paradigm tissue-specific stem cell. Am J Pathol 2006;169:338–346.

26 Al-Awqati Q, Oliver JA: Stem cells in the kidney. Kidney Int 2002;61:387–395.

27 Asnaghi V, Ferrari G, Rastaldi MP, Gabellini D, Dell'Antonio G, Maestroni A, Rugglieri D, Luzi L, Piemonti L, Zerbini G: Bone marrow-derived stem cells repopulate glomerular and tubular kidney components. Effects of hyperglycemia (abstract). J Am Soc Nephrol 2004;15:29.

28 Chen J, Park H-C, Addabbo F, Ni J, Pelger E, Li H, Plotkin M, Goligorsky MS: Kidney-derived mesenchymal stem cells contribute to vasculogenesis, angiogenesis, and endothelial repair. Kidney Int 2008;74:879–889.
29 Kunter U, Rong S, Djuric Z, Boor P, Muller-Newen G, Yu D, Floege J: Transplanted mesenchymal cells accelerate glomerular healing in experimental glomerulonephritis. J Am Soc Nephrol 2006;17:2202–2212.
30 Caplan A: Why are MSCs therapeutic? New data: new insight. J Pathol 2009;217:318–324.
31 Chapel A, Bertho JM, Bensidhoum M, et al: Mesenchymal stem cells home to injured tissues when co-infused with hematopoietic cells to treat a radiation-induced multi-organ failure system. J Gene Med 2003;5:1028–1038.
32 Poulsom R, Alison MR, Cook T, Jeffery R, Ryan E, Forbes SJ, Hunt T, Wyles S, Wright NA: Bone marrow stem cells contribute to healing of the kidney. J Am Soc Nephrol 2003;14:S48–S54.
33 Zerbini G, Piemonti L, Maestroni A, Dell'Antonio G, Bianchi G: Stem cells and the kidney: a new therapeutic tool? J Am Soc Nephrol 2006;17:S123–S126.
34 LeBleu V, Sugimoto H, Mundel TM, Gerami-Naini B, Finan E, Miller CA, Gattone VH, Lu L, Shield CF, Folkman J, Kalluri R: Stem cell therapies benefit Alport's syndrome. J Am Soc Nephrol 2009;20:2359–2370.
35 Imasawa T, Nagasawa R, Utsunomiya Y, Kawamura T, Zhong Y, Makita N, et al: Bone marrow trans plantation attenuated murine IgA nephropathy: role of a stem cell disorder. Kidney Int 1999;56:1809–1817.
36 Ninichuk V, Gross O, Segerer S, Hoffman R, Radomska E, Buchstaller A, Huss R, Akis N, Schlondorff D, Anders H-J: Multipotent mesenchymal stem cells reduce interstitial fibrosis but do not delay progression of chronic renal disease in collagen 4A3-deficient mice. Kidney Int 2006;70:121–129.
37 Morigi M, Imberti B, Zoja C, Corna D, Tomasoni S, Abate M, Rottoli D, Angioletti S, Benigni A, Perico N, Alison M, Remuzzi G: Mesenchymal stem cells are renotropic: helping to repair the kidney and improve function in acute renal failure. J Am Soc Nephrol 2004;15:1794–1804.
38 Little M, Bertram JF: Is there such a thing as a renal stem cell? J Am Soc Nephrol 2009;20:2112–2117.
39 Humphreys BD: Slow-cycling cells in renal papilla: Stem cells awaken? J Am Soc Nephrol 2009;20:2277–2279.
40 Oliver JA, Kinakis FH, Friedlander J, Sampogna RV, Martens TP, Liu C, Efstratiadis A, Al-Awqati Q: Proliferation and migration of label-retaining cells of the kidney papilla. J Am Soc Nephrol 2009;20:2315–2327.
41 Ito T, Suzuki A, Imai E, Okabe M, Hori M: Bone marrow is a reservoir of repopulation mesangial cells during mesangial remodeling. J Am Soc Nephrol 2001;12:2625–2635.
42 Imai E, Ito T: Can bone marrow cells differentiate into renal cells?. Pediatr Nephrol 2002;17:790–794.
43 Ito T, Suzuki A, Okabe M, Imai E, Hori M: Application of bone marrow-derived stem cells in experimental nephrology. Exp Nephrol 2001;9:444–450.
44 Poulsom R, Forbes SJ, Hodivala-Dilke K, Ryan E, Wyles S, Navaratnarasah S, Jeffery R, Hunt T, Alison M, Cook T, Pusey C, Wright NA: Bone marrow contributes to renal parenchymal turnover and regeneration. J Pathol 2001;195:229–235.
45 Gupta S, Verfaillie C, Chmielewski D, Krem S, Eidman K, Connaire J, Heremans Y, Lund T, Blackstad M, Jiang Y, Luttun A, Rosenberg M: Isolation and characterization of kidney-derived stem cells. J Am Soc Nephrol 2006;17:3028–3040.
46 Bussolati B, Bruno S, Grange C, Buttiglieri S, Deregibus MC, Cantino D, Camussi G: Isolation of renal progenitor cells from adult human kidney. Am J Pathol 2005;166:545–555.
47 Herrera MB, Bussolati B, Bruno S, Fonsato V, Romanazzi GM, Camussi G: Mesenchymal stem cells contribute to the renal repair of acute tubular epithelial injury. Int J Mol Med 2004;14:1035–1041.
48 Li J, Deane JA, Campanale NV, Bertram JF, Ricardo SD: The contribution of bone marrow-derived cells to the development of renal interstitial fibrosis. Stem Cells 2007;25:697–706.
49 Ross EA, Williams MJ, Hamazaki T, Terada N, Clapp WL, Adin C, Ellison GW, Jorgensen M, Batich CD: Embryonic stem cells proliferate and differentiate when seeded into kidney scaffolds. J Am Soc Nephrol 2009;20:2338–2347.
50 Maeshima A, Yamashita S, Nojima Y: Identification of renal progenitor-like tubular cells that participate in the regeneration processes of the kidney. J Am Soc Nephrol 2003;14:3138–3146.
51 Denkel B, Zngi L, Shezen E, Reich-Zeliger S, Eventov-Friedman S, Katchman H, Jacob-Hirsch J, Amariglio N, Rechavi G, Margalit R, Reisner Y: Isolation and characterization of nontubular sca-1+lin-multipotent stem/progenitor cells from adult mouse kidney. J Am Soc Nephrol 2006;17:3300–3314.
52 Kitamura S, Yamasaki Y, Kinomura M, Sugaya T, Sugiyama H, Maeschima Y, Makino H: Establishment and characterization of renal progenitor like cells from S3 segment of nephron in rat adult kidney. FASEB J 2005;19:1789–1797.
53 Wong CY, Cheong SK, Mok PL, Leong CF: Differentiation of human mesenchymal cells into mesangial cells in post-glomerular injury murine model. Pathology 2008;40:52–57.
54 Yokoo T, Ohashi T, Shen JS, Sakurai K, Miyasaki Y, Utsunomiya Y, et al: Human mesenchymal stem cells in rodent whole-embryo culture and reprogrammed to contribute to kidney tissues. Proc Nat Acad Sci USA 2005;102:3296–3300.
55 Teng J, Turbat-Herrera EA, Herrera GA: The role of translational research advancing the understanding of the pathogenesis of light chain (LC)-mediated glomerulopathies. Pathol Int 2007;57:398–412.
56 Keeling J, Teng J, Herrera GA: AL-amyloidosis and light chain deposition disease light chains induce divergent phenotypic transformations of human mesangial cells. Lab Invest 2004;84:1322–1338.
57 Keeling J, Herrera GA: Matrix metalloproteinases and mesangial remodeling in light chain-related glomerular damage. Kidney Int 2005;68:1590–1603.
58 Keeling J, Herrera GA: An in vitro model of light chain deposition disease. Kidney Int 2009;75:634–645.

59 Herrera GA, Turbat-Herrera EA, Teng J: Repair of damaged mesangium by rat mesenchymal stem cells using a 6D living cell model (abstract). Amyloid 2010;17:44.

60 Teng J, Herrera GA: Replacement of injured rat mesenchymal mesangial cells by glomerulopathic light chains (GLCs) by rat mesenchymal cells (RMSCs): a 6D living cell model of mesangial repair (abstract). J Am Soc Nephrol 2008;19:435A.

61 Koh CJ, Atala A: Tissue engineering, stem cells, and cloning: opportunities for regenerative medicine. J Am Soc Nephrol 2004;15:1113–1125.

62 Rabelink TJ, van Kooten C: Stem cell therapy for glomerular disease. J Am Soc Nephrol 2006;17:2086–2088.

63 Zhu H, Mitsuhashi N, Klein A, Barsky LW, Weinberg K, Barr ML, Demetriou A, Wu GD: The role of hyaluronan CD44 in mesenchymal stem cell migration in the extracellular matrix. Stem Cells 2006;24:928–935.

64 Herrera MB, Bussolati B, Bruno S, Morando L, Mauriello-Romanazzi G, Sanavio F, Stamenkovic I, Biancone L, Camussi G: Exogenous mesenchymal stem cells localize to the kidney by means of CD44 following acute tubular injury. Kidney Int 2007;72:430–441.

65 Yakoo T, Kawamura T, Kobayash E: Stem cells for kidney repair: useful tool for acute renal failure. Kidney Int 2008;74:847–849.

66 Togel FE, Westenfelder C: Mesenchymal stem cells: a new therapeutic tool for AKI. Nephrol 2010;6:179–183.

67 Hopkins C, Li J, Rae F, Little MH: Stem cell options for kidney disease. J Pathol 2009;217:265–281.

Guillermo A. Herrera, MD
Nephrocor
1700 North Desert Drive
Tempe, AZ 85 281 (USA)
Tel. +1 602 4436 119, Fax +1 602 4436 102 ext. 3106, E-Mail nephronpath@gmail.com

Author Index

Subject Index